The Receptors

The Tachykinin Receptors,
edited by ***Stephen H. Buck,*** 1994
The Beta-Adrenergic Receptors,
edited by ***John P. Perkins,*** 1991
Adenosine and Adenosine Receptors,
edited by ***Michael Williams,*** 1990
The Muscarinic Receptors,
edited by ***Joan Heller Brown,*** 1989
The Serotonin Receptors,
edited by ***Elaine Sanders-Bush,*** 1988
The Alpha-2 Adrenergic Receptors,
edited by ***Lee Limbird,*** 1988
The Opiate Receptors,
edited by ***Gavril W. Pasternak,*** 1988
The Alpha-1 Adrenergic Receptors,
edited by ***Robert R. Ruffolo, Jr.,*** 1987
The GABA Receptors,
edited by ***S. J. Enna,*** 1983

The Tachykinin Receptors

Edited by

Stephen H. Buck
Biopharm Consultists, Tucson, AZ

Humana Press
Totowa, New Jersey

999 Riverview Drive, Suite 208
Totowa, New Jersey 07512

Printed in the United States of America. 10 9 8 7 6 5 4 3 2 1

Library of Congress Cataloging in Publication Data

The Tachykinin receptors / edited by Stephen H. Buck.
p. cm. -- (Receptors)
Includes index.
ISBN 0-89603-266-3
1. Tachykinins--Receptors. I. Buck, Stephen H. II. Series.
[DNLM: 1. Receptors, Tachykinin. QU 55 T117 1994]
QP552.T33T33 1994
599'.0192456--dc20
DNLM/DLC
for Library of Congress 94-13436
CIP

Preface

The Tachykinin Receptors is the first treatise to cover one of the most fascinating areas of neuropeptide biology—the receptors for the tachykinin family of peptides. Receptors for these biological mediators were surmised to exist for decades because of the diverse pharmacological effects of the tachykinins, which followed theoretical ligand–receptor interaction criteria. The existence of tachykinin receptors has been biochemically confirmed over the last decade with the introduction of numerous types of receptor binding assays and ligands that have facilitated such assays. In addition, the elegant report in 1987 of Masu et al. (*Nature* **329,** 836–838) of the primary structure of a substance P receptor protein of the G-protein superfamily has heralded a new era in which distinct receptors for the various natural tachykinin agonists have been structurally identified. These tachykinin receptors are structurally similar to and apparently function in a manner similar to those for many classical nonpeptide neurotransmitters. Furthermore, potent and selective antagonists of some of the tachykinin receptors have now been identified, including the very recent and interesting examples that are nonpeptides themselves, yet still completely block the receptor interactions of the much larger tachykinin peptides.

The Tachykinin Receptors contains chapters contributed by many of the leading international scientists in the field, one that is advancing very rapidly, perhaps more so than for any other neuropeptide family, and continues to attract many new and talented participants. One hopes that this book will be of assistance to these scientists and to all others who desire a single source of critical information on the great progress being made in understanding the receptors for the tachykinin peptides. It has been a pleasure to work with the Series Editor for *The Receptors*, David Bylund; with the publisher, Humana Press; and with the many authors who I am fortunate to have as my colleagues and friends.

As happens early in any area of science where new biological entities are rapidly being discovered in many laboratories, nomenclature for the tachykinin peptides and their receptors can be a confusing issue. The International Tachykinin Nomenclature Committee has continued to study the situation carefully under the past chairpersonship of Elizabeth Burcher of the University of New South Wales, Australia, and under the current chairpersonship of R. Michael Snider of Pfizer Inc., Groton, Connecticut. Since the Committee has yet to make recommendations, we have arbitrarily settled on a single nomenclature for the sake of consistency. For guidance on the tachykinin nomenclature issue, the reader is directed to *Substance P and Neurokinins* (J. L. Henry et al., eds., 1987, Springer-Verlag, New York), to any of the recent review articles cited throughout this volume, and to the abstracts (*Regul. Pept.* **Suppl. 1**, 1992) of the International Symposium on Substance P and Related Peptides that was held November 3–6, 1992, in Shizuoka, Japan.

Stephen H. Buck

Contents

Contributors

PAUL BLOUNT • *Department of Anatomy and Neurobiology, Washington University School of Medicine, St. Louis, MO*
NORMAN D. BOYD • *Department of Pharmacology and Experimental Therapeutics, Boston University School of Medicine, Boston, MA*
STEPHEN H. BUCK • *Biopharm Consultists, Tucson, AZ*
ELIZABETH BURCHER • *School of Physiology and Pharmacology, University of New South Wales, Kensington, Australia*
THAN-VINH DAM • *Douglas Hospital Research Centre and Department of Psychiatry, McGill University, Verdun, Canada*
PEDRO D'ORLÉANS-JUSTE • *Department of Pharmacology, University of Sherbrooke Medical School, Sherbrooke, Canada*
JIAN-ZHONG GUO • *Department of Pharmacology, Tokyo Medical and Dental University, Tokyo, Japan*
JUDITH M. HALL • *Pharmacology Group, Biomedical Sciences Division, King's College, London*
SCOTT L. HARBESON • *Alkermes Inc., Cambridge, MA*
MARCEL HIBERT • *Marion Merell Dow Research Institute, Strasbourg Center, Strasbourg, Cedex, France*
RAYMOND G. HILL • *Merck Sharp and Dohme Research Labs, Neuroscience Research Centre, Harlow, UK*
JAN HOFLACK • *Marion Merell Dow Research Institute, Strasbourg Center, Strasbourg, Cedex, France*
RUMIKO HOSOKI • *Department of Pharmacology, Tokyo Medical and Dental University, Tokyo, Japan*
LESLIE L. IVERSEN • *Merck Sharp and Dohme Research Labs, Neuroscience Research Centre, Harlow, UK*
REINHARD K. KAGE • *Department of Pharmacology and Experimental Therapeutics, Boston University School of Medicine, Boston, MA*

James E. Krause • *Department of Anatomy and Neurobiology, Washington University School of Medicine, St. Louis, MO*

Susan E. Leeman • *Department of Pharmacology and Experimental Therapeutics, Boston University School of Medicine, Boston, MA*

Fred Lembeck • *Institute for Experiemntal and Clinical Pharmacology, University of Graz, Austria*

John A. Lowe III • *Department of Chemistry, Pfizer Inc., Groton, CT*

Carlo Alberto Maggi • *Pharmacology Department, A. Menarini Pharmaceuticals, Florence, Italy*

John E. Maggio • *Department of Biological Chemistry and Molecular Pharmacology, Harvard Medical School, Boston, MA*

Patrick W. Mantyh • *Molecular Neurobiology Lab, VA Medical Center, Minneapolis, MN*

Stafford McLean • *Department of Neuroscience, Pfizer Inc., Groton, CT*

Christian J. Mussap • *School of Physiology and Pharmacology, University of New South Wales, Kensington, Australia*

Shigehiro Nakajima • *Department of Pharmacology, University of Illinois at Chicago, IL*

Yasuko Nakajima • *Department of Anatomy and Cell Biology, University of Illinois, Chicago, IL*

Masanori Otsuka • *Department of Pharmacology, Tokyo Medical and Dental University, Tokyo, Japan*

James W. Putney • *Lab of Cellular and Molecular Pharmacology, NIEHS-NIH, Research Triangle Park, NC*

Remi Quirion • *Douglas Hospital Research Centre and Department of Psychiatry, McGill University, Verdun, Canada*

Domenico Regoli • *Department of Pharmacology, University of Sherbrooke Medical School, Sherbrooke, Canada*

Noureddine Rouissi • *Department of Pharmacology, University of Sherbrooke Medical School, Sherbrooke, Canada*

Paolo Rovero • *Instituto di Mutagenesi e Differenziamento, Pisa, Italy*

Bruce S. Sachais • *Department of Anatomy and Neurobiology, Washington University School of Medicine, St. Louis, MO*

Jennifer A. Stephenson • *Pharmaceutical Research Institute, Bristol-Myers Squibb Co., Noble Park, Australia*

Hidenori Suzuki • *Department of Pharmacology, Tokyo Medical and Dental University, Tokyo, Japan*

Susanne Trumpp-Kallmeyer • *Marion Merrell Dow Research Institute, Strasbourg Center, Strasbourg, Cedex, France*

Paul L. M. van Giersbergen • *Marion Merrell Dow Research Institute, Strasbourg Center, Strasbourg, Cedex, France*

Steven R. Vigna • *Departments of Cell Biology and Medicine, Duke University Medical Center, Durham, NC*

Mitsuhiko Yanagisawa • *Department of Pharmacology, Tokyo Medical and Dental University, Tokyo, Japan*

Koichi Yoshioka • *Department of Pharmacology, Tokyo Medical and Dental University, Tokyo, Japan*

PART I

HISTORICAL PERSPECTIVES OF TACHYKININS

CHAPTER 1

History of Tachykinin Peptides

John E. Maggio and Patrick W. Mantyh

1. Introduction

A scientific highlight of the first half of the twentieth century was the discovery and characterization of many biologically active molecules. Many large (hormones, neuropeptides) and small (classical transmitters) intercellular messengers were described during this period. Among these was substance P, which claims the longest scientific history and is arguably the most thoroughly characterized of the brain–gut peptides. It is the oldest neuropeptide in the sense that it was the first active compound from neural tissue that was later shown to be a peptide. For 50 years after its discovery, substance P (SP) was generally believed to be the only tachykinin in mammals. The relatively recent discovery of other mammalian tachykinins has raised many important new questions about the physiological role of this peptide family and has reawakened interest in the field.

The large volume of scientific literature on tachykinins (several thousand original publications) reflects both the current interest and the long history of the field. Other reviews and symposia (Erspamer and Anastasi, 1966; Skrabanek and Powell, 1977, 1980, 1983a,b;

The Tachykinin Receptors Ed.: S. H. Buck © 1994 Humana Press Inc., Totowa, NJ

Von Euler and Pernow, 1977; Nicoll et al., 1980; Oehme et al., 1981; Porter and O'Connor, 1982; Sandberg and Iversen, 1982; Jessell, 1983; Pernow, 1983; Skrabanek, 1984; Jordan and Oehme, 1985; Sandberg 1985; Segawa, et al., 1985; Henry et al., 1987; Maggio, 1988; Maggio and Manthyh, 1989) give information on certain aspects of tachykinin history that cannot be considered in depth in this chapter. The exhaustive literature surveys of Skrabanek and Powell (Skrabanek and Powell, 1977, 1980, 1983a) are especially valuable as a guide to older literature. This chapter concentrates on the discovery and early characterization of the tachykinin peptides; other relevant topics are considered elsewhere in this volume.

2. Discovery and Early Studies of Tachykinins

2.1. *Discovery of Substance P*

In 1931, while investigating the tissue distribution of the newly characterized transmitter acetylcholine, Von Euler and Gaddum (1931) noted in acid ethanol extracts of equine intestine a hypotensive and spasmogenic activity that differed from acetylcholine in that it contracted the rabbit jejunum in the presence of atropine. The activity of this preparation, first referred to as P on kymograph tracings, was distinct from the biologically active principles known at that time. The active component was soon termed substance P (P for preparation), and early experiments suggested that SP was peptide or protein in nature (Gaddum and Schild, 1934; Von Euler, 1942).

Most of the substantial research effort invested in tachykinins over the next three decades concentrated on the chemical properties, pharmacological activities, and tissue distribution of SP (for reviews, *see* Pernow, 1953, 1963, 1983; Lembeck and Zetler, 1962, 1971). Although the SP of that time was a biological extract of variable (and often low) purity, and research relied heavily on bioassays, many of the major findings of this era proved to be substantially correct. That SP was identical in brain and gut, found in all vertebrates, present and active in the gastrointestinal tracts of many species, and unevenly distributed in the nervous system were all findings made long before SP was completely purified and sequenced (Von Euler, 1936;

Gernandt, 1942; Pernow, 1953, 1963, 1983; Amin et al., 1954; Eliasson et al., 1956; Von Euler and Ostlund, 1956; Grabner et al., 1959; Lembeck and Zetler, 1962, 1971). In retrospect, both the quality and the quantity of research on SP before its structure was known were very impressive.

Up through the 1950s, SP was prepared by acidic extraction of tissue, often followed by precipitations to remove lipids and proteins (Von Euler, 1942; Pernow, 1951; Amin et al., 1954; Meinardi and Craig, 1966; Skrabanek and Powell, 1980). These preparations undoubtedly contained, in addition to SP, other bioactive peptides, notably other tachykinins. By the 1960s, purification protocols employing such techniques as alumina chromatography, ion exchange, and paper electrophoresis had advanced to the point of yielding preparations in which SP itself was probably the major component (Boissonnas et al., 1963; Zuber, 1963; Meinardi and Craig, 1966; Pernow, 1983; Skrabanek and Powell, 1980), but another decade passed before the peptide finally surrendered its sequence.

By the mid-1950s, the elementary pharmacology of SP had been established. Thus, SP preparations were found to be active on gastrointestinal tissues of fish, birds, and amphibians, as well as mammals (Pernow, 1953; Von Euler and Ostlund, 1956; Lembeck and Zetler, 1962). The activity of SP was not blocked by antagonists of acetylcholine, histamine, or serotonin (Douglas et al., 1951; Gaddum, 1953; Pernow, 1953), suggesting a direct action on the smooth muscle. Substance P stimulated peristalsis in vivo (Gernandt, 1942), but this activity was inhibited by hexamethonium in vitro (Beleslin and Varagic, 1958) indicating that indirect action (via nerve fibers releasing acetylcholine) was also involved.

Studies of the tissue distribution of SP biological activity had also been completed in considerable detail by the mid-1950s. The original rabbit jejunum and blood pressure assays (Von Euler, 1936; Von Euler and Pernow, 1977) were in most cases replaced by the guinea pig ileum (Erspamer, 1949; Pernow, 1951). Substance P activity was extracted from the brains and intestines of many vertebrates (lampreys, cartilaginous and bony fishes, amphibians, reptiles, birds, mammals), and its occurrence in subregions of those tissues was extensively mapped (Gernandt, 1942; Douglas et al., 1951; Pernow, 1953; Amin et al., 1954; Gaddum and Szerb, 1961; Von Euler and Pernow, 1977). The bioassay mappings of this period have been generally confirmed by radioimmunoassay (RIA), particularly when the presence of other tachykinins is taken into account.

The biological activities and distribution of SP led to early speculations about its physiological role, some of which remain relevant today. The observations that SP is concentrated in the intestinal wall and that it is a very powerful spasmogen of gastrointestinal tissue led Von Euler (1936) to suggest that SP might play a role in controlling intestinal motility. The observation that SP concentrations were much higher in dorsal than in ventral roots of the spinal cord (Lembeck, 1953; Pernow, 1953; Amin et al., 1954) led Lembeck (1953) to suggest that SP might serve as a primary sensory transmitter. Both of these suggestions later found considerable experimental support.

2.2. Discovery of Nonmammalian Tachykinins

2.2.1. Eledoisin

In 1949, while screening for biogenic amines in extracts of the posterior salivary glands of the Mediterranean octopus *Eledone moschata*, Erspamer (1949) discovered a hypotensive, sialagogic, and spasmogenic activity that could not be ascribed to any known substances. The active substance, initially named moschatine (Erspamer, 1949) and shortly after renamed eledoisin, had two notable properties: It was present in very high concentrations and was recognized as a polypeptide with certain similarities to SP (Anastasi and Erspamer, 1962; Erspamer and Anastasi, 1962; Erspamer, 1949, 1971).

In 1962, the structure of eledoisin was established (Erspamer and Anastasi, 1962). The very high concentrations of eledoisin in the glands (0.01% by weight, 10^3-fold higher than the concentrations of SP in mammalian tissues) made it the first tachykinin to be sequenced. At that time, peptide sequencing was a heroic undertaking in that the quantity of material required was immense by modern standards. From over 10,000 pairs of posterior salivary glands (5.37 kg) obtained from 1450 kg of live animals, Erspamer and Anastasi (Anastasi and Erspamer, 1962; Erspamer and Anastasi, 1962) purified enough homogeneous eledoisin to determine its correct structure. Its structure (Table 1) has the notable feature of blocked termini (it lacks free amino and carboxy ends) that would later be found in many bioactive peptides. It was recognized that eledoisin, although very similar to SP in pharmacological activities and protease sensitivity, could nevertheless be readily distinguished from the mammalian peptide by parallel bioassays (Erspamer and Falconieri-Erspamer, 1962). That is, despite the qualitative similarity of their activities, there were important quantitative differences in the relative potencies of the two peptides in various mammalian systems (Erspamer and Anastasi, 1962; Erspamer, 1971).

2.2.2. Physalaemin

In the same year that eledoisin was sequenced, Erspamer and coworkers noted a potent eledoisin-like bioactivity in the skin of the South American frog, *Physalaemus biligonigerus* (Erspamer, 1971). This spasmogenic and hypotensive principle, named physalaemin, was purified, sequenced, and synthesized in 1964 (Erspamer et al., 1964). The sequence of physalaemin, like its biological activities, showed considerable homology with that of eledoisin (Table 1); qualitative similarities and quantitative distinctions between physalaemin, eledoisin, and SP were observed in parallel bioassay. With their structures known, many analogs of physalaemin and eledoisin could be synthesized for systematic pharmacological study (Erspamer and Melchiorri, 1973). One of the more important findings was that most of the biological activity of the peptides seemed to reside in the conserved carboxy-terminal sequence (Erspamer and Melchiorri, 1973; Bertaccini, 1976). Because physalaemin and eledoisin and their analogs showed a characteristic fast onset of action on tissues of the gut (as compared with the slower acting bradykinins), the term *tachykinin* was coined (Erspamer and Anastasi, 1966) for the peptides of the group.

2.2.3. Other Amphibian Tachykinins

The third member of the family was sequenced in 1970 (Anastasi and Falconieri-Erspamer, 1970). The new tachykinin, phyllomedusin, shared with physalaemin and eledoisin both a similar carboxy-

Table 1
Tachykinin Sequences

Tachykinin	Structure[a]
Mammals	
Substance P[b]	H-Arg-Pro-Lys-Pro-Gln-Gln-PHE-PHE-GLY-LEU-MET-NH_2
Neurokinin A/Substance K[c]	H-His-Lys-Thr-Asp-Ser-PHE-Val-GLY-LEU-MET-NH_2
Neurokinin B/Neuromedin K[d]	H-Asp-Met-His-Asp-Phe-PHE-Val-GLY-LEU-MET-NH_2
Neuropeptide gamma[e]	H-Asp-Ala-Gly-His-Gly-Gln-Ile-Ser-His-Lys-Arg-His-Lys-Thr-Asp-Ser-PHE-Val-GLY-LEU-MET-NH_2
Birds	
Chicken Substance P[f]	H-Arg-Pro-Arg-Pro-Gln-Gln-PHE-Phe-GLY-LEU-MET-NH_2
Chicken Neurokinin A/Substance K[f]	H-His-Lys-Thr-Asp-Ser-PHE-Val-GLY-LEU-MET-NH_2
Molluscs	
Eledoisin[g]	pGlu-Pro-Ser-Lys-Asp-Ala-PHE-Ile-GLY-LEU-MET-NH_2
Amphibians	
Physalaemin[h]	pGlu-Ala-Asp-Pro-Asn-Lys-PHE-Tyr-GLY-LEU-MET-NH_2
Phyllomedusin[i]	pGlu-Asn-Pro-Asn-Arg-PHE-Ile-GLY-LEU-MET-NH_2
Kassinin[j]	H-Asp-Val-Pro-Lys-Ser-Asp-Gln-PHE-Val-GLY-LEU-MET-NH_2
Lys^5Thr^6-physalaemin[k]	pGlu-Ala-Asp-Pro-Lys-Thr-PHE-Tyr-GLY-LEU-MET-NH_2
Hylambatin[k]	H-Asp-Pro-Pro-Asp-Pro-Asp-Arg-PHE-Tyr-GLY-Met-MET-NH_2
Glu^2Pro^5-kassinin[l]	H-Asp-Glu-Pro-Lys-Pro-Asp-Gln-PHE-Val-GLY-LEU-MET-NH_2
Uperolein[m]	pGlu-Pro-Asp-Pro-Asn-Ala-PHE-Tyr-GLY-LEU-MET-NH_2
Entero-hylambatin[n]	H-Asp-Pro-Pro-Asn-Pro-Asp-Arg-PHE-Tyr-GLY-MET-Met-NH_2
Entero-kassinin[n]	H-Asp-Glu-Pro-Asn-Ser-Asp-Gln-PHE-Ile-GLY-LEU-MET-NH_2

SP-like peptide[n,o]	
	H-Asp-Ile-Pro-Lys-Pro-Asp-Gln-PHE-Phe-GLY-LEU-MET-NH_2
PG-KI[p]	pGlu-Pro-His-Pro-Asp-Glu-PHE-Val-GLY-LEU-MET-NH_2
PG-KII[p]	pGlu-Pro-Asn-Pro-Asp-Glu-PHE-Val-GLY-LEU-MET-NH_2
PG-KIII[p]	pGlu-Pro-His-Pro-Asn-Glu-PHE-Val-GLY-LEU-MET-NH_2
PG-SPI[p]	pGlu-Pro-Asn-Pro-Asp-Glu-PHE-Phe-GLY-LEU-MET-NH_2
PG-SPII[p]	pGlu-Pro-Asn-Pro-Asn-Glu-PHE-Phe-GLY-LEU-MET-NH_2
Ranatachykinin A[q]	H-Lys-Pro-Ser-Pro-Asp-Arg-PHE-Tyr-GLY-LEU-MET-NH_2
Ranatachykinin B[q]	H-Tyr-Lys-Ser-Asp-Ser-PHE-Tyr-GLY-LEU-MET-NH_2
Ranatachykinin C[q]	H-His-Asn-Pro-Ala-Ser-PHE-Ile-GLY-LEU-MET-NH_2
Ranatachykinin D[q]	H-Lys-Pro-Asn-Pro-Glu-Arg-Phe-Tyr-Ala-Pro-MET-NH_2
Ranakinin[r]	H-Lys-Pro-Asn-Pro-Glu-Arg-PHE-Tyr-GLY-LEU-MET-NH_2
Frog Neurokinin B/Neuromedin K[r]	H-Asp-Met-His-Asp-Phe-PHE-Val-GLY-LEU-MET-NH_2
Frog Neurokinin A/Substance K[s]	H-His-Lys-Leu-Asp-Ser-PHE-Ile-GLY-LEU-MET-NH_2
Insects	
Locutatachykinln I[t]	H-Gly-Pro-Ser-Gly-PHE-Tyr-GLY-Val-Arg-NH_2
Locustatachykinin II[t]	H-Ala-Pro-Leu-Ser-Gly-PHE-Tyr-GLY-Val-Arg-NH_2
Locustatachykinin III[t]	H-Ala-Pro-Gln-Ala-Gly-PHE-Tyr-GLY-Val-Arg-NH_2
Locustatachykinin IV[t]	H-Ala-Pro-Ser-Leu-Gly-PHE-His-GLY-Val-Arg-NH_2
Reptiles	
Alligator Substance P[u]	H-Arg-Pro-Arg-Pro-Gln-Gln-PHE-Phe-GLY-LEU-MET-NH_2
Alligator Neuropeptide gamma[u]	
	H-Asp-Ala-Gly-Tyr-Gly-Gln-Ile-Ser-His-Lys-Arg-His-Lys-Thr-Asp-Ser-PHE-Val-GLY-LEU-MET-NH_2
Alligator Neurokinin A/Substance K[u]	H-His-Lys-Thr-Asp-Ser-PHE-Val-GLY-LEU-MET-NH_2

(continued)

Table 1 *(continued)*

Tachykinin	Structure[a]
Fish	
Scyliorhinin I[v]	H-Ala-Lys-Phe-Asp-Lys-PHE-Tyr-GLY-LEU-MET-NH_2
Scyliorhinin II[v]	H-Ser-Pro-Ser-Asn-Ser-Lys-Cys-Pro-Asp-Gly-Pro-Asp-Cys-PHE-Val-GLY-LEU-MET-NH_2 (S–S bridge between Cys residues)
Des [Ser[1]Pro[2]] Scyliorhinin II[w]	H-Ser-Asn-Ser-Lys-Cys-Pro-Asp-Gly-Pro-Asp-Cys-PHE-Val-GLY-LEU-MET-NH_2 (S–S bridge between Cys residues)
Carassin[x]	H-Ser-Pro-Ala-Asn-Ala-Gln-Ile-Thr-Arg-Lys-Arg-His-Lys-Ile-Asn-Ser-PHE-Val-GLY-LEU-MET-NH_2
Trout Substance P[y]	H-Lys-Pro-Arg-Pro-His-Gln-PHE-Phe-GLY-LEU-MET-NH_2
Cod Substance P[y]	H-Lys-Pro-Arg-Pro-Gln-Gln-PHE-Ile-GLY-LEU-MET-NH_2
Trout/Cod Neurokinin A/Substance K[y]	H-His-Lys-Ile-Asn-Ser-PHE-Val-GLY-LEU-MET-NH_2
Dogfish Substance P[z]	H-Lys-Pro-Arg-Pro-Gly-Gln-PHE-Phe-GLY-LEU-MET-NH_2
Skate Neurokinin A/Substance K[z]	H-His-Lys-Leu-Gly-Ser-PHE-Val-GLY-LEU-MET-NH_2
His[3]-Scyliorhinin I[z]	H-Ala-Lys-His-Asp-Lys-PHE-Tyr-GLY-LEU-MET-NH_2
Canonical tachykinin sequence	-PHE- X -GLY-LEU-MET-NH_2

[a]H-, Free NH_2-terminal α-amino group; -NH_2, COOH-terminal α-carboxamide; pGlu-, NH_2-terminal pyroglutamic acid moiety.
[b]Chang et al., 1971.

[c]Kimura et al., 1983. Alternate names: neurokinin α, α-neurokinin, neuromedin L.
[d]Kimura et al., 1983. Alternate names: neurokinin β, β-neurokinin.
[e]Kage et al., 1988. Some species have Leu replacing Ile^7.
[f]Conlon et al., 1988a.
[g]Erspamer et al., 1962.
[h]Erpsamer et al., 1964.
[i]Anastasi et al., 1970.
[j]Anastasi et al., 1977.
[k]Nakajima et al., 1980. Alternate name: rugosa uperolein II.
[l]Yasuhara et al., 1981. Alternate name: hylambates-kassinin.
[m]Anastasi et al., 1975.
[n]Melchiorri and Negri, 1984.
[o]Presumed sequence.
[p]Simmaco et al., 1990.
[q]Kozawa et al., 1991.
[r]O'Harte et al., 1991.
[s]Wang et al., 1 992b.
[t]Schoofs et al., 1990a,b.
[u]Wang et al., 1992a.
[v]Conlon et al., 1986.
[w]Conlon and Thim, 1988.
[x]Conlon et al., 1991.
[y]Jensen and Conlon, 1992.
[z]Waugh et al., 1993.

terminal sequence and a similar spectrum of bioactivities, although again quantitative differences were evident. These facts led Erspamer (1971) to predict before SP was sequenced that the mammalian peptide would share with the nonmammalian tachykinins the carboxy-terminal sequence -Phe-X-Gly-Leu-Met-NH_2. More amphibian tachykinins were subsequently isolated (Anastasi et al., 1975; Anastasi et al., 1977; Nakajima et al., 1980; Yasuhara et al., 1981; Melchiorri and Negri, 1984), and the process may be expected to continue. By 1985, 10 amphibian tachykinins had been described; by 1993 the number had swelled to 22 (Table 1). Generally, they have been named for the genus from which each was first isolated. Thus, the dodecapeptide tachykinin, kassinin (X = Val), was first isolated from the skin of the African rhacophorid frog, *Kassina senegalensis* (Anastasi et al., 1977).

Studies of the nonmammalian tachykinins have been essential to advances in our understanding of the roles of tachykinins in mammals. In particular, the quantitative differences in potency within a family of structurally related peptides whose biological activities are qualitatively similar led directly to the discovery that SP was not the only mammalian tachykinin.

2.3. Rediscovery of Substance P

In the late 1960s, while attempting to isolate a corticotropin-releasing factor from bovine hypothalamus, Leeman and Hammerschlag (1967) found a peptide sialagogue that was not blocked by cholinergic or adrenergic antagonists. Lembeck and Starke (1968), on the basis of their own experiments and the earlier work of Vogler et al. (1963), suggested that the hypothalamic sialogen might indeed be SP. The sialagogic peptide was purified to homogeneity (by gel filtration, cation-exchange chromatography, and electrophoresis), recognized as SP, and sequenced in 1971 (Chang and Leeman 1970; Chang et al., 1971). Later it was shown (Studer et al., 1973) to be identical to the SP extracted from equine intestine by methods originally used four decades earlier by Von Euler and Gaddum (1931).

Substance P was rediscovered yet again in the early 1970s by Otsuka et al. (1972a), who were searching for sensory neurotransmitters. They found in extracts of bovine dorsal roots a peptide that excited spinal motoneurons (Otsuka et al., 1972a); subsequently SP was shown to be at least partly responsible for this biological activity (Otsuka et al., 1972b).

The availability of pure synthetic SP fostered an explosion of research interest in SP, and by 1978, SP was by one tally the third most active field of research in the life sciences (Garfield, 1980). In the midst of this activity on SP, the interest in the other tachykinins waned. From 1971 until 1983, SP was generally assumed to be the only mammalian tachykinin, although the assumption was implicit rather than explicit. In many ways, it was like the implicit assumption made for the four preceding decades that the activities of the biological extracts known as SP could be attributed to a single chemical entity.

Even research that was and is above reproach can be profitably reevaluated in light of later findings. In the case of the tachykinins, such an important later finding was that SP is not in fact the only tachykinin in mammals. Interpretation of some of the earlier work is complicated by the fact that the experimental techniques employed lacked the specificity to distinguish rigorously one mammalian tachykinin from another. Thus, at least some of the biological actions originally ascribed to SP were later shown to be mediated by other members of the tachykinin family.

3. Structure–Activity Relationships

The sequences of the naturally occuring tachykinins described to date are given in Table 1. The peptides in the family share not only the spectrum of biological activities that initially defined the group (Erspamer and Anastasi, 1966), but also the homologous structural features that now define it. It is the highly conserved carboxy-terminal sequence -Phe-X-Gly-Leu-Met-NH_2 that defines the tachykinins, although the authors have included in Table 1 several peptides that do not strictly satisfy that definition. To a first approximation, most of the biological activities of the tachykinins depend on this defining sequence. This conclusion is based on measurements of the activities of many tachykinin fragments and analogs in many systems (Erspamer and Melchiorri, 1973; Bertaccini, 1976; Bury and Mashford, 1976; Couture et al., 1979; Lee et al., 1982; Sandberg and Iversen, 1982; Pernow, 1983; Watson et al., 1983; Sandberg, 1985). The amino-terminal portion of the peptides is neither without activity nor without effect on the activity of carboxy-terminal fragments, but in general carboxyl-terminal fragments of six or more amino acids retain tachykinin bioactivity, whereas amino-terminal fragments do not. Removal of the methionineamide, or even of the carboxy-terminal amide group, abolishes activity, although the amide may be replaced with other uncharged functional groups without such a loss (Anastasi and Falconieri-Erspamer, 1970; Lee et al., 1982; Sandberg, 1985). In some systems, the partial COOH-terminal sequences of SP have been reported to be more potent than the full-length peptide (Bertaccini, 1976; Bury and Mashford, 1976; Couture et al., 1979; Lee et al., 1982; Sandberg and Iversen, 1982; Pernow, 1983; Sandberg, 1985).

As noted, although the tachykinins share a similar spectrum of biological activities, they are distinguished by quantitative differences in their relative potencies in various bioassays. A major determinant of that specificity (Erspamer, 1981; Lee et al., 1982; Sandberg, 1985) is the nature of the amino acid at position X (Table 1), which may have an aromatic or aliphatic side chain. The general interpretation of these potency differences was that multiple types of SP receptors existed in mammalian tissues (Teichberg et al., 1981; Gater et al., 1982; Lee et al., 1982; Sandberg and Iversen, 1982), although, clearly, other explanations were possible; a classification of receptors based solely on agonist potency must always be regarded as preliminary.

It is hard to exclude the possibility that one or more of the intermediates in preprotachykinin processing or tachykinin metabolism is an active peptide in its own right. Such a claim has been advanced for neuropeptide K, a 36-amino acid peptide with the neurokinin A/substance K (NKA/SK) sequence at its carboxy terminus (Tatemoto et al., 1985; Theodorsson-Norheim et al., 1985).

Because this peptide has an intact tachykinin sequence -Phe-X-Gly-Leu-Met-NH_2, it would be expected to have some tachykinin biological activity; whether neuropeptide K is more than an intermediate in NKA/SK biosynthesis (Deacon et al., 1987) remains to be established. The carboxyl-terminal octapeptide of NKA/SK (H-TDS-FVGLM-NH_2) is another peptide with an intact tachykinin sequence. Like neuropeptide K, it is found in tissue extracts (Too et al., 1989); whether this octapeptide is more than a breakdown product of NKA/SK remains to be established. Neuropeptide K and NKA/SK octapeptide have been arbitrarily omitted from Table 1.

4. Discovery of Novel Mammalian Tachykinins

Shortly after the golden anniversary of the discovery of SP was celebrated, a new discovery with very important implications for tachykinin research was reported. In 1983, four independent research groups using three different approaches reported that mammalian tissues contain not only SP, but also other tachykinins (Kangawa et al., 1983; Kimura et al., 1983, 1985; Maggio et al., 1983; Nawa et al., 1983; Minamino et al., 1984; Maggio, 1985). The peptides were given various names by their various discoverers *(see below)*.

As discussed above, there are quantitative differences between the potencies of the various tachykinins in mammalian bioassays, a finding usually interpreted as evidence for multiple types of SP receptors in mammalian tissues (Teichberg et al., 1981; Gater et al., 1982; Lee et al., 1982; Piercey et al., 1982; Sandberg and Iversen, 1982). Our group (Maggio et al., 1983) considered another possible explanation of the fact that SP (X = Phe) is much less potent in many mammalian bioassays than aliphatic nonmammalian tachykinins. We hypothesized that the SP-E receptor (*see* Iversen chapter) was not a receptor for SP at all, but was instead a receptor for an undiscovered mammalian tachykinin, similar in structure to kassinin or eledoisin. The possibility that SP was not the only mammalian tachykinin had already been raised by others in symposia or in print. The credit for this idea belongs most appropriately to Erspamer, who recognized both before and after structural elucidation of SP that mammalian SP was "more likely a family of polypeptides" (Erspamer, 1949; Erspamer and Melchiorri, 1973).

To search for the hypothetical aliphatic mammalian tachykinin postulated as the natural ligand for SP-E receptors, we adopted an immunochemical approach (Maggio et al., 1983). A radioimmunoassay that was developed showed high cross-reactivity with aliphatic tachykinins (i.e., X is a branched aliphatic residue in -Phe-X-Gly-Leu-Met-NH_2) and low crossreactivity with aromatic tachykinins such as SP (X = Phe) or physalaemin (Maggio et al., 1983; Maggio and Hunter, 1984). The assay revealed in extracts of bovine spinal cord a positively charged, chymotrypsin-sensitive substance with carboxyl-terminal kassinin-like immunoreactivity, chromatographically distinct from SP. Reverse-phase HPLC showed that this novel peptide, for which the name substance K (SK) was proposed, was distinct from the known tachykinins in both the oxidized and reduced forms (Maggio et al., 1983). The spasmogenic activity of SK (in the presence of classical blockers) followed the slower (than SP) time course typical of kassinin (Erspamer, 1981) and was blocked by a tachykinin antagonist (Maggio et al., 1983).

The new mammalian tachykinin that had been named substance K *(above)* was first sequenced by Kimura et al. (1983). This group had been searching for novel neuropeptides in extracts of porcine spinal cord (Kimura et al., 1985) using spasmogenic activity on the guinea pig ileum as an assay. After purification by sizing gels and HPLC, the structures of two novel tachykinins were reported; a cationic peptide named neurokinin α, which later proved identical to bovine SK (Nawa et al., 1983; Hunter and Maggio, 1984) and an anionic peptide named neurokinin β. Both of these novel neuropeptides were aliphatic tachykinins with carboxy-termini identical to that of kassinin (-Phe-Val-Gly-Leu-Met-NH_2).

At about the same time, Kangawa et al. (1983) were employing a similar strategy in searching for novel neuropeptides from porcine spinal cord. This group reported the structure of a novel anionic tachykinin homologous to kassinin and named it neuromedin K (NMK). This peptide proved to be identical to neurokinin β sequenced by Kimura et al. (1983). Among the other peptides sequenced by these workers was a cationic tachykinin named neuromedin L (Minamino et al., 1984), which proved to be identical to neurokinin α and substance K.

The elucidation of the sequence of cDNAs encoding SP by Nawa et al. (1983) revealed a novel tachykinin sequence that was named substance K. This peptide proved to be identical with neurokinin α, neuromedin L, and substance K described above.

In retrospect, it seems remarkable that the new mammalian tachykinins were not discovered earlier. The original SP preparation of Von Euler and Gaddum (1931) surely contained more than one tachykinin. The same was true of Chang and Leeman's hypothalamic extract (Chang and Leeman, 1970) of four decades later, up to the ion-exchange chromatography step that would have separated SP (the most potent sialagogue) from SK and NMK.

Reluctant to wait for convergent evolution to settle on a single name for each of the novel mammalian tachykinins, nomenclature committees (Jordan and Oehme, 1985; Henry et al., 1987) have attempted to reduce confusion by changing the names chosen by the groups who discovered the peptides. Thus, neurokinin A (NKA) was proposed to replace substance K, neurokinin α, and neuromedin L; neurokinin B (NKB) was proposed to replace neurokinin β and neuromedin K. Thus far attempts to rename substance P have met with little success.

5. Other Tachykinins

A strategy employed for isolation and characterization of novel mammalian tachykinins was purification based on immunoreactivity or bioassay activity. These approaches, separately or in combination, have in recent years yielded the sequences of many naturally occurring tachykinins. Swelled by recent contributions from several groups, notably Conlon and coworkers (*see* Table 1), the number of known sequences had approached 50 by the end of 1992. Tachykinin-like immunoreactive or bioactive components have been reported in all phyla from coelenterates to vertebrates, suggesting that tachykinins may be among the oldest neuropeptides in a phylogenetic sense, as well as in the sense of scientific history.

References

Amin, A. H., Crawford, T. B. B., and Gaddum, J. H. (1954) The distribution of substance P and 5-hydroxy-tryptamine in the central nervous system of the dog. *J. Physiol. Lond.* **126,** 596–618.

Anastasi, A. and Erspamer, V. (1962) Occurrence and some properties of eledoisin in extracts of posterior salivary glands of Eledone. *Br. J. Pharmacol.* **19,** 326–336.

Anastasi, A., Erpsamer, V., and Endean, R. (1975) Structure of uperolein a physalaemin-like endecapeptide occuring in the skin of *Uperoleia rugosa* and *Uperoleia marmorata. Experential* **31,** 394–395.

Anastasi, A. and Falconieri-Erspamer, G. (1970) Occurrence of phyllomedusin a physalaemin-like decapeptide in the skin of *Phyllomedusa bicolor. Experential* **26,** 866,867.

Anastasi, A., Montecucchi, P., Erspamer, V., and Visser, J. (1977) Amino acid composition and sequence of kassinin, a tachykinin dodecapeptide from the skin of the African frog *Kassina senegalensis. Experential* **33,** 857–858.

Beleslin D. and Varagic, V. (1958) The effect of substance P on the peristaltic reflex of the isolated guinea-pig ileum. *Br. J. Pharmacol.* **13,** 321–325.

Bertaccini, G. (1976) Active peptides of nonmammalian origin. *Pharmacol. Rev.* **28,** 127–177.

Boissonnas, R. A., Franz, J., and Sturmer, E. (1963) On the chemical characterization of substance P. *Ann. NY Acad. Sci.* **104,** 376,377.

Bury, R. W. and Mashford, M. L. (1976) Biological activity of C-terminal partial sequences of substance P. *J. Med. Chem.* **19,** 854–856.

Chang, M. M. and Leeman, S. E. (1970) Isolation of a sialogogic peptide from bovine hypothalamic tissue and its characterization as substance P. *J. Biol. Chem.* **245,** 4784–4790.

Chang, M. M., Leeman, S. E., and Niall, H D. (1971) Amino acid sequence of substance P. *Nature New Biol.* **232,** 86,87.

Conlon, J. M., Deacon, C. F., O'Toole, L., and Thim, L. (1986) Scyliorhinin I and Il: Two novel tachykinins from dogfish gut. *FEBS Lett.* **200,** 116.

Conlon, J. M. and Thim, L. (1988) Isolation of the tachykinin des[Ser1Pro2] scyliorhinin ll from the intestine of the ray, *Torpedo marmorata. Gen. Comp. Endocrinol.* **71,** 383–388.

Conlon, J. M., Katsoulis, S., Schmidt, W. E., and Thim, L. (1988a) [Arg^3]substance P and neurokinin A from chicken small intestine. *Regul. Pept.* **20,** 171–180.

Conlon, J. M., O'Harte, F., Peter, R. E., and Kah, O. (1991) Carassin: A tachykinin that is structurally related to neuropeptide-gamma of the goldfish. *J. Neurochem.* **56,** 1432–1436.

Couture, R., Fournier, A., Magnan, J., St.-Pierre, S., and Regoli, D. (1979) Structure-activity studies on substance P. *Can. J. Physiol. Pharmacol.* **57,** 1427–1436.

Deacon, C. F., Agoston, D. V., Nau, R., and Conlon, J. M. (1987) Conversion of neuropeptide K to neurokinin A and vesicular colocalization of neurokinin A and substance P in the guinea-pig small intestine. *J. Neurochem.* **48,** 141–146.
Douglas, W. W., Feldberg, W., Paton, W. D. M., and Schachter, M. S. (1951) Distribution of histamine and substance P in the wall of dogs digestive tract. *J. Physiol. Lond.* **115,** 163–176.
Eliasson, R., Lie, L., and Pernow, B. (1956) A comparative study of substance P from intestine and brain. *Br. J. Pharmacol.* **11,** 137–140.
Erspamer, V. (1949) Ricerche preliminaria sulla moschatina. *Experential* **5,** 79–81.
Erspamer, V. (1971) Biogenic amines and active polypeptides of the amphibian skin. *Annu. Rev. Pharmacol.* **11,** 327–350.
Erspamer, V. (1981) The tachykinin peptide family. *Trends Neurosci.* **4,** 267–269.
Erspamer, V. and Anastasi, A. (1962) Structure and pharmacological actions of eledoisin, the active endecapeptide of the posterior salivary glands of *Eledone. Experiential* **18,** 58,59.
Erspamer, V. and Anastasi, A. (1966) Polypeptides active on plain muscle in the amphibian skin, in *Hypotensive Peptides* (Erdos, E. G., Back, N., Sicuteri, F., and Wilde, A. F., eds.), Springer-Verlag, New York, pp. 63–75.
Erspamer, V., Anastasi, A., Bertaccini, G., and Cei, J. M. (1964) Structure and pharmacological actions of physalaemin, the main active polypeptide of the skin of *Physalaemus fuscumaculatus. Experiential* **20,** 489,490.
Erspamer, V. and Falconieri-Erspamer, G. (1962) Pharmacological actions of eledoisin on extravascular smooth muscle. *Br. J. Pharmacol.* **19,** 337–354.
Erspamer, V. and Melchiorri, P. (1973) Active peptides of the amphibian skin and their synthetic analogs. *Pure Appl. Chem.* **35,** 463–494.
Gaddum, J. H. (1953) Tryptamine receptors. *J. Physiol. Lond.* **119,** 363–368.
Gaddum, J. H. and Schild, H. (1934) Depressor substances in extracts of intestine. *J. Physiol. Lond.* **83,** 1–14.
Gaddum, J. H. and Szerb, J. C. (1961) Assay of substance P on goldfish intestine in a microbath. *Br. J. Pharmacol.* **17,** 451–463.
Garfield, E. (1980) ABCs of cluster mapping. Part 1. Most active fields in the life sciences 1978. *Curr. Contents* **40,** 5–12.
Gater, P. R., Jordon, C. C., and Owen, D. G. (1982) Relative activities of substance P related peptides in the guinea-pig ileum and rat parotid gland, in vitro. *Br. J. Pharmacol.* **75,** 341–348.
Gernandt B. (1942) Untersuchung über die biologische Wirkung der Substanz P. *Acta Physiol. Scand.* **3,** 270–274.
Grabner, K., Lembeck, F., and Neuhold, K. (1959) Substanz P im Gehirn verschiedener Species. *Naunyn Schmiedeberg's Arch. Exp. Pathol. Pharmacol.* **236,** 331–334.
Henry, J. L., Couture, R., Cuello, A. C., Pelletier, G. Quirion, R., and Regoli, D., eds. (1987) *Substance P and Neurokinins.* Springer-Verlag, New York.
Hunter, J. C. and Maggio, J. E. (1984) Pharmacological characterization of a novel tachykinin isolated from mammalian spinal cord. *Eur. J. Pharmacol.* **97,** 159,160.
Jensen, J. and Conlon, J. M. (1992) Substance-P-related and neurokinin-A-related peptides from the brain of the cod and trout. *Eur. J. Biochem.* **206,** 659–664.

Jessell, T. M. (1983) Substance P in the nervous system, in *Handbook of Psychopharmacology. Neuropeptides,* vol. 16 (Iversen, L. L., Iversen, S. D., and Snyder, S. H., eds.), Plenum, New York, pp. 1–105.

Jordan, C. C. and Oehme, P., eds. (1985) *Substance P, Metabolism and Biological Actions.* Taylor & Francis, London.

Kage, R., McGregor, G. P., Thim, L., and Conlon, J. M. (1988) Neuropeptidegamma: A peptide isolated from rabbit intestine that is derived from gammaprepro-tachykinin. *J. Neurochem.* **50,** 1412–1417.

Kangawa, K., Minamino, N., Fukuda, A., and Matsuo, H. (1983) Neuromedin K: A novel mammalian tachykinin identified in porcine spinal cord. *Biochem. Biophys. Res. Commun.* **114,** 533–540.

Kimura, S., Ogawa, T., Goto, K., Sugita, Y., Munekata, E., and Kanazawa, I. (1985) Endogenous ligands for tachykinin receptors in mammals, in *Substance P: Metabolism and Biological Actions* (Jordan, C. C. and Oehme, P., eds.), Taylor & Francis, London, pp. 33–43.

Kimura, S., Okada, M., Sugita, Y., Kanazawa, I., and Munekata, E. (1983) Novel neuropeptides, neurokinins α and β, isolated from porcine spinal cord. *Proc. Jpn. Acad. Ser. B. Phys. Biol. Sci.* **59,** 101–104.

Kozowa, H., Hino, J., Minamino, N., Kangawa, K., and Matsuo, H. (1991) Isolation of four novel tachykinins from frog *(Rana catesbeiana)* brain and intestine. *Biochem. Biophys. Res. Comm.* **177,** 588–595.

Lee, C.-M., Iversen, L., Hanley, M. R., and Sandberg, B. E. B. (1982) The possible existence of multiple receptors for substance P. *Naunyn Schmiedeberg's Arch. Pharmacol.* **318,** 281–287.

Leeman, S. E. and Hammerschlag, R. (1967) Stimulation of salivary secretion by a factor from hypothalamic tissue. *Endocrinology* **81,** 803–810.

Lembeck, F. (1953) Zur Frage der zentralen Ubertragung afferenter Impulse. Ill. Das Vorkommen und die Bedeutung der Substanz P in den dorsalen Wurzeln des Ruckenmarks. *Naunyn Schmiedeberg's Arch. Exp. Pathol. Pharmacol.* **219,** 197–213.

Lembeck, F. and Starke, K. (1968) Substanz P und Speichelsekrection. *Naunyn Schmeideberg's Arch. Exp. Pathol. Pharmacol.* **259,** 375–385.

Lembeck, F. and Zetler, G. (1962) Substance P. A polypeptide of possible physiological significance especially within the nervous system. *Int. Rev. Neurobiol.* **4,** 159–215.

Lembeck, F. and Zetler, G. (1971) Substance P, in *Pharmacology of Naturally Occurring Polypeptides and Lipid-Soluble Acids* (Walker, J. M., ed.), Pergamon, Oxford, UK, p. 2972.

Maggio, J. E. (1988) Tachykinins. *Annu. Rev. Neurosci.* **11,** 13–28.

Maggio, J. E. (1985) Kassinin in mammals: the newest tachykinins. *Peptides* **6 (Suppl. 3),** 237–243.

Maggio, J. E. and Hunter, J. C. (1984) Regional distribution of kassinin-like immunoreactivity in rat central and peripheral tissues and the effect of capsaicin. *Brain Res.* **307,** 370–373.

Maggio, J. E. and Mantyh, P. W. (1989) Gut tachykinins, in *Handbook of Physiology* (Makhlouf, G., ed.), American Physiological Society, Bethesda, MD, pp. 661–690.

Maggio, J. E., Sandberg, B. E. B., Bradley, C. V., Iversen, L. L., Santikarn, S., Williams, D. H., Hunter, J. C., and Hanley, M.R. (1983) Substance K: A novel tachykinin in mammalian spinal cord, in *Substance P—Dublin, 1983* (Skrabanek, P. and Powell, D., eds.), Boole, Dublin, Ireland, pp. 20,21.

Meinardi, H. and Craig, L. C. (1966) Studies of substance P, in *Hypotensive Peptides* (Erdos, E. G., Back, N., Sicuteri, F., and Wilde, A. F., eds.), Springer-Verlag, New York, pp. 594–606.

Melchiorri, P. and Negri, L. (1984) Evolutionary aspects of amphibian peptides, in *Evolution and Tumour Pathology of the Neuroendocrine System* (Falkmer, S., Håkanson, R., and Sundler, F., eds.), Elsevier, New York, pp. 231–244.

Minamino, N., Kangawa, K., Fukuda, A., and Matsuo, H. (1984) Neuromedin L: A novel mammalian tachykinin identified in porcine spinal cord. *Neuropeptides* **4,** 157–166.

Nakajima, T., Yasuhara, T., Erspamer, V., Falconieri-Erspamer, G., Negri, L., and Endean, R. (1980) Physalaemin- and bombesin-like peptides in the skin of the Australian leptodactylid frog Uperoleia rugosa. *Chem. Pharm. Bull.* **28,** 689–695.

Nawa, H., Hirose, T., Takashima, H., Inayama, S., and Nakanishi, S. (1983) Nucleotide sequences of cloned cDNAs for two types of bovine substance P precursor. *Nature* **306,** 32–36.

Nicoll, R. A., Schenker, C., and Leeman, S. E. (1980) Substance P as a neurotransmitter candidate. *Annu. Rev. Neurosci.* **3,** 227–268.

Oehme, P., Bienert, M., Hecht, K., and Bergmann, J. (1981) Substanz P. Ausgewählte Probleme der Chemie, Biochemie, Pharmacologie, Physiologie und Pathophysiologie. *Beitr. Wirkstofforsch.* **12,** 1–185.

O'Harte, F., Burcher, E., Lovas, S., Smith, D. D., Vaudry, H., and Conlon, J. M. (1991) Ranakinin: A novel NK1 tachykinin receptor agonist isolated with neurokinin B from the brain of the frog Rana ridibunda. *J. Neurochem.* **57,** 2086–2091.

Otsuka, M. S., Konishi, S., and Takahashi, T. (1972a) A further study of motoneuron depolarizing peptide extracted from dorsal roots of bovine spinal nerves. *Proc. Jpn. Acad.* **48,** 747–752.

Otsuka, M. S., Konishi, S., and Takahashi, T. (1972b) The presence of a motoneuron depolarizing peptide in bovine dorsal roots of spinal nerves. *Proc. Jpn. Acad.* **48,** 342–346.

Pernow, B. (1951) Substance P distribution in the digestive tract. *Acta Physiol. Scand.* **24,** 97–102.

Pernow, B. (1953) Studies on substance P: Purification, occurrence and biological actions. *Acta Physiol. Scand. Suppl.* **105,** 1–90.

Pernow, B. (1963) Pharmacology of substance P. *Ann. NY Acad. Sci.* **104,** 393–402.

Pernow, B. (1983) Substance P. *Pharmacol. Rev.* **35,** 85–141.

Piercey, M. F., Dobry, P. J. K., Einspahr, F. J., Schroeder, L. A., and Masiques, N. (1982) Use of substance P fragments to differentiate substance P receptors of different tissue. *Regul. Pept.* **3,** 337–349.

Porter, R. and O'Connor, M., eds. (1982) *Substance P in the Nervous System.* Pitman, London (Ciba Found. Symp. 91).

Sandberg, B. E. B. (1985) Structure-activity relationships for substance P: A review, in *Substance P: Metabolism and Biological Actions* (Jordan, C. C. and Oehme, P., eds.), Taylor & Francis, London, pp. 65–81.
Sandberg, B. E. B. and Iversen, L. L. (1982) Substance P. *J. Med. Chem.* **25,** 1090–1095.
Schoofs, L., Holman, G. M., Hayes, T. K., Nachman, R. J., and De Loof, A. (1990a) Locustatachykinin I and II, two novel insect neuropeptides with homology to peptides of the vertebrate tachykinin family. *FEBS Lett.* **261,** 397–401.
Schoofs, L., Holman, G. M., Hayes, T. K., Kochansky, J. P., Nachman, R. J., and De Loof, A. (1990b) Locustatachykinin III and IV: Two novel insect neuropeptides with homology to peptides of the vertebrate tachykinin family. *Regul. Pept.* **31,** 199–212.
Segawa, T., Nakata, Y., and Inoue, A. (1985) Precursors for substance P, in *Substance P: Metabolism and Biological Actions* (Jordan, C. C. and Oehme, P., eds.), Taylor & Francis, London, pp. 1–12.
Simmaco, M., Severini, C., De Biase, D., Barra, D., Bossa, F., Roberts, J. D., Melchiorri, P., and Erspamer, V. (1990) Six novel tachykinin-and bombesin-related peptides from the skin of the Austrialian frog *Pseudophryne guntheri. Peptides* **11,** 299–304.
Skrabanek, P. (1984) Eye of newt and toe of frog: substance P and the charmed pot of neuropeptides. *Ir. J. Med. Sci.* **153,** 47–59.
Skrabanek, P. and Powell, D. (1977) *Substance P.* Vol. 1. Eden, Montreal, Canada.
Skrabanek, P. and Powell, D. (1980) *Substance P.* Vol. 2. Eden, Montreal, Canada.
Skrabanek, P. and Powell, D. (1983a) *Substance P.* Vol. 3. Eden, Montreal, Canada.
Skrabanek, P. and Powell, D. (1983b) *Substance P—Dublin 1983.* Boole, Dublin, Ireland.
Studer, R.O., Trzeciak, H., and Lergier, W. (1973) Isolierung und Aminosauresequenz Von Substanz P aus Pferdedarm. *Helv. Chim. Acta* **56,** 860–866.
Tatemoto, K., Lundberg, J. M., Jornvall, H., and Mutt, V. (1985) Neuropeptide K: Isolation, structure and biological activities of a novel brain tachykinin. *Biochem. Biophys. Res. Commun.* **128,** 947–953.
Teichberg, V. I., Cohen, S., and Blumberg, S. (1981) Distinct classes of substance P receptors revealed by a comparison of the activities of substance P and some of its segments. *Regul. Pept.* **1,** 323–333.
Theodorsson-Norheim, E., Brodin, E., Norheim, I, and Rosell, N. (1985) Antisera raised against elodoisin and kassinin detect immunoreactive material in rat tissue extracts: Tissue distribution and chromotographic characterization. *Regul. Pept.* **9,** 229–244.
Theodorsson-Norheim, E., Norheim, I., Oberg, K., Brodin, E., Lundberg, J. M., Tatemoto, K., and Lindgrenn, P. G. (1985) Neuropeptide K: A major tachykinin in plasma and tumor tissues from carcinoid patients. *Biochem. Biophys. Res. Commun.* **131,** 77–83.
Too, H. P., Cordova, J. L., and Maggio, J. E. (1989) Heterogeneity of tachykinin peptides in the rat spinal cord and dorsal root ganglia. *Peptides* **10,** 25–30.

Vogler, K., Haefely, W., Hurlimann, A., Studer, R. O., Lergier, W., Strassle, R., and Berneis, K. H. (1963) A new purification procedure and biological properties of substance P. *Ann. NY Acad. Sci.* **104,** 378–389.

Von Euler, U. S. (1936) Untersuchungen uber Substanz P, die atropinfester, darmerregende und gefbasserweiternde Substanz aus Darm und Gehirn. *Naunyn Schmiedeberg's Arch. Exp. Pathol. Pharmacol.* **181,** 181–197.

Von Euler, U. S. (1942) Herstellung und Eigenschaften Von Substanz P. *Acta Physiol. Scand.* **4,** 373–375.

Von Euler, U. S. and Gaddum, J. H. (1931) An unidentified depressor substance in certain tissue extracts. *J. Physiol. Lond.* **72,** 74–87.

Von Euler, U. S. and Ostlund, E. (1956) Occurrence of a substance P-like polypeptide in fish intestine and brain. *Br. J. Pharmacol.* **11,** 323–325.

Von Euler, U. S. and Pernow, B. (1977) *Substance P.* Raven, New York (Nobel Symp Ser., no. 37).

Wang, Y., O'Harte, F., and Conlon, J. M. (1992a) Structural characterization of tachykinins (neuropeptide gamma, neurokinin A and substance P) from a reptile, *Alligator mississipiensis. Gen. Comp. Endocrinol.* **88,** 277–286.

Wang, Y., Badgery-Parker, T., Lovas, S., Chartrel, N., Vaudry, H., Burcher, E., and Conlon, J. M. (1992b) Primary structure and receptor-binding properties of a neurokinin A-related peptide from frog gut. *Biochem. J.* **287,** 827–832.

Watson, S. P., Sandberg, B. E. B., Hanley, M. R., and Iversen, L. L. (1983) Tissue selectivity of substance P alkyl esters suggesting multiple receptors. *Eur. J. Pharmacol.* **87,** 77–84.

Waugh, D., Hazon, N., Balment, R. J., and Conlon, J. M. (1993) Structures and biological activities of novel tachykinins from fish brain. *Thirteenth Am. Peptide Symp.*, in press.

Yasuhara, T., Nakjima, T., Falconieri-Erspamer, G., and Erspamer, V. (1981) New tachykinins Glu2 Pro5-kassinin (Hylambates-kassinin) and hylambatin in the skin of the African rhacophorid frog *Hylambates maculatus. Biomed. Res.* **2,** 613–617.

Zuber, H. (1963) Isolation of substance P from bovine brain. *Ann. NY Acad. Sci.* **104,** 391–392.

CHAPTER 2

History of Tachykinin Receptors

Leslie L. Iversen

1. Introduction

The rapid advances in molecular pharmacology in recent years have made the concept of multiple receptors and receptor subtypes for neurotransmitters and neuropeptides a commonplace. We are no longer surprised to learn that on closer examination the muscarinic receptor exists in five distinct molecular forms, encoded by separate genes (Hulme et al., 1990), or that the dopamine receptor similarly has five subtypes, some of which in turn exist in a variety of isoforms because of variable mRNA splicing (Iversen, 1992), or the existence of genetically determined minor variants (Van Tol et al., 1992). That the varied biological actions of the three mammalian tachykinins should be mediated by three different receptors is thus easily accepted—indeed, one would anticipate that such a complex family of related neuropeptides might be associated with a larger number of receptor subtypes.

This was not the situation 10 years ago, however, when the existence of neurokinin A (NKA) and neurokinin B (NKB) in mammalian tissues had not yet been demonstrated. Substance P (SP) was

The Tachykinin Receptors Ed.: S. H. Buck © 1994 Humana Press Inc., Totowa, NJ

the only known mammalian tachykinin, and although it was clear that it exerted a multitude of biological effects, there was no compelling reason to suppose that these were mediated by more than one class of receptor. This chapter will briefly review developments in this field over the past decade and prospects for the future.

2. Multiple Tachykinin Receptor Subtypes: Evidence from Bioassay Data

2.1. Early Data

In the 1970s and early 1980s as SP and related tachykinins were first chemically synthesized and the pure materials became more widely available, several laboratories published the results of studies in which the biological activities of the various peptides were compared. Most were agreed that whereas all of the peptides shared the common tachykinin C-terminal sequence

-Phe-X-Gly-Leu-MetNH2

and had a similar spectrum of biological effects, their potencies in particular test systems were markedly influenced by the N-terminal portion of the molecule.

Piercey et al. (1982), from the Upjohn group, following a different line of argument, pursued their earlier studies on the biological effects of N-terminal fragments of SP with the proposal that substance P receptors be classified as "SP-1" and "SP-2" to indicate those insensitive or sensitive to N-terminal fragments, respectively. It is clear that peptides related to the N-terminus of substance P do exert biological effects, and these are often opposite to those of SP (Hall and Stewart, 1983). High-affinity binding sites for the N-terminal heptapeptide, SP(1–7), have also been identified in brain (Igwe et al., 1990). Nevertheless, it is probably misleading to think of the effects of N-terminal fragments of SP as being mediated by "tachykinin receptors" as such molecules no longer possess the common tachykinin C-terminal sequence.

For our own laboratory in the early 1980s, the most influential study comparing tachykinin peptides was that from Erspamer's group

(Falconieri-Erspamer et al., 1980). In this paper, the tachykinins physalaemin, SP, and eledoisin were compared with the newly discovered frog skin peptide, kassinin, using more than 20 different bioassays. The conclusion was clear: In nearly all of the bioassays SP and physalaemin diplayed similar relative potencies. Eledoisin and kassinin, however, were up to 50 times more potent than either physalaemin or SP in some bioassays, equipotent in others, and in some cases (e.g., blood pressure effects) had only 5–20% the potency of physalaemin or SP. Teichberg et al. (1981) compared the potencies of SP and a range of C-terminal fragments in different bioassays. They found that, whereas in guinea pig ileum the potency of SP was similar to that of its hepta- and octapeptide fragments, SP was markedly less potent than these fragments in bovine pupillary sphincter or guinea pig urinary bladder. They concluded that these results suggested the existence of different subclasses of SP receptors. Couture and Regoli (1982) compared the effects of SP and related peptides on a variety of smooth muscle bioassays. They showed that there was incomplete cross-desensitization between SP or physalaemin (which showed complete cross-desensitization to each other) and eledoisin. They concluded, however, "It is too early to affirm that these different patterns indicate the existence of different receptor types; the discovery and use of specific and competitive inhibitors will be required to prove it."

2.2. The "SP-P" and "SP-E" Classification

In our laboratory in Cambridge, England, C. M. Lee undertook a comparative study of SP and several related tachykinins in guinea pig ileum and rat vas deferens bioassays. The key findings were that whereas SP, physalaemin, eledoisin, and kassinin were approximately equipotent in guinea pig ileum, the latter two peptides were markedly more potent than SP or physalaemin in rat vas deferens (Table 1). Like Couture and Regoli (1982), we also found that eledoisin and SP failed to show complete cross-tachyphylaxis in the guinea pig ileum. It was suggested that two classes of receptors might be distinguished, the "SP-P" (physalaemin), which responded potently to SP and physalaemin and the "SP-E" (eledoisin), which responded

Table 1
SP-P and SP-E Receptors, Bioassay Data

	EC_{50}, n*M*	
Test substance	Guinea pig ileum	Rat vas deferens
Substance P	1.4	3500.0[a]
Physalaemin	0.9	5833.0[a]
Eledoisin	1.6	40.7[a]
Kassinin	1.3	19.3[a]
SP-*O*-Methyl	1.5	>10000.0[b]

[a]Data calculated from Lee et al. (1982).
[b]Data calculated from Watson et al. (1983).

most potently to eledoisin or kassinin (Lee et al., 1982). Watson (1983) went on to show that although there were differences in the rate of degradation of [^{3}H] SP in rat vas deferens and guinea pig ileum preparation incubated with the radiolabeled peptide in vitro, these differences were not sufficiently great to account for the marked difference in potency of SP in these two tissues. Watson et al. (1983) also strengthened the original hypothesis by demonstrating that SP-*O*-methyl ester exhibited a marked selectivity for SP-P receptors at which the synthetic peptide was almost equipotent with SP, while it had barely detectable activity on SP-E sites (Table 1). Watson et al. (1983) also showed that an SP-P pattern of agonist selectivity could be demonstrated in guinea pig ileum, vas deferens, and bladder, and in rat bladder, whereas an SP-E pattern was found in rat vas deferens and duodenum, and in mouse and hamster bladder.

Similar findings were made by others. Early support for the SP-P and SP-E classification came from a group at ICI Pharmaceuticals (Growcott et al., 1983) who found that eledoisin selectively protected a subgroup of tachykinin receptors from phenoxybenzamine alkylation in guinea pig urinary bladder in vitro. Regoli and colleagues in Montreal (Regoli et al., 1984a–c) reported on a large series of experiments comparing tachykinin peptide responses in a range of different bioassays. Although partly adapting the SP-P and

SP-E classification, they eventually proposed their own classification of "SP-P," "SP-A," and "SP-B" receptors for sites preferring SP, NKA, and NKB, respectively (Dion et al., 1987).

The postulated existence of SP-E receptors for which a kassinin-like peptide might be the preferred agonist, rather than SP, prompted our laboratory to initiate a search for such a hypothetical "substance K" (*see* Maggio, Chapter 1). Following the discovery of the novel mammalian peptides, NKA (= substance K) and NKB (= neuromedin K), it was reported that both of these did indeed exhibit SP-E-type pharmacology (Hunter and Maggio, 1984; Kimura et al., 1984; Nawa et al., 1984; Holzer-Petsche et al., 1985).

2.3. *The "SP-N" Receptor*

Shortly after the idea of two receptor classes had been assimilated—if not universally accepted—a third class was proposed by Laufer et al. (1985). The quality of the supporting data was so convincing that this so-called "SP-N" receptor subtype gained almost instant recognition. Laufer et al. took advantage of the fact that the SP response of the isolated guinea pig ileum has both a neurally mediated and a direct smooth muscle component. Substance P acts on postganglionic cholinergic neurons in the gut to cause acetylcholine release, which in turn causes contraction. This neurally mediated response to SP had usually been abolished by the addition of atropine in previous studies of SP receptors on smooth muscle. Laufer et al. deliberately set out to study the neural component, which they were able to do by either desensitizing the smooth muscle receptors by exposure to SP-*O*-methyl ester, or by using the antagonist, [Arg^{6}, D-Trp^{7,9},MePhe^{8}]SP(6–11). The neuronal receptor proved to be much more sensitive to NKB than to SP or other tachykinins (Table 2). The suggestion that the neuronal receptor for SP in guinea pig ileum differed from the smooth muscle receptor had earlier been made by the Glaxo group (Hawcock et al., 1982) on the basis of their finding that the synthetic peptide, [D-Pro^{2},D-Phe^{7},D-Trp^{8}]SP, acted as an agonist at the neuronal sites, but had weak antagonist properties at the smooth muscle SP receptors. The Israeli group went on to develop the synthetic peptide agonist senktide (succinyl[Asp^{6},MePhe^{8}]SP

Table 2
Potencies of Tachykinins at SP-N (NK_3) Receptor in Guinea Pig Ileum

Tachykinin	EC_{50}, nM	
	Neuronal[a]	Muscle[b]
Substance P	60.0	2.5
Neurokinin A	46.1	20.8
Neurokinin B	1.9	2.7

[a]Data from Laufer et al. (1985). Neuronal receptors were studied in isolated tissues previously exposed to SP-*O*-Methyl ester (100 n*M*) for 2 min to desensitize smooth muscle receptors. Muscle receptors were studied in the presence of atropine (1 μ*M*) to block neuronally mediated responses.

(6–11)), which acts as a potent agonist at the SP-N sites, with virtually no effects on SP-P or SP-E sites (Wormser et al., 1986). Senktide remains one of the most selective agonist tools available to distinguish among the tachykinin receptor subtypes *(see below)*.

2.4. The "NK" Receptor Classification

At the 1986 Substance P Symposium in Montreal, it was agreed after a lengthy debate to adopt the nomenclature "NK-1," "NK-2," and "NK-3" (i.e., NK_1, NK_2, NK_3) to replace the previous SP-P, SP-E, and SP-N terminology and the alternative proposed by Regoli and colleagues. This nomenclature has gained wide acceptance, at least among pharmacologists. McKnight et al. (1988a) observed an unusual pattern of agonist specificity in guinea pig trachea; the order of potency of the tachykinins was that seen at NK_2 receptors: neuropeptide K > NKA > NKB > SP. However, the synthetic peptide, [Glp^6,L-Pro^9]SP(6–11) ("septide"), which is generally NK_1 receptor selective, was even more potent than the corresponding D-Pro^9 compound in guinea pig trachea. Other synthetic peptides appeared to behave anomalously. These results led McKnight and colleagues to postulate the possible existence of a fourth tachykinin receptor subtype, the "NK_4," to explain the findings. Others, however, have examined the pharmacology of the tachykinin responses in guinea pig trachea and come to the conclusion that there is a mixture of NK_1 and NK_2 sites that explains the complex pattern of peptide specificity (Advenier et al., 1987; Dion et al., 1987; Maggi et al., 1991). The active degradation of tachykinins and related peptides by peptidases in the isolated trachea may contribute to the unusual rank order of agonist potencies observed in this preparation (Hagan et al., 1991). More recently, however, Burcher et al. (1991) studied the actions of neuropeptide gamma, the most potent in causing bronchospasm in human isolated bronchus, and concluded that it acted via a "nonclassical NK_2 receptor."

2.5. The Development of Selective Agonists

Until recently, there have been few tachykinin receptor antagonists available, and the peptide antagonists that were developed originally by Folkers and others (Folkers et al., 1986) proved largely nonselective in their receptor specificity (Regoli et al., 1984c). Although the classification of receptors has traditionally relied on the use of selective antagonists, the development of the concept of multiple tachykinin receptors rested entirely on differences in the relative potencies of agonists. The development of a series of highly selective, synthetic, agonist peptides during the past decade was thus particularly important in strengthening the concept of multiple receptors and in providing valuable research tools.

Several laboratories recognized the importance of the conformation of the peptide chain in the tachykinins and were able to develop receptor-selective ligands by synthesizing conformationally restricted analogs. The C-terminal residue, Gly^9, proved to be of particular importance, and the introduction of L-Pro or D-Pro at this point (Piercey et al., 1985; Lavielle et al., 1986; Petitet et al., 1991) or a lactam bridge (Cascieri et al., 1986; Hagan et al., 1991) yielded analogs with improved receptor selectivity. Regoli and Drapeau and colleagues in Montreal undertook a systematic program of work synthesizing and testing ever more potent peptide agonist ligands, and some of the best of the existing selective agonists come from their work (Drapeau et al., 1987a,b; Rovero et al., 1989). Some of the analogs now available are listed in Table 3.

3. Heterogeneity of Radioligand Binding Sites for Tachykinins

In addition to pharmacological studies, which revealed differences in the rank order of potencies of agonists in different tissues, the application of radioligand binding techniques to the study of tachykinin receptors soon provided evidence for heterogeneity.

The first successful radioligand binding studies, using ^{125}I- or ^{3}H-labeled SP were performed with NK_1 sites in salivary glands (Liang and Cascieri, 1981; Lee et al., 1983), brain (Hanley et al., 1980; Cascieri and Liang, 1983; Perrone et al., 1983; Torrens et al., 1983), and gut (Watson and Iversen, 1984).

Shortly thereafter, when ^{125}I-eledoisin was assessed as a radioligand, groups at Merck Research Laboratories and in Paris discovered that this radioligand labeled a different population of high-affinity sites in rat brain membranes (Beaujouan et al., 1984; Cascieri et al., 1985; Torrens et al., 1985). Although the eledoisin binding site in brain was at first thought to represent the "SP-E" receptor (i.e., NK_2), it later became clear that the NK_3 receptor accounts for most of the eledoisin binding in brain. Thus, the NK_3 agonist NKB was found to have preferential affinity for the eledoisin binding sites in brain (Torrens et al., 1984; Lee et al., 1986).

Further evidence for distinct populations of tachykinin receptors emerged from autoradiographic studies, mapping the distribution of ^{125}I-SP and ^{125}I-eledoisin binding sites in brain and other tissues (Rothman et al., 1984; Ninkovic et al., 1985; Quirion and Dam, 1985; Beaujouan et al., 1986; Burcher et al., 1986;). These studies clearly demonstrated anatomically distinct patterns of distribution of the binding sites for these and other tachykinin receptor radioligands.

4. Evidence from Receptor Cloning and the Use of Selective Antagonists

The definitive proof that more than one tachykinin receptor existed came from Kyoto in Japan with the successful cDNA cloning of three receptor genes from mammalian tissues that appear to correspond to the NK_1, NK_2, and NK_3 receptor types (Masu et al., 1987;

Table 3
Selective Agonists for Tachykinin Receptors

Agonist	Selectivity	Reference
[L-Pro^9]SP(6–11) = SEPTIDE	NK_1	Piercey et al. (1985)
[L-Pro^9]SP(1–11)	NK_1	Lavielle et al. (1986)
GlpPhePhe(R)Gly[ANC-2]LeuMetNH$_2$ = L-363,581	NK_2/NK_3	Cascieri et al. (1984)
succinyl[Asp^6,MePhe^8]SP(6–11) = SENKTIDE	NK_3	Wormser et al. (1986)
[MePhe^7]NKB(4–10)	NK_3	Drapeau et al. (1987b)
[Sar^9,Met(O$_2$)^{11}]SP	NK_1	Drapeau et al. (1987a)
[β-Ala^8]NKA(4–10)	NK_2	Rovero et al. (1989)

Nakanishi, 1991). These discoveries validated the conclusions reached from the earlier work summarized above.

Meanwhile, the selective antagonists that tachykinin pharmacology had long needed were finally developed, and during the past few years there have been rapid advances in this regard. Whereas earlier peptide-based antagonists were largely nonselective, newer peptides developed in this laboratory (McKnight et al., 1988b, 1991) and by the Glaxo group (Ward et al., 1990) offered new competitive antagonists with selectivity for NK_2 (L-659,877) and NK_1 (GR71251) receptors, respectively. The Menarini group in Italy also developed linear peptides that acted as selective antagonists at NK_2 sites, and they used one of these compounds (MEN 10,376) along with the Merck antagonist, L-659,877, to study NK_2 receptors in various tissues. They concluded that there might be two subvarieties of this receptor, the NK_{2A} (defined as more sensitive to MEN 10,376 than L-659,877) and the NK_{2B} with the reverse antagonist selectivity (Maggi et al., 1990, 1992).

More recently, the important discovery of potent and highly selective nonpeptide antagonists for NK_1 (Garret et al., 1991; Snider et al., 1991; Appell et al., 1992) and for NK_2 receptors (Emonds-Alt et al., 1992) has further accelerated progress in this field. One of the surprising features to emerge from studies of these new compounds is that striking species differences exist in the sensitivity of both NK_1 and NK_2 sites to these antagonists (Gitter et al., 1991; Advenier et al., 1992).

5. Future Prospects

Developments have been rapid in the tachykinin receptor field in the last few years, with the first successful cloning of tachykinin receptors and the discovery of nonpeptide antagonists. It is already being suggested, on pharmacological grounds, that the NK_1 and NK_2 receptors may be capable of further subclassification. Is it possible, for example, that the striking differences in NK_1 receptor sensitivity to nonpeptide antagonists among different species reveals the possible existence of two distinct classes of NK_1 receptor? If so, these differences seem to rest on very minor differences between the molecules, as guinea pig and rat NK_1 receptors differ only in a very few amino acid residues (Gorbulev et al., 1992). Petitet et al. (1992) have also suggested the possible existence of NK_1 subtypes in guinea pig ileum, on the grounds that SP-*O*-Me and septide were relatively weak in displacing ^{3}H-[Pro^9]SP from NK_1 binding sites, but were almost equipotent with SP in evoking contractile responses.

It is worth noting that the techniques used to clone the tachykinin receptors so far identified may have favored the discovery of those that are coupled to inositol phospholipid degradation via phospholipase C. A further series of tachykinin receptors that are coupled to adenylate cyclase or to other intracellular coupling mechanisms may remain to be discovered. It will come as no surprise if further members of the tachykinin receptor family emerge in the next few years.

References

Advenier, C., Naline, E., Drapeau, G., and Regoli, D. (1987) Relative potencies of neurokinins in guinea pig. *Eur. J. Pharmacol.* **139,** 133–137.

Advenier, C., Rouissi, N., Nguyen, Q. T., Emonds-Alt, X., Brelière, J-C., Neliat, G., Naline, E., and Regoli, D. (1992) Neurokinin A (NK_2) receptor revisited with SR 48968, a potent nonpeptide antagonist. *Biochem. Biophys. Res. Comm.* **184,** 1418–1424.

Appell, K. C., Fragale, B. J., Losgig, J., Singh, S., and Tomczuk, B. E. (1992) Antagonists that demonstrate species differences in neurokinin-1 receptors. *Mol. Pharmacol.* **11,** 774–778.

Beaujouan, J. C., Torrens, Y., Viger, A., and Glowinski, J. (1984) A new type of tachykinin binding site in the rat brain characterized by specific binding of a labeled eledoisin derivative. *Mol. Pharmacol.* **26,** 248–254.

Beaujouan, J. C., Torrens, Y., Saffroy, M., and Glowinski, J. (1986) Quantitative autoradiographic analysis of the distribution of binding sites for [^{125}I]Bolton Hunter derivatives of eledoisin and substance P in the rat brain. *Neuroscience* **18,** 857–875.

Burcher, E., Buck, S. H., Lovenberg, W., and O'Donohue T. L. (1986) Characterization and autoradiographic localization of multiple tachykinin binding sites in gastrointestinal tract and bladder. *J. Pharm. Exp. Ther.* **236,** 819–831.

Burcher, E., Alouan, L. A., Johnson, P. R. A., and Black, J. L. (1991) Neuropeptide gamma, the most potent contractile tachykinin in human isolated bronchus, acts via a "nonclassical" NK_2 receptor. *Neuropeptides* **20,** 79–82.

Cascieri, M. A. and Liang, T. (1983) Characterization of the substance P receptor in rat brain cortex membranes and the inhibition of radioligand binding by guanine nucleotides. *J. Biol. Chem.* **258,** 5158–5164.

Cascieri, M. A., Chicchi, G. G., and Liang, T. (1985) Demonstration of two distinct tachykinin receptors in rat brain cortex. *J. Biol. Chem.* **260,** 1501–1507.

Cascieri, M. A., Chicchi, G. G., Freidinger, R. M., Dylion Colton, C., Perlow, D. S., Williams, B., Curtis, N. R., McKnight, A. T., Maguire, J. J., Veber, D. F., and Liang, T. (1986) Conformationally constrained tachykinin analogs which are selective ligands for the eledoisin binding site. *Mol. Pharmacol.* **29,** 34–38.

Couture, R. and Regoli D. (1982) Mini review: smooth muscle pharmacology of substance *P. Pharmacology* **24,** 1–25.

Dion, S., D'Orléans-Juste, P., Drapeau, G., Rhaleb, N.-E., Rouissi, N., Tousignant, C., and Regoli, D. (1987) Characterization of neurokinin receptors in various isolated organs by the use of selective agonists. *Life Sciences* **41,** 2269–2278.

Drapeau, G., d'Orléans-Juste, P., Dion, S., Rhaleb, N.-E., and Regoli, D. (1987a) Specific agonists for neurokinin B receptors. *Eur. J. Pharmacol.* **136,** 401–403.

Drapeau, G., d'Orléans-Juste, P., Dion, S., Rhaleb, N.-E., Rouissi, N.-E., and Regoli, D. (1987b) Selective agonists for substance P and neurokinin receptors. *Neuropeptides* **10,** 43–54.

Emonds-Alt, X., Vilain, P., Goulaouic, P., Proietto, V., Van Broeck, D., Advenier, C., Naline, E., Neliat, G., Le Fur, G., and Breliere, J. C. (1992) A potent and selective nonpeptide antagonist of the neurokinin A (NK_2) receptor. *Life Sciences* **50,** PL101–PL106.

Falconieri-Erspamer, G., Erspamer, V., and Piccinelli, D. (1980) Parallel bioassay of physalaemin and kassinin, a tachykinin dodecapeptide from the skin of the African frog kassina senegalensis. *Naunyn Schmiedeberg's Arch. Pharmacol.* **311,** 61–65.

Folkers, K., Rosell, S., Chu, J.-Y., Lu, L.-A., Tang, P.-F.L., and Ljungqvist, A. (1986) Design and synthesis of antagonists of substance P. *Acta Chem. Scand.* **B40,** 295–302.

Garret, C., Carruette, A., Fardin, V., Moussaoui, S., Peyronel, J.-F., Blanchard, J.-C., and Laduron, P.M. (1991) Pharmacological properties of a potent and selective nonpeptide substance P antagonist. *Proc. Natl. Acad. Sci. USA* **88,** 10,208–10,212.

Gitter, B. D., Waters, D. C., Bruns, R. F., Mason, N. R., Nixon, J. A., and Jeffry Howbert, J. (1991) Species differences in affinities of nonpeptide antagonists for substance P receptors. *Eur. J. Pharmacol.* **197,** 237,238.

Gorbulev, A., Akhundova, A., Luzuis, H., and Fahrenholz, F. (1992) Molecular cloning of substance P receptor cDNA from guinea-pig uterus. *Biochim. Biophys. Acta* **1131,** 99–102.

Growcott, J. W., Jamieson, A., Tarpey, A. V., and Topham, L. D. (1983) Further evidence for multiple tachykinin receptors. *Eur. J. Pharmacol.* **86,** 59–64.

Hagan, R. M., Ireland, S. J., Jordan, C. C., Beresford, I. J. M., Deal, M. J., and Ward, R. (1991) Receptor-selective, peptidase-resistant agonists at neurokinin NK-1 and NK-2 receptors: new tools for investigating neurokinin function. *Neuropeptides* **19,** 127–135.

Hall, M. E. and Stewart, J. M. (1983) Substance P and behavior: opposite effects of N-terminal and C-terminal fragments. *Peptides* **4,** 763–768.

Hanley, M. R., Sandberg, B.E. B., Lee, C. M., Iversen, L. L., and Wade, R. (1980) Specific binding of ^{3}H-substance P to rat brain membranes. *Nature (Lond.)* **286,** 810–812.

Hawcock, A. B., Hayes, A. G., and Tyers, M. B. (1982) Agonist effects of [D-Pro^2,D-Phe^7,D-Trp^9]substance P—evidence for different receptors. *Eur. J. Pharmacol.* **80,** 135–138.

Holzer-Petsche, U., Schimek, E., Amann, R., and Lembeck, F. (1985) *In vivo* and *in vitro* actions of mammalian tachykinins. *Naunyn Schmiedeberg's Arch. Pharmacol.* **330,** 130–135.

Hulme, E. C, Birdsall, N. J. M., and Buckley, N. J. (1990) Muscarinic receptor subtypes. *Annu. Rev. Pharmacol.* **30,** 633–673.

Hunter, J. C. and Maggio, J. E. (1984) Pharmacological characterisation of a novel tachykinin isolated from mammalian spinal cord. *Eur. J. Pharmacol.* **97,** 159,160.

Igwe, O. J., Kim, D. C., Seybold, V. S., and Larson, A. A. (1990) Specific binding of substance P aminoterminal heptapeptide [SP(1–7)] to mouse brain and spinal cord membranes. *J. Neurosci.* **10,** 3653–3663.

Iversen, L. L. (1992) Which D4 do you have? *Nature (Lond.)* **358,** 109.

Kimura, S., Katsutoshi, G., Ogawa, T., Sugita, Y., and Kanazaw, I. (1984) Pharmacological characterization of novel mammalian tachykinins, neurokinin α and neurokinin β. *Neurosci. Res.* **2,** 97–104.

Laufer, R., Wormser, U., Friedman, Z. Y., Gilon, C., Chorev, M., and Selinger, Z. (1985) Neurokinin B is a preferred agonist for a neuronal substance P receptor and its action is antagonized by enkephalin. *Proc. Natl. Acad. Sci. USA* **82,** 7444–7448.

Lavielle, S., Chassaing, G., Julien, S., Marquet, A., Bergström, L., Beaujouan, J.-C., Torrens, Y., and Glowinski, J. (1986) Specific recognition of SP or NKB receptors by analogs of SP substituted at positions 8 and 9. *Eur. J. Pharmacol.* **125,** 461,462.

Lee, C.-M., Iversen, L. L., Hanley, M. R., and Sandberg, B. E. B. (1982) The possible existence of multiple receptors for substance P. *Naunyn Schmiedeberg's Arch. Pharmacol.* **318,** 281–287.
Lee, C.-M., Javitch, J. A., and Snyder, S.H. (1983) ^{3}H-substance P binding to salivary gland membranes. *Mol. Pharmacol.* **23,** 563–569.
Lee, C.-M., Campbell, N. J., Williams, B. J., and Iversen, L. L. (1986) Multiple tachykinin binding sites in peripheral tissues and in brain. *Eur. J. Pharmacol.* **130,** 209–217.
Liang, T. and Cascieri, M. A. (1981) Substance P receptor on parotid cell membranes. *J. Neurosci.* **1,** 1133–1141.
McKnight, A. T., Maguire, J. J., Varney, M. A., Williams, B. J., and Iversen, L.L. (1988a) Characterisation of tachykinin receptors using selectivity of agonists and antagonists: evidence for an NK-4 type. *Regul. Pept.* **22,** 126.
McKnight, A. T., Maguire, J. J., Williams, B. J., Foster, A. C., Tridgett, R., and Iversen, L. L. (1988b) Pharmacological specificity of synthetic peptides a antagonists at tachykinin receptors. *Regul. Pept.* **22,** 127.
McKnight, A. T., Maguire, J. J., Elliott, N. J., Fletcher, A. E., Foster, A. C., Tridgett, R., Williams, B. J., Longmore, J., and Iversen, L. L. (1991) Pharmacological specificity of novel, synthetic, cyclic peptides as antagonists at tachykinin receptors. *Br. J. Pharmacol.* **104,** 355–360.
Maggi, C. A., Patacchini, R., Guiliani, S., Rovero, P., Dion, S., Regoli, D., Giachetti, A., and Meli, A. (1990) Competitive antagonists discriminate between NK_2 tachykinin receptor subtypes. *Br. J. Pharmacol.* **100,** 588–592.
Maggi, C. A., Patacchini, R., Quartara, L., Rovero, P., and Santicioli, P. (1991) Tachykinin receptors in the guinea-pig isolated bronchi. *Eur. J. Pharmacol.* **197,** 167–174.
Maggi, C. A., Eglezos, A., Quartara, L., Patacchini, R., and Giachetti, A. (1992) Heterogeneity of NK-2 tachykinin receptors in hamster and rabbit smooth muscles. *Regul. Pept.* **37,** 85–93
Masu, Y., Nakayama, K., Tamaki, H., Harada, Y., Kuno, M., and Nakanishi, S. (1987) cDNA cloning of bovine substance-K receptor through oocyte expression system. *Nature* **329,** 836–838.
Nakanishi, S. (1991) Mammalian tachykinin receptors. *Annu. Rev. Neurosci.* **14,** 123–136.
Nawa, H., Doteuchi, M., Igano, K., Inouye, K., and Nakanishi, S. (1984) Substance K: a novel mammalian tachykinin that differs from substance P in its pharmacological profile. *Life Sci.* **34,** 1153–1160.
Ninkovic, M., Beaujouan, J.C., Torrens, Y., Saffroy, M., Hall, M. D., and Glowinski, J. (1985) Differential localization of tachykinin receptors in rat spinal cord. *Eur. J. Pharmacol.* **106,** 463,464.
Perrone, M. H., Diehl, R. E., and Haubrich, D. R. (1983) Binding of [^{3}H] substance P to putative substance P receptors in rat brain membranes. *Eur. J. Pharmacol.* **95,** 131–133.
Petitet, F., Beaujouan, J.-C., Saffroy, M., Torrens, Y., Chassaing, G., Lavielle, S., Besseyre, J., Garret, C., Curruette, A., and Glowinski, J. (1991) Further demonstration that [Pro^9]-substance P is a potent and selective ligand of NK-1 tachykinin receptors. *J. Neurochem.* **56,** 879–889.

Petitet, F., Saffroy, M., Torrens, Y., Lavielle, S., Chassaing, G., Loeuillet, D., Glowinski, J., and Beaujouan, J.-C. (1992) Possible existence of a new tachykinin receptor subtype in the guinea pig ileum. *Peptides* **13,** 383–388.

Piercey, M. F., Dobry, P. J. K., Einspahr, F. J., Schroeder, L. A., and Masiques, N. (1982) Use of substance P fragments to differentiate substance P receptors of different tissues. *Regul. Pept.* **3,** 337–349.

Piercey, M. F., Dobry-Schreur, P. J. K., Masiques, N., and Schroeder, L. A. (1985) Stereospecificity of $SP_1$and SP_2 substance P receptors. *Life Sci.* **36,** 777–780.

Quirion, R. and Dam, T.-V. (1985) Multiple tachykinin receptors in guinea pig brain. High densities of substance K (neurokinin A) binding sites in the substantia nigra. *Neuropeptides* **6,** 191–204.

Regoli, D., d'Orleans-Juste, P., Escher, E., and Mizrahi, J. (1984a) Receptors for substance P. I. The pharmacological preparations. *Eur. J. Pharmacol.* **97,** 161–170.

Regoli, D., Mizrahi, J., d'Orleans-Juste, P., and Escher, E. (1984b) Receptors for substance P. II. Classification by agonist fragments and homologues. *Eur. J. Pharmacol.* **97,** 171–177

Regoli, D., Escher, E., Drapeau, G., d'Orleans-Juste, P., and Mizrahi, J. (1984c) Receptors for substance P. III Classification by competitive antagonists. *Eur. J. Pharmacol.* **97,** 179–189.

Rothman, R. B., Danks, J. A., Herkenham, M., Cascieri, M. A., Chicchi, G. G., Liang, T., and Pert, C.B. (1984) Autoradiographic localization of a novel peptide binding site in rat brain using the substance P analog, eledoisin. *Neuropeptides* **4,** 343–349.

Rovero, P., Pestellini, V., Rhaleb, N.-E., Dion, S., Rouissi, N., Tousignant, C., Télémaque, S., Drapeau, G., and Regoli, D. (1989) Structure-activity studies of Neurokinin A. *Neuropeptides* **13,** 263–270.

Snider, R. M., Constantine. J. W., Lowe III, J. A., Longo, K. P., Lebel, W. S., Woody, H. A., Drozda, S. E., Desai, M. C., Vinick, F. J., Spencer, R. W., and Hess, H.-J. (1991) A potent nonpeptide antagonist of the substance P (NK_1) receptor. *Science* **251,** 435–437.

Teichberg, V. I., Cohen, S., and Blumberg, S. (1981) Distinct classes of substance P receptors revealed by a comparison of the activities of substance P and some of its segments. *Regul. Pept.* **1,** 327–333.

Torrens, Y., Beaujouan, J.C., Viger, A. and Glowinski, J. (1983) Properties of a ^{125}I-substance P derivative binding to synaptosomes from various brain structures and the spinal cord of the rat. *Naunyn Schmiedeberg's Arch. Pharmacol.* **324,** 134–139.

Torrens, Y., Lavielle, S., Chassaing, G., Marquet, A., Glowinski, J., and Beaujouan, J. C. (1984) Neuromedin K, a tool to further distinguish two central tachykinin binding sites. *Eur. J. Pharmacol.* **102,** 381,382.

Torrens, Y., Beaujouan, J. C., and Glowinski, J. (1985) Pharmacological characterisation of two tachykinin binding sites in the rat cerebral cortex. *Neuropeptides* **6,** 59–70.

Van Tol, H. M. H., Wu, C. M., Guan, H.-C., Ohara, K., Bunzow, J. R., Civelli, O., Kennedy, J., Seeman, P., Niznik, H. B., and Jovanovic, V. (1992) Multiple dopamine D4 receptor variants in the human population. *Nature (Lond.)* **358,** 149–152.

Ward, P., Ewan, G. B., Jordan, C. C., Ireland, S. J., Hagan, R. M., and Brown, J. R. (1990) Potent and highly selective neurokinin antagonists. *J. Med. Chem.* **33,** 1848–1851.

Watson, S. P. (1983) Rapid degradation of [^{3}H]-substance P in guinea-pig ileum and rat vas deferens *in vitro. Br. J. Pharmacol.* **79,** 543–552.

Watson, S. P. and Iversen, L. L. (1984) ^{3}H-substance P binding to guinea-pig ileum longitudinal smooth muscle membranes. *Regul. Pep.* **8,** 273–281.

Watson, S. P., Sandberg, B. E. B., Hanley, M. R., and Iversen, L. L. (1983) Tissue selectivity of substance P alkyl esters: suggesting multiple receptors. *Eur. J. Pharmacol.* **87,** 77–84.

Wormser, U., Laufer, R., Hart, Y., Chorev, M., Gilon, C., and Selinger, Z. (1986) Highly selective agonists for substance P receptor subtypes. *EMBO J.* **5,** 2805–2808.

Part II

Characterization of Tachykinin Receptors

CHAPTER 3

Biochemical Methods and Assays

Paul L. M. van Giersbergen
and Stephen H. Buck

1. Introduction

The tachykinins are a family of peptides that share the common C-terminal sequence Phe-X-Gly-Leu-Met-NH_2. These peptides exert their biological effects via three pharmacologically and molecular biologically well-defined receptors (*see* other chapters in this volume), the NK_1, NK_2, and NK_3 receptors. Substance P (SP) binds preferentially to the NK_1 receptor, neurokinin A (NKA) to the NK_2 receptor, and neurokinin B (NKB) to the NK_3 receptor. Among other things, the number of receptor sites in tissues/cells and the affinities of compounds/peptides can be measured readily in receptor binding assays that exist for each individual tachykinin receptor. In this chapter, we will describe the development of these assays and the various factors that can influence their outcome. For reasons of clarity, the current tachykinin receptor nomenclature has been used throughout the chapter.

The Tachykinin Receptors Ed.: S. H. Buck © 1994 Humana Press Inc., Totowa, NJ

2. Early Binding Assays

The first tachykinin receptor binding assay was performed by Nakata et al. (1978). These investigators used synaptic membranes prepared from various parts of the rabbit brain and spinal cord. Substance P was labeled with tritium and used as the radioligand. No association and dissociation experiments were performed. Incubations were carried out on ice for 1 min, which makes it highly unlikely that steady-state was reached. Nevertheless, the saturation curve shown and the reported dissociation constant, K_D of 2.74 nM, are within the range of curves and values later reported by others. Rosenthal (1967) transformation of the data did not result in a linear plot, although a straight line was forced through the data points. The data might be better fit to a two-site model. In addition, the rank order of potency of several C-terminal fragments of SP to inhibit [^{3}H]SP binding is not consistent with the rank order later established in binding assays (*see*, e.g., Buck and Shatzer, 1988). Nakata et al. (1978) reported that SP(5–11) and SP(6–11) are more potent than SP(3–11) and SP(4–11) in inhibiting binding of [^{3}H]SP. However, later it was demonstrated in both peripheral tissues and in brain that SP progressively loses affinity for the NK_1 receptor when more and more amino acids are removed from its N-terminus; thus SP(3–11) is more potent than SP(6–11) (Beaujouan et al., 1982; Cascieri and Liang, 1983; Buck and Shatzer, 1988). The peptides, physalaemin and eledoisin, that normally display appreciable affinity for [^{3}H]SP-labeled sites, were inactive up to a concentration of 10 μM. The fact that the experiments were not performed at steady-state might well explain, at least in part, these aberrant results.

In subsequent studies, binding was performed on whole cells. Dispersed acinar cells from guinea pig pancreas (Jensen and Gardner, 1979; Sjödin et al., 1980) and rat parotid gland (Liang and Cascieri, 1980; Putney et al., 1980) were used. Kinetic experiments showed that the binding of the radioligand, ^{125}I-physalaemin (Jensen and Gardner, 1979; Putney et al., 1980), ^{125}I-BH (BH)-SP (Liang and Cascieri, 1980) or ^{125}I-Tyr^8-SP (Sjödin et al., 1980), reached steady state between 15–30 min at ambient or higher temperatures.

The association was considerably slower when determined at 4°C (Jensen and Gardner, 1979; Sjödin et al., 1980). In none of these studies were saturation experiments performed. The IC_{50} of SP to inhibit radioligand binding varied between 1–5 n*M* among the different laboratories. In contrast to this rather small variation are the results obtained in inhibition studies with physalaemin and eledoisin. In the studies in which ^{125}I-physalaemin was employed, physalaemin and SP were equipotent and more potent than eledoisin, whereas physaleamin and eledoisin were equipotent and displayed lower affinity than SP when other radioligands were used. It is unlikely that this difference is caused by the different radioligand used, assuming that all ligands recognize a similar binding site. In fact, we now know that physalaemin and SP have higher affinity for the NK_1 receptor than eledoisin, regardless of the radioligand used (Viger et al., 1983; Park et al., 1984; Buck and Shatzer, 1988).

An IC_{50} value of 3 n*M* for SP to inhibit [^{3}H]SP binding from rat brain membranes was also reported by Hanley et al. (1980). In this study, it was shown that the K_D, as obtained from Rosenthal analysis of a saturation curve (0.38 n*M*), correlated well with the kinetic K_D (0.46 n*M*). Since rapid degradation of [^{3}H]SP was observed at 37°C, several peptidase inhibitors were included in the incubation buffer (20 m*M* HEPES, pH 7.4). Experiments were performed on ice, and it was shown that steady state is not reached within 1 min. The saturation curve showed that the nonspecific binding increased very rapidly with increasing radioligand concentrations. At selected radioligand concentrations, specific binding was as high as 50–60% of total binding. These authors also showed that better results were obtained when separating bound from free radioligand using filtration as compared to centrifugation. Although the C-terminal fragments of SP were more potent in the studies of Hanley et al. (1980) as compared to Nakata et al. (1978), it was again demonstrated that SP(5–11) and SP(6–11) were more potent than the longer fragments. An explanation for these puzzling results remains elusive.

A different approach to tachykinin receptor binding was taken by the group of Lembeck. In their initial paper (Lembeck et al., 1977), the release of SP from rat brain synaptosomes was studied. SP could be released by electrical field stimulation or high potassium in a Ca^{2+} dependent manner. In addition, it was demonstrated that these synaptosomes could synthesize SP and retain the peptide in a phosphatidylserine bound state. This binding of SP to phospholipids was studied in more detail in a subsequent paper (Mayer et al., 1979). The "binding" of ^{125}I-Tyr^{8}-SP to the lipid extractable fraction of synaptic vesicles was pH dependent, reversible, and saturable. High-affinity binding was demonstrated in saturation and kinetic experiments. Radioligand binding could be inhibited by SP and C-terminal fragments (with the proper rank order of potency!), but μM concentrations needed to be used! This is a peculiar result when taking into account the high-affinity of the radioligand in saturation experiments. In addition, rather than inhibiting ^{125}I-Tyr^{8}-SP binding, physaleamin increased total binding of the radioligand. Thus, it is obvious that Lembeck and colleagues were not studying the true NK_1 receptor. This was also recognized by these investigators, and they subsequently looked for a possible role of a protein in SP binding. Although the hypothesis of a phospholipid storage site for SP was maintained, the involvement of proteins in SP binding was clearly demonstrated using trypsin, which completely eliminated ligand binding (Saria et al., 1980).

In these early reports, high-affinity binding sites for SP were shown to be present in a variety of tissues. Except for the inhibition studies with truncated SP fragments, the results reported do not vary much and are in good agreement with today's results. There is still no good explanation for the aberrant data obtained in the inhibition experiments, but several methodological problems exist regarding membrane preparation and composition of incubation buffer that could have influenced the outcome of these experiments (*see* Section 3.1.). These early studies all suffer from a poor ratio between total binding and nonspecific binding. This could be due to several factors, including choice of ligand, tissue preparation, high non-specific binding, degradation of the ligand, the way bound and unbound ligand are separated, the composition of the incubation buffer, and so on.

3. Tachykinin Receptor Assays

3.1. Membrane Preparation and Incubation Buffer

It should be mentioned that there are only small differences among the protocols for the three receptors. Thus, if a method for one tachykinin receptor is developed, it can usually be applied to the two other receptors. Membrane preparation and the composition of the incubation buffer are basically the same, but the radioligand and usually the receptor source are different, of course. Several different tachykinin receptor assay protocols have been used. In the early days of tachykinin receptor binding, each laboratory developed its own protocol, and in many cases these protocols, with some minor modifications, are still used. In general terms, it can be stated that the preparation of membranes consists of dissection of the tissue of interest, homogenization and centrifugation, followed by one or more resuspending/centrifugation steps. The final pellet is then resuspended in the incubation buffer. No comparative studies have appeared regarding membrane preparation. It is thus very difficult to compare the many different protocols that all vary to a lesser or greater extent. The method originally described by Lee et al. (1983) for [^{3}H]SP binding to rat salivary gland is the method that is by far the most widely used. This method can be applied for binding to all three tachykinin receptors as was shown by Buck et al. (1984). The tissue is homogenized in ice-cold Tris-HCl (50 m*M*, pH 7.4) containing 120 m*M* NaCl and 5 m*M* KCl using a polytron. After centrifugation at 48,000*g* for 10 min, the pellet is resuspended in ice-cold Tris-HCl (50 m*M*, pH 7.4) containing 300 m*M* KCl and 10 m*M* EDTA and incubated on ice for 30 min. The purpose of these two steps is to remove any endogenously bound peptide or other proteins that might interfere with access to the receptor. In addition, endogenous cations are removed to inhibit metallopeptidases that could degrade the radioligand or the peptide ligands. After centrifugation as above, the pellet is washed twice with plain Tris-HCl (50 m*M*, pH 7.4). The final pellet is resuspended in incubation buffer. During the preparation of the membranes, trypsin and chymotrypsin inhibitors are occasionally added to protect the receptor from proteolytic degradation (Bahouth and Musacchio, 1985). Trypsin has repeatedly been shown to markedly reduce tachykinin receptor binding (Saria et al., 1980; Mohini et al., 1985).

A variety of buffers, including 50 m*M* Tris-citrate (Nakata et al., 1978), 50 m*M* Tris-HCl (Cascieri and Liang, 1983; Lee et al., 1983; Perrone et al., 1983; Buck et al., 1984, Park et al., 1984; Watson and Iversen, 1984), 20 m*M* HEPES (Hanley et al., 1980), 20–25 m*M* HEPES of high ionic strength (Bahouth et al., 1984; Laufer et al., 1986; Guard et al., 1990; Aharony et al., 1991), Krebs-Ringer phosphate (Viger et al., 1983; Torrens et al., 1984), and 50 m*M* Tris-acetate (Foster and Tridgett, 1988; McKnight et al., 1991) have been employed in tachykinin receptor binding assays. In one report (Cascieri and Liang, 1983), it was stated that nonspecific binding was lower when 50 m*M* Tris-HCl was used as compared to a Krebs-Ringer phosphate buffer. However, although Tris-based buffers are more commonly used, the results obtained with the different buffers are all very similar. What is important is that tachykinin receptor binding should be performed in a buffer at a concentration of 20–50 m*M* and at physiological pH. The pH dependence of tachykinin binding has not been extensively studied, but binding of [^{3}H]SP appears to be constant in the pH range of 6.9–8.4 (Park et al., 1984). A similar pH optimum for [^{3}H]SP binding to chick brain membranes was reported by Too and Hanley (1988). These investigators also found that at pH 2 and 11 specific binding could no longer be detected. The binding of another radioligand, [^{3}H]physalaemin, was maximal between pH 8.0–9.0, but experiments were, nevertheless, performed at pH 7.4 (Mohini et al., 1985).

The ionic strength of HEPES based buffers is an important factor in the outcome of the experiments as was demonstrated by Bahouth and coworkers (Bahouth et al., 1985; Bahouth and Musacchio, 1985; Mohini et al., 1985). Although physalaemin is more potent than SP as a sialagogue (Liang and Cascieri, 1981), this peptide was reported to have a lower affinity than SP for the NK_1 receptor (Nakata et al., 1978; Liang and Cascieri, 1980; Beaujouan et al., 1982) when radiolabeled SP was used, but higher affinity than SP when binding experiments were performed using ^{125}I-physalaemin (Jensen and Gardner, 1979; Putney et al., 1980). This observation prompted Bahouth and coworkers (Bahouth et al., 1984, 1985; Bahouth and Musacchio, 1985; Mohini et al., 1985) to study the binding of several radioligands under a variety of conditions. Using rat salivary gland and [^{3}H]SP, it was demonstrated that SP had 40 times higher affinity than physalaemin when [^{3}H]SP binding was done in low ionic strength buffer (0.35*M* sucrose and 25 m*M* HEPES, pH 7.4) as was also used by Hanley et al. (1980). However, in high ionic strength buffer (0.25*M* sodium sulfate and 25 m*M* HEPES, pH 7.4), SP and physalaemin are equipotent in inhibiting [^{3}H]SP binding (Bahouth and Musacchio, 1985). In low ionic strength buffer, the affinity of [^{3}H]SP is abnormally high, $K_D = 0.14$ n*M*. With increasing concentration of sodium sulfate, the K_D value for [^{3}H]SP increases (Bahouth and Musacchio, 1985).

In order to optimize binding of the radioligand to the receptor protein, several substances are added to the incubation buffer. Since peptide ligands, including radioligands, are used, peptidase inhibitors need to be included. Substance P is highly susceptible to proteolytic degradation and can be cleaved by a variety of enzymes including enkephalinase and angiotensin converting enzyme (*see,* for review, Mussap et al., 1993). As has been clearly demonstrated by several groups (Lee et al., 1983; Tousignant et al., 1989; Guard et al., 1990), no one peptidase inhibitor can completely block the degradation of a tachykinin peptide radioligand. Therefore, a cocktail of peptidase inhibitors is added to the incubation buffer. Commonly used peptidase inhibitors are thiorphan (an enkephalinase inhibitor), phosphoramidon (an enkephalinase inhibitor), bacitracin (a wide-spectrum protease inhibitor), chymostatin (a serine protease inhibitor), leupeptin (protease inhibitor), and bestatin (aminopeptidase inhibitor). As pointed out by Lee et al. (1983), rather than stimulating [^{3}H]SP binding, some thiol-peptidase inhibitors (*N*-ethylmaleimide and *p*-chloromercuriphenylsulfonate) markedly reduce specific binding. In their study, Hanley et al. (1980) used thiol-peptidase inhibitors and, in addition to the absence of divalent cations in their incubation buffer, this may explain the poor level of specific binding obtained. Bovine serum albumin is routinely added at a concentration of 0.5% (w/v) to prevent sticking of the peptide to the tubes and pipet tips. In this respect, peptides stick much less to polypropylene tubes than to polystyrene tubes. However, not all polypropylene tubes are the same, and it is advisable to test tubes from different suppliers. Binding of peptide radioligands to tachykinin receptors is markedly enhanced by divalent cations like magnesium and manganese, and these ions are typically present in a concentration of 1–5 m*M* in the incubation buffer. For more details on the effects of cations on binding of radioligands to tachykinin receptors, *see* Section 4. of this chapter.

3.2. Association and Dissociation

The kinetics of tachykinin receptor binding have been studied extensively. In general, the effects of temperature on the binding kinetics are tested at 4, 20, and 37°C. At increasing temperatures, association is faster and more specific binding is obtained (Jensen and Gardner, 1979; Viger et al., 1983; Bergström et al., 1987a,b). At 37°C, however, a short-lasting steady state is attained (Bergström et al., 1987a,b) because of the rapid degradation of the peptide radioligand (Lee et al., 1983). For this reason, tachykinin binding experiments are performed at room temperature. At room temperature, most radioligands reach steady-state within 30 min and binding remains unchanged for up to 1–2 h (Jensen and Gardner, 1979; Guard et al., 1990). Binding for some other ligands, such as ^{125}I-iodohistidyl-neurokinin A (Van Giersbergen et al., 1991), BH-scyliorhinin II (Mussap and Burcher, 1990), and ^{125}I-BH [Sar^9,Met(O$_2$)^{11}]substance P (Lew et al., 1990), is considerably slower. It should be kept in mind that the association rate is directly proportional to the radioligand concentration (*see,* e.g., Aharony et al., 1991; Van Giersbergen et al., 1991). The rate of association is not only dependent on temperature, type of radioligand, and radioligand concentration, but may also depend on the tissue used (Lew et al., 1990) and the ionic strength of the incubation buffer (Bahouth and Musacchio, 1985).

The rate of dissociation is influenced by the same factors that affect the association rate. In all studies, association obeys the laws of pseudo first-order kinetics. This is, however, not true for the dissociation of the radioligand from the tachykinin receptor. In the majority of papers, a biphasic dissociation has been found. Dissociation, which is generally induced by addition of an excess of "cold radioligand" rather than by dilution, from NK_1 (Viger et al., 1983; Bahouth and Musacchio, 1985), NK_2 (Bergström et al., 1987a; Van Giersbergen et al., 1991), and NK_3 (Beaujouan et al., 1984; Bergström et al., 1987b; Guard et al., 1990) receptors has been reported to be biphasic, a rapid phase followed by a much slower phase. Most studies that do show a monophasic dissociation have used only a limited number of the total time points in the analysis of data thereby probably missing the second slow phase (Jensen and Gardner, 1979; Hanley et al., 1980; Beaujouan et al., 1982; Bahouth et al., 1985; Too and Hanley, 1988). Only the dissociation of [^{3}H]physalaemin (Mohini et al., 1985) and of BH-NH-senktide (Laufer et al., 1986) from rat brain membranes appears to be monophasic. The determination of the kinetic K_D (the dissociation rate, k_{-1}, divided by the association rate, k_{+1}) is seriously hampered by biphasic dissociation. Plotting k_{obs}, as obtained from association experiments performed at different radioligand concentrations, against the radioligand concentration allows the determination of k_{-1} and k_{+1}. The kinetic K_D thus obtained has been shown to correlate well with the dissociation constant derived from saturation experiments (Aharony et al., 1991; Van Giersbergen et al., 1991). The biphasic dissociation is difficult to explain. Several possibilities have been reviewed by Cuatrecasas and Hollenberg (1976).

3.3. Choice of Ligand

Over the years, many radioligands have been synthesized for the three tachykinin receptors; however, only a few have received a broad acceptance and are commonly used. For receptor binding, there is a choice of two radiolabels, tritium or iodine-125. In terms of usefulness in binding assays, ^{125}I has the advantage that much higher specific activity is obtained. This renders these radioligands powerful tools to study (tachykinin) receptors in tissues with low expression of these receptors. Because of the rapid decay of ^{125}I, the radioligand needs to be used within a few weeks of synthesis. In addition, it should always be verified that the introduction of the rather large iodine molecule does not change the biological properties of the ligand. If, however, a tissue contains a high density of the appropriate receptor and can be readily obtained in large quantity, then a ^{3}H-labeled radioligand can be used. ^{3}H-radioligands have a much longer shelf life and the biological properties of the ligand usually are not significantly changed. Several tachykinin peptides, either naturally occurring or synthesized analogs, have been labeled with ^{3}H or ^{125}I. Without exception, all the tachykinin receptor peptide radioligands published are agonists. A recent development is the use of radiolabeled nonpeptide tachykinin receptor antagonists

3.3.1. NK_1 Receptor Radioligands

The first report of tachykinin receptor binding used [^{3}H]SP (Nakata et al., 1978) synthesized according to the method of Segawa et al. (1976). Due to the low specific binding observed by Nakata et al. (1978) and Hanley et al. (1980), other investigators turned to ^{125}I-labeled peptides. We now know that the observed low specific binding with [^{3}H]SP is not related to the radioligand itself, but to a variety of methodological problems. The group of Snyder (Lee et al., 1983) elegantly demonstrated that under nearly optimal conditions, the commercially available [^{3}H]SP is a useful radioligand. A popular method of labeling peptides with ^{125}I involves use of chloramine T. This requires the presence of a tyrosine residue in the peptide and since no tyrosine is present in SP, Tyr^{8}-SP was used by Mayer et al. (1979) to prepare ^{125}I-Tyr^{8}-SP. This radioligand, which is equipotent with SP on the guinea pig ileum (Mayer et al., 1979), is occasionally used in receptor binding studies (it is commercially available), but is not the radioligand of choice. Another tyrosine-containing analog of SP, [Tyr^{1}, Nle^{11}]-SP, was iodinated by Bahouth et al. (1984). This radioligand bound to NK_1 receptors with high-affinity, but specific binding was only 45% of total binding and was, therefore, not used in subsequent studies. Another popular method to iodinate peptides is to conjugate the ^{125}I-BH (BH) reagent to the peptide, which results in a ^{125}I-*N*(1)acylated peptide (Bolton and Hunter, 1972, 1973).

This method was first used by Liang and Cascieri (1980) and Michelot et al. (1980) to label SP. Both groups demonstrated that the coupling of the BH reagent to SP did not reduce the biological activity of the peptide. In fact, Liang and Cascieri (1980) demonstrated that cold BH-SP was more potent than SP in inhibiting BH-SP binding to parotid cells and in inducing salivation in rats. In their study, these investigators also successfully used BH-SP to study the binding of tachykinin peptides to rat parotid cells. It was rapidly recognized that BH-SP is a very useful radioligand and, at present, it is the most widely used. Probably because physalaemin contains a tyrosine moiety, this peptide has been labeled with ^{125}I and, as described before, ^{125}I-physalaemin was used to study NK_1 receptors present on parotid acinar cells (Putney et al., 1980) and pancreatic acinar cells (Jensen and Gardner, 1979). Since there are no obvious advantages in using ^{125}I-physalaemin over BH-SP, it is logical that the endogenous mammalian tachykinin be used, even though Mohini et al. (1985) demonstrated that tritiated physalaemin is a high-affinity ligand with low nonspecific binding. Recently, selective NK_1 receptor agonists, [Sar^9,$Met(O_2)^{11}$]-SP and [Pro^9]-SP, were radioiodinated with the ^{125}I-BH reagent (Lew et al., 1990) or tritiated (Petitet et al., 1988, 1991). Both radioligands, although more selective, appear not to be better than the existing ones in defining NK_1 binding sites, and Lew et al. (1990) have reinforced this conclusion. Because of its higher specific activity, 137 Ci/mmol vs on average, 30 Ci/mmol for the previously synthesized tritiated SP radioligands, [4,5-^3H-Leu^{10}]-substance P (Aharony et al, 1991) might become a useful radioligand. A new era of tachykinin receptor binding could be brought about by developing radioligands using the recently discovered nonpeptide antagonists (for review, *see* Watling, 1992). In fact, two reports have recently appeared demonstrating that tachykinin NK_1 receptors can be identified by autoradiography (McLean et al., 1991) and by binding (Cascieri et al., 1992) using the radiolabeled nonpeptide antagonists, [^3H]CP96,345 and ^{125}I-L-703,606, respectively. A novel approach to receptor binding was reported by Anton et al. (1991). These investigators have biotinylated an N-terminally extended analog of SP. This compound retains biological activity and was used to detect SP binding sites on human IM-9 lymphoblast cells using high resolution streptavidin-peroxidase immunostaining. This interesting approach needs further validation, however.

3.3.2. NK_2 Receptor Radioligands

From functional experiments, the possible existence of multiple tachykinin receptors was suggested in the early 1980s (for details, *see* Chapter 2). In 1984, the "grinders and binders" caught up with the functional classification of the tachykinin receptors. Using ^{125}I-BH eledoisin (BH-E) and BH-SP, Torrens et al. (1984) and Beaujouan et al. (1984) were the first to clearly demonstrate the existence of two different tachykinin receptor binding sites in the rat brain. We now know that the site labeled by BH-E corresponds to the NK_3 receptor. In the same year, Buck et al. (1984) demonstrated the presence of a third tachykinin receptor, the NK_2 receptor, using ^{125}I-BH neurokinin A. This radioligand has since been used in several binding studies (Buck and Burcher, 1985; Burcher and Buck, 1986; Burcher et al., 1986; Lee et al., 1986; Foster and Tridgett, 1988). However, Bolton-Hunter derivatization of NKA was shown to markedly increase the affinity of this peptide for the NK_3 receptor thereby reducing its specificity (Bergström et al., 1987b). Early in 1988, this radioligand was replaced by [^{125}I]iodohistidyl^1-NKA, which has high specific activity, relatively good selectivity for the NK_2 receptor, and specific binding is routinely more than 90% of total binding. This commercially available product is the radioligand of choice for the NK_2 receptor. Although NK_1 receptor binding methodology could be applied to the NK_2 receptor, the NK_2 receptor has received less attention than the NK_1 receptor; the immediate availability of a good radioligand may explain the lack of other radioligands. Only [^{3}H]NKA has been synthesized and tested as another possible radioligand (Bergström et al., 1987a). This radioligand has rather low-affinity (K_D = 13.3 nM), which reduces its usefulness for the study of NK_2 receptors in tissues with a low receptor number. As mentioned for the NK_1 receptor, radiolabeling of a nonpeptide NK_2 antagonist might prove to be very useful.

3.3.3. NK_3 Receptor Radioligands

Several radioligands for the NK_3 receptor have been synthesized. Little receptor binding has been done with most of them. BH-E was the first radioligand used (Buck et al., 1984; Cascieri and Liang, 1984; Torrens et al., 1984). Eledoisin poorly distinguishes between NK_2 and NK_3 receptors. It is, therefore, not useful in peripheral tissues that contain a mixture of tachykinin receptors. Because of the almost complete absence of NK_2 receptors in brain, BH-E is a good radioligand to label brain NK_3 receptors and is frequently used as such. Attempts to use radiolabeled analogs of the endogenous ligand for the NK_3 receptor, NKB, have been unsuccessful in autoradiographic studies because of high nonspecific binding (Buck et al., 1986). This may be due to the rather water insoluble nature of NKB. In search for a NK_3 receptor radioligand, the specific and potent NK_3 receptor agonist, senktide, was conjugated with the ^{125}I-BH reagent (Laufer et al., 1986). ^{125}I-BH-senktide potently contracts the guinea pig ileum thus showing that the labeled peptide retains its full biological activity (Laufer et al., 1986). In receptor binding experiments, ^{125}I-BH-senktide binds with high-affinity and specificity to the rat brain NK_3 receptor. The level of specific binding obtained was between 70 and 80% of total binding (Laufer et al., 1986), which is not ideal for a tachykinin receptor radioligand. In fact, the poor ratio between nonspecific and specific binding hampered the use of this radioligand in autoradiographic studies (Mantyh et al., 1989). The tritiated version of senktide has also been synthesized and used (Guard et al., 1990). Guard and coworkers (1990) were able to detect NK_3 receptors in membranes from guinea pig brain and ileal longitudinal muscle-myenteric plexus. Unfortunately, no data on total binding achieved and percentage of specific binding were reported. Scyliorhinin II, a peptide isolated from the dogfish, *Scyliorhinus canicula* (Conlon et al., 1986), which preferentially binds to the NK_3 receptor (Buck and Krstenansky, 1987), was radiolabeled and successfully used in receptor binding and autoradiographic experiments (Mussap and Burcher, 1990). These studies were performed on rat brain tissues. Recently, we have used ^{125}I-BH-scyliorhinin II to demonstrate the presence of tachykinin receptor-like binding sites in the dogfish brain and stomach (Van Giersbergen et al., 1991). In both tissues, specific binding was more than 85% of total binding. Thus, at present the commercially available radioligands, BH-E and [^{3}H]senktide, are the ligands of choice to study the central and peripheral NK_3 receptors, respectively. However, the labeled dogfish peptide appears to be a good alternative.

3.4. Centrifugation or Filtration

Separation of bound from free ligand is a crucial step in receptor binding. During this process, the dissociation of the radioligand from the receptor should be avoided. Two methods can be used, vacuum filtration through filters or centrifugation. Filtration is the method of choice since it is much easier. Only under certain conditions, such as very fast dissociation of the ligand from the receptor, unacceptable high nonspecific binding of the ligand to the filter, or a ligand with low-affinity ($K_D > 10$ nM), does centrifugation need to be used (*see also* Bylund and Yamamura, 1990). This, of course, does not mean that centrifugation cannot be used in other cases. Tachykinin receptor binding assays do not suffer from the drawbacks that exclude vacuum filtration. It is, therefore, not surprising that from the beginning of the study of these receptors by binding, vacuum filtration has been used by practically all laboratories. As mentioned before, binding of the radioligand to the filter, which may comprise most of the nonspecific binding, may be a problem. Thus, for each radioligand the choice of filter type is important. Nakata et al. (1978) and Hanley et al. (1980) used Millipore filters. As was demonstrated by Cascieri and Liang (1983), the use of these filters and of several other types led to considerable adsorption of the radioligand by the filters and resulted in high nonspecific binding. Since then, Whatman glass fiber filters have been almost exclusively used. These filters have the advantage over Millipore filters in that they allow much faster filtration (Perrone et al., 1983), which also reduces nonspecific binding. Perrone et al. (1983) compared GF/A, GF/B, GF/C, and GF/F Whatman filters and found that nonspecific binding was lowest when [^{3}H]SP was filtered through GF/F filters. At present, regardless of radioligand used, it appears that GF/B filters are the most widely used in the three different tachykinin receptor assays. To further reduce nonspecific binding, the Whatman filters can be presoaked in different solutions. For the NK_1 receptor assay, polyethylenimine at 0.1% is typically used. For the NK_2 and NK_3 receptor assays, the GF/B filters are generally presoaked in ice-cold 0.5% bovine serum albumin, but a combination of polyethylenimine (0.3%) and Triton X-100 (0.5%) is also used.

4. Influence of Cations

The influence of cations on tachykinin NK_1 receptor binding has been extensively studied. Very few reports have been published that describe the effects of these ions on NK_2 receptor binding and, to our knowledge, none regarding the NK_3 receptor.

The first report on this subject was by Saria et al. (1980) who studied the binding of ^{125}I-Tyr^8-substance P to rat brain synaptic vesicles. It was demonstrated that Ca^{2+} and Mg^{2+} (1–10 m*M*) caused a concentration-dependent, parallel leftward shift of the plot as obtained from Rosenthal analysis of saturation experiments performed in the absence or presence of these ions. Thus, no effect on K_D was found, but the number of binding sites was decreased by the ions. These results are in sharp contrast to more recent data. In 1983, two groups (Cascieri and Liang, 1983; Lee et al., 1983) independently discovered that divalent cations increase binding of radiolabeled substance P. Binding of BH-SP to rat cortical membranes was minimal in the absence of divalent cations, and the addition of Mn^{2+}, Ca^{2+} or Mg^{2+} resulted in a 4–28-fold increase of binding, with Mn^{2+} being more potent than the other two ions (Cascieri and Liang, 1983). A similar rank order of potency, $Mn^{2+} > Ca^{2+} > Mg^{2+}$, was also demonstrated by Lee et al. (1983) for binding of [^{3}H]SP to rat salivary gland membranes. However, in the absence of divalent cations there was appreciable binding and these ions increased binding of [^{3}H]SP by only 60%. It appears that there is differential sensitivity to the effects of divalent cations in different tissues of the rat. Zinc and cobalt reduced [^{3}H]SP binding, whereas sodium had no effect at concentrations up to 150 m*M*. The enhanced binding elicited by the divalent cations is associated with an increased B_{max} but no change in K_D (Lee et al., 1983).

The effects of the different ions, monovalent or divalent, are more complex when HEPES is used as buffer. As was reported by Lee et al. (1983), Mn^{2+}, Mg^{2+}, and Ca^{2+} slightly (20–35%) increased the binding of [^{3}H]SP to salivary gland membranes in high ionic strength HEPES buffer (Bahouth and Musacchio, 1985). This effect was mediated by an effect on the B_{max} rather than on receptor affinity. The effect of divalent cations on [^{3}H]SP binding under low ionic strength was not studied. Previously, however, these authors (Bahouth et al., 1984) had shown that the affinity of ^{3}H-Tyr^8 physalaemin in low ionic strength buffer was markedly reduced by Mn^{2+}, but not by Mg^{2+} or Ca^{2+}. In high ionic strength buffer, divalent cations slightly increased the binding of [^{3}H]physalaemin to rat salivary gland membranes by increasing the number of receptor sites detected. Mg^{2+} was more potent than Ca^{2+} or Mn^{2+}. When [^{3}H]physalaemin was used to detect rat brain NK_1 receptors, the stimulating effect of the divalent cations was more pronounced (Mohini et al., 1985). Saturation experiments with and without Mn^{2+} indicated it increased both receptor affinity and density (Mohini et al., 1985). In the absence of ions, the binding of [^{3}H]physalaemin to rat brain membranes was negligible. It should be mentioned that significantly more binding of [^{3}H]physalaemin was found if, during the membrane preparation, the membranes were exposed to high salt concentrations (Mohini et al., 1985). Augmenting the ionic strength of the 20 m*M* HEPES buffer by addition of sodium, potassium, lithium, or ammonium salts resulted in a marked increase of binding (Mohini et al., 1985). As was previously demonstrated for [^{3}H]SP binding (Lee et al., 1983; Cascieri and Liang, 1983), the sensitivity of the rat brain NK_1 receptor, as compared to the salivary gland receptor with respect to modulation by ions, is also different when [^{3}H]physalaemin is used as the radioligand. Thus, in the absence of ions almost no binding is observed to the brain receptor, monovalent ions increase binding to the brain receptor, but decrease binding to the salivary gland receptor, and the stimulatory effect of divalent ions is more pronounced in membranes prepared from rat brain (Lee et al., 1983; Liang and Cascieri, 1983; Mohini et al., 1985; Bahouth and Musacchio, 1985; Too and Hanley, 1988).

In an initial study, Buck and Shatzer (1988) demonstrated that divalent cations, in the efficacy rank order, $Mn^{2+} > Mg^{2+} > Ca^{2+}$, increased the binding of ^{125}I-iodohistidyl-neurokinin A to NK_2 receptors in the hamster urinary bladder. The three cations were equipotent with a maximum effect at 2–3 m*M*. The presence of Mn^{2+} resulted in higher affinity of the radioligand and in an increase in B_{max}. A differential sensitivity to divalent cations has also been observed for the NK_2 receptor. We have shown that in SKLKB82#3 cells, a murine fibroblast into which had been transfected a cDNA encoding the NK_2 receptor isolated from the bovine stomach (Henderson et al., 1990), binding of ^{125}I-iodohistidyl-neurokinin A is minimal in the absence of Mn^{2+} (Van Giersbergen et al., 1991). Addition of this cation produced a more than tenfold increase in binding in SKLKB82#3 cells (Van Giersbergen et al., 1991) as compared to a 3–4-fold increase in the hamster urinary bladder (Buck and Shatzer, 1988). Mg^{2+} and Ca^{2+} were less efficacious than Mn^{2+} in SKLKB82#3 cells.

Little is known as to how the different ions affect the binding of tachykinin radioligands to their receptors. The stimulation of binding by the divalent cations occurs regardless of the radioligand used and generally with the same potency/efficacy rank order. Since these effects have been demonstrated for a large number of G-protein-coupled receptors (Blume, 1978; Chang and Snyder, 1980; U'Prichard and Snyder, 1980; Goodman et al., 1982), an effect of these cations on the conformation of the radioligand or a direct effect on the receptor binding site can be excluded as possible mechanisms. A possible interaction with the coupling of the receptor to its G-protein has been suggested (*see* Section 5.). Experimental evidence that the cations do not affect the primary binding site of the NK_1 receptor was reported by Mohini Sharma and Musacchio (1987). These investigators demonstrated that the sulfhydryl reagent, *N*-ethylmaleimide, blocked the effects of divalent cations and of guanine nucleotides on physalaemin-protected NK_1 receptors, demonstrating that these modulators of binding act on a site of the receptor that contains SH-groups and is different from the physalaemin binding site.

5. Effect of Guanine Nucleotides

Guanine nucleotides are known to regulate the receptor binding of several peptide (Snyder, 1979; Childers and Snyder, 1980) and amine (Creese et al., 1979; U'Prichard and Snyder, 1980; Yamada et al., 1980; Ehlert et al., 1981) hormones/neurotransmitters. Therefore, the effects of these substances on tachykinin receptor binding were investigated by several groups. Lee et al. (1983) demonstrated that binding of [^{3}H]SP to rat salivary gland membranes could be inhibited by the addition of the guanine nucleotides, GTP, GDP, GMP-PNP (GppNHp), but not by GMP and the adenine nucleotides, ATP and ADP. Maximal inhibition of specific binding was 60% with an IC_{50} for GTP and GDP of approximately 7 μM. Saturation experiments performed in the presence or absence of GTP revealed that GTP reduced the B_{max} and almost doubled the K_D (Lee et al., 1983). In the same year, Cascieri and Liang (1983) found that guanine nucleotides, but not a series of other nucleotides like ATP, cAMP, NAD, and CTP, inhibited binding of BH-SP to rat brain cortex membranes. In this tissue, as compared to the salivary gland, the nucleotides were markedly more efficacious inhibiting more than 85% of specific binding with a similar potency. Thus, these data indicate that the NK_1 receptor is coupled to a G-protein.

Other groups have confirmed the partial inhibition of guanine nucleotides of binding of radiolabeled SP to NK_1 receptors in rat salivary gland (Bahouth et al., 1985; Bahouth and Musacchio, 1985; Mohini et al., 1985) and the full inhibition in brain membranes (Too and Hanley, 1988). In contrast, when [^{3}H]physalaemin was used as radioligand, only partial inhibition of binding was obtained by these nucleotides in both brain (Mohini et al., 1985) and salivary gland (Bahouth et al., 1984) membrane preparations. Thus, the effects of the guanine nucleotide may vary with the tissue and the radioligand used. It has been consistently demonstrated that guanine nucleotides reduced the B_{max} with or without changing the K_D (Lee et al., 1983; Bahouth and Musacchio, 1985; Mohini et al., 1985; Aharony et al., 1991). Typically, addition of guanine nucleotides induces a conversion of high-affinity binding to low-affinity binding, thereby decreasing the affinity of agonists, but not altering the B_{max}. This is more evident using an antagonist as radioligand. Until recently, there were no tachykinin antagonist radioligands. Therefore, Luber-Narod et al. (1990) used BH-SP and modified their binding assay in such a way that they could detect low-affinity binding, which is undetectable using the classical filtration assay. It was found that, using a low radioligand concentration, Gpp(NH)p, a nonhydrolyzable analog of GTP, completely inhibited BH-SP binding to rat salivary membranes, whereas the inhibition of the binding of a high radioligand concentration was incomplete. Binding of BH-SP that remained even at high Gpp(NH)p concentration was interpreted as binding to a low-affinity state. In addition, Gpp(NH)p caused a classical rightward shift of the SP displacement curve (Luber-Narod et al., 1990). Further evidence for high- and low-affinity states of the NK_1 receptor was recently published by Cascieri et al. (1992). This paper represents the first tachykinin receptor binding study utilizing a radiolabeled antagonist. The inhibition by SP of binding of ^{125}I-L-703606 to the human NK_1 receptor expressed in Chinese hamster ovary cells was best resolved by a two-site model. Addition of Gpp-(NH)p eliminated the high-affinity site resulting in a monophasic SP displacement curve. Gpp(NH)p inhibited BH-SP binding to the human NK_1 receptor, but did not reduce ^{125}I-L-703606 binding (Cascieri et al., 1992).

The NK_2 and NK_3 receptors are also coupled to a G-protein as evidenced by inhibition of binding by guanine nucleotides (Laufer et al., 1986; Buck and Shatzer, 1988; Foster and Tridgett, 1988; Guard et al., 1990). As was demonstrated for the NK_1 receptor, Gpp(NH)p decreased the B_{max} and the K_D was either unaffected (Buck and Shatzer, 1988) or increased (Laufer et al., 1986). Differential sensitivity of NK_2 receptor binding to Gpp(NH)p in different tissues was also observed (Van Giersbergen et al., 1991). The use of guanine nucleotides only indirectly demonstrates the regulation of agonist binding by G-proteins. Therefore, Macdonald and Boyd (1989) studied the binding of BH-SP in a reconstitution assay employing a receptor preparation deficient in G-proteins. Alkaline treatment of rat salivary gland membranes resulted in a total loss of high-affinity BH-SP binding. Upon addition of purified G-proteins, G_o or G_i, the binding of BH-SP was restored. This restoration could be blocked by simultaneous incubation with Gpp(NH)p. This report provides clear evidence that coupling of the NK_1 receptor to a G-protein results in high-affinity binding. The nature of the endogenous G-protein(s) that is coupled to the NK_1 receptor remains to be elucidated. Stimulation of the NK_1 receptor leads to an increase in phosphatidylinositol turnover (Mantyh et al., 1984) and not to an increased production of cAMP (Lee et al., 1983). This, together with the fact that pertussis toxin and cholera toxin do not influence the binding to the NK_1 receptor (Macdonald and Boyd, 1989), excludes the involvement of G_o or G_i.

The inhibition of peptide agonist binding to tachykinin receptors by guanine nucleotides can be modified by divalent cations. In the presence of magnesium, the IC_{50} of GTP and GDP to inhibit [^{3}H]SP binding to rat salivary gland is approximately sevenfold higher (Lee et al., 1983). This may be explained by the fact that magnesium stimulates the breakdown of these nucleotides to the inactive GMP and guanosine (Lee et al., 1983). In contrast, the potency of Gpp(NH)p was substantially enhanced by magnesium. This latter finding, using Gpp(NH)p and manganese, was also reported by Bahouth and Musacchio (1985) in the same preparation and for binding of ^{125}I-iodohistidyl-neurokinin A to the NK_2 receptor in the hamster urinary bladder (Buck and Shatzer, 1988). Thus, divalent cations might increase the affinity of the guanine nucleotides for their binding site. If and how this observation is related to the binding stimulation properties of divalent cations is not clear.

6. Appendix: Binding Protocols

The following is a standard protocol for tachykinin receptor binding. This method is taken from Lee et al. (1983) and, with some modifications, is used in several laboratories. The radioligands of choice and the tissues that contain a high receptor number for the three tachykinin receptors are listed in Table 1.

6.1. Membrane Preparation

The tissue (Table 1) is weighed, minced with scissors (peripheral tissues), and homogenized with a Polytron (setting 6 for 15 s) in 15 mL of 50 m*M* Tris-HCl buffer (pH 7.4, 4°C) containing 120 m*M* NaCl and 5 m*M* KCl. The homogenate is centrifuged at 48,000*g* for 15 min at 4°C. The pellet is resuspended (Polytron as above) in 15 mL 50 m*M* Tris-HCl buffer (pH 7.4, 4°C) containing 10 m*M* EDTA and 300 m*M* KCl and incubated for 30 min on ice. The homogenate is centrifuged as above and the pellet washed twice by resuspension and again centrifugated in 50 m*M* Tris-HCl buffer (pH 7.4, 4°C). The final pellet is resuspended in incubation buffer (*see* Section 6.2.) at a concentration 10 times the final incubation tube concentration (Table 1) and allowed to set at room temperature for 15 min before use.

6.2. Incubation and Filtration

The binding assay is carried out in duplicate or triplicate in 12 × 75 mm polypropylene cell culture tubes in incubation buffer. The incubation tubes receive 150 µL of incubation buffer consisting of 50 m*M* Tris-HCl (pH 7.4, room temperature) containing 0.1% bovine serum albumin, 2 m*M* $MnCl_2$, 40 µg/mL bacitracin, 4 µg/mL leupeptin, 4 µg/mL chymostatin, and 1 µ*M* thiorphan, 50 µL of radioligand diluted in incubation buffer, and 50 µL of test compound diluted in incubation buffer. The final incubation volume is 0.5 mL. The assay is started by the addition of 250 µL membrane homogenate, vortexing, and incubation of the tubes at room temperature for 30–120 min depending on receptor and radioligand used (Table 1). The incubation is terminated by addition of 3 mL 50 m*M* Tris-HCl (pH 7.4, 4°C)

Table 1
Standard Radioligands and Tissues
in Tachykinin Receptor Binding Assays

Receptor	Radioligand	[Radioligand], n*M*	Tissue	[Tissue], mg ww[a]/tube	Incubation time, h
NK_1	^{3}H-substance P	2	Rat salivary gland	10	1
		2	Rat cortex	10	1
	BH[b]-substance P	0.1	Rat salivary gland	1	1
		0.1	Rat cortex	1	1
		0.1	Guinea pig lung	4	1
NK_2	^{125}I-iodohistidyl-NKA	0.1	Hamster bladder	2	2
		0.1	Rat bladder	4	2
		0.1	Rat duodenum	8	2
NK_3	BH-eledoisin	0.1	Rat cortex	14	2

[a]ww = Wet weight.
[b]BH = ^{125}I-BH.

and rapid filtration under vacuum through GF/B Whatman filters. The filters are rapidly washed with two times 3 mL of the same buffer and transferred to appropriate tubes/vials for detection of radioactivity. Before use, the filters are presoaked in 0.1% polyethylenimine (NK_1 receptor) or 0.5% bovine serum albumin (NK_2 and NK_3 receptor). Specific binding is defined as the difference in binding between tubes containing no test compound and tubes containing 1 μ*M* of unlabeled radioligand. The protein content can be determined on an aliquot taken during the last tissue wash and resuspended. Using one of many computer programs available, the binding parameters, IC_{50}, K_i and Hill slope can be determined. The pros and cons of the different programs have been reviewed by Unnerstall (1990).

References

Aharony, D., Catanese, C. A., and Woodhouse, D.P. (1991) Binding of the novel ligand [4,5-^{3}H-Leu^{10}]substance P to high-affinity receptors on guinea pig lung membranes: Modulation by GTP analogs and sulfhydryl modifying agents. *J. Pharmacol. Exp. Ther.* **259,** 146–155.

Anton, P. A., Reeve, Jr, J. R., Vidrich, A., Mayer, E., and Shanahan, F. (1991) Development of a biotinylated analog of substance P for use as a receptor probe. *Lab. Invest.* **64,** 703–708.

Bahouth, S. W., Stewart, J. M., and Musacchio, J. M. (1984) Specific binding of a ^{125}I-labeled substance P analog to rat submaxillary gland. *J. Pharmacol. Exp. Ther.* **230,** 116–123.

Bahouth, S. W., Lazaro, D. M., Brundish, D. E., and Musacchio, J. M. (1985) Specific binding of [^{3}H-Tyr^8]physalaemin to rat submaxillary gland substance P receptor. *Mol. Pharmacol.* **27,** 38–45.

Bahouth, S. W. and Musacchio, J. M. (1985) Specific binding of [^{3}H]substance P to the rat submaxillary gland. The effects of ions and guanine nucleotides. *J. Pharmacol. Exp. Ther.* **234,** 326–336.

Beaujouan, J. C., Torrens, Y., Herbet, A., Daguet, M.-C., Glowinski, J., and Prochiantz, A. (1982) Specific binding of an immunoreactive and biologically active ^{125}I-labeled substance P derivative to mouse mesencephalic cells in primary culture. *Mol. Pharmacol.* **22,** 48–55.

Beaujouan, J. C., Torrens, Y., Viger, A., and Glowinski, J. (1984) A new type of tachykinin binding site in the rat brain characterized by specific binding of a labeled eledoisin derivative. *Mol. Pharmacol.* **26,** 248–254.

Bergström, L., Beaujouan, J. C., Torrens, Y., Saffroy, M., Glowinski, J., Lavielle, S., Chassaing, G., Marquet, A., D'Orleans-Juste, P., Dion, S., and Regoli, D. (1987a) ^{3}H-neurokinin A labels a specific tachykinin-binding site in the rat duodenal smooth muscle. *Mol. Pharmacol.* **32,** 764-771.

Bergström, L., Torrens, Y., Saffroy, M., Beaujouan, J. C., Lavielle, S., Chassaing, G., Morgat, J. L., Glowinski, J., and Marquet, A. (1987b) [^{3}H]Neurokinin B and ^{125}I-Bolton Hunter eledoisin label identical tachykinin binding sites in the rat brain. *J. Neurochem.* **48,** 125–133.

Blume, A. J. (1978) Interactions of ligands with opiate receptors of brain membranes: Regulation by ions and nucleotides. *Proc. Natl. Acad. Sci. USA* **75,** 1713–1717.

Bolton, A. E. and Hunter, W. M. (1972) A new method for labelling protein hormones with radioiodine for use in the radioimmunoassay. *J. Endocrinol.* **55,** 30,31.

Bolton, A. E. and Hunter, W. M. (1973) The labelling of proteins to high specific radioactivities by conjugation to a ^{125}I-containing acylating agent. *Biochem. J.* **133,** 529–538.

Buck, S. H. and Burcher, E. (1985) The rat submaxillary gland contains predominantly P-type tachykinin binding sites. *Peptides* **6,** 1079–1084.

Buck, S. H., Burcher, E., Shults, C. W., Lovenberg, W., and O'Donohue, T. L. (1984) Novel pharmacology of substance K-binding sites: A third type of tachykinin receptor. *Science* **226,** 987–989.

Buck, S. H., Helke, C. J., Burcher, E., Shults, C. W., and O'Donohue, T. L. (1986) Pharmacologic characterization and autoradiographic distribution of binding sites for iodinated tachykinins in rat central nervous system. *Peptides* **7,** 1109–1120.

Buck, S. H. and Krstenansky, J. L. (1987) The dogfish peptides scyliorhinin I and scyliorhinin II bind with differential selectivity to mammalian tachykinin receptors. *Eur. J. Pharmacol.* **144,** 109–111.

Buck, S. H. and Shatzer, S. A. (1988) Agonist and antagonist binding to tachykinin peptide NK-2 receptors. *Life Sci.* **42,** 2701–2708.

Burcher, E. and Buck, S. H. (1986) Multiple tachykinin binding sites in hamster, rat and guinea pig urinary bladder. *Eur. J. Pharmacol.* **128,** 165–177.

Burcher, E., Buck, S. H., Lovenberg, W., and O'Donohue, T. L. (1986) Characterization and autoradiographic localization of multiple tachykinin binding sites in gastrointestinal tract and bladder. *J. Pharmacol. Exp. Ther.* **236,** 819–831.

Bylund, D. B. and Yamamura, H. I. (1990) Methods for receptor binding, in *Methods in Neurotransmitter Receptor Analysis* (Yamamura, H. I., Enna, S. J., and Kuhar, M. J., eds.), Raven, New York, pp. 37–68.

Cascieri, M. A., Ber, E., Ming Fong, T., Sadowski, S., Bansal, A., Swain, C., Seward, E., Frances, B., Burns, D., and Strader, C. D. (1992) Characterization of the binding of a potent, selective, radioiodinated antagonist to the human neurokinin-1 receptor. *Mol. Pharmacol.* **42,** 458–463.

Cascieri, M. A. and Liang, T. (1983) Characterization of the substance P receptor in rat brain cortex membranes and the inhibition of radioligand binding by guanine nucleotides. *J. Biol. Chem.* **258,** 5158–5164.

Cascieri, M. A. and Liang, T. (1984) Binding of [^{125}I]Bolton hunter conjugated eledoisin to rat brain cortex membranes—Evidence for two classes of tachykinin receptors in the mammalian central nervous system. *Life Sci.* **35,** 179–184.

Chang, R. S. L. and Snyder, S. H. (1980) Histamine H1-receptor binding sites in guinea pig brain membranes: Regulation of agonist interactions by guanine nucleotides and cations. *J. Neurochem.* **34,** 916–922.

Childers, S. R. and Snyder, S. H. (1980) Differential regulation by guanine nucleotides of opiate agonist an antagonist receptor interactions. *J. Neurochem.* **34,** 583–594

Conlon, J. M., Deacon, C. F., O'Toole, L., and Thim, L. (1986) Scyliorhinin I and II: Two novel tachykinins from dogfish gut. *FEBS Lett.* **200,** 111–116.

Creese, I., Usdin, T. B., and Snyder, S. H. (1980) Dopamine receptor binding regulated by guanine nucleotides. *Mol. Pharmacol.* **16,** 69–76.

Cuatrecasas, P. and Hollenberg, M. D. (1976) Membrane receptors and hormone action. *Adv. Prot. Chem.* **30,** 251–451.

Ehlert, F. J., Roeske, W .R., and Yamamura, H. I. (1981) Muscarinic receptor: Regulation by guanine nucleotides, ions and N-ethylaleimide. *Fed. Proc.* **40,** 153–159.

Foster, A. C. and Tridgett, R. (1988) Comparison of the binding of radiolabelled neurokinin A and eledoisin in rat cortex synaptic membranes. *Br. J. Pharmacol.* **94,** 602–608.

Goodman, R. R., Cooper, M. J., Gavish, M., and Snyder, S. H. (1982) Guanine nucleotide and cation regulation of the binding of [^{3}H]cyclohexyladenosine and [^{3}H]diethylphenylxanthine to adenosine A1 receptors in brain membranes. *Mol. Pharmacol.* **21,** 329–335.

Guard, S., Watson, S. P., Maggio, J. E., Phon Too, H., and Watling, K. J. (1990) Pharmacological analysis of [^{3}H]-senktide binding to NK_3 tachykinin receptors in guinea pig ileum longitudinal muscle-myenteric plexus and cerebral cortex membranes. *Br. J. Pharmacol.* **99,** 767–773.

Hanley, M. R., Sandberg, B. E. B., Lee, C. M., Iversen, L. L., Brundish, D. E., and Wade, R. (1980) Specific binding of ^{3}H-substance P to rat brain membranes. *Nature* **286,** 810–812.

Henderson, A. K., Lai, J., Buck, S. H., Fujiwara, Y., Singh, G., Yamamura, m. S., Nakanishi, S., Roeske, W.R., and Yamamura, H. I. (1990) A cloned neurokinin A receptor mediates phosphatidylinositol hydrolysis in a transfected murine fibroblast. *Life Sci.* **47,** PL-7–PL-12.

Jensen, R. T. and Gardner, J. D. (1979) Interaction of physalaemin, substance P, and eledoisin with specific membrane receptors on pancreatic acinar cells. *Proc. Natl. Acad. Sci. USA* **76,** 5679–5683.

Laufer, R., Gilon, C., Chorev, M., and Selinger, Z. (1986) Characterization of a neurokinin B receptor site in rat brain using a highly selective radioligand. *J. Biol. Chem.* **261,** 10,257–10,263.

Lembeck, F., Mayer, N., and Schindler, G. (1977) Substance P in rat brain synaptosomes. *Naunyn Schmiedeberg's Arch. Pharmacol.* **301,** 17–22.

Lee, C.-M., Javitch, J. A. and Snyder, S. H. (1983) ^{3}H-substance P binding to salivary gland membranes. Regulation by guanyl nucleotides and divalent cations. *Mol. Pharmacol.* **23,** 563–569.

Lee, C.-M., Campbell, N. J., Williams, B. J., and Iversen, L. L. (1986) Multiple tachykinin binding sites in peripheral tissues and in brain. *Eur. J. Pharmacol.* **130,** 209–217.

Lew, R., Geraghty, D. P., Drapeau, G., Regoli, D., and Burcher, E. (1990) Binding characteristics of [^{125}I]Bolton-Hunter [Sar^9,Met(O$_2$)^{11}]substance P, a new selective radioligand for the NK_1 receptor. *Eur. J. Pharmacol.* **184,** 97–108.

Liang, T. and Cascieri, M. A. (1980) Specific binding of an immunoreactive and biologically active ^{125}I-labeled N(1)acylated substance P derivative to parotid cells. *Biochem. Biophys. Res. Commun.* **96,** 1793–1799.

Liang, T. and Cascieri, M. A. (1981) Substance P receptor on parotid cell membranes. *J. Neurosci.* **1,** 1133–1141.

Luber-Narod, J., Boyd, N. D., and Leemean, S. E. (1990) Guanine nucleotides decrease the affinity of substance P binding to its receptor. *Eur. J. Pharmacol.* **188,** 185–191.

Macdonald, S. G. and Boyd, N. D. (1989) Regulation of substance P receptor affinity by guanine nucleotide-binding proteins. *J. Neurochem.* **53,** 264–272.

Mantyh, P. W., Gates, T., Mantyh, C. R., and Maggio, J. E. (1989) Autoradiographic localization and characterization of tachykinin receptor binding sites in the rat brain and peripheral tissues. *J. Neurosci.* **9,** 258–279.

Mantyh, P. W., Pinnock, R. D., Downes, C. P., Goedert, M., and Hunt, S. P. (1984) Correlation between inositol phospholipid hydrolysis and substance P receptors in rat CNS. *Nature* **309,** 795–797.

Mayer, N., Lembeck, F., and Gamse, R. (1979) Substance P: Characteristics of binding to synaptic vesicles of rat brain. *Naunyn Schmiedeberg's Arch. Pharmacol.* **306,** 45–51.

McKnight, A. T., Maguire, J. J., Elliott, N. J., Fletcher, A. E., Foster, A. C., Tridgett, R., Williams, B. J., Longmore, J., and Iversen, L. L. (1991) Pharmacological specificity of novel, synthetic, cyclic peptides as antagonists at tachykinin receptors. *Br. J. Pharmacol.* **104,** 355–360.

McLean, S., Ganong, A. H., Seeger, T. F., Bryce, D. K., Pratt, K. G., Reynolds, L. S., Siok, C. J., Lowe, III, J. A., and Heym, J. (1991) Activity and distribution of binding sites in brain of a nonpeptide substance P (NK1) receptor antagonist. *Science* **251,** 437–439.

Michelot, R., Gozlan, H., Beaujouan, J. C., Besson, M. J., Torrens, Y., and Glowinski, J. (1980) Synthesis and biological activities of substance P iodinated derivatives. *Biochem. Biophys. Res. Commun.* **95,** 491–498.

Mohini Sharma, P. and Musacchio, J. M. (1987) N-ethylmaleimide blocks the modulatory effects of divalent cations and guanine nucleotides on the brain substance P receptor. *Eur. J. Pharmacol.* **138,** 9–19.

Mohini, P., Bahouth, S. W., Brundish, D. E., and Musacchio, J. M. (1985) Specific labeling of rat brain substance P receptor with [^{3}H]physalaemin. *J. Neurosci.* **5,** 2078–2085.

Mussap, C. J., Geraghty, D. P., and Burcher, E. (1993) Tachykinin receptors: A radioligand binding perspective. *J. Neurochem.* **60,** 1987–2009.

Mussap, C. J. and Burcher, E. (1990) [^{125}I]-Bolton-Hunter scyliorhinin II: A novel, selective radioligand for the tachykinin NK_3 receptor in rat brain. *Peptides* **11,** 827–836.

Nakata, Y., Kusaka, Y., Segawa, T., Yajima, H., and Kitagawa, K. (1978) Substance P: Regional distribution and specific binding to synaptic membranes in rabbit central nervous system. *Life Sci.* **22,** 259–268.

Park, C. H., Massari, V. J., Quirion, R., Tizabi, Y., Shults, and O'Donohue, T. L. (1984) Characteristics of ^{3}H-substance P binding sites in rat brain membranes. *Peptides* **5,** 833–836.

Perrone, M. H., Diehl, R. E., and Haubrich, D. R. (1983) Binding of [^{3}H]substance P tp putative substance P receptors in rat brain membranes. *Eur. J. Pharmacol.* **95,** 131–133.

Petitet, F., Beaujouan, J. C., Saffroy, M., Torrens, Y., Chassaing, G., Lavielle, S., Besseyre, J., Garret, C., Carruette, A., and Glowinski, J. (1991) Further demonstration that [Pro^9]-substance P is a potent and selective ligand of NK-1 tachykinin receptors. *J. Neurochem.* **56,** 879–889.

Petitet, F., Saffroy, M., Torrens, Y., Glowinski, J., Beaujouan, J. C., Lavielle, S., Chassaing, G., and Marquet, A. (1988) ^{3}H[Pro^9]SP, a new selective ligand for NK-1 binding sites. *Regul. Pept.* **22,** 147.

Putney, J. W., Jr., Van De Walle, C. M., and Wheeler, C. S. (1980) Binding of ^{125}I-physalaemin to rat parotid acinar cells. *J. Physiol.* **301,** 205–212.

Rosenthal, H. E. (1967) Graphical method for the determination and presentation of binding parameters in a complex system. *Anal. Biochem.* **20,** 525–532.

Saria, A., Mayer, N., Lembeck, F., and Pabst, M. (1980) Regional distribution and biochemical properties of ^{125}I-Tyr^8-substance P binding sites in synaptic vesicles. *Naunyn Schmiedeberg's Arch. Pharmacol.* **311,** 151–157.

Segawa, T., Nakata, Y., Nakamura, K., Yajima, H., and Kitagawa, K. (1976) Substance P in the central nervous system of rabbits: Uptake system differs from putative transmitters. *Jpn. J. Pharmacol.* **26,** 757–760.

Sjödin, L., Brodin, E., Nilsson, G., and Conlon, T. P. (1980) Interaction of substance P with dispersed pancreatic acinar cells from the guiea pig. Binding of radioiodinated peptide. *Acta. Physiol. Scand.* **109,** 97–105.

Snyder, S. H. (1979) Peptide and neurotransmitter receptors in the brain: Regulation by ions and guanyl nucleotides, in *Central Regulation of the Endocrine System* (Fuxe, K., Hökfelt, T., and Luft, R., eds.), Plenum, New York, pp. 109–117.

Torrens, Y., Lavielle, Chassaing, G., Marquet, A., Glowinski, J., and Beaujouan, J. C. (1984) Neuromedin K, a tool to further distinguish two central tachykinin binding sites. *Eur. J. Pharmacol.* **102,** 381,382.

Too, H.-P. and Hanley, M. R. (1988) Solubilization and characterization of substance P-binding sites from chick brain membranes. *Biochem. J.* **252,** 545–551.

Tousignant, C., Guillemette, G., Drapeau, G., and Regoli, D. (1989) Comparison of binding assay and biological activity on a NK-1 system with new selective agonists. *Neuropeptides* **14,** 275–283.

Unnerstall, J. R. (1990) Computer-assisted analysis of binding data, in *Methods in Neurotransmitter Receptor Analysis* (Yamamura, H. I., Enna, S. J., and Kuhar, M. J., eds.), Raven, New York, pp. 37–68.

U'Prichard, D. C. and Snyder, S. H. (1980) Interactions of divalent cations and guanine nucleotides at alpha$_2$ adrenergic receptor binding sites in bovine brain membranes. *J. Neurochem.* **34,** 385–394.

Van Giersbergen, P. L. M., Conlon, J. M., and Buck, S. H. (1991) Binding sites for tachykinin peptides in the brain and stomach of the dogfish, *Scyliorhinus canicula*. *Peptides* **12,** 1161–1163.

Van Giersbergen, P. L. M., Shatzer, S. A., Henderson, A. K., Lai, J., Nakanishi, S., Yamamura, H. I., and Buck, S. H. (1991) Characterization of a tachykinin peptide NK_2 receptor transfected into murine fibroblast B82 cells. *Proc. Natl. Acad. Sci. USA* **88,** 1661–1665.

Viger, A., Beaujouan, J. C., Torrens, Y., and Glowinski, J. (1983) Specific binding of a ^{125}I-substance P derivative to rat brain synaptosomes. *J. Neurochem.* **40,** 1030–1039.

Watling, K. J. (1992) Nonpeptide antagonists herald new era in tachykinin research. *Trends Pharmacol. Sci.* **13,** 266–269.

Watson, S. P. and Iversen, L. L. (1984) ^{3}H-substance P binding to guinea pig ileum longitudinal smooth muscle membranes. *Regul. Pept.* **8,** 273–281.

Yamada, S., Yamamura, H. I., and Roeske, W. R. (1980) The regulation of cardiac α_1-adrenergic receptors by guanine nucleotides and by muscarinic cholinergic agonists. *Eur. J. Pharmacol.* **63,** 239–241.

CHAPTER 4

Agonist and Antagonist Receptor Binding

Stafford McLean and John A. Lowe III

1. Introduction

Substance P (SP), neurokinin A (NKA), and neurokinin B (NKB) are members of the tachykinin family of peptides that share the carboxy terminal sequence Phe-X-Gly-Leu-Met-NH_2. These mammalian tachykinins elicit their effects through three different receptors known as NK_1, NK_2, and NK_3. Pharmacological and biochemical studies suggest that substance P is the preferred endogenous ligand for the NK_1 receptor, whereas neurokinin A and neurokinin B have highest affinity for NK_2 and NK_3 receptors, respectively (Buck and Burcher, 1986; Maggio, 1988; Quirion and Dam, 1988; Guard and Watson, 1991). The characterization of the NK_1, NK_2, and NK_3 receptors as distinct sites was confirmed with the cloning and sequencing of separate cDNAs encoding each of the receptors (Masu et al., 1987; Sasai and Nakanishi, 1989; Yokota et al., 1989; Gerard et al., 1990; Hershey and Krause, 1990; Shigemoto et al., 1990; Hershey et al., 1991). The deduced amino acid sequence

The Tachykinin Receptors Ed.: S. H. Buck

and hydropathy analysis suggest that all three of the tachykinin receptors contain seven transmembrane spanning domains and sequence similarity with other members of the superfamily of G-protein-coupled receptors. Maximal sequence homology among NK_1, NK_2, and NK_3 receptors is found in the putative core transmembrane domains, whereas homology is lowest at the amino and carboxy terminal ends (Nakanishi, 1991).

Complete characterization of tachykinin receptors has been hampered by the limited receptor selectivity of the endogenous peptides and the difficulty in finding tissue expressing a homogenous receptor population. Efforts to overcome these problems have focused on the development of more selective peptide analogs (*see* Chapters 12 and 13 of this volume) and, more recently, the expression of a single receptor type in either COS or CHO cells (Ingi et al., 1991; Cascieri et al., 1992; Takeda et al., 1992). The use of molecular biological techniques to construct chimeric receptors, as well as receptors with single point mutations, will facilitate identification of agonist and antagonist binding sites (Gether et al., 1992; Fong et al., 1992b).

The characterization of antagonist binding sites presupposes the availability of receptor selective molecules that act as full antagonists. Fortunately, the recent development of nonpeptide tachykinin receptor antagonists that are selective (>1000-fold) for NK_1 and NK_2 receptors has circumvented many of the problems plaguing peptide analogs and will facilitate biochemical characterization of tachykinin receptors. CP-96,345, disclosed as the first nonpeptide antagonist of NK_1 receptors, was discovered by empirical screening using [^{3}H]substance P binding to bovine caudate membranes (Snider et al., 1991). Since this initial disclosure, several nonpeptide antagonists selective for either NK_1 or NK_2 receptors have been described (Garret et al., 1991; Appell et al., 1992; Emonds-Alt et al., 1992; Hirschmann et al., 1992; McLean et al., 1992; Tabart et al., 1992; Venepalli et al., 1992) and are shown in Fig. 1. In this chapter, we will discuss the binding of agonists and particularly antagonists to tachykinin receptors, with emphasis on the NK_1 receptor.

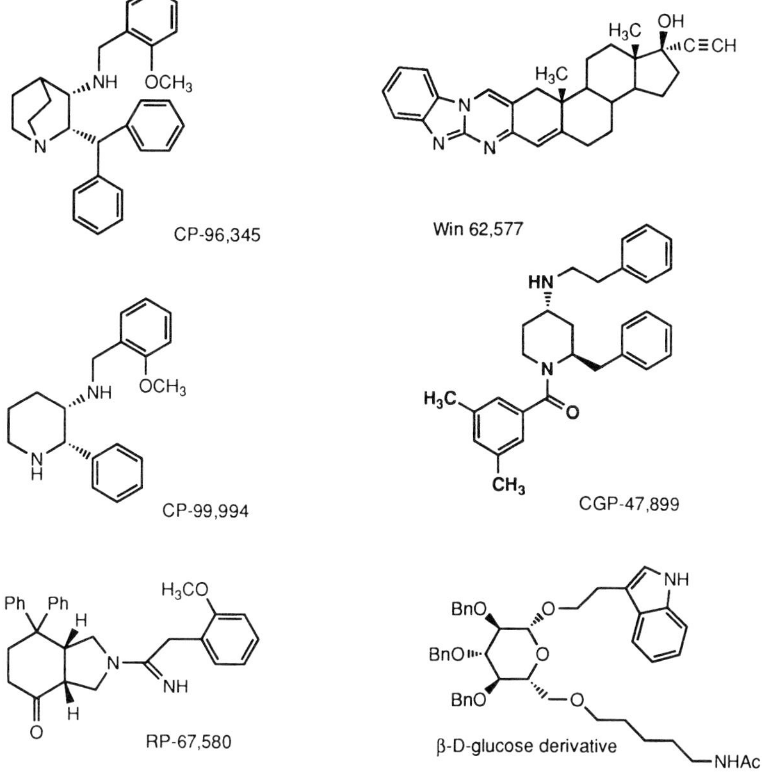

Fig. 1. Structure of NK_1 receptor antagonists.

2. NK_1 Receptors

2.1. Characterization of the NK_1 Receptor

Binding studies characterizing NK_1 receptors often have used [^{3}H]substance P or ^{125}I-Bolton-Hunter (BH)-coupled substance P as the radioligand (*see* Quirion et al., 1991 for a review). At the concentrations generally used (0.05–1.0 n*M*), both ligands preferentially label NK_1 receptors with 65–96% specific binding (Burcher et al., 1991; Quirion et al., 1991). Characterization of the binding site as the NK_1 receptor requires that the rank order of affinities in inhibiting the radioligand be as follows: substance P ≥ physalaemin > eledoisin ≥ neurokinin A > neurokinin B (Buck and Burcher, 1986). However, substance P, neurokinin A, and neurokinin B exhibit less than 100-fold selectivity for their "preferred" receptor whether measured in binding assays (Buck et al., 1984; Chassaing et al., 1992; Lavielle et al., 1988) or functional assays (Regoli et al., 1991). Thus, tissues enriched in NK_2 or NK_3 receptors may be labeled by radiolabeled substance P.

Several peptide analogs of substance P with amino acid substitutions to enhance selectivity have been synthesized (Folkers et al., 1984; Folkers et al., 1986; Lavielle et al., 1986a; Drapeau et al., 1987; Lavielle et al., 1988; Regoli et al., 1988). Two examples of selective NK_1 receptor agonists are [Pro^9]-substance P and Sar^9-Met-$(O_2)^{11}$-substance P that lack activity in biological assays for NK_2 and NK_3 receptor activation (Drapeau et al., 1987; Petitet et al., 1991), and at concentrations of 10 μM fail to inhibit the binding of NK_2 or NK_3 selective ligands (Lavielle et al., 1986b; Bergstrom et al., 1987). These ligands have been radiolabeled, i.e., [^{3}H]Pro^9-substance P and ^{125}I-BH-Sar^9-Met$(O_2)^{11}$-substance P, and appear to bind selectively to NK_1 receptors (Lew et al., 1990; Tousignant et al., 1990, 1991; Petitet et al., 1991). However, the lability of peptides to enzymatic degradation raises concerns about potential differences in the distribution and/or levels of endogenous peptidases that may affect binding constants leading to the mistaken attribution of receptor subtypes. Similarly, agonists are sensitive to the affinity state of the receptor, that is, whether it is coupled or uncoupled from its respective G-protein, making it difficult to determine whether a two-component displacement curve represents different affinity states of a single receptor or evidence for receptor subtypes. The value of selective antagonists for pharmacologically and biochemically characterizing receptors and elucidating the presence or absence of receptor subtypes is well illustrated by the efforts to classify α_1-adrenergic subtypes (Minneman, 1988). Thus, the recent discovery of several nonpeptide NK_1 receptor antagonists, representing a structurally diverse series of compounds, provides a formidable arsenal for characterizing the pharmacology and function of the NK_1 receptor.

2.2. NK_1 Receptor Antagonists

The nonpeptide NK_1 receptor antagonists bind to the NK_1 receptor with a wide range of affinities (Table 1). The data are derived from inhibition of ^{125}I-BH-SP binding to the human NK_1 receptor transfected into CHO cells or to the NK_1 receptor in membranes prepared from guinea pig striatum with several exceptions as noted. Although the affinity of the compounds for the NK_1 receptor is dependent upon the species examined, as discussed later in the chapter, the compounds do not distinguish between guinea pig and human NK_1 receptors. The prototypic compound, CP-96,345 [(2S,3S)-*cis*-2-(diphenylmethyl)-*N*-[(2-methoxyphenyl)-methyl]-1-azabicyclo[2.2.2]octan-3-amine], has high-affinity for the NK_1 receptor comparable to that of the preferred endogenous ligand, substance P. In contrast to substance P or spantide, a D-amino acid substituted analog of substance P that acts as an "antagonist" (Folkers et al., 1984), CP-96,345 lacks affinity for NK_2 or NK_3 receptors labeled by ^{125}I-NKA and ^{125}I-BH-eledoisin, respectively. The SAR which led to CP-96,345 is outlined in Fig. 2, where the importance of three key elements for high-affinity binding to the NK_1 receptor is illustrated:

1. The 2-methoxybenzylamino side chain;
2. The benzhydryl group; and
3. The bridgehead nitrogen (Lowe et al., 1992).

Substitutions on the benzylamino side chain have significant effects on binding to the NK_1 receptor. Potent activity was obtained with substitutions in the ortho position, whereas meta and para substitutions led to considerable loss in affinity. The electron-donating methoxy group in the ortho position yielded the most active compound with even closely related groups, such as ethoxy and ethyl, significantly lowering receptor affinity. A narrow SAR was demonstrated for the benzhydryl group with substitutions on the phenyl rings or replacement of the benzhydryl with a single benzyl group leading to a significant loss in activity. Removal of the bridgehead nitrogen also produced a dramatic loss in affinity. A three-point binding model for the interaction of CP-96,345 with the NK_1 receptor was

Table 1
Selectivity of Compounds for the NK_1 Tachykinin Receptor[a]

Competitor	NK_1, IC_{50}, n*M*	NK_2, IC_{50}, n*M*	NK_3, IC_{50}, n*M*
CP-99,994	0.48 ± .07	>10,000	>10,000
CP-96,345	0.58 ± .13	>10,000	>10,000
(±) RP 67580	83.6 ± 13.9	>10,000	>10,000
Win-62,577[b]	>25,000[b]	?	?
CGP 47,899[c]	10[c]	?	?
β-D-glucose analog[d]	60[d]	?	?
Spantide	353 ± 31	4930 ± 640	>10,000
Substance P	0.31 ± 0.15	221 ± 84	68 ± 21

[a]Values are from competition assays using 0.1 n*M* of ^{125}I-BH-SP and CHO cells transfected with the human NK_1 receptor, or ^{125}I-BH-eledoisin binding to rat cerebral cortex (NK_3), or 0.05 n*M* of ^{125}I-NKA binding to hamster bladder (NK_2) except where otherwise noted. IC_{50} values were determined using the iterative curve fitting program, Lundon2.

[b]Appell et al., 1992.

[c]Schilling et al., 1992.

[d]Hirschmann et al., 1992.

proposed based on these results (Lowe et al., 1992). The three points are:

1. The ion-pairing site interaction with the bridgehead nitrogen;
2. The accessory binding site interaction with the benzhydryl group; and
3. The specificity site interaction with the 2-methoxybenzylamino side chain.

Given the importance of the bridgehead nitrogen for binding and the report that the NK_1 receptor binding site for CP-96,345 resides in the extracellular loop regions above transmembrane segments V and VI (Gether et al., 1992), it may be assumed that the ion-pairing site interaction occurs in this domain of the receptor. The accessory binding site has been proposed by Ariens to recognize hydrophobic double ring systems, exemplified by the benzhydryl group, which are

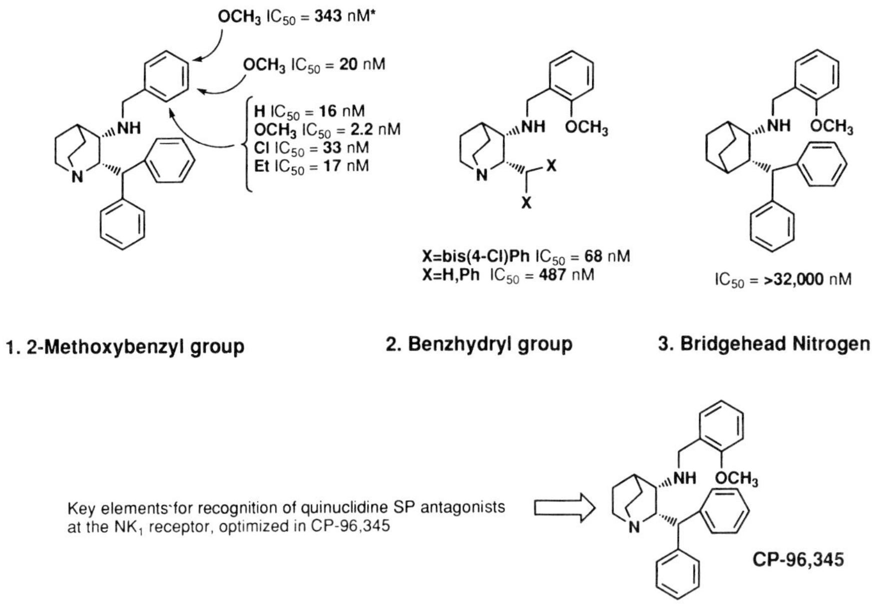

Fig. 2. Structure–activity relationship of CP-96,345 binding to the human NK_1 receptor. IC_{50} values are from inhibition of [^{3}H]SP binding to IM-9 cells.

a common structural feature of G-protein-coupled receptor antagonists (Ariens et al., 1979).

Molecular modeling of the relative spatial orientation of the benzylamino group and the benzhydryl group of CP-96,345 indicated that the benzhydryl group forces one of its phenyl rings to adopt an orientation parallel to the benzylamino group (Desai et al., 1992). This suggested that the position of the inner phenyl ring of the benzhydryl group relative to the benzylamino side chain might be important for activity. To test this hypothesis, molecules in which a single phenyl ring could be locked into place relative to the benzylamino group were synthesized. CP-99,994 [(2S,3S)-*cis*-3-(2-methoxybenzylamino)-2-phenylpiperidine] represents the result of these efforts (McLean et al., 1992). CP-99,994 has high-affinity and selectivity for the NK_1 receptor, comparable to that of CP-96,345. It blocks the activity of substance P or capsaicin in in vitro and in vivo models (McLean et al., 1992) with approximately twofold greater potency than CP-96,345. CP-99,994 retains the ring nitrogen postulated to be important for ion-pairing with the NK_1 receptor and the 2-methoxy benzyl amino side chain (Lowe et al., 1992), but lacks the double ring motif provided by the benzhydryl group of CP-96,345, indicating that a single phenyl ring is sufficient to perform this function at the binding site in the NK_1 receptor.

Subsequent to the initial studies reported for CP-96,345, two other nonpeptide antagonists have appeared. RP-67,580 [(3aR,7aR)-7,7-diphenyl-2-[1-imino-2-(2-methoxyphenyl)ethyl]perhydroisoindol-4-one], also discovered by SAR optimization of an empirical screening lead, has greatly reduced affinity for the human NK_1 receptor relative to CP-96,345. However, as discussed later, RP-67,580 has the highest affinity for the rat NK_1 receptor (Garret et al., 1991). The structural similarities of RP-67,580 and CP-96,345 are noteworthy and suggest that RP-67,580 may provide complementary SAR information. Although based on binding to the rat NK_1 receptor, the binding affinity of RP-67,580 and its analogs, outlined in Fig. 3, suggests the importance of the 2-methoxy and amidine groups for potency (Tabart et al., 1992). Interestingly, substitutions on the phenyl rings reduced affinity for the receptor similar to the effects observed for substitutions on the benzhydryl group of CP-96,345.

A series of heterocyclic derivatives of the androstane steroid skeleton that are NK_1 receptor antagonists has also been reported (Venepalli et al., 1992). The SAR for these compounds is summarized in Fig. 4, where the importance of the groups at the C-17 position of the steroid nucleus can be seen. The most potent compound from this effort gave an IC_{50} value of 50 n*M* at the rat brain NK_1 receptor. The structural similarity of these heterosteroids with the previous compounds is not apparent; however, it is worth noting that these compounds do not interact with the human NK_1 receptor (Appell et al., 1992).

Two other compounds reported to be NK_1 receptor antagonists are CGP-47,899 (Schilling et al., 1992) and a β-D-glucose based nonpeptide compound (Hirschmann et al., 1992). Using [^{3}H]SP bind-

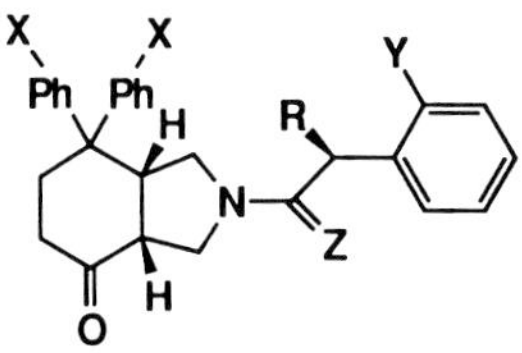

X	Y	Z	R	SP IC_{50}, nM*
H	H	O	H	181
H	OCH_3	O	H	75
H	$OCH_2CH_2NMe_2$	O	H	40
H	H	O	CH_3	35
H	H	O	OH	113
H	OCH_3	O	CH_3	12
4-CH_3	H	O	H	393
H	H	NH	H	36
H	OCH_3	NH	H	17#

Fig. 3. Structure–activity of RP67580 binding to rat NK_1 receptor. IC_{50} values are from inhibition of [^{3}H]SP binding to brain membranes (data from Tabart et al., 1992).

ing to bovine retina, CGP-47,899 (K_i = 10 n*M*) was identified and shown to block substance P-induced contraction of the guinea pig ileum. Its affinity for NK_2 and NK_3 receptors was not reported. Hirschmann et al. (1992) described an acetyl derivative of a peptidomimetic compound that possessed activity at somatostatin and β_2-adrenergic receptors, but also had an IC_{50} value of 120 n*M* at the NK_1 receptor. As the authors noted, *N*-acetylation quite remarkably abolished activity at somatostatin and β_2-adrenergic receptors, yielding a compound that bound selectively to the NK_1 receptor with an IC_{50} value of 60 n*M*. Antagonist activity was demonstrated by blockade of substance P-mediated inositol phosphate production. The authors speculated that the benzyl rings on the glucose scaffold might bind to the NK_1 receptor in a manner similar to the phenyl groups of CP-96,345 and RP-67,580. The benzyl groups of

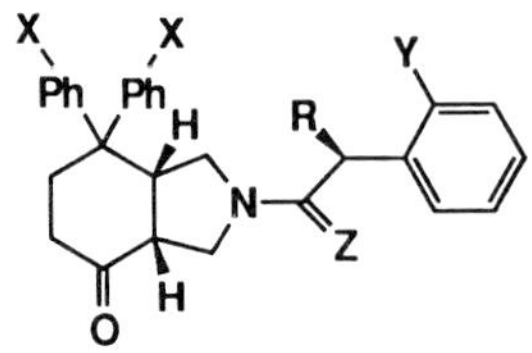

X	Y	Z	R	SP IC_{50}, nM
H	H	O	H	181
H	OCH_3	O	H	75
H	$OCH_2CH_2NMe_2$	O	H	40
H	H	O	CH_3	35
H	H	O	OH	113
H	OCH_3	O	CH_3	12
4-CH_3	H	O	H	393
H	H	NH	H	36
H	OCH_3	NH	H	17

Fig. 4. Structure–activity relationship of heterosteroid analog binding to rat NK_1 receptor. IC_{50} values are from displacement of [^{3}H]SP binding to brain membranes (data from Venepelli et al., 1992).

CGP-47,899, which has some structural similarities to CP-99,994, may also be important in its binding to the NK_1 receptor. These latter compounds indicate the breadth and variety of structural leads available for pursuit of novel nonpeptide NK_1 antagonists.

2.3. Receptor Characterization Using Nonpeptide Antagonists

CP-96,345 was radiolabeled with tritium (Lowe et al., 1991) to provide a high-affinity and selective radiolabeled antagonist to characterize the NK_1 receptor. [^{3}H]CP-96,345 bound with high-affinity (K_d = 0.22 n*M*) to a single site in guinea pig striatal membranes (McLean et al., 1991). Using a ligand concentration of 1.0 n*M*, kinetic analysis of the time course of association at 22°C indicated that specific binding of [^{3}H]CP-96,345 reached a plateau at 5 min with a k_{obs} value of 0.54 min^{-1} (Fig. 5A). The dissociation of [^{3}H]CP-96,345, precipitated by the addition of excess concentrations of CP-99,994 gave a k_{-1} value of 0.08 min^{-1} M^{-1} for a $t_{1/2}$ of 8.6 min (Fig. 5B). This agrees very well with the dissociation constant obtained using an ^{125}I-labeled analog of CP-96,345 (Cascieri et al., 1992). Using the equation $k_{+1} = (k_{obs} - k_{-1})/[L]$ yielded an association constant of 4.6×10^8 min^{-1} M^{-1}. The K_D value calculated from association and dissociation rate constant data was 0.17 nM, in good agreement with the K_D value calculated from saturation isotherm experiments.

In competition studies, the rank order of potency for inhibiting [^{3}H]CP-96,345 was substance P ≥ physalaemin ≥ Sar^9-Met(O_2)^{11}-substance P > eledoisin > NKA ≥ NKB, consistent with previous radioligand binding studies of the NK_1 receptor using ^{125}I-BH-SP (Cascieri and Liang, 1983; Lavielle et al., 1988; Tousignant et al., 1991). The rank order of compounds for inhibiting the binding of [^{3}H]CP-96,345 mirrored previous work, but the K_i values obtained were reduced approximately 100-fold relative to those observed using radiolabeled agonists. In fact, NKA, NKB, and eledoisin at 10 μM did not significantly inhibit the binding of [^{3}H]CP-96,345 despite their demonstrated agonist activity at the NK_1 receptor (Regoli et al., 1988). In contrast, the affinity estimates for antagonists such as [D-Pro^2, D-Trp^{7,9}]-substance P or CP-96,345 were consistent with previous reports (Cascieri and Liang, 1983; Snider et al., 1991). Cascieri et al., (1992) using CHO cells transfected with the human NK_1 receptor, compared the potency of agonists and antagonists for inhibiting the binding of ^{125}I-SP and an ^{125}I-labeled analog of CP-96,345 and obtained similar results. The potencies of NKA and NKB against the ^{125}I-labeled analog of CP-96,345 were reduced relative to those obtained using ^{125}I-SP, whereas CP-96,345 had equivalent IC_{50} values against both radioligands. Although CP-96,345 interacts with the NK_1 receptor in a competitive manner (Snider et al., 1991), the results of the binding studies suggest that [^{3}H]SP and [^{3}H]CP-96,345 interact with the NK_1 receptor in subtly different ways, as would be expected based on their respective agonist and antagonist properties. Moreover, the binding data combined with results emerging from molecular biological studies of the NK_1 receptor suggest that the agonists and antagonists may bind to

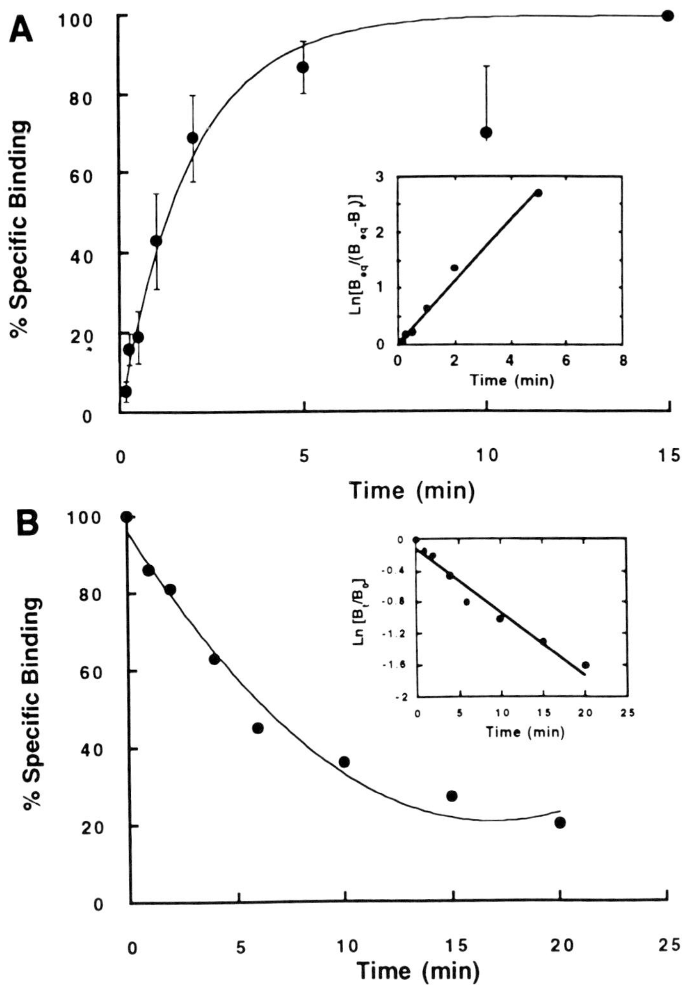

Fig. 5. Kinetic analysis of the binding of [^{3}H]CP-96,345 binding to guinea pig brain membranes. **(A)** Association rate of [^{3}H]CP-96,345 (1.0 n*M*) at 22°C. Each point represents the mean of three separate experiments with each point determined in triplicate. Inset, linear transformation of the association data using the pseudo first order equation $\ln([B_{eq}]/[B_{eq}-B_t]) = K_{obs}t$ where B_{eq} represents the amount of radioligand bound at equilibrium and B_t is the amount of ligand bound at time *t*. **(B)** Dissociation of [^{3}H]CP-96,345 from guinea pig brain membranes incubated with [^{3}H]CP-96,345 (1.0 n*M*) for 20 min at 22°C. Dissociation was initiated by the addition of 1 µ*M* of CP-99,994. Inset, linear transformation of the dissociation data using the first order rate equation $\ln (B/B_0)=-k_{-1}t$ where *B* represents the amount bound at any time *t* and B_0 is the amount bound at the time dissociation was initiated.

entirely different or even mutually exclusive sites on the receptor. Gether et al. (1992) constructed chimeric NK_1-NK_3 receptors in which they replaced the C-terminal portion (TM V to TM VII) of the NK_1 receptor with the corresponding portion of the NK_3 receptor, resulting in a loss of CP-96,345 binding. In contrast, the affinity of substance P for this chimeric receptor was not dramatically altered. This is consistent with the results of Yokota et al. (1992) who concluded that the high-affinity binding site for substance P is determined by the middle portion of the receptor around transmembrane segments II–IV (*see* Chapter 7 of this volume). This is clearly an exciting area and will require extensive SAR studies using a variety of chimeric, as well as point mutated, receptors to elucidate the molecular determinants of binding to the NK_1 receptor. We might note, however, that substance P's significant affinity for the NK_3 receptor suggests considerable overlap in the SP binding site between these two receptors.

Analysis of the primary amino acid sequence of the NK_1 receptor suggests that it belongs to the family of G-protein-coupled receptors having seven hydrophobic, transmembrane spanning segments. Like other members of this family, such as the adrenergic receptors, the binding of agonists to the NK_1 receptor is modulated by guanine nucleotides (Lee et al., 1983; Bahouth and Musacchio, 1985; Cascieri and Liang, 1983; MacDonald and Boyd, 1989). The G-protein-linked receptors are thought to exist in two affinity states, the proportion of which varies according to the concentration of guanine nucleotides (DeLean et al., 1980). Guanine nucleotides generally decrease the affinity of the agonist–receptor complex, but not the number of sites (Rodbell et al., 1980); however, changes in B_{max} values for the NK_1 receptor have been reported (Lee et al., 1983; Bahouth and Musacchio, 1985). In contrast to agonist binding, the binding of antagonists is not influenced by guanine nucleotides (DeLean et al., 1980). The effect of GTP on the binding of [^{3}H]SP and [^{3}H]CP-96,345 to the NK_1 receptor in guinea pig striatal membranes was examined. As shown in Fig. 6, GTP had no effect on the binding of [^{3}H]CP-96,345, but at 1 m*M* produced a maximal (80%) inhibition of [^{3}H]SP binding. The apparent IC_{50} value of 20

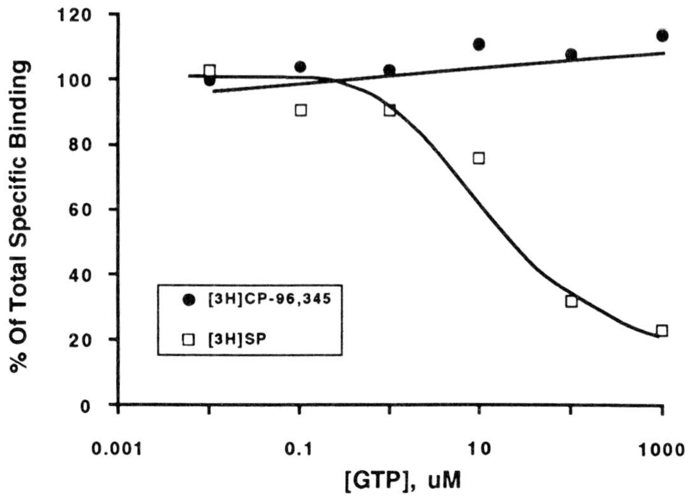

Fig. 6. Effect of GTP on the binding of [^{3}H]CP-96,345 (1.0 n*M*) and [^{3}H]SP (1.0 n*M*) to guinea pig brain membranes.

μ*M* for inhibition of [^{3}H]SP binding is similar to that obtained previously (Lee et al., 1983; Bahouth and Musacchio, 1985). Because [^{3}H]CP-96,345 bound to the NK_1 receptor independently of G-protein coupling, it was possible to investigate whether agonists interacted with two affinity states of the NK_1 receptor. We first examined the inhibition of [^{3}H]CP-96,345 binding to the guinea pig striatum by substance P and Sar^9-Met$(O_2)^{11}$-substance P and found that the binding was best fit by a two-site model. Next, in order to relate the data to functional activation of the NK_1 receptor, CHO cells transfected with the human NK_1 receptor (provided by J. Krause) were used since agonist stimulation of the NK_1 receptor activates phosphatidylinositol turnover in these cells (Ingi et al., 1991; Takeda et al., 1992). Again, the data were best fit by a two site model (p <.001) in which 60–70% of the sites were high-affinity (K_i = 1.0 n*M*) and 30–40% were low-affinity ($K_i \geq 200$ n*M*) (Fig. 7). The K_i value of substance P for the high-affinity site is consistent with the EC_{50} value (~1.0 n*M*) for substance P-induced phosphatidylinositol turnover (S. H. Zorn, personal communication). Cascieri et al. (1992)

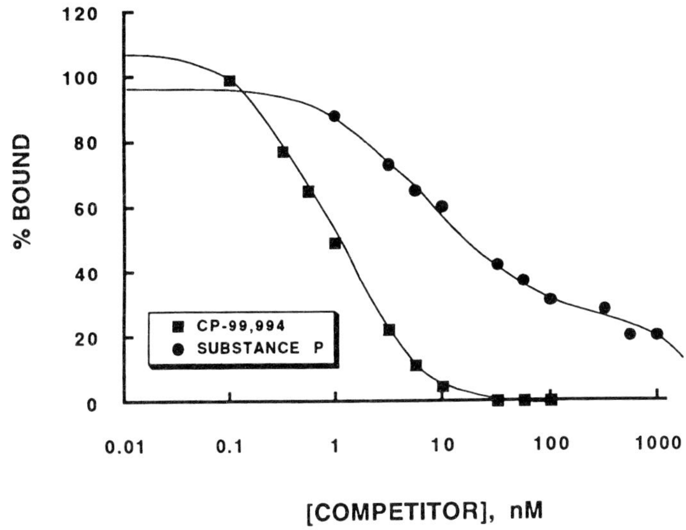

Fig. 7. Inhibition of [^{3}H]CP-96,345 binding to CHO cells stably transfected with the human NK_1 receptor. Membranes prepared from the CHO cells were incubated with [^{3}H]CP-96,345 (1.0 n*M*) for 20 min at 22°C in the presence of increasing concentrations of substance P or CP-99,994.

found a similar proportion of high- and low-affinity NK_1 sites and demonstrated that >90% of the high-affinity sites could be converted to low-affinity by the addition of 100 μ*M* Gpp(NH)p. Our estimates of the K_i values for substance P binding to the high- and low-affinity sites are considerably reduced relative to those obtained by Cascieri et al. (1992), but in agreement with the values reported by Luber-Narod et al. (1990).

2.4. Species Differences in the Binding of NK_1 Receptor Antagonists

A species difference in the affinity of CP-96,345 for the NK_1 receptor was first noted by Snider et al. (1991) who found that CP-96,345 bound with subnanomolar affinity to bovine brain, but was 35-fold less active in inhibiting [^{3}H]SP binding to rat brain. The species variation in binding to the NK_1 receptor has been confirmed and extended to other species (Beresford et al., 1991; Bryce and McLean, 1991; Gitter et al., 1991; Watling et al., 1991) with CP-96,345 exhibiting reduced affinity for the NK_1 receptor in chickens, rats, and mice and significantly higher affinity in gerbil, guinea pig, hamster, rabbit, dog, monkey, and human. A species dependent enzymatic degradation of radiolabeled [^{3}H]SP is unlikely to explain these results, since: 1. the K_D for [^{3}H]SP does not vary between guinea pig and rat, and 2. the assays were carried out in the presence of several peptidase inhibitors. Furthermore, saturation analysis using [^{3}H]CP-96,345 and [^{3}H]SP binding to mouse, rat, guinea pig, and marmoset brains revealed a 10- to 50-fold reduction in K_d values for [^{3}H]CP-96,345 binding to rat and mouse, but no species variation in the binding of [^{3}H]SP (Table 2).

The binding of several nonpeptide antagonists to the rat and human NK_1 receptor transfected into CHO cells is summarized in Table 3. The variation in selectivity for the rat and human receptor is striking. CP-96,345 and CP-99,994 are approximately 2–3 orders of magnitude more selective for the human NK_1 receptor. In contrast, the heterosteroid, Win-62,577, although binding to the rat NK_1 receptor with moderate affinity, does not inhibit ^{125}I-BH-SP binding to the human receptor. (±)RP-67,580 lies in-between these extremes, with an approximately 10-fold separation in affinity for the rat and human NK_1 receptor. The availability of these compounds with their spectrum of species affinities should provide the pharmacological tools for characterizing the antagonist binding site in these respective receptors.

The dramatic species difference in binding to the NK_1 receptor appears remarkable given that the rat and human NK_1 receptor differ by only 22 amino acids that are generally conservative substitutions and are distributed throughout the protein sequence (Takeda et al., 1991). Molecular biology studies using site-directed mutagenesis and chimeric receptors have identified two residues (116 and 290) that largely account for the marked differences in the binding of CP-96,345 and RP-67,580 to the rat and human NK_1 receptor (Fong et al., 1992b; Sachais et al., 1993). Fong et al. (1992b) concluded that these residues do not interact directly with the compounds, but alter the

Table 2
Results of Saturation Isotherms
Using Membranes Prepared from the Caudate Nucleus
of Several Species and [^{3}H]Substance P and [^{3}H]CP-96,345[a]

	[^{3}H]Substance P	[^{3}H]CP-96,345
Species	K_d ± SEM, n*M*	K_d ± SEM, n*M*
Mouse	1.81 ± 0.90	28.7 ± 6.9
Rat	2.17 ± 1.14	26.6 ± 9.4
Guinea pig	2.38 ± 0.84	2.24 ± 0.26
Marmoset	0.61 ± 0.15	0.64 ± 0.26

[a]The results are from 5–7 experiments performed in triplicate.

Table 3
Competition Assays Using ^{125}I-BH-Substance P (0.1 n*M*)
Binding to the Rat and Human NK_1 Receptor
Stably Transfected into Chinese Hamster Ovary Cells

	IC_{50} ± SEM, n*M*	
Competitor	Human NK_1 receptor	Rat NK_1 receptor
CP-96,345	0.58 ± 0.13	19 ± 5
CP-99,994	0.48 ± 0.07	127 ± 26
(±) RP 67,580	83.6 ± 13.9	6.6 ± 1.3
WIN 62,577[a]	>25,000	~50

[a]Data from Appell et al., 1992

local conformation of the receptor leading to changes in antagonist binding.

Whether the species difference in antagonist binding represents a species homolog of the NK_1 receptor or a receptor subtype is not entirely clear. The binding studies have generally focused on binding to brain tissue from various animals and have not investigated antagonist binding in different tissues from the same species. This is important as receptor subtypes are conventionally defined by ligands exhibiting different potencies in different tissues of the same species. Appell et al. (1992) examined the binding of ^{125}I-BH-SP to human astrocytoma cells and urinary bladder, guinea pig brain and ileum, and rat brain and duodenum. ^{125}I-BH-SP binding did not differ among these tissues nor did the affinity of the rat selective heterosteroids and cyanines distinguish between CNS and peripheral tissues. These data and the finding that within a species only one mRNA encodes the NK_1 receptor (different isoforms have been described [Fong et al., 1992a]) suggest that species variation, and not NK_1 receptor subtypes, account for the observed antagonist selectivity. However, recent evidence proposing another tachykinin receptor or subtype in the guinea pig ileum (Chassaing et al., 1992) suggests that a more extensive investigation using a variety of tissues and several different ligands be undertaken.

The availability of high-affinity NK_1 receptor selective antagonists and agonists coupled with the extensive molecular biology efforts underway bodes well for progress in understanding the molecular mechanisms underlying binding to the NK_1 receptor and the resulting signal transduction. These tools should also help to clarify the presence of receptor subtypes or species variants of a single receptor.

3. NK_2 Receptors

3.1. Characterization of the NK_2 Receptor

In rat brain, ^{125}I-BH-reagent-coupled eledoisin binding was inhibited by tachykinins with the following rank order of potency: neurokinin B > kassinin = eledoisin > physalaemin > neurokinin A (substance K) > substance P (Buck et al., 1984), consistent with binding to what is now called the NK_3 receptor (Henry, 1987). In contrast, in membranes from rat duodenum or mouse bladder the rank order in inhibiting ^{125}I-BH-eledoisin binding was neurokinin A > kassinin > eledoisin > neurokinin B > substance P (Buck et al., 1984; Burcher and Buck, 1986) suggesting that ^{125}I-BH-eledoisin labels different sites in the CNS and periphery. Based on these results and the finding that a nearly identical competition profile was obtained against ^{125}I-BH-neurokinin A and ^{125}I-BH-eledoisin, it was concluded that a site distinct from the NK_1 and NK_3 receptors present in brain was labeled. Originally referred to as the SP-K receptor, it is now termed the NK_2 receptor (Henry, 1987). Unfortunately, ^{125}I-BH-NKA is not selective for NK_2 receptors as addition of the ^{125}I-BH group to NKA markedly enhances its affinity for NK_3 sites (Bergstrom et al., 1987). Also, in tissues enriched in NK_1 receptors ^{125}I-BH-NKA binds to the NK_1 receptor (Geraghty et al., 1992).

Direct iodination of the histidine in neurokinin A, ^{125}I-iodohistidyl-neurokinin A, provides a high specific activity radioligand whose binding characteristics are presumably less affected than the ^{125}I-BH derivative of NKA. Saturation experiments using a variety of different tissues, including cell lines transfected with the NK_2 receptor, give K_D values of 0.5–1.0 n*M* with Hill coefficients near unity, consistent with binding to a single site (Burcher et al., 1986; Van Giersbergen et al., 1991). However, Ingi et al. (1991), using monkey kidney COS cells transfected with rat NK_2 receptors, found high- and low-affinity binding sites for ^{125}I-iodohistidyl-NKA with association rates that were indistinguishable, but different dissociation rate constants. Addition of guanine nucleotides markedly diminished the number of high-affinity sites without affecting the number of low-affinity sites. The presence of high- and low-affinity sites may explain the biphasic dissociation curves that have been observed for the binding of radiolabeled agonists to the NK_2, as well as the NK_1 and NK_3, receptors (Torrens et al., 1983; Beaujouan et al., 1984; Guard et al., 1990) and suggest that the ligand is interacting with more than one population of sites. Although the guanine nucleotide sensitive and insensitive sites could represent distinct receptors, a more parsimonious explanation is that the high-affinity site represents the NK_2 receptor coupled with a G-protein, whereas the low-affinity site is the G-protein dissociated form of the receptor. Clearly, a radiolabeled antagonist would be useful in sorting out these possibilities.

3.2. NK_2 Receptor Antagonists

Biochemical characterization of the NK_2 receptor has been limited by the paucity of pharmacological tools, in particular receptor selective antagonists. Recently, several peptide antagonists have been synthesized that have greater than 100 n*M* affinity for the NK_2 receptor (Table 4; Buck et al., 1990; Dion et al., 1990; Rovero et al., 1990a; McKnight et al., 1991; Advenier et al., 1992; McElroy et al., 1992). An exciting development is the discovery of SR48968, the first nonpeptide NK_2 receptor antagonist (Advenier et al., 1992; Emonds-Alt et al., 1992). SR48968 has high-affinity (K_i = 0.5 n*M*) and selectivity (>10,000) for the NK_2 receptor in rat duodenal membranes, acting as a full antagonist to block NKA elicited effects on various isolated smooth muscle preparations. At the International Substance P and Related Peptides meeting held in November, 1992, the binding of [^{3}H]SR-48968 and [^{3}H]GR100679 was reported. Both compounds bound with high-affinity (<1.0 n*M*) and selectivity to the NK_2 receptor with the appropriate pharmacology (Emonds-Alt et al., 1993; Hagan et al., 1992). Detailed kinetic studies on the binding of these compounds were not reported; however, Hagan et al. (1992), using autoradiography, visualized discrete, specific binding of [^{3}H]GR100,679 to brain sections from neonatal rats. They were unable to obtain any specific binding in adult brains. The presence of NK_2 receptors in brain is controversial with several groups reporting specific binding to presumed NK_2 sites (Mantyh, 1984; Quirion and Dam, 1985; Dam et al., 1990), while others have suggested that this represents binding to NK_1 or NK_3 sites (Bergstrom et al., 1987; Foster and Tridgett, 1988; Saffroy et al., 1988). Furthermore, mRNA encoding NK_2 receptors is reported to be absent in brain (Sasai and Nakanishi, 1989). However, the blockade by selective NK_2 antagonists of turning behavior elicited by central infusion of NK_2 receptor selective agonists suggests the presence of NK_2 receptors in brain (Elliott et al., 1991; Poncelet et al., 1993; *see* Chapter 5 of this volume for further discussion of the presence of NK_2 receptors in brain).

Table 4
Amino Acid Sequences of Peptide Antagonists and the Structure of SR48968[a]

Compound	Sequence/structure	Activity	Tissue selectivity
MEN 10,207	H-Asp-Tyr-D-Trp-Val-D-Trp-D-Trp-Arg-NH_2	Antagonist	RPA>HT
MDL 28,564	H-Asp-Ser-Phe-Val-Gly-Leuψ(CH_2NH)Leu-NH_2	Agonist/antagonist?	RPA/HT
L-659,877	cyclo(Gln-Trp-Phe-Gly-Leu-Met)	Antagonist	HT>RPA
GR100,679	CyCO-Gly-Ala-D-Trp-Phe-*N*Me_2[b]	Antagonist	HT>RPA
R-396	AcLeu-Asp-Gln-Trp-Phe-Gly-NH_2	Antagonist	RPA>HT
SR48968	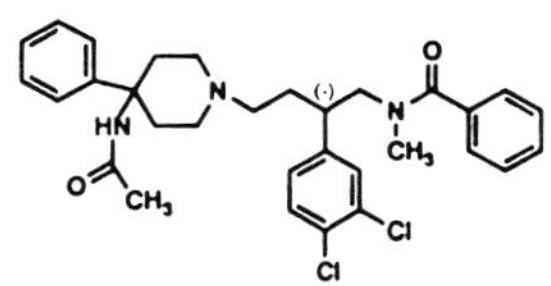	Antagonist	RPA>HT

[a]Activity is derived from bioassays described in the text. Tissue selectivity refers to potency in blocking agonist effects in the rabbit pulmonary artery and hamster trachea.

[b]Cy = cyclohexyl.

3.3. NK_2 Receptor Heterogeneity

The data suggesting the presence of NK_2 receptor subtypes, or perhaps stated more conservatively, species variants of the NK_2 receptor, have been derived primarily from functional assays. These results have been reviewed in detail elsewhere in this volume and so will be considered here only briefly. The compounds listed in Table 4 are at least 100-fold selective for the NK_2 relative to the NK_1 and NK_3 receptors in functional and radioligand binding assays (Buck et al., 1990; Dion et al., 1990; Rovero et al., 1990a,b; McKnight et al., 1991; Advenier et al., 1992). Thus, differences in rank order of potency in different tissue preparations is unlikely to be accounted for by contamination with NK_1 or NK_3 receptors. In the rabbit pulmonary artery, NK_2 receptor-mediated contraction was antagonized with a rank order of potency of MEN-10,207 > L-659,877 > R-396 (Maggi et al., 1990; Patacchini et al., 1991). In contrast, contraction of the hamster trachea by neurokinin A was blocked by L-659,877 ≥ R-396 > MEN-10,207. SR48968, the recently described nonpeptide antagonist (Advenier et al., 1992; Emonds-Alt et al., 1992), was the most potent antagonist in both tissues; however, it had approximately 100-fold greater potency in the rabbit pulmonary artery. The activity of MDL-28,564 in these two preparations was striking. In the rabbit pulmonary artery, it was a full agonist; however, in hamster trachea or bladder MDL-28,564 lacked appreciable agonist activity and blocked the response to neurokinin A (Buck et al., 1990; Patacchini et al., 1991).

It is difficult to carry out radioligand binding assays because of the small amount of smooth muscle present in many of the tissues used in the bioassays. However, sufficient tissue is available from the hamster bladder and bovine stomach to perform ligand binding assays. In these experiments, NKA did not distinguish between species, similar to the results from functional assays. In contrast, MEN-10,207 had 40-fold higher affinity for the NK_2 receptor in bovine stomach than in hamster bladder. L-659,877 had the opposite profile, inhibiting ^{125}I-NKA binding to hamster bladder with 50-fold higher affinity than bovine stomach. Similar results were obtained using mouse fibroblasts transfected with the bovine NK_2 receptor (Van Giersbergen et al., 1991). Although these studies clearly demonstrate that the NK_2 receptors in different species exhibit a different pharmacology, there is no evidence within a species that NK_2 receptors differ in their pharmacology. In fact, the available data suggest that within a species the apparent affinities do not differ across tissues (*see* Advenier et al., 1992). Thus, similar to the NK_1 receptor, the evidence to date suggests species homologs of the NK_2 receptor. The finding that the mouse, rat, bovine, and human NK_2 receptors have 85–95% amino acid sequence identity is consistent with a single species of NK_2 receptor (Masu et al., 1987; Sasai et al., 1989; Gerard et al., 1990; Sundelin et al., 1992).

4. NK_3 Receptors

The existence of different types of tachykinin receptors was hypothesized based on the different rank order of potencies of substance P/physalaemin and kassinin/eledoisin, depending on the tissue used, and on a lack of cross-desensitization between substance P and eledoisin (Lee et al., 1982; Watson et al., 1983). Radioligand binding studies using ^{125}I-BH-coupled eledoisin binding to rat cortical membranes confirmed the presence of two distinct sites (Beaujouan et al., 1984; Buck et al., 1984; Cascieri and Liang, 1984). Competition studies revealed a site with a pharmacology different from the site labeled by ^{125}I-BH-substance P with eledoisin and the endogenous peptide, neurokinin B, exhibiting greater potency in inhibiting ^{125}I-BH-eledoisin binding than ^{125}I-BH-substance P binding. As discussed previously, when the binding of ^{125}I-BH-eledoisin was examined in different tissues two pharmacologically distinct sites were labeled. The binding site found in the periphery is currently referred to as the NK_2 receptor, while the site in the cerebral cortex is the NK_3 receptor (Buck et al., 1984; *see* Henry, 1987 concerning nomenclature). [^{3}H]NKB bound to a single site with a pharmacology and distribution identical to ^{125}I-BH-eledoisin (Bergstrom et al., 1987) and, of the naturally occurring tachykinins, neurokinin B binds with the highest affinity suggesting that it is the preferred endogenous ligand for the NK_3 receptor (Buck et al., 1984; Torrens et al., 1984; Cascieri et al., 1985). In the search for more selective ligands for the NK_3 receptor, Mussap and Burcher (1990) radioiodinated scyliorhinin II, a peptide isolated from dogfish gut (Conlon et al., 1986), which was reported to be 1000-fold selective for the NK_3 receptor (Beaujouan et al., 1988). Unfortunately, the ^{125}I-BH-coupled scyliorhinin analog, although having high-affinity for the NK_3 receptor, also bound to the NK_1 receptor in rat submandibular gland (Mussap and Burcher, 1991). To date, the most selective ligand for labeling the NK_3 receptor is senktide. Senktide, succinyl-[Asp^6,MePhe^8]substance P_{6-11}, was synthesized by Wormser et al. (1986) and has at least four orders of magnitude selectivity for the NK_3 receptor. Characterization of senktide binding has used ^{125}I-BH-senktide and, more recently, [^{3}H]senktide (Laufer et al., 1986; Guard et al., 1990). Interestingly, affinity constants obtained for ^{125}I-BH-senktide and [^{3}H]senktide differed by an order of magnitude. Addition of the bulky BH reagent to the peptide may alter its binding properties as reported previously for neurokinin A (Bergstrom et al., 1987). Alternatively, since brain membranes from rat and guinea pig were used to characterize ^{125}I-BH-senktide and [^{3}H]senktide binding, respectively, this may reflect species variants in the NK_3 receptor. Differences in NK_3 receptor pharmacology between rats and guinea pigs have been observed in the binding of [^{3}H]senktide to cortical membranes, as well as in agonist potency and efficacy in depolarizing the superior cervical ganglia (Renzetti et al., 1991; Seabrook et al., 1992). An additional possibility is that the use of low concentrations of iodinated probes may result in labeling only the high-affinity form of the receptor. This would be consistent with the monophasic dissociation of ^{125}I-BH-senktide and the curvilinear dissociation curves obtained with [^{3}H]senktide (Laufer et al., 1986; Guard et al., 1990). Existence of a high- and low-affinity form of the NK_3 receptor can be hypothesized given that the binding of radiolabeled agonists to the NK_3 receptor was reported to be sensitive to guanine nucleotides (Cascieri et al., 1985; Laufer et al., 1986; Guard et al., 1990). Clearly, the availability of an antagonist insensitive to the G-protein-linked state of the receptor would help in interpreting the contradictory data (Cascieri et al., 1985; Beaujouan et al., 1984).

The need for additional pharmacological tools to investigate the NK_3 receptor is further emphasized by recent molecular biology studies. Buell et al. (1992) cloned the human NK_3 receptor and reported a 74 and 68% homology with the human NK_1 and NK_2 receptors, respectively. Cross-species comparison of the rat and human sequences indicated 86% homology, which is considerably less than the 95% identity of the primary structure of rat and human NK_1 receptors. This comparison is noteworthy since Buell et al. (1992) found [^{3}H]senktide binding was inhibited with a pharmacology consistent with labeling the NK_3 receptor, i.e., senktide = NKB > eledoisin > NKA = SP. However, the IC_{50} values were 10–50-fold less than previously reported (Guard et al., 1991). The twofold higher concentrations of [^{3}H]senktide used by Buell et al. might account for a portion of these elevated values. Nonetheless, the reported IC_{50} of 424 n*M* for eledoisin at the human NK_3 receptor may explain an earlier report that NK_3 receptors were not detected in monkey and human CNS using ^{125}I-BH-eledoisin (Dietl and Palacios, 1991). In contrast, [^{3}H]senktide was reported to bind to monkey spinal cord (Guard and Watson, 1991), and NK_3 message is present in human brain (Buell et al., 1992). Confirming the presence of NK_3 receptors in human brain and resolving a possible species difference in the receptor are important tasks for the future.

5. Conclusions

The availability of nonpeptide antagonist ligands for the NK_1 and, more recently, the NK_2 receptor has facilitated biochemical characterization of these receptors. As efforts to produce new nonpeptide tachykinin antagonists continue, clarification of tachykinin pharmacology and of the relevance of tachykinins to disease states should be possible. Advances in molecular genetics coupled with the availability of these antagonists should provide greater understanding of the structure and function of tachykinin receptors and may ultimately lead to mechanistic insights that will facilitate the design of new antagonists. The eventual application of this work to the treatment of diseases involving tachykinins will afford even greater significance to these studies.

References

Advenier, C., Rouissi, N., Nguyen, Q. T., Emonds-Alt, X., Breliere, J.-C., Neliat, G., Naline, E., and Regoli, D. (1992) Neurokinin A (NK_2) receptor revisited with SR 48968, a potent non-peptide antagonist. *Biochem. Biophys. Res. Commun.* **184,** 1418–1424.

Appell, K. C., Fragale, B. J., Loscig, J., Singh, S., and Tomczuk, B. E. (1992) Antagonists that demonstrate species differences in neurokinin-1 receptors. *Mol. Pharmacol.* **41,** 772–778.

Ariens, E. J., Beld, A. J., Rodrigues de Miranda, J. F., and Simomis, A. M. (1979) *The Receptors: A Comprehensive Treatise.* Plenum, New York, pp. 33–91.

Bahouth, S. W. and Musacchio, J. M. (1985) Specific binding of [^{3}H]substance P to the rat submaxillary gland. The effects of ions and guanine nucleotides. *J. Pharmacol. Exp. Ther.* **234,** 326–336.

Beaujouan, J. C., Torrens, Y., Saffroy, M., and Glowinski, J. (1988) Neuropeptide K, scyliorhinin I and scyliorhinin II: new tools in the tachykinin receptor field. *Eur. J. Pharmacol.* **151,** 353,354.

Beaujouan, J. C., Torrens, Y., Viger, A., and Glowinski, J. (1984) A new type of tachykinin binding site in the rat brain characterized by specific binding of a labeled eledoisin derivative. *Mol. Pharmacol.* **26,** 248–254.

Beresford, I. J. M., Birch, P. J., Hagan, R. M., and Ireland, S. J. (1991) Investigation into species variants in tachykinin NK_1 receptors by use of the nonpeptide antagonist, CP-96,345. *Br. J. Pharmacol.* **104,** 292,293.

Bergstrom, L., Beaujouan, J. C., Torrens, Y., Saffroy, M., Glowinski, J., Lavielle, S., Chassaing, G., Marquet, A., d'Orleans-Juste, P., Dion, S., and Regoli, D. (1987) ^{3}H-neurokinin A labels a specific tachykinin-binding site in the rat duodenal smooth muscle. *Mol. Pharmacol.* **32,** 764–771.

Bergstrom, L., Torrens, Y., Saffroy, M., Beaujouan, J. C., Lavielle, S., Chassaing, G., Morgat, J. L., Glowinski, J., and Marquet, A. (1987) [^{3}H]neurokinin B and ^{125}I-BH eledoisin label identical tachykinin binding sites in the rat brain. *J. Neurochem.* **48,** 125–133.

Bryce, D. K. and McLean, S. (1991) Differential binding of CP-96,345 across species: possible heterogeneity of the NK-1 tachykinin receptor. *Soc. Neurosci. Abs.* **17,** 804.

Buck, S. H. and Burcher, E. (1986) The tachykinins: a family of peptides with a brood of "receptors." *Trends Pharmacol. Sci.* **7,** 65–68.

Buck, S. H., Burcher, E., Shults, C. W., Lovenberg, W., and O'Donohue, T. L. (1984) Novel pharmacology of substance K-binding sites: a third type of tachykinin receptor. *Science* **226,** 987–989.

Buck, S. H., Harbeson, S. L., Hassmann, C. F., Shatzer, S. A., Rouissi, N., Nantel, F., and Van Giersbergen, P. L. (1990) [Leu^9 ψ(CH_2NH)Leu^{10}]-neurokinin A (4-10) (MDL 28,564) distinguishes tissue tachykinin peptide NK_2 receptors. *Life. Sci.* **47,** PL37–PL41.

Buell, G., Schulz, M. F., Arkinstall, S. J., Maury, K., Missotten, M., Adami, N., Talabot, F., and Kawashima, E. (1992) Molecular characterisation, expression and localisation of human neurokinin-3 receptor. *FEBS Lett.* **299,** 90–95.

Burcher, E. and Buck, S. H. (1986) Multiple tachykinin binding sites in hamster, rat and guinea pig urinary bladder. *Eur. J. Pharmacol.* **128,** 165.

Burcher, E., Buck, S. H., Lovenberg, W., and O'Donohue, T. L. (1986) Characterization and autoradiographic localization of multiple tachykinin binding sites in gastrointestinal tract and bladder. *J. Pharmacol. Exp. Ther.* **236,** 819–831.

Burcher, E., Mussap, C. J., Geraghty, D. P., McClure-Sharp, J. M., and Watkins, D. J. (1991) Concepts in characterization of tachykinin receptors. *Ann. NY Acad. Sci.* **632,** 123–136.

Cascieri, M. A., Ber, E., Fong, T. M., Sadowski, S., Bansal, A., Swain, C., Seward, E., Frances, B., Burns, D., and Strader, C. D. (1992) Characterization of the binding of a potent, selective, radioiodinated antagonist to the human neurokin-1 receptor. *Mol. Pharmacol.* **42,** 458–463.

Cascieri, M. A., Chicchi, G. G., and Liang, T. (1985) Demonstration of two distinct tachykinin receptors in rat brain cortex. *J. Biol. Chem.* **260,** 1501–1507.

Cascieri, M. A. and Liang, T. (1983) Characterization of the substance P receptor in rat brain cortex membranes and the inhibition of radioligand binding by guanine nucleotides. *J. Biol. Chem.* **258,** 5158–5164.

Cascieri, M. A. and Liang, T. (1984) Binding of 1^{25}I-BH conjugated eledoisin to rat brain cortex membranes-evidence for two classes of tachykinin receptors in the mammalian central nervous system. *Life. Sci.* **35,** 179–184.

Chassaing, G., Lavielle, S., Brunissen, A., Carruette, A., Garret, C., Petitet, F., Saffroy, M., Beaujouan, J. C., Torrens, Y., and Glowinski, J. (1992) [Pro^9]SP and [pGlu^6,Pro^9]SP(6-11) interact with two different receptors in the guinea pig ileum as demonstrated with new SP antagonists. *Neuropeptides* **23,** 73–79.

Conlon, J. M., Deacon, C. F., O'Toole, L., and Thim, L. (1986) Scyliorhinin I and II: two novel tachykinins from dogfish gut. *FEBS Lett.* **200,** 111–116.

Dam, T.-V., Takeda, Y., Krause, J. E., Escher, E., and Quirion, R. (1990) γ-Preprotachykinin-(72-92)-peptide amide: an endogenous γ-preprotachykinin I gene derived peptide which preferentially binds to neurokinin-2 receptors. *Proc. Natl. Acad. Sci. USA* **87,** 246–250.

DeLean, A., Stadel, J. M., and Lefkowitz, R. J. (1980) A ternary complex model explains the agonist-specific binding properties of the adenylate cyclase coupled β-adrenergic receptor. *J. Biol. Chem.* **255,** 7108–7117.

Desai, M. C., Lefkowitz, S. L., Thadio, P. F., Longo, K. P., and Snider, R. M. (1992) Discovery of a potent substance P antagonist: recognition of the key molecular determinant. *J. Med. Chem.* **135,** 4911–4913.

Dietl, M. M. and Palacios, J. (1991) Phylogeny of tachykinin receptor localization in the vertebrate central nervous system: apparent absence of neurokinin-2 and neurokinin-3 binding sites in the human brain. *Brain Res.* **539,** 211–222.

Dion, S., Rouissi, N., Nantel, F., Drapeau, G., Regoli, D., Naline, E., and Advenier, C. (1990) Receptors for neurokinins in human bronchus and urinary bladder are of the NK-2 type. *Eur. J. Pharmacol.* **178,** 215–219.

Drapeau, G., d'Orleans-Juste, P., Dion, S., Rhaleb, N. E., Rouissi, N. E., and Regoli, D. (1987) Selective agonists for substance P and neurokinin receptors. *Neuropeptides* **10,** 43–54.

Elliott, P. J., Mason, G. S., Stephens-Smith, M., and Hagen, R. M. (1991) Behavioural and biochemical responses following activation of midbrain dopamine pathways by receptor selective neurokinin agonists. *Neuropeptides* **19,** 119–126.

Emonds-Alt, X., Advenier, C., Le Fur, G., and Breliere, J. C. (1993) SR48968, a neurokinin A (NK_2) receptor antagonist. *Regul. Pept.* **suppl. 1,** S4.

Emonds-Alt, X., Vilain, P., Goulaouic, P., Proietto, V., Van Broeck, D., Advenier, C., Naline, E., Neliat, G., Le Fur, G., and Breliere, J. C. (1992) A potent and selective nonpeptide antagonist of the neurokinin A (NK_2) receptor. *Life. Sci.* **50,** PL101–PL106.

Folkers, K., Hakanson, R., Horig, J., Jie-Cheng, X., and Leander, S. (1984) Biological evaluation of substance P antagonists. *Br. J. Pharmacol.* **83,** 449–456.

Folkers, K., Rosell, S., Chu, J. Y., Lu, L. A., Tang, P. F., and Ljungqvist, A. (1986) Design and synthesis of antagonists of substance P. *Acta Chem. Scand. (B)* **40,** 295–302.

Fong, T. M., Anderson, S. A., Yu, H., Huang, R. R., and Strader, C. D. (1992a) Differential activation of intracellular effector by two isoforms of human neurokinin-1 receptor. *Mol. Pharmacol.* **41,** 24–30.

Fong, T. M., Yu, H., Huang, R.-R., and Strader, C. D. (1992b) Molecular basis for the species selectivity of the neurokinin-1 receptor antagonists CP-96,345 and RP67580. *J. Biol. Chem.* **267,** 25,668–25,671.

Foster, A. C. and Tridgett, R. (1988) Comparison of the binding of radiolabelled neurokinin A and eledoisin in rat cortex synaptic membranes. *Br. J. Pharmacol.* **94,** 602–608.

Garret, C., Carruette, A., Fardin, V., Moussaoui, S., Peyronel, J. F., Blanchard, J. C., and Laduron, P. M. (1991) Pharmacological properties of a potent and selective nonpeptide substance P antagonist. *Proc. Natl. Acad. Sci. USA* **88,** 10,208–10,212.

Geraghty, D. P., Mussap, C. J., and Burcher, E. (1992) Radioiodinated substance P, neurokinin A, and eledoisin bind predominantly in NK_1 receptors in guinea pig lung. *Mol. Pharmacol.* **41,** 147–153.

Gerard, N. P., Eddy, R. L., Jr., Shows, T. B., and Gerard, C. (1990) The human neurokinin A (substance K) receptor. Molecular cloning of the gene, chromosome localization, and isolation of the cDNA from tracheal and gastric tissues. *J. Biol. Chem.* **266,** 1354.

Gether, U., Johansen, T. E., Snider, R. M., Lowe III, J. A., Yokota, Y., Nakanishi, S., and Schwartz, T. W. (1992) Binding epitopes for peptide and nonpeptide ligands on the NK_1 (substance P) receptor. *Regul. Pept.* **suppl. 1,** S6.

Gitter, B. D., Waters, D. C., Bruns, R. F., Mason, N. R., Nixon, J. A., and Howbert, J. J. (1991) Species differences in affinities of nonpeptide antagonists for substance P receptors. *Eur. J. Pharmacol.* **197,** 237,238.

Guard, S. and Watson, S. P. (1991) Tachykinin receptor types: classification and membrane signalling mechanisms. *Neurochem. Intl.* **18,** 149–165.

Guard, S., Watson, S. P., Maggio, J. E., Too, H. P., and Watling, K. J. (1990) Pharmacological analysis of [^{3}H]-senktide binding to NK3 tachykinin receptors in guinea-pig ileum longitudinal muscle-myenteric plexus and cerebral cortex membranes. *Br. J. Pharmacol.* **99,** 767–773.

Hagan, R. M., Beresford, I. J. M., Elliott, P. J., Ball, D. I., Sheldrick, R. L. G., Ireland, S. J., Pritchard, J. M., and McElroy, A. B. (1992) NK_2 antagonists and the role of NK_2 receptors. *Regul. Pept.* **suppl. 1,** S8.

Henry, J. L. (1987) Discussions of nomenclature for tachykinins and tachykinin receptors, in *Substance P and Neurokinins* (Henry, J. S., Couture, R., Cuello, A. C., Pelletier, G., Quirion, R., and Regoli, D., eds.), Springer-Verlag, New York, xvii.

Hershey, A. D., Dykema, P. E., and Krause, J. E. (1991) Organization, structure and expression of the gene encoding the rat substance P receptor. *J. Biol. Chem.* **266,** 4366–4374.

Hershey, A. D. and Krause, J. E. (1990) Molecular characterization of a functional cDNA encoding the rat substance P receptor. *Science* **247,** 958–962.

Hirschmann, R., Nicolaou, K. C., Pietranico, S., Salvino, J., Leahy, E. M., Sprengeler, P. A., Furst, G., Smith, A. B., Strader, C. D., Cascieri, M. A., Candelore, M. R., Donaldson, C., et al. (1992) Nonpeptidal peptidomimetics with a β-D-glucose scaffolding. A partial somatostatin agonist bearing a close structural relationship to a potent substance P antagonist. *J. Am. Chem. Soc.* **114,** 9217,9218.

Ingi, T., Kitajima, Y., Minamitake, Y., and Nakanishi, S. (1991) Characterization of ligand-binding properties and selectivities of three rat tachykinin receptors by transfection and functional expression of their cloned cDNAs in mammalian cells. *J. Pharmacol. Exp. Ther.* **259,** 968–975.

Laufer, R., Gilon, C., Chorev, M., and Selinger, Z. (1986) Characterization of a neurokinin B receptor site in rat brain using a highly selective radioligand. *J. Biol. Chem.* **261,** 10,257–10,263.

Lavielle, S., Chassaing, G., Julien, S., Besseyre, J., and Marquet, A. (1986a) Influence of the amino acids of substance P in the recognition of its receptor: affinities of synthesized substance P analogues for the specific ^{125}I-BHsubstance P binding site on rat brain synaptosomes. *Neuropeptides* **7,** 191–200.

Lavielle, S., Chassaing, G., Julien, S., Marquet, A., Bergstrom, L., Beaujouan, J. C., Torrens, Y., and Glowinski, J. (1986b) Specific recognition of SP or NKB receptors by analogues of substance P substituted at positions 8 and 9. *Eur. J. Pharmacol.* **125,** 461,462.

Lavielle, S., Chassaing, G., Ploux, O., Loeuillet, D., Besseyre, J., Julien, S., Marquet, A., Convert, O., Beaujouan, J.-C., Torrens, Y., Bergstrom, L., Saffroy, M., et al. (1988) Analysis of tachykinin binding site interactions using constrained analogues of tachykinins. *Biochem. Pharmacol.* **37,** 41–49.

Lee, C.-M., Iversen, L. L., Hanley, M. R., and Sandberg, B. E. B. (1982) The possible existence of multiple receptors for substance P. *Naunyn Schmiedeberg's Arch. Pharmacol.* **318,** 281–287.

Lee, C.-M., Javitch, J. A., and Snyder, S. H. (1983) ^{3}H-substance P binding to salivary gland membranes. *Mol. Pharmacol.* **23,** 563–569.

Lew, R., Geraghty, D. P., Drapeau, G., Regoli, D., and Burcher, E. (1990) Binding characteristics of [^{125}I]BH [Sar^{9},Met(O^{2})^{11}]substance P, a new selective radioligand for the NK_1 receptor. *Eur. J. Pharmacol.* **184,** 97–108.

Lowe, J. A., Drozda, S. E., Snider, R. M., Longo, K. P., Zorn, S. H., Morrone, J., Jackson, E. R., McLean, S., Bryce, D. K., Bordner, J., Nagahisa, A., Kanai, Y., et al. (1992) The discovery of (2S,3S)-*cis*-(diphenylmethyl)-*N*-[(2-methoxy–phenyl)methyl]-1-azabicyclo[2.2.2]-octan-3-amine as a novel, nonpeptide substance P antagonist. *J. Med. Chem.* **35,** 2591–2600.

Lowe, J. A., Drozda, S. E., Snider, R. M., Longo, K. P., and Bordner, J. (1991) Preparation and radiolabelling of CP-96,345, the first nonpeptide substance P antagonist. *Biorg. Med. Chem. Lett.* **1,** 129–132.

Luber-Narod, J., Boyd, N. D., and Leeman, S. E. (1990) Guanine nucleotides decrease the affinity of substance P binding to its receptor. *Eur. J. Pharmacol.* **188,** 185–191.

MacDonald, S. G. and Boyd, N. D. (1989) Regulation of substance P receptor affinity by guanine nucleotide-binding proteins. *J. Neurochem.* **53,** 264–272.

Maggi, C. A., Patacchini, R., Giuliani, S., Rovero, P., Dion, S., Regoli, D., Giachetti, A., and Meli, A. (1990) Competitive antagonists discriminate between NK_2 tachykinin receptor subtypes. *Br. J. Pharmacol.* **100,** 589–592.

Maggio, J. E. (1988) Tachykinins. *Annu. Rev. Neurosci.* **11,** 13–28.

Mantyh, C. R. (1984) The autoradiographic distribution of kassin and substance K binding sites is different from the distribution of substance P binding sites in brain. *Eur. J. Pharmacol.* **102,** 361–364.

Masu, Y., Nakayama, K., Tamaki, H., Harada, Y., Kuno, M., and Nakanishi, S. (1987) cDNA cloning of bovine substance-K receptor through oocyte expression system. *Nature* **329,** 836–838.

McElroy, A. B., Clegg, S. P., Deal, M. J., Ewan, G. B., Hagan, R. M., Ireland, S. J., Jordan, C. C., Porter, B., Ross, B. C., and Ward, P. (1992) Highly potent and selective heptapeptide antagonists of the neurokinin NK-2 receptor. *J. Med. Chem.* **35,** 2582–2591.

McKnight, A. T., Maguire, J. J., Elliott, N. J., Fletcher, A. E., Foster, A. C., Tridgett, R., Williams, B. J., Longmore, J., and Iversen, L. L. (1991) Pharmacological specificity of novel, synthetic cyclic peptides as antagonists at tachykinin receptors. *Br. J. Pharmacol.* **104,** 355–360.

McLean, S., Ganong, A., Seymour, P. A., Snider, R. M., Desai, M., Rosen, T., Bryce, D. K., Longo, K. P., Reynolds, L. S., Robinson, G., Schmidt, A. W., Siok, C., et al. (1992) Pharmacology of CP-99,994; a nonpeptide antagonist of the tachykinin NK_1 receptor. *Regul. Pept.* **suppl. 1,** S120.

McLean, S., Ganong, A. H., Seeger, T. F., Bryce, D. K., Pratt, K. G., Reynolds, L. S., Siok, C. J., Lowe, J. A., and Heym, J. (1991) Activity and distribution of binding sites in brain of a nonpeptide substance P (NK1) receptor antagonist. *Science* **251,** 437–439.

Minneman, K. P. (1988) α_1-Adrenergic receptor subtypes, inositol phosphates, and sources of cell Ca^{2+}. *Pharmacol Rev.* **40,** 87–119.

Mussap, C. J. and Burcher, E. (1990) [^{125}I]-BH scyliorhinin II: a novel, selective radioligand for the tachykinin NK_3 receptor in rat brain. *Peptides* **11,** 827–836.

Mussap, C. J. and Burcher, E. (1991) Multiple tachykinin receptors in rat submandibular gland. *Ann. NY Acad. Sci.* **632,** 447–451.

Nakanishi, S. (1991) Mammalian tachykin receptors. *Annu. Rev. Neurosci.* **14,** 123–136.

Patacchini, R., Astolfi, M., Quartara, L., Rovero, P., Giachetti, A., and Maggi, C. A. (1991) Further evidence for the existence of NK_2 tachykinin receptor subtypes. *Br. J. Pharmacol.* **104,** 91–96.

Petitet, F., Beaujouan, J. C., Saffroy, M., Torrens, Y., Chassaing, G., Lavielle, S., Besseyre, J., Garret, C., Carruette, A., and Glowinski, J. (1991) Further demonstration that [Pro^9]-substance P is a potent and selective ligand of NK-1 tachykinin receptors. *J. Neurochem.* **56,** 879–889.

Poncelet, M., Gueudet, C., Emonds-Alt, X., Breliere, J. C., Le Fur, G., and Soubrie, P. (1993) Turning behaviour induced in mice by a neurokinin A receptor agonist: stereoselective blockade by SR 48968, a no peptide receptor antagonist. *Neurosci. Lett.* **149,** 40–42.

Quirion, R. and Dam, T.-V. (1985) Multiple tachykinin receptors in guinea pig brain. High densities of substance K (neurokinin A) binding sites in substantia nigra. *Neuropeptides* **6,** 191–204.

Quirion, R. and Dam, T.-V. (1988) Multiple neurokinin receptors: recent developments. *Regul. Pept.* **22,** 18–25.

Quirion, R., Dam, T.-V. and Guard, S. (1991) Selective neurokinin receptor radioligands. *Ann. NY Acad. Sci.* **632,** 137–144.

Regoli, D., Drapeau, G., Dion, S., and Couture, R. (1988) New selective agonists for neurokinin receptors: pharmacological tools for receptor characterization. *Trends Pharmacol. Sci.* **9,** 290–295.

Renzetti, A. R., Barsacchi, P., Criscuoli, M., and Lucacchini, A. (1991) Characterization of NK-3 binding sites in rat and guinea pig cortical membranes by the selective ligand [^{3}H]senktide. *Neuropeptides* **18,** 107–114.

Rodbell, M. (1980) The role of hormone receptors and GTP-regulatory proteins in membrane transduction. *Nature* **284,** 17–22.

Rovero, P., Pestellini, V., Maggi, C. A., Patacchini, R., Regoli, D., and Giachetti, A. (1990a) A highly selective NK-2 tachykinin receptor antagonist containing D-tryptophan. *Eur. J. Pharmacol.* **175,** 113–115.

Rovero, P., Pestellini, V., Patacchini, R., Giuliani, S., Maggi, C. A., Meli, A., and Giachetti, A. (1990b) Synthesis and biological activity of NK-2 selective tachykinin antagonists containing d-tryptophan. *Peptides* **11,** 619,620.

Sachais, B. S., Snider, R. M., Lowe III, J. A., and Krause, J. E. (1993) Molecular basis for the species selectivity of the substance P antagonist CP-96,345. *J. Biol. Chem.* **268,** 2319–2323.

Saffroy, M., Beaujouan, J. C., Torrens, Y., Besseyre, J., Bergstrom, L., and Glowinski, J. (1988) Localization of tachykinin binding sites (NK1, NK2, NK3 ligands) in the rat brain. *Peptides* **9,** 227–241.

Sasai, Y. and Nakanishi, S. (1989) Molecular characterization of rat substance K receptor and its mRNA. *Biochim. Biophys. Acta* **165,** 695–702.

Schilling, W., Bittiger, H., Brugger, F., Criscione, L., Hauser, K., Ofner, S., Olpe, H. R., Vassout, A., and Veenstra, S. (1992) Approaches towards the design and synthesis of nonpeptidic substance P antagonists. *XIIth International Symposium on Medicinal Chemistry* Sept. 13–17, ML-11.3–ML-11.0.

Seabrook, G. R., Main, M., Bowery, B., Wood, N., and Hill, R. G. (1992) Differences in neurokinin receptor pharmacology between rat and guinea pig superior cervical ganglia. *Br. J. Pharmacol.* **105,** 923–928.

Shigemoto, R., Yokota, Y., Tsuchida, K., and Nakanishi, S. (1990) Cloning and expression of a rat neuromedin K receptor cDNA. *J. Biol. Chem.* **265,** 623–628.

Snider, R. M., Constantine, J. W., Lowe, J. A., Longo, K. P., Lebel, W. S., Woody, H. A., Drozda, S. E., Desai, M. C., Vinick, F. J., Spencer, R. W., and Hess, H-J. (1991) A potent nonpeptide antagonist of the substance P (NK_1) receptor. *Science* **251,** 435–437.

Sundelin, J. B., Provvedini, D. M., Wahlestedt, C. R., Laurell, H., Pohl, J. S., and Peterson, P. A. (1992) Molecular cloning of the murine substance K and substance P receptor genes. *Eur. J. Biochem.* **203,** 625–631.

Tabart, M., Peyronel, J. F., Truchon, A., Moutonnier, C., Dubroeucq, M. C., Fardin, V., Carruette, A., and Garret, C. (1992) Perhydroisoindolone derivatives: a new class of potent and selective nonpeptide NK_1 antagonists. *XIIth International Symposium on Medicinal Chemistry,* Sept. 13–17, P-190.A-P-190.0 (Abstract).

Takeda, Y., Blount, P., Sachais, B. S., Hershey, A. D., Raddatz, R., and Krause, J. E. (1992) Ligand binding kinetics of substance P and neurokinin A receptors stably expressed in Chinese hamster ovary cells and evidence for differential stimulation of inositol 1,4,5-trisphosphate and cyclic AMP second messenger responses. *J. Neurochem.* **59,** 740–745.

Takeda, Y., Chou, K. B., Takeda, J., Sachais, B. S., and Krause, J. E. (1991) Molecular cloning, structural characterization and functional expression of the human substance P receptor. *Biochem. Biophys. Res. Commun.* **179,** 1232–1240.

Torrens, Y., Beaujouan, J. C., Viger, A., and Glowinski, J. (1983) Properties of a ^{125}I-SP derivative binding to synaptosomes from various brain structures and the spinal cord of the rat. *Naunyn Schmiedeberg's Arch. Pharmacol.* **324,** 134–139.

Torrens, Y., Lavielle, S., Chassaing, G., Marquet, A., Glowinski, J., and Beaujouan, J. C. (1984) Neuromedin K, a tool to further distinguish two central tachykinin binding sites. *Eur. J. Pharmacol.* **102,** 381,382.

Tousignant, C., Guillemette, G., Drapeau, G., Telemaque, S., Dion, S., and Regoli, D. (1990) ^{125}I-BH[Sar^9, Met(O_2)^{11}]-SP, a new selective ligand for the NK-1 receptor in the central nervous system. *Brain Res.* **524,** 263–270.

Tousignant, C., Guillemette, G., and Regoli, D. (1991) Binding sites for [^{3}H][Sar^9, Met(O_2)^{11}]substance P in rat brain and guinea pig ileum. *Brain Res.* **560,** 1–11.

Van Giersbergen, P. L., Shatzer, S. A., Henderson, A. K., Lai, J., Nakanishi, S., Yamamura, H. I., and Buck, S. H. (1991) Characterization of a tachykinin peptide NK_2 receptor transfected into murine fibroblast B82 cells. *Proc. Natl. Acad. Sci. USA* **88,** 1661–1665.

Venepalli, B. R., Aimone, L. D., Appell, K. C., Bell, M. R., Dority, J. A., Goswami, R., Hall, P. L., Kumar, V., Lawrence, K. B., and Logan, M. E. (1992) Synthesis and substance P receptor binding activity of androstano[3,2-b]pyrimido[1,2-*a*]benzimidazoles. *J. Med. Chem.* **35,** 374–378.

Watling, K. J., Guard, S., Howson, W., and Walton, L. (1991) Inhibition of binding to tachykinin NK_1 receptor sites in rat, guinea-pig, and human brain by the nonpeptide substance P antagonist (±)CP-96,345. *Br. J. Pharmacol.* **104,** 27P.

Watson, S. P., Sandberg, B. E. B., Hanley, M. R., and Iversen, L. L. (1983) Tissue selectivity of substance P alkyl esters: suggesting multiple receptors. *Eur. J. Pharmacol.* **87,** 77–84.

Wormser, U., Laufer, R., Hart, Y., Chorev, M., Gilon, C., and Selinger, Z. (1986) Highly selective agonists for substance P receptor subtypes. *EMBO J.* **5,** 2805–2808.

Yokota, Y., Akazawa, C., Ohkubo, H., and Nakanishi, S. (1992) Delineation of structural domains involved in the subtype specificity of tachykinin receptors through chimeric formation of substance P/substance K receptors. *EMBO J.* **11,** 3585–3591.

Yokota, Y., Sasai, Y., Tanaka, K., Fujiwara, T., Tsuchida, K., Shigemoto, R., Kakizuka, A., Ohkubo, H., and Nakanishi, S. (1989) Molecular characterization of a functional cDNA for rat substance P receptor. *J. Biol. Chem.* **264,** 17,649–17,652.

CHAPTER 5

Comparative Distribution of Receptor Types in the Mammalian Brain

Than-Vinh Dam and Rémi Quirion

1. Introduction

The tachykinins (or neurokinins [NK]) are a group of structurally related peptides found in many regions of central nervous system, as well as in various peripheral tissues (Erspamer, 1981; Pernow, 1983; Maggio, 1988). Mammalian NKs are derived from the preprotachykinin I and II genes (PPT-I and PPT-II), two different but related genes (Nakanishi, 1987; Krause et al., 1989). Substance P (SP), neurokinin A (NKA), and the NKA-derived peptides, neuropeptide K, neuropeptide γ (NPγ), and NKA (3–10) are encoded by mRNAs resulting from PPT-I gene transcription, whereas neurokinin B (NKB) is derived from the maturation of the PPT-II gene (Nakanishi, 1987; Krause et al., 1989). These various NKs produce their physiological effects by activating specific membrane receptors.

The existence of at least three major types of NK receptors is now well established (Buck and Burcher, 1986; Regoli et al., 1987; Maggio, 1988; Quirion and Dam, 1988; Guard and Watson, 1991;

The Tachykinin Receptors Ed.: S. H. Buck © 1994 Humana Press Inc., Totowa, NJ

Nakanishi, 1991). Following an initial characterization based on the use of specific bioassays and series of agonists and antagonists, the NK_1, NK_2, and NK_3 receptors were cloned from a variety of tissues (Masu et al., 1987; Yokota et al., 1989; Hershey and Krause, 1990; Shigemoto et al., 1990; *see also* various chapters in this volume).

All three receptor types belong to the G-protein-coupled rhodopsin superfamily. SP, NKA (and extended forms), and NKB act as preferential, but not exclusive, endogeneous ligand for the NK_1, NK_2, and NK_3 receptor class, respectively (Quirion and Dam, 1988; Guard and Watson, 1991; Quirion et al., 1991; Regoli et al., this volume). Consequently, the development of type-selective agonists and antagonists was critical in order to properly assess the respective functional roles of each receptor class (for example, Regoli et al., this volume). These molecules in their radiolabeled forms (^3H or ^{125}I) allow for the detailed characterization of binding parameters and autoradiographic distribution of each NK receptor type in brain and peripheral tissues.

Using such highly selective radioligands, various groups (Guard et al., 1990; Lew et al., 1990; Mussap and Burcher, 1990; Burcher et al., 1991; McLean et al., 1991; Snider et al., 1991; Tousignant et al., 1992), including ours (Dam et al., 1990a,b,c; Guard et al., 1991; Guard et al., 1993; Quirion and Dam, 1993) reported on the binding parameters and autoradiographic distribution of the NK_1, NK_2, and NK_3 receptor types. The use of selective radioligands is highly preferable as radiolabeled probes derived from endogeneous ligands generally lack the adequate selectivity required to study a single receptor type (Quirion et al., 1991 for further discussion of this issue). Accordingly, in the present study, the comparative autoradiographic distribution of brain NK_1, NK_2, and NK_3 receptors was investigated using $[^3H][Sar^9,Met(O_2)^{11}]$-substance P as NK_1 ligand (Dam et al., 1990a), ^{125}I-NPγ as putative NK_2 probe (Dam et al., 1990b; Guard et al., 1993), and $[^3H]$senktide as NK_3 ligand (Guard et al., 1990; Dam et al., 1990c). The respective anatomical distribution of these three receptor types was investigated in rat, mouse, and guinea pig brains in order to determine potential species differences. This is especially relevant in view of the differential effectiveness of various

NK-related agonists and antagonists in these species (Buck et al., 1990; Maggi et al., 1990, 1991; Fardin and Garret, 1991; Gitter et al., 1991; Snider et al., 1991; Rouissi et al., 1993). Detailed information on the discrete localization of these various NK receptors in peripheral tissues is summarized elsewhere in this volume (Burcher et al.).

2. Technical Considerations

Detailed descriptions of methods and incubation assay conditions have previously been reported for each radioligand (Dam et al., 1990a,b,c; Quirion and Dam, 1993). Rationales for the preferred use of selective radioligands (Quirion et al., 1991) and their respective binding parameters and ligand selectivity pattern are well established (Guard et al., 1990, 1991; Dam et al., 1990a,b,c).

Male Sprague-Dawley rat (200–250 g), male CD mouse (20–25 g), and male Hartley guinea pig (400–450 g) brain sections are prepared as described by us (Dam et al., 1990b; Quirion and Dam, 1993 for detailed protocols). Briefly, animals are sacrificed by decapitation and brains are quickly removed and frozen in 2-methylbutane (–20°C) before storage at –80°C until needed. Brains are then cut into 20-µm sections and thaw-mounted onto gelatin-coated slides (*see* Quirion and Dam, 1993 for a detailed protocol). Following an overnight stay at 4°C (under vacuum), slide-mounted sections are stored at –80°C until use.

For receptor autoradiography, tissue sections are incubated with either 3 n*M* $[^3H][Sar^9,Met(O_2)^{11}]$-substance P (32.5 Ci/mmol, NEN-Dupont, Wilmington, DE) as selective NK_1 ligand, 25 p*M* ^{125}I-neuropeptide γ (1000 Ci/mmol, a generous gift from J. Krause, St. Louis, MO) as NK_2 probe, or 3.0 n*M* $[^3H]$senktide (55.2 Ci/mmol, NEN-Dupont) as a selective NK_3 ligand in a buffer containing 50 m*M* Tris-HCl, pH 7.4 at 25°C, 0.02% bovine serum album, 0.2 mg/100 mL chymostatin, 0.4 mg/100 mL leupeptin, 4 mg/100 mL bacitracin, and 3 m*M* $MnCl_2$ for 90 min at room temperature. Specific binding is determined as the difference in labeling observed in the presence and absence of 1 µ*M* SP, NPγ, or eledoisin for NK_1, NK_2, or NK_3 receptors, respectively.

3. Distribution of NK_1 Receptors

The discrete autoradiographic localization of NK_1 receptors in the mammalian brain has been reported first using iodinated SP (Shults et al., 1982, 1984; Beaujouan et al., 1986; Buck et al., 1986; Dam and Quirion, 1986; Danks et al., 1986; Saffroy et al., 1988; Mantyh et al., 1989), [^{3}H]substance P (Quirion et al., 1983), or ^{125}I-physalaemin (Wolf et al., 1985). More recently, selective NK_1 radioligands have been used, including [^{3}H][Sar^9,Met(O_2)^{11}]-SP (Dam et al., 1990a), ^{125}I-Bolton-Hunter [Sar^9,Met(O_2)^{11}]SP (Lew et al., 1990), and [^{3}H]CP-96,345 (McLean et al., 1991; Snider et al., 1991). Globally, the observed distribution of the NK_1 receptor type is rather similar using these various probes except for the better resolution and signal-to-noise ratio obtained with most iodinated radioligands. Species differences are also noted regarding the discrete localization of NK_1 receptors in the mammalian brain (*see below,* and Dietl and Palacios, 1991). This, coupled to the fact that some NK_1 receptor antagonists are more potent in mouse and rat than in other species (and vice versa; Beresford et al., 1991; Fardin and Garret, 1991; Gitter et al., 1991; Garret et al., 1991; Rouissi et al., 1991, 1993; Regoli et al., this volume), suggests that it may not be appropriate to speculate about functional role of brain NK_1 receptors on the basis of the detailed distribution observed in a single species.

3.1. NK_1 Receptors in the Rat Brain

The distribution in the rat brain of NK_1 receptors obtained using the selective agonist, [^{3}H][Sar^9,Met(O_2)^{11}-SP, is shown in Fig. 1. Additional information was reported elsewhere (Dam et al., 1990a), as well as a very detailed account of the comparative localization of specific ^{125}I-BH-SP binding site/putative NK_1 sites and SP-like immunoreactive materials (Shults et al., 1984). The respective discrete localization of [^{3}H][Sar^9,Met(O_2)^{11}]-SP and ^{125}I-BH-SP binding sites are similar except for a few areas in which the labeling seen with the selective tritiated ligand is not as evident as that of ^{125}I-BH-SP (e.g., pyramidal cell layer of the hippocampus, lamina X and motoneurons of the spinal cord). This may be due to the presence of rather

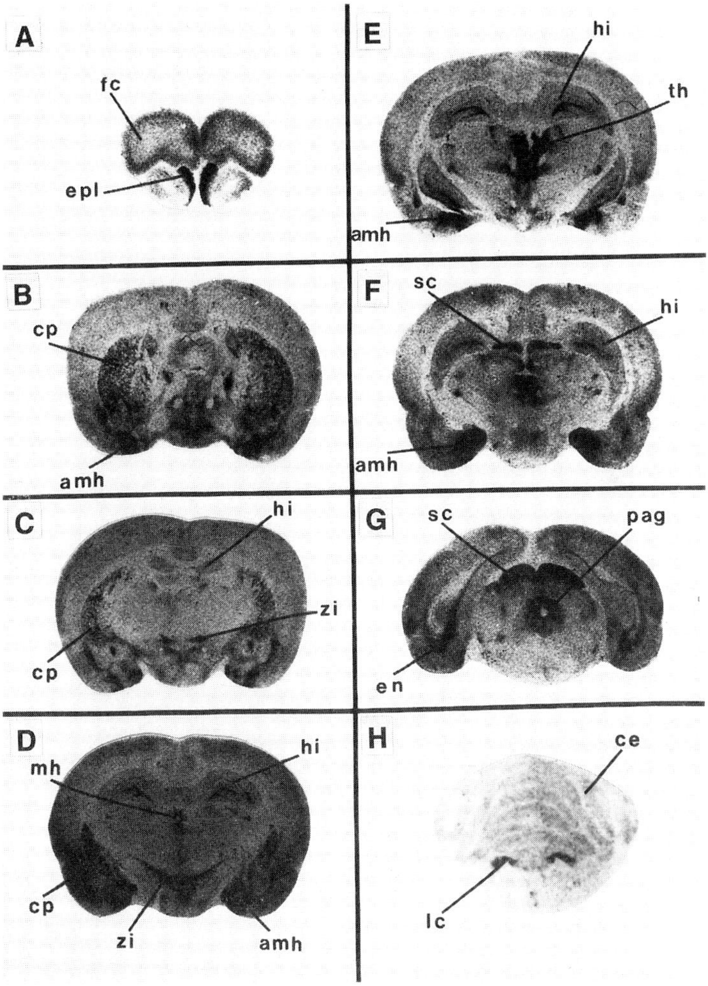

Fig. 1. Photomicrographs of the autoradiographic distribution of [^{3}H]-[Sar^9,Met(O$_2$)^{11}]SP/NK$_1$ binding sites in coronal rat brain sections. Sections were incubated in the presence of 3.0 n*M* [^{3}H][Sar^9,Met(O$_2$)^{11}]SP as described in the text. Abbreviations used: amh, amygdalo-hippocampal area; ce, cerebellum; cp, caudate putamen; en, entorhinal cortex; epl, external plexiform layer of the olfactory bulb; fc, frontal cortex; hi, hippocampus; lc, locus coeruleus; mh, medial habenula; pag, paraqueductal grey matter; sc, superior colliculus; th, thalamus; zi, zona incerta.

low quantities of NK_1 sites in these areas requiring the use of an iodinated probe for optimal characterization. Alternatively, it could be related to the potential recognition of other NK receptor classes or subtypes by ^{125}I-BH-SP; this radioligand potentially lacks adequate selectivity under certain assay conditions (Dam et al., 1990a; Yashpal et al., 1990).

NK_1 sites are widely, but discretely, distributed in the rat brain (Fig. 1). High levels of specific labeling are most evident in the external plexiform layer of the olfactory bulb, the amygdalo-hippocampal area, the granular cell layer of the dentate gyrus, the zona incerta, the medial habenula, some discrete thalamic and hypothalamic nuclei, the superior colliculus, the locus ceruleus, the parabrachial nucleus, lobules 9 and 10 of the cerebellum, and the hypoglossal nucleus. Moderate densities of NK_1 sites are seen in the striatal and septal areas, the anterior hypothalamus, the inferior colliculus, and the periaqueductal area.

Cortical laminae are generally not enriched with specific NK_1 labeling except in the superficial layers of the frontal cortex and the entorhinal, endopiriform, and occipital cortices (Fig. 1). Most other regions are not enriched with specific [^{3}H][Sar^9,Met(O$_2$)^{11}]-SP binding; the substantia nigra is devoid of specific labeling (Fig. 1G). While this chapter is not aimed at extrapolating on the functional significance of the observed distribution of NK_1 sites, the presence of high densities of specific labeling in structures associated with pain pathways lends additional support to the involvement of this peptide as a pain neurotransmitter (*see* Pernow, 1983 for a detailed review). In contrast, the absence of specific NK_1 labeling in the substantia nigra is still very puzzling and difficult to relate to the modulatory effects of neurokinins on mesencephalic dopaminergic neurons (Quirion et al., 1983; Quirion and Dam, 1988). Recent *in situ* hybridization studies aimed at investigating the locus of synthesis of the NK_1 receptor revealed a rather good correlation with NK_1 receptor autoradiography with the substantia nigra not expressing significant amounts of NK_1 receptor mRNA (Elde et al., 1990; Gerfen, 1991).

3.2. NK_1 Receptors in the Mouse Brain

At first glance, the autoradiographic distribution of specific [^{3}H][Sar^9,Met(O$_2$)^{11}] SP/NK_1 binding sites in the mouse brain (Fig. 2) is rather similar to that observed in the rat (Fig. 1). For example, as in the rat brain, the external plexiform layer of the olfactory bulb, the medial habenula, the amygdalo-hippocampal area, some midline thalamic nuclei, and the superior colliculus of the mouse brain are enriched with specific NK_1 binding sites, while the substantia nigra is not (Fig. 2). However, important differences are seen in cortical, hippocampal, and striatal areas (*see also* Dietl and Palacios, 1991). The mouse cortex and hippocampal formation are basically devoid of specific NK_1 labeling (Fig. 2), whereas certain cortical areas and the dentate gyrus of the hippocampus of the rat brain are enriched with NK_1 sites (Fig. 1). On the other hand, NK_1 receptors are more abundant on mouse than rat striatal areas (Figs. 1B, 2B).

The functional significance of these findings remains to be fully explored. However, it is clear that the distribution of the NK_1 receptor type is not identical in the rat and mouse brain. This may be relevant as most behavioral studies have been performed thus far in the rat (*see* Pernow, 1983 and this volume), whereas future experiments in transgenic animals over- (or under-) expressing the NK_1 receptor are likely to be performed in mice. Additionally, the existence of abundant, well-characterized, inbred strains of mice could be used as a model to investigate potential relationships among the expression of the NK_1 receptor in the brain, a given phenotype, and the chromosomal localization of the receptor gene.

3.3. NK_1 Receptors in the Guinea Pig Brain

A comparative evaluation of the autoradiographic distribution of NK_1 receptors in the rat and guinea pig brain recently became most relevant as functional and receptor binding data clearly demonstrated that, although nonpeptide antagonists such as CP-96,345 were rather weak blockers in the rat, they potently blocked NK_1-related effects in other species, including the guinea pig (Gitter et al., 1991; McLean

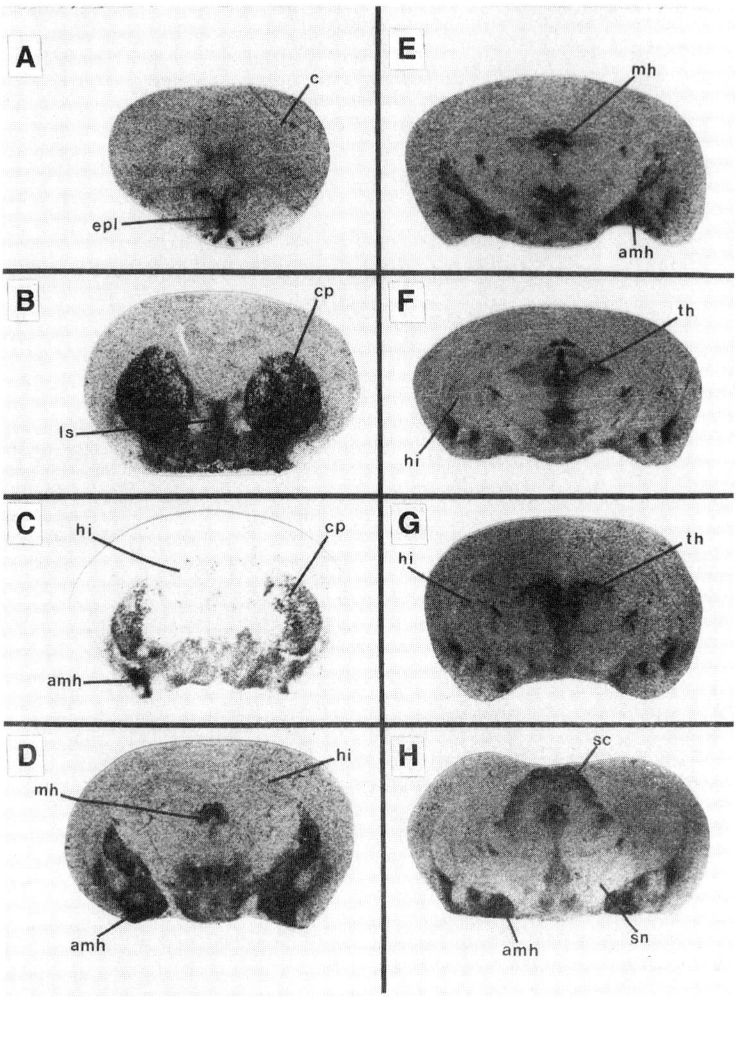

Fig. 2. Photomicrographs of the autoradiographic distribution of [^{3}H][Sar^9,Met(O$_2$)^{11}]SP/NK$_1$ binding sites in coronal mouse brain sections. Sections were incubated in the presence of 3.0 n*M* [^{3}H][Sar^9,Met(O$_2$)^{11}]SP as described in the text. Abbreviations used: amh, amygdalo-hippocampal area; c, cortex; cp, caudate putamen; epl, external plexiform layer of the olfactory bulb; hi, hippocampus; ls, lateral septum; mh, medial habenula; sc, superior colliculus; sn, substantia nigra; th, thalamus.

et al., 1991; Snider et al., 1991; Rouissi et al., 1991, 1993). The opposite situation is observed using RP67,580 (Garret et al., 1991). It remains to be established if these species differences are related to the existence of genuine NK_1 receptor subtypes across species or to the presence of species-dependent isoforms of a unique receptor (Tousignant et al., 1992; Rouissi et al, 1993; Regoli et al., this volume). In this context, the recent demonstration that differences between 5-HT_{1B} (present in rat and mouse only) and 5-HT_{1D} (other species) receptors are owing to a single amino acid substitution in the binding pocket reveals that great care must be taken when classifying receptor subtypes on the basis of data obtained in different species (Oksenberg et al., 1992).

We had some difficulties in obtaining adequate signal-to-noise ratio using $[^3H][Sar^9,Met(O_2)^{11}]$-SP as NK_1 radioligand in the guinea pig brain. This may relate to the somewhat lower affinity of this selective agonist in guinea pig vs rat preparations (Rouissi et al., 1993). Accordingly, autoradiographic data obtained using $[^3H]$SP (Dam and Quirion, 1986) are reproduced and discussed here (Fig. 3). Their pharmacological characterization as NK_1 receptors was reported elsewhere (Dam and Quirion, 1986).

The most striking comparative feature of the localization of NK_1 receptors in the guinea pig brain relates to the presence of very high densities of specific labeling in the hippocampal formation and various cortical areas (Fig. 3). This is especially evident since the mouse hippocampus is not enriched with NK_1 labeling (Fig. 2), whereas only the granular cell layer of the dentate gyrus of the rat brain contained a significant quantity of NK_1 sites (Fig. 1). In the guinea pig hippocampal formation, all subfields of the Ammon's horn, as well as various laminae, are highly enriched with specific $[^3H]$SP binding (Figs. 3D, 3E). This may suggest that SP, in a species-dependent manner, can modulate the activity of various neuronal populations of the hippocampus.

Other areas most enriched with specific $[^3H]$SP/NK_1 sites in the guinea pig brain included, as in the rat and mouse brain, the septal nucleus, the amygdalo-hippocampal area, the medial habenula, the superior colliculus, the entorhinal cortex, and to a lesser extent, the

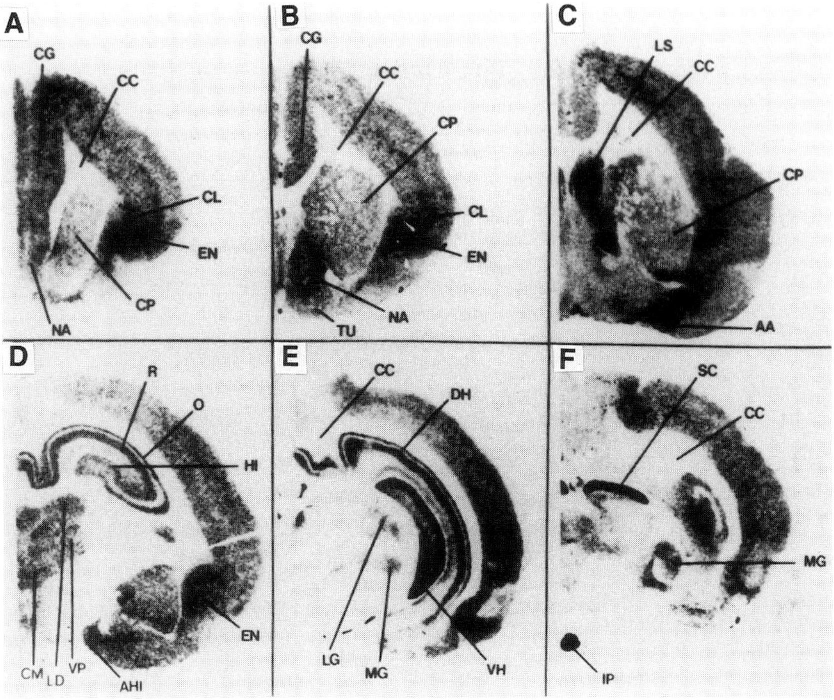

Fig. 3. Photomicrographs of the autoradiographic distribution of [^{3}H]SP/NK_1 binding sites in guinea pig brain sections. Sections were incubated in the presence of 0.5 n*M* [^{3}H]SP as described in the text. Abbreviations used: aa, amygdala; ahi, amygdalo-hippocampal area; cc, corpus callosum; cg, cingulate cortex; cl, claustrum; cm, centro-medial nucleus of the thalamus; cp, caudate putamen; dh, dorsal hippocampus; en, endopiriform nucleus; hi, hippocampus; ip, interpeduncular nucleus; ld, latero-dorsal nucleus of the thalamus; lg, lateral geniculate; ls, lateral septum; mg, medial geniculate; na, nucleus accumbens; o, oriens layer of the hippocampus; r, radiatum layer of the hippocampus; sc, superior colliculus; sn, substantia nigra; tu, olfactory tubercule; vh, ventral hippocampus; vp, ventro-posterior nucleus of the thalamus. Reproduced from Dam and Quirion (1986) with permission.

caudal striatum (Fig. 3). The interpeduncular nucleus is also highly enriched with specific [^{3}H]SP binding sites, whereas the substantia nigra, as in all other species studied thus far, is devoid of NK_1 sites (Quirion and Dam, 1985, 1988; Dietl and Palacios, 1991). It is evident that, although the global distribution of NK_1 receptors is rather similar in rat, mouse, and guinea pig brains, important differences are seen in higher nonvegetative areas such as the striatum, the hippocampal formation, and the cortex. It is thus critical to explore several species before concluding a given role for this receptor class, especially in these brain regions.

4. Distribution of NK_2 Receptors

The existence of genuine NK_2 receptors in the CNS has been controversial. Early on, we (Quirion and Dam, 1985, 1988; Yashpal et al., 1990) and others (Mantyh et al., 1984) provided evidence for the presence of this receptor type in the rat brain, whereas others failed (Buck et al., 1986; Bergstrom et al., 1987; Saffroy et al., 1988). It now appears that these controversial results were related to the use of radioligands that did not display either sufficient affinity or selectivity for the NK_2 receptor type, as well as to the existence of rather low concentrations (compared to NK_1 and NK_3 sites; except during ontogeny, Dam et al., 1988) of this receptor class in the CNS (Dam et al., 1990b; Quirion et al., 1991, Guard et al., 1993). However, it is now clear that this receptor class is expressed in the rat brain, the corresponding mRNA being present, albeit in low amounts, in various regions, including the spinal cord and the striatum (Poosch et al., 1991; Takeda and Krause, 1991). Specific NK_2-related behavioral effects have also been characterized following the injections of selective NK_2 agonists and antagonists in the brain (Elliot et al., 1991; Fleetwood-Walker et al., 1990; Overton et al., 1992; Tschope et al., 1992).

4.1. NK_2 Receptors in the Rat Brain

The extended form of NKA, NPγ (γ-preprotachykinin-[72–92]) peptide amide, was used here as a radioligand as previously reported (Dam et al., 1990b; Guard et al., 1993). While this probe is not as selective as previously thought (Dam et al., 1990b), it clearly labels a specific class of NK_2 receptors (NK_{2B}?) in the rat brain, this on the basis of the comparative potency of selective NK_1-, NK_2-, and NK_3-related molecules to compete for specific ^{125}I-NPγ binding in the rat brain (Guard et al., 1993). For example, senktide failed to compete for ^{125}I-NPγ binding, but the NK_2 antagonist, L659,877 (but not MEN 10,207), potently inhibited binding in areas such as the CA_1 to CA_3 subfields of the hippocampal formation and some thalamic nuclei. In other regions, ^{125}I-NPγ labeling was competed by rather low concentrations of the NK_1 agonist, [Sar^9,$Met(O_2)^{11}$]SP (Guard et al., 1993).

The autoradiographic distribution of the total population of specific ^{125}I-NPγ binding sites is shown in Fig. 4. Of these, specific labeling depicted in Fig. 4A (prefrontal cortex) and Fig. 4C (hippocampal formation and thalamus) clearly represent labeling of the NK_2 type. In other regions (Fig. 2B, striatum, Fig. 2E, locus ceruleus), specific ^{125}I-NPγ binding is likely to the NK_1 receptor (Guard et al., 1993). Thus, in the rat brain NK_2 sites are directly distributed, being most abundant in various subfields of the hippocampal formation (except the dentate gyrus) and the centro-lateral and ventro-lateral thalamic nuclei. In preliminary experiments, we observed that the radiolabeled form of the NK_2 receptor antagonist, [^{3}H]SR-48968A (Emonds-Alt et al., 1992; a gift from Dr. Breliere, Sanofi, France), also labeled the hippocampal formation and midline thalamic nuclei of the rat brain, supporting data obtained with ^{125}I-NPγ. However, signal-to-noise ratio was poor, not allowing for a detailed investigation of the localization of [^{3}H]SR-48968A binding in areas not as enriched with putative NK_2 receptor sites (Dam and Quirion, unpublished results). Assay conditions and/or alternate ligand preparation will thus have to be modified to allow for a general use of this NK_2 receptor blocker.

4.2. NK_2 Receptors in the Guinea Pig Brain

As in the rat brain, specific ^{125}I-NPγ binding sites are mostly concentrated in various subfields of the hippocampal formation in the guinea pig brain (Fig. 5). Other areas enriched with labeling include the frontal cortex (Fig. 5A), the amygdaloid body (Fig. 5D), and certain thalamic nuclei (Fig. 5C,E). Competition studies remain to be performed in the guinea pig brain, but it is likely that only specific ^{125}I-NPγ labeling seen in the hippocampal formation and thalamus represents genuine NK_2 receptors (*see* Section 4.1.), although putative species differences cannot be excluded at this time.

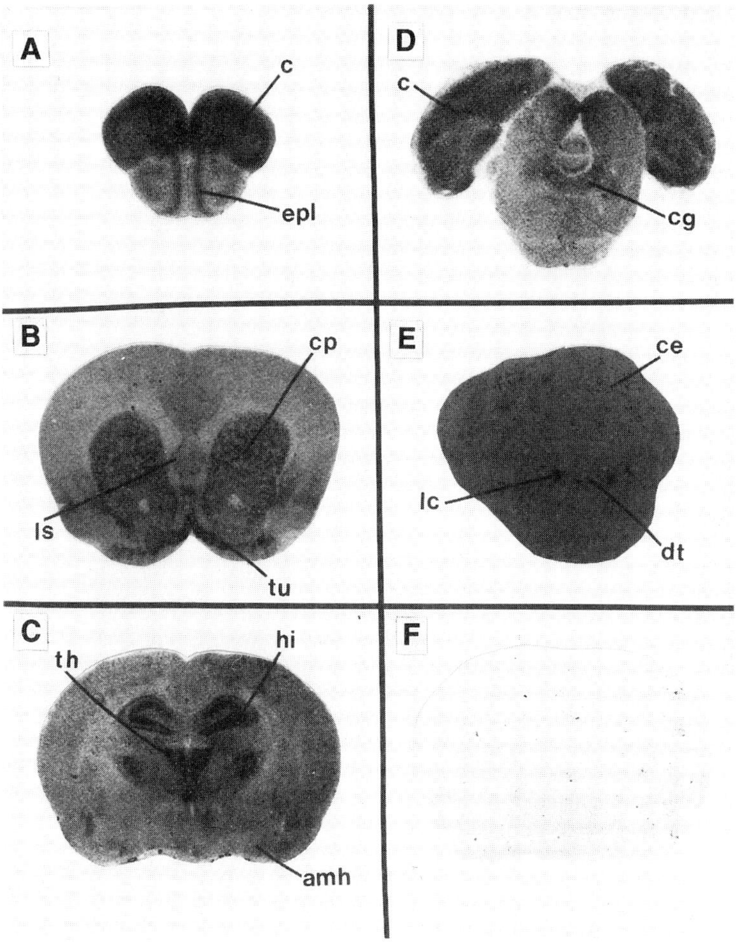

Fig. 4. Photomicrographs of the autoradiographic distribution of ^{125}I-NPγ/putative NK_2 binding sites in coronal rat brain sections. Sections were incubated in the presence of 25 p*M* ^{125}I-NPγ as described in the text. Nonspecific labeling seen in the presence of 1.0 μ*M* NKA is shown in F. Abbreviations used: amh, amygdalo-hippocampal area; c, cortex; ce, cerebellum; cg, dorsal tegmentum; cp, caudate putamen; dt, dorsal tegmental nuclei; epl, external plexiform layer of the olfactory bulb; hi, hippocampus; lc, locus ceruleus; ls, lateral septum; th, thalamus; tu, olfactory tubercule.

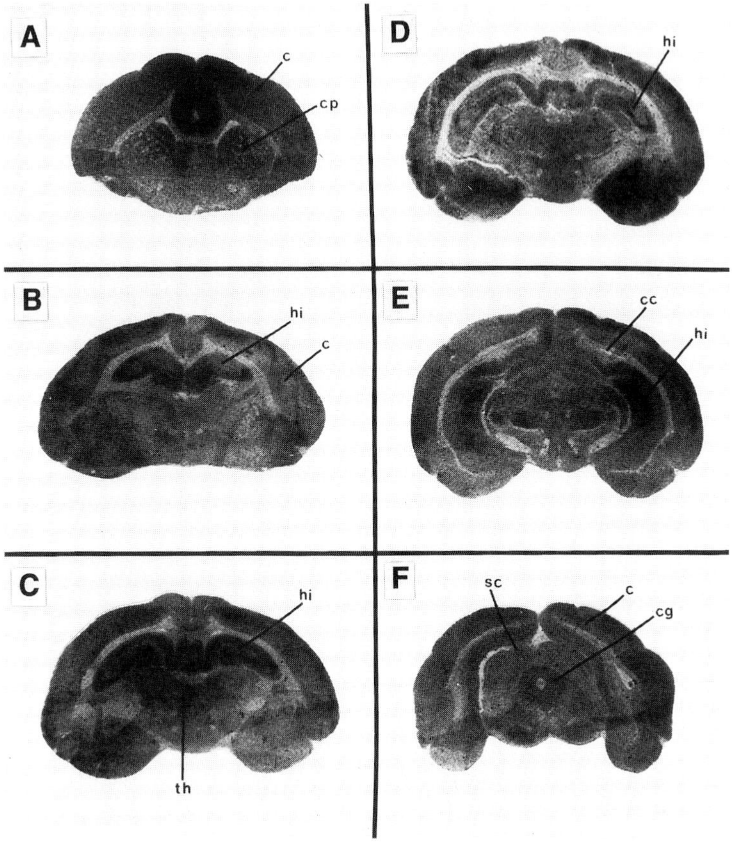

Fig. 5. Photomicrographs of the autoradiographic distribution of ^{125}I-NPγ/putative NK_2 binding sites in coronal guinea pig brain sections. Sections were incubated in the presence of 25 p*M* ^{125}I-NPγ as described in the text. Abbreviations used: c, cortex; cc, corpus callosum; cg, central grey matter; cp, caudate putamen; hi, hippocampus; sc, superior colliculus, and th, thalamus.

It appears that NK_2 receptors are expressed and translated in the mammalian brain. However, the distribution of this receptor class is rather limited, being especially concentrated in the hippocampus and the thalamus. These regions could be key targets in relation to NK_2 receptor-mediated events in the CNS.

5. Distribution of NK_3 Receptors

Brain NK_3 receptors were originally characterized using radiolabeled NKs, including neurokinin B (Bergstrom et al., 1987b), eledoisin (Cascieri and Liang, 1984; Beaujouan et al., 1984, 1986; Danks et al., 1986; Dam et al.,1988; Saffroy et al., 1988; Mantyh et al., 1989), and scyliorhinin II (Mussap and Burcher, 1990). However, it is now clear that these endogeneous peptides lack adequate selectivity for NK_3 receptor binding studies (Burcher et al., 1991; Quirion et al., 1991). Thus far, only senktide in its tritiated (but not iodinated) form appears to be a highly selective, useful NK_3 receptor ligand (Dam et al., 1990c; Guard et al., 1990, 1991).

5.1. NK_3 Receptors in the Rat Brain

The discrete localization of specific [^{3}H]senktide/NK_3 receptors in the rat brain is shown in Fig. 6. Its most striking feature relates to the presence of very high densities of specific labeling in midcortical laminae throughout the cortex. This is clearly unique to this receptor class as the NK_1 and NK_2 types possess a differential profile of cortical distribution *(see above)*. Additionally, high densities of NK_3 sites are seen in the supraoptic area, zona incerta, amygdalo-hippocampal area, medial habenula, and interpeduncular nucleus. The substantia nigra (pars compacta) and the ventral tegmental area are also reasonably enriched with specific NK_3 sites (Fig. 6F). Other areas expressing NK_3 binding sites include the external plexiform layer of the olfactory bulb, basolateral nucleus of the amygdala, superior colliculus, central gray matter, and, to a lesser extent, the dorsal hippocampal formation and the striatum. Most other areas are not enriched with specific [^{3}H]senktide binding (Fig. 6). Other anatomical features of the distribution of NK_3 binding sites have been previously discussed (Dam et al., 1990c; Stoessl and Hill, 1990). The presence of specific NK_3 binding sites in areas enriched with dopaminergic cell bodies (A9 and A10) suggests that this receptor class may be responsible for the modulatory effects of NKs on dopaminergic neurons observed in these structures (Elliott et al., 1991; Stoessl et al., 1991; Overton et al., 1992).

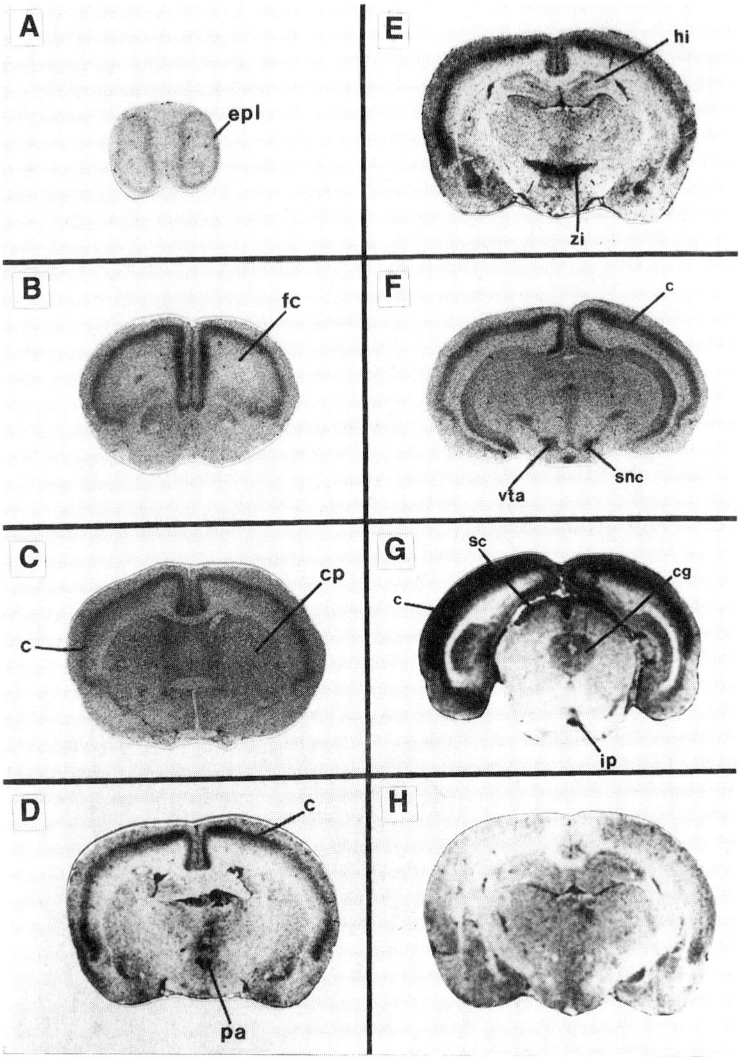

Fig. 6. Photomicrographs of the autoradiographic distribution of [^{3}H]-senktide/NK$_3$ binding sites in coronal rat brain sections. Sections A–G depict the total binding in the presence of 3 n*M* [^{3}H]senktide, whereas H shows binding remaining in the presence of a saturating concentration (1.0 μ*M*) of eledoisin used to define nonspecific labeling (same level as section E). Abbreviations used: c, cortex; cg, central greymatter; cp, caudate putamen; epl, external plexiform layer of the olfactory bulb; fc, frontal cortex; hi, hippocampus; ip, interpeduncular nucleus; pa, paraventricular nucleus of the hypothalamus; sc, superior colliculus; snc, substantia nigra, pars compacta; vta, ventral tegmental areas; zi, zona incerta.

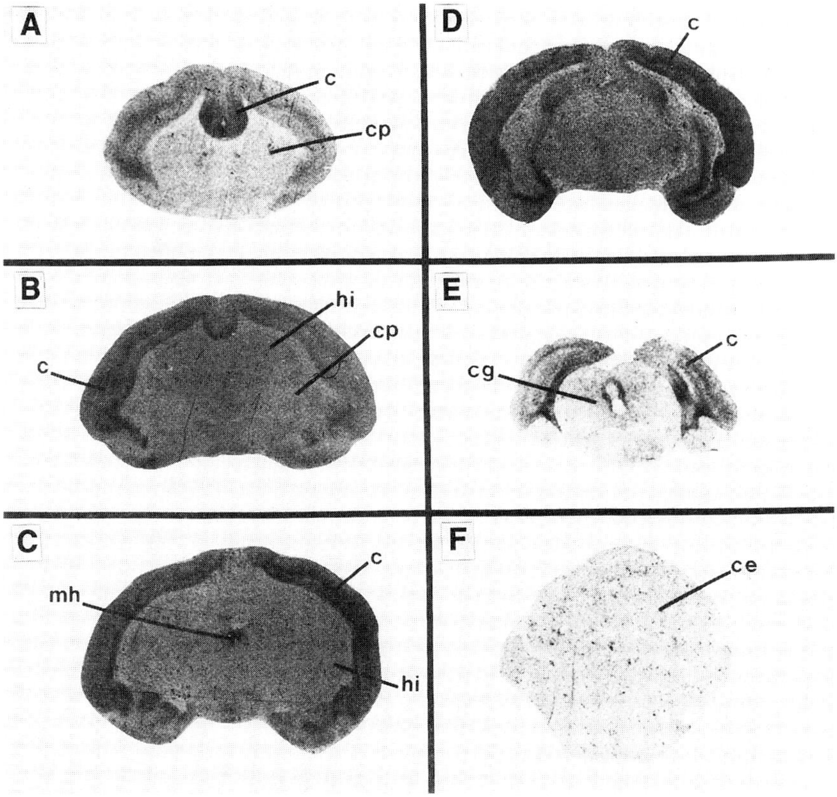

Fig. 7. Photomicrographs of the autoradiographic distribution of [^{3}H]-senktide/NK_3 binding sites in coronal guinea pig brain sections. Sections were incubated in the presence of 3.0 n*M* [^{3}H]senktide as described in the text. Abbreviations used: c, cortex; ce, cerebellum; cg, central grey matter; cp, caudate putamen; hi, hippocampus; mh, medial habenula.

5.2. NK_3 Receptors in the Guinea Pig Brain

As in the rat brain, the guinea pig brain cortex is highly enriched with specific [^{3}H]senktide binding sites (Fig. 7). Cortical NK_3 labeling is especially abundant in mid- and deep cortical laminae (Fig. 7B,C). However, in contrast to the rat, most other areas, except for the medial habenula (Fig. 7C) and the caudal hippocampus (Fig. 7D), contained very low levels of specific labeling, including the striatum and various thalamic and hypothalamic nuclei (Fig. 7). It thus appears that species differences may also exist in regard to the discrete distribution of NK_3 receptors in mammalian brain, especially in subcortical areas. In fact, Dietl and Palacios (1991) obtained similar results and suggested that NK_3 receptors were found only in the brain of lower mammals. However, at the time of their study, selective radioligands were not available. Consequently, the genuine existence of this receptor class in primate brain should be reevaluated using highly selective radioligands such as [^{3}H]senktide.

6. Conclusions

It is now clear that NK_1, NK_2, and NK_3 receptor types are present and differentially distributed in the mammalian brain. This is also supported by the recent demonstration of the expression of their respective mRNAs in the CNS (Hershey et al., 1991). Moreover, the autoradiographic results presented here, as well as earlier studies (Dam and Quirion, 1986; Dietl and Palacios, 1991), revealed the existence of species differences in the level of expression and distribution of a given NK receptor type in the brain. Accordingly, it is difficult to extrapolate on potential clinical applications on the basis of data obtained in one species only. Finally, the future availability of highly selective agonists and nonpeptide antagonists as radioligands and, hopefully, receptor monoclonal antibodies should allow for detailed ultrastructural investigations of the precise cellular and subcellular localization of each NK receptor type in the mammalian brain.

Acknowledgments

The support of the Medical Research Council of Canada is gratefully acknowledged. T.-V. Dam holds a Fellowship award from the Medical Research Council of Canada, and R. Quirion is a "Chercheur-Boursier Mérite Exceptionel" of the "Fonds de la Recherche en Santé du Quebec." The expert secretarial assistance of Joan Currie is acknowledged.

References

Beaujouan, J. C., Torrens, Y., Viger, A., and Glowinski, J. (1984) A new type of tachykinin binding site in the rat brain characterized by the specific binding of a labeled eledoisin derivative. *Mol. Pharmacol.* **26,** 248–254.

Beaujouan, J.C., Torrens, Y., Saffroy, M., and Glowinski, J. (1986) Quantitative autoradiographic analysis of the distribution of binding sites for ^{125}I-Bolton-Hunter derivatives of eledoisin and substance P in the rat brain. *Neuroscience* **18,** 857–875.

Beresford, I. J. M., Birch, P. J., Hagan, R. M., and Ireland, S. J. (1991) Investigation into species variants in tachykinin NK_1 receptors by the use of the non-peptide antagonist, CP-96,345. *Br. J. Pharmacol.* **104,** 292,293.

Bergstrom, L., Beaujouan, J. C., Torrens, Y., Saffroy, M., Glowinski, J., Lavielle, S., Chassiang, G., Marquet, A., D'Orleans-Juste, P., Dion, S., and Regoli, D. (1987a) [^{3}H]neurokinin A labels a specific tachykinin binding site in the rat duodenal smooth muscle. *Mol. Pharmacol.* **32,** 764–771.

Bergstrom, L., Torrens, Y., Saffroy, M., Beaujouan, J. C., Lavielle, S., Chassaing, G., Morgat, J. L., Glowinski, J., and Marquet, A. (1987b) ^{3}H-neurokinin B and ^{125}I-Bolton Hunter eledoisin label identical tachykinin binding sites in the rat brain. *J. Neurochem.* **48,** 125–133.

Buck, S. H. and Burcher, E. (1986) The tachykinins: a family of peptides with a brood of "receptors." *Trends Pharmacol. Sci.* **7,** 65–68.

Buck, S. H., Helke, C. J., Burcher, E., Shults, C. W., and O'Donohue, T. L. (1986) Pharmacologic characterization and autoradiographic distribution of binding sites for iodinated tachykinins in the rat central nervous system. *Peptides* **7,** 1109–1120.

Buck, S. H., Harbeson, S. L., Hassmann III, C. F., Shatzer, S. A., Rouissi, N., Nantel, F., and van Giersbergen, P. L. M. (1990) [Leu^9(CH_2NH)Leu^{10}]-neurokinin A (4–10) (MDL 28,564) distinguishes tissue tachykinin peptide NK-2 receptors. *Life Sci.* **47,** 37–41.

Burcher, E., Mussap, C. J., Geraghty, D. P., and McClure-Sharp, J. M., and Watkins, D. J. (1991) Concepts in characterization of tachykinin receptors. *Ann. NY Acad. Sci.* **632,** 123–136.

Cascieri, M. A. and Liang, T. (1984) Binding of [^{125}I]-Bolton Hunter conjugated eledoisin to rat brain cortex membranes—evidence for two classes of tachykinin receptors in the mammalian central nervous system. *Life Sci.* **35,** 179–184.

Dam, T. V. and Quirion, R. (1986) Pharmacological characterization and Autoradiographic localization of substance P receptors in guinea pig brain. *Peptides* **7,** 855–864.

Dam, T. V., Escher, E., and Quirion, R. (1988) Evidence for the existence of three classes of neurokinin receptors in brain. Differential ontogeny of neurokinin-1, neurokinin-2, and neurokinin-3 binding sites in rat cerebral cortex. *Brain Res.* **453,** 372–376.

Dam, T. V., Martinelli, B., and Quirion, R. (1990a) Autoradiographic distribution of brain neurokinin-1/substance P receptors using a highly selective ligand [^{3}H][Sar^9,Met(O_2)^{11}]-substance P. *Brain Res.* **531,** 333–337.

Dam, T. V., Takeda, Y., Krause, J. E., Escher, E., and Quirion, R. (1990b) γ-preprotachykinin (72-92) peptide amide: an endogenous preprotachykinin-I-gene derived peptide that preferentially binds to neurokinin-2 receptor. *Proc. Natl. Acad. Sci. USA* **87,** 246–250.

Dam, T. V., Escher, E., and Quirion, R. (1990c) Visualization of neurokinin-3 receptor sites in rat brain using the highly selective ligand [^{3}H]senktide. *Brain Res.* **506,** 175–178.

Danks, J. A., Rothman, R. B., Cascieri, M., Chicchi, G. G., Liang, T., and Herkenham, M. (1986) A comparative autoradiographic study of the distribution of substance P and eledoisin binding sites in rat brain. *Brain Res.* **385,** 273–291.

Dietl, M. M. and Palacios, J. M. (1991) Phylogeny of tachykinin receptor localization in the vertebrate central nervous system: apparent absence of neurokinin-2 and neurokinin-3 binding sites in the human brain. *Brain Res.* **539,** 211–222.

Elde, R., Schalling, M., Ceddatelli, S., Nakanashi, S., and Hokfelt, T. (1990) Localization of neuropeptide receptor mRNA in rat brain: initial observations using probes for neurotensin and substance P receptors. *Neurosci. Lett.* **120,** 134–138.

Elliot, P. J., Mason, G. S., Stephens-Smith, M., and Hagan, R. M. (1991) Behavioural and biochemical responses following activation of midbrain dopamine pathways by receptor selective neurokinin agonists. *Neuropeptides* **19,** 119–126.

Emonds-Alt, X., Vilain, P., Goulaouic, P., Proitto, V., Van Broeck, D., Advenier, C., Naline, E., Neliat, G., Le Fur, G., and Breliere, J. C. (1992) A potent and selective non-peptide antagonist of the neurokinin A (NK_2) receptor. *Life Sci.* **50,** PL101–PL104.

Erspamer, V. (1981) The tachykinin peptide family. *Trends Neurosci.* **4,** 267–269.

Fardin, V. and Garret, C. (1991) Species differences between [^{3}H]substance P binding in rat and guinea pig shown by the use of peptide agonists and antagonists. *Eur. J. Pharmacol.* **201,** 231–234.

Fleetwood-Walker, S. M., Mitchell, R., Hope, P. J., El-Yassir, N., Maloney, V., and Bladon, C. (1990) The involvement of neurokinin receptor subtypes in somatosensory processing in the superficial dorsal horn of the cat. *Brain Res.* **519,** 169–182.

Garret, C., Carruette, A., Fardin, V., Moussaoui, S., Peyronel, J.-F., Blanchard, J.-C., and Laduron, P. M. (1991) Pharmacological properties of a potent and selective nonpeptide substance P antagonist. *Proc. Natl. Acad. Sci. USA* **88,** 10,208–10,212.

Gerfen, C.R. (1991) Substance P (neurokinin-1) receptor mRNA is selectively expressed in cholinergic neurons in the striatum and basal forebrain. *Brain Res.* **556,** 165–170.

Gitter, B. D., Waters, D. C., Bruns, R. F., Mason, N. R., Nixon, J. A., and Howbert, J. J. (1991) Species differences in affinities of non-peptide antagonists for substance P receptors. *Eur. J. Pharmacol.* **197,** 237,238.

Guard, S., Watson, S. P., Maggio, J E., Too, P. H., and Watling, K. J. (1990) Pharmacological analysis of [^{3}H]senktide binding to NK_3 tachykinin receptors in guinea pig ileum longitudinal muscle myenteric plexus and cerebral cortex membranes. *Br. J. Pharmacol.* **99**, 767–773.

Guard, S., Dam, T. V., Watson, S. P., Martinelli, B., Watling, K. J., and Quirion, R. (1991) [^{3}H]-[Sar^9,Met(O$_2$)^{11}]-SP and [^{3}H]-succinyl-[Asp^6, MePhe^8]-SP(6–11)-

senktide: New selective radioligands for NK_1 and NK_3 tachykinin receptors. *Biotechnol. Update* **6,** 4–8.

Guard, S. and Watson, S. P. (1991) Tachykinin receptor subtypes: Classification and membrane signalling mechanisms. *Neurochem. Int.* **18,** 149–165.

Guard, S., Zargenar, R., Takeda, Y., Krause, J. E., Quirion, R., and Watling, K. J. (1993) Demonstration of the presence of NK_2 tachykinin receptor binding sites in rat brain using [^{125}I] neuropeptide gamma. *Eur. J. Pharmacol.*, submitted.

Hershey, A. D. and Krause, J. E. (1990) Molecular characterization of a functional cDNA encoding the rat substance P receptor. *Science* **247,** 958–962.

Hershey, A. D., Polenzani, L., Woodward, R. M., Miledi, R., and Krause, J. E. (1991) Molecular and genetic characterization, functional expression, and mRNA expression patterns of a rat substance P receptor. *Ann. NY Acad. Sci.* **632,** 63–78.

Krause, J. E., MacDonald, M. R., and Takeda, Y. (1989) The polyprotein nature of substance P precursors. *BioEssays* **10,** 62–69.

Lew, R., Geraghty, D. P., Drapeau, G.,Regoli, D., and Burcher, E. (1990) Binding characteristics of ^{125}I-Bolton-Hunter [Sar^9,Met(O_2)^{11}]-substance P, a new selective radioligand for the NK_1 receptor. *Eur. J. Pharmacol.* **184,** 97–108.

Maggi, C. A., Patacchini, R., Giuliani, S., Rovero, P., Dion, S., Regoli, D., Giachetti, A., and Meli, A. (1990) Competitive antagonists discriminate between NK_2 tachykinin receptor subtypes. *Br. J. Pharmacol.* **100,** 588–592.

Maggi, C. A., Patacchini, R., Astoll, M., Rovero, P., Giuliani, S., and Giachetti, A. (1991) NK_2 receptor agonists and antagonists. *Ann. NY Acad. Sci.* **632,** 184–191.

Maggio, J. E. (1988) Tachykinins. *Annu. Rev. Neurosci.* **11,** 13–28.

Mantyh, P. W., Maggio, J. M., and Hunt, S. P. (1984) The autoradiographic distribution of kassinin and substance K binding sites is different from the distribution of substance P binding sites in the rat brain. *Eur. J. Pharmacol.* **102,** 361–364.

Mantyh, P.W., Gates, T., Mantyh, C.R., and Maggio, J.E. (1989) Autoradiographic localization and characterization of tachykinin receptor binding sites in the rat brain and peripheral tissues. *J. Neurosci.* **9,** 258–279.

Masu, Y., Nakayama, K, Tamaki, H., Harada, Y., Kuno, M., and Nakanishi, S. (1987) cDNA cloning of bovine substance-K receptor through oocyte expression system. *Nature (Lond.)* **328,** 836–838.

McLean, S., Ganong, A. H., Seeger, Th. F., Bryce, D. K., Pratt, K. G., Reynolds, L. S., Siok C. J., Lowe III, J. A., and Heym, J. (1991) Activity and distribution of binding sites in brain of a nonpeptide substance P (NK_1) receptor antagonist. *Science* **251,** 437–439.

Mussap, C. J. and Burcher, E. (1990) [^{125}I]-Bolton-Hunter Scyliorhinin II: a novel, selective radioligand for the tachykinin NK_3 receptor in rat brain. *Peptides* **11,** 827–836.

Nakanishi, S. (1987) Substance P precursor and kininogen: their structures, gene organization and regulation. *Physiol. Rev.* **67,** 1117–1142.

Nakanishi, S. (1991) Mammalian tachykinin receptors. *Annu. Rev. Neurosci.* **14,** 123–136.

Oksenberg, D., Marsters, S., O'Dowd, B. F., Jin, H., Havlik, S., Peroutka, S. J., and Ashkenai, A. (1992) A single amino-acid difference confers major pharmacological variation between human and rodents 5-HT_{1B} receptors. *Nature* **360,** 161–163.

Overton, P., Elliot, P. J., Hagan, R. M., and Clark, D. (1992) Neurokinin agonists differentially affect A9 and A10 dopamine cells in the rat. *Eur. J. Pharmacol.* **214,** 165,166.

Pernow, B. (1983) Substance P. *Pharmacol. Rev.* **35,** 85–141.

Poosch, M. S., Goebel, D. J., and Bannon, M. J. (1991) Distribution of neurokinin receptor gene expression in the rat brain. *Soc. Neurosci. Abst.* **17,** 320.13.

Quirion, R., Shults, C. W., Moody, T. W., Pert, C. B., Chase, T. N., and O'Donohue, T. L. (1983) Autoradiographic distribution of substance P receptors in rat central nervous system. *Nature* **303,** 714–716.

Quirion, R. and Dam, T. V. (1985) Multiple tachykinin and substance P receptors, in *Substance P: Metabolism and Biological Actions* (Jordan, C. C. and Oehme, P., eds.), Taylor and Francis, London, pp. 45–64.

Quirion, R. and Dam, T. V. (1988) Multiple neurokinin receptors: recent developments. *Regul. Peptides* **22,** 18–25.

Quirion, R., Dam, T. V., and Guard, S. (1991) Selective neurokinin receptor radioligands. *Ann. NY Acad. Sci.* **632,** 137–144.

Quirion, R. and Dam, T. V. (1993) Radioligand binding studies of the three major classes of neurokinin receptors, in *Methods in Neuroscience*, vol. 11 (Conn, P. M., ed.), Academic, New York, in press.

Regoli, D., Drapeau, G., Dion, S., and D'Orleans-Juste, P. (1987) Recent developments in neurokinin pharmacology. *Life Sci.* **40,** 100–117.

Rouissi, N., Gitter, B. D., Waters, D. C., Howbert, J. J., Nixon, J. A., and Regoli D. (1991) Selectivity and specificity of new, non-peptide, guinuclidine antagonists of substance P. *Biochem. Biophys. Res. Commun.* **176,** 894–901.

Rouissi, N., Claing, A., Nicolau, M., Jukic, D., D'Orleans-Juste, P., and Regoli, D. (1993) Substance P (NK_1 receptor) antagonists: *in vivo* and *in vitro* activities in rats and guinea pigs. *Life Sci.* **52,** 1141–1147.

Saffroy, M., Beaujouan, J. C., Torrens, Y., Besseyre, J., Bergstrom, L., and Glowinski, J. (1988) Localization of tachykinin binding sites (NK_1, NK_2, NK_3 ligands) in the rat brain. *Peptides* **9,** 227–241.

Shigemoto, R., Yokota, Y., Tsuchida, K., and Nakanishi, S. (1990) Cloning and expression of a rat neuromedin K receptor cDNA. *J. Biol. Chem.* **265,** 623–628.

Shults, C. W., Quirion, R., Jensen, R. T., Moody, T. W., O'Donohue, T. L., and Chase, T. N. (1982) Autoradiographic localization of substance P receptors using [^{125}I]substance P. *Peptides* **3,** 1073–1075.

Shults, C. W., Quirion, R., Chronwall, B., Chase, T. N., and O'Donohue, T. L. (1984) A comparison of the anatomical distribution of substance P and substance P receptors in the rat central nervous system. *Peptides* **5,** 1097–1128.

Snider, R. M., Constantine, J. W., Lowe, J. A. III, Longo, K. P., Lebel, W. S., Woody, II. A., Drozda, S. E., Desai, M. C., Vinick, F. J., Spencer, R. W., and Hess, H. (1991) A potent nonpeptide antagonist of the substance P (NK_1) receptor. *Science* **251,** 435–437.

Stoessl, A. J. and Hill, D. R. (1990) Autoradiographic visualization of NK-3 tachykinin binding sites in the rat brain, utilizing [^{3}H]senktide. *Brain Res.* **534,** 1–7.

Stoessl, A. J., Szezutkowski, E., Glenn, B., and Watson, I. (1991) Behavioral effects of selective tachykinin agonists in midbrain dopamine neurons. *Brain Res.* **565,** 254–262.

Takeda, Y. and Krause, J. E. (1991) Pharmacological and molecular biological studies on the diversity of rat tachykinin NK_2 receptor subtypes in rat CNS, duodenum, vas deferens, and urinary bladder. *Ann. NY Acad. Sci.* **632,** 479–481.

Tousignant, C., Chretien, L., Guillemette, G., and Regoli, D. (1992) Characterization of NK_1 receptors in guinea pig and rat brain membranes with NK_1 peptides and a non-peptide antagonist. *Brain Res.* **596,** 243–250.

Tschope, C., Picard, P., Culman, J., Prat, A., Itoi, K, Regoli, D., Unger, T., and Couture, R. (1992) Use of selective antagonists to dissociate the central cardiovascular and behavioral effects of tachykinins on NK_1 and NK_2 receptors in the rat. *Br. J. Pharmacol.* **107,** 750–755.

Wolf, S. S., Moody, T. W., Quirion, R., and O'Donohue, T. L. (1985) Biochemical characterization and autoradiographic localization of central substance P receptors using ^{125}I-physalaemin. *Brain Res.* **332,** 299–307.

Yashpal, K., Dam, T. V., and Quirion, R. (1990) Quantitative autoradiographic distribution of multiple neurokinin binding sites in rat spinal cord. *Brain Res.* **506,** 259–266.

Yokota, Y., Sasai, Y., Tanaka, K, Fujiwara, T., Tsuchida, K., Shigemoto, R., Kakizuka, A., and Ohkubo, H., and Nakanishi, S. (1989) Molecular characterization of a functional cDNA for rat substance P receptor. *J. Biol. Chem.* **264,** 17,649–17,652.

CHAPTER 6

Autoradiographic Localization of Receptors in Peripheral Tissues

Elizabeth Burcher, Christian J. Mussap, and Jennifer A. Stephenson

1. Introduction

1.1. Radioligands—General Considerations

Autoradiography incorporates the fundamental principles of radioligand binding and photographic techniques. The ideal requisites for a peptide radioligand used in autoradiography are:

1. High-affinity;
2. Absolute receptor selectivity;
3. Low nonspecific binding;
4. Resistance to enzymatic degradation; and
5. Biological activity.

Thus, iodinated radioligands should demonstrate biological activity comparable to the endogenous peptide in order to validate extrapolation from binding sites to receptors. Tachykinin radioligands developed to date have largely failed to fulfill all the above-

The Tachykinin Receptors Ed.: S. H. Buck

mentioned criteria with the problem of lack of selectivity proving a major stumbling block to unambiguous receptor localization. High nonspecific binding can be another major concern, particularly in the case of some NK_3 preferring radioligands *(see below)*, where this problem is still to be overcome.

The "trademark" Phe-X-Gly-Leu-Met-NH_2 C-terminus common to all tachykinins underscores the lack of receptor selectivity inherent to peptide-derived radioligands. This is disappointing since iodinated tachykinin radioligands often possess extremely high specific binding resulting in autoradiographs that appear near-perfect technically. Lack of selectivity has necessitated tedious characterization of binding (in slide-mounted tissue sections) using a range of concentrations of competing agents. Localization of a single receptor/receptor subtype can be achieved by coincubation with an appropriate receptor-blocking concentration of a highly selective agonist–antagonist, but resulting density of specific binding can be very low. Whenever possible, studies should employ the most selective radioligands currently available.

Iodinated radioligands are more convenient to use than tritiated radioligands because of their considerably higher specific activity. The very much shorter exposure times required for ^{125}I enable faster "workup" of methods and more rapid results. This is offset somewhat by the higher resolution attainable with [^{3}H]-labeled radioligands. Iodinated radioligands also have the advantage of being easily prepared in the laboratory using commercially available [^{125}I]NaI or ^{125}I-Bolton-Hunter reagent thus more cheaply and conveniently producing large amounts of radioligand. We find this an important attribute since it enables experimental iodination of novel selective analogs without the significant financial outlay needed for custom tritiation, which may prove ultimately futile if unacceptably high nonspecific binding prevails. However, disadvantages of iodinated radioligands include the necessity to demonstrate biological activity, which usually involves synthesis of the nonradioactive form of the radioligand.

Susceptibility of (endogenous) peptide radioligands to peptidase degradation is another issue deserving of comment. Radioligand degradation can be readily monitored using HPLC, and cocktails of peptidase inhibitors (e.g., chymostatin, leupeptin, thiorphan, phosphoramidon) are routinely used in homogenate binding studies. Although cryocut tissue sections are likely to possess only minute quantities of peptidases, inclusion of the relevant inhibitors is nonetheless recommended, especially in tissues (e.g., lung) enriched with peptidases. Development of nonpeptide radioligands (of high specific activity, absolute selectivity, and very low nonspecific binding) will clearly avoid this problem and generate more confidence in the reliability of autoradiographic studies.

1.2. Radioligands for Tachykinin Receptors

The first tachykinin radioligand successfully used in autoradiographic studies was [^{3}H]substance P that delineated SP (NK_1) binding sites in rat brain (Quirion et al., 1983). However, [^{3}H]SP was largely superseded by the high specific activity ligand ^{125}I-Bolton-Hunter SP (BHSP) thus enabling more rapid turnover of experiments and greater ease in radioligand preparation using ^{125}I-Bolton-Hunter conjugate. Similarly, iodinated derivatives of eledoisin (BHELE) and neurokinin B (BHNKB) (for labeling of NK_3 sites) and neurokinin A (BHNKA) (for NK_2 sites) were soon developed for use in homogenate binding and autoradiographic studies (for reviews, *see* Burcher et al., 1991; Quirion et al., 1991; Mussap et al., 1993). Novel radoligands based on selective agonists have now been developed for the NK_1 receptor. These include ^{125}I-Bolton-Hunter [Sar^9,Met(O_2)^{11}]-SP (BHSarSP) (Lew et al., 1990), [^{3}H]-[Pro^9]-SP (Petitet et al., 1991), and [^{3}H][Sar^9,Met(O_2)^{11}]-SP (Quirion et al., 1991).

Although possessing low nonspecific binding, the NK_2 preferring radioligand, BHNKA, is somewhat unselective with significant interaction at central NK_3 sites (Buck et al., 1986; Foster and Tridgett, 1988). This illustrates how addition of the bulky Bolton-Hunter group is capable of distorting the binding characteristics of the native molecule, and how accompanying functional studies with the unlabeled chemical equivalent of the radioligand may be required to control for such aberrations. [^{125}I]iodohistidyl-NKA (INKA) was developed as an alternative, selective NK_2 radioligand (Buck and Shatzer, 1988; Burcher et al., 1989; Mussap and Burcher, 1993). Over the years, INKA proved most useful in autoradiographic studies, exhibiting 95% specific binding in most tissues and yielding high-resolution autoradiographs. However, INKA binding to NK_1 sites in guinea pig lung has been conclusively demonstrated (Burcher et al., 1989; Geraghty et al., 1992; *see* Section 2.1.), negating claims of high selectivity. We therefore synthesized ^{125}I-[Lys^5,Tyr(I$_2$)^7,MeLeu^9,Nle^{10}]-NKA(4–10) (Burcher et al., 1993), which has greatly enhanced selectivity at the NK_2 receptor and ~85% specific binding in rat bladder, gut, and vas deferens homogenates (Badgery-Parker et al., 1993). However, in our experience this radioligand is less suitable for studies in guinea pig tissues, where it displays high nonspecific binding (Badgery-Parker et al., 1993), possibly reflecting species differences in receptor sequence. Current tachykinin radioligands suitable for autoradiography are listed in Table 1.

Since the discovery of neuropeptide K and neuropeptide γ(NPγ) in the 1980s, there has been some interest in determining whether these two N-terminally extended forms of NKA elicit their potent actions (including smooth muscle contraction and salivation) via novel receptor types. Efforts to produce a radioligand based on neuropeptide K have been unsuccessful, with very low yield of purified ligand and exceedingly high nonspecific binding (Mantyh et al., 1989; Burcher and Mussap, unpublished results). NPγ can be iodinated (possesses 3 histidine residues), but this may result in a number of different products, necessitating further separation of the various labeled constituents. Moreover, recent studies suggest that NPγ is not altogether selective, with capacity to interact at NK_1 sites in addition to high-affinity at NK_2 sites (Quirion et al., 1991).

The autoradiographic localization and characterization of NK_3 sites in the central nervous system (CNS) has routinely been performed with BHELE (Buck et al., 1986; *see* Dam and Quirion, this volume). However, the notorious lack of selectivity of this radioligand, most evident in periphery (Burcher and Buck, 1986; *see* Burcher et al., 1991; Mussap et al., 1993), has posed difficulties for researchers

Table 1
Commonly Used Tachykinin Radioligands

Radioligand	Selectivity			Nonspecific binding
[^{3}H]SP	NK-1	–	good	>40%
BHSP	NK-1	–	good	~15%
BHSarSP	NK-1	–	excellent	~10%
[^{3}H][Pro^9]SP	NK-1	–	excellent	30%
[^{3}H]NKA	NK-2	–	good	>30%
INKA	NK-2	–	good	≤5%
^{125}I-[Lys^5,Tyr(I$_2$)^7, MeLeu^9,Nle^{10}] NKA(4–10)	NK-2	–	excellent	>10%[a]
BHNKA	NK-2	–	moderate	~5%
[^{3}H]NKB	NK-3	–	good	>60%
BHELE	NK-3	>	NK-2 ≥ NK-1	15–30%
BHSCYII	NK-3	>	NK-1 >> NK-2	15%[b]

[a]In rat and human, but not guinea pig.
[b]In CNS; higher nonspecific binding in periphery.
BH, ^{125}I-Bolton-Hunter.

interested in visualizing elusive peripheral NK$_3$ sites. Iodinated NKB and selective analogs have proved totally unsuitable for this purpose due to very high nonspecific binding (*see* Mussap, 1992; Mussap et al., 1993). We introduced ^{125}I-Bolton-Hunter scyliorhinin II (BHSCYII) as an alternative, higher affinity ligand with low nonspecific binding at central NK$_3$ sites (Mussap and Burcher, 1990). Although substantially more selective than BHELE, BHSCYII does exhibit a limited capacity for NK$_1$ binding (Mussap and Burcher, 1991; Mussap, 1992). Both [^{3}H]NKB and the highly selective [^{3}H]-senktide have been used with success in CNS autoradiographic studies (*see* Dam and Quirion, this volume), but are untested in periphery and remain subject to problems of tritiated ligands (discussed earlier). Iodinated radioligands derived from new, nonpeptide NK$_3$ antagonists (when available) may sidestep these longstanding obstacles to high resolution visualization of NK$_3$ receptors in the periphery.

1.3. Methodology and Technical Considerations

Most autoradiographic studies are performed in vitro. Frozen tissue sections (unfixed or very lightly fixed) are cut on a cryostat and thaw-mounted onto cleaned and gelatin-coated (subbed) glass microscope slides. Subbing of slides facilitates tissue adhesion. For radiolabeling, frozen tissue sections are thawed and incubated with radioligand. At binding equilibrium, the reaction is terminated by repeated washing in buffer to remove excess radioligand and reduce nonspecific binding. The appropriate peptidase inhibitor cocktail, as well as the optimum incubation time and washing conditions for each tissue and radioligand combination, should be established in preliminary studies.

Autoradiographs can be generated by tight apposition of radiolabeled sections to photosensitive film. Differential tissue quenching is a problem in the CNS using tritiated, but not iodinated, ligands. Binding over discrete regions of brain and spinal cord is readily visualized with film, and consecutive CNS sections have the advantage of being almost identical in appearance. In contrast, peripheral tissue autoradiography often requires higher resolution to visualize binding over distinct cellular structures. Consequently, the more complex process of emulsion-dipping (Herkenham and Pert, 1982) or apposition of labeled sections to emulsion-coated coverslips (Kuhar, 1985a) is usually required.

Although enhanced resolution of binding sites is achieved using photographic emulsion techniques, both the direct-dipping and emulsion-coated coverslip methods are themselves fraught with technical difficulties. Repeated manipulation of delicate coverslips (for the initial coating, then apposition and final development and staining) inevitably results in breakages, and image fogging occurs with even slight overexposures. Because the emulsion-coated coverslip is at some distance from the bound radioligand, image resolution is not as sharp as that achieved using direct dipping. However, with the latter technique immersion into molten (42°C) emulsion further impairs tissue morphology. A classic example is the salivary gland, where labeled sections dipped in emulsion often exhibit considerable tissue degradation, with light field micrographs quite unsuitable for publication. In most instances, this problem can be avoided by photographing an adjacent unlabeled section treated with a histologic stain appropriate for the labeled feature(s) in question. According to convention, binding is depicted with alternate dark- and lightfield photomicrographs of the same labeled section, with a darkfield photomicrograph of an adjacent section illustrating nonspecific binding.

An additional problem arising from direct dipping of slides is rapid dissociation of the bound radioligand, manifest as an aggregation of silver grains at the base of the section. Our experience has been that this may be largely avoided by fixation of sections in paraformaldehyde vapor at 65°C, thereby forming covalent links between bound (peptide) radioligand and receptor proteins. Loss of precipitated silver grains may also occur if pH and temperature are not stringently controlled throughout the development and staining. Acidic conditions are most detrimental, so selection of stains has been limited to pyronin Y and methylene blue (and their nonacidic derivatives); in our hands, hematoxylin-eosin is unsuitable.

2. Autoradiographic Localization of Receptors in Peripheral Tissues

2.1. Respiratory Tract

In the respiratory tract, tachykinins occur in neurons, including primary afferent fibers (Lundberg et al., 1984), from which they can be released, possibly as a result of an axon reflex (*see* Barnes et al., 1991). Tachykinins are potent in eliciting contraction of bronchial smooth muscle in several species (guinea pig, rabbit, ferret, hamster, and human), but not dog or rat. In addition, tachykinins stimulate glandular and epithelial secretion, mucociliary clearance, and plasma extravasation, and activate inflammatory cells (*see* Frossard and Advenier, 1991). These actions have suggested a role in the pathogenesis of asthma, and characterization of the responsible receptors has been actively pursued in view of important therapeutic implications.

Using BHSP, NK_1 binding sites have been demonstrated in the airways of several species. Most work has been carried out in the guinea pig, where dense BHSP binding was described over tracheal and bronchial smooth muscle (Carstairs and Barnes, 1986; Hoover and Hancock, 1987; Watkins and Burcher, 1987; Burcher et al., 1989). In the tracheal lamina propria, BHSP binding sites are found in association with small blood vessels and glands (Hoover and Hancock, 1987; Burcher et al., 1989). In the rabbit, there is no specific binding of BHSP over tracheal smooth muscle, but binding sites are found over bronchial smooth muscle with receptor density increasing markedly in the smaller airways (Black et al., 1990). This distribution of binding sites correlates with functional data implicating NK_1 receptors in airway smooth muscle contraction (in guinea pig and rabbit), mucus secretion, and bronchial blood flow (*see* Frossard and Advenier, 1991).

In the human lung, one report exists of BHSP binding sites over smooth muscle and glands (Carstairs and Barnes, 1986). However, other groups found that BHSP and BHSarSP show no specific binding to human bronchial smooth muscle, but do label pulmonary arterial endothelium small blood vessels and glands (Goldie, 1990; Strigas and Burcher, 1993). This lack of NK_1 sites on smooth muscle is in accordance with the inability of NK_1 receptor agonists to contract human airways. BHSP binding sites also occur over epithelium, glands, and blood vessels in human nasal mucosa (*see* Barnes et al., 1991). In human trachea, localization of tachykinin receptors over the trachealis muscle, glands, and epithelium, but not submucosal blood vessels, has been reported using an immunohistochemical technique involving an antiidiotypic antibody (Fischer et al., 1992). Since the antibody recognized both NK_1 and NK_2 receptors, further characterization was not possible.

In spite of ample evidence for the presence of NK_2 receptors mediating contraction of bronchial smooth muscle in many species, including humans (*see* Frossard and Advenier, 1991), autoradiographic visualization of NK_2 binding sites has been unsuccessful until recently *(see below)*. Although we earlier found weak specific binding of INKA to bronchial smooth muscle in sections of guinea pig major bronchi, characterization studies in sections and membranes showed that the radioligand was actually binding to NK_1 receptors (Burcher et al., 1989; Geraghty et al., 1992). A similar situation applies to the rabbit. Here, specific binding of INKA over bronchial smooth muscle was apparent with distribution similar to that of BHSP; however, characterization of binding in membranes revealed that INKA was binding predominantly to NK_1 sites (Black et al., 1992). These studies highlight the lack of absolute selectivity of this radioligand (noted above) and indicate that NK_2 receptors on airway smooth muscle may have unusual characteristics (Burcher et al., 1991; Geraghty et al., 1992; Black et al., 1992). Using our novel, selective radioligand, ^{125}I-[Lys^5,Tyr(I$_2$)^7,MeLeu^9,Nle^{10}]-NKA(4–10), we have now demonstrated specific binding to NK_2 preferring sites in homogenates of guinea pig lung (Zeng et al., 1993), and autoradiographic studies have now shown specific binding on guinea pig bronchial smooth muscle (Strigas and Burcher, 1993). To date, there have been no reports of NK_2 binding sites on human respiratory smooth muscle.

There have been few reports of NK_3 receptors in peripheral tissues, and functional, binding, and autoradiographic studies have all failed to provide evidence for NK_3 sites in the mammalian respiratory tract (Mussap and Burcher, 1990; Geraghty et al., 1992; *see* Frossard and Advenier, 1991).

2.2. *Gastrointestinal Tract*

Tachykinins are potent in contracting the gastrointestinal smooth muscle of all vertebrates; in some cases, contraction is partly mediated via neuronal mechanisms (*see* Bartho and Holzer, 1985). In addition, tachykinins promote glandular secretion and can increase gastrointestinal blood flow. High levels of SP- and other tachykinin-like immnoreactivity occurs in gut, and immunohistochemical techniques reveal that most immunoreactivity is in enteric neurons rather than in primary afferent, capsaicin-sensitive nerves (*see* Bartho and Holzer, 1985; Holzer, 1988).

In accordance with these findings, the mammalian gastrointestinal tract is enriched with NK_1 and NK_2 receptors, primarily on smooth muscle. In the rat, dense binding of BHSP and other NK_1 preferring ligands is found over circular muscle in the stomach, small intestine, and especially colon, as well as over ganglia of the myenteric and submucous plexuses (Burcher et al., 1986). Binding sites for BHNKA (characterized as NK_2 sites in parallel binding studies) are densely localized over circular muscle of stomach, duodenum, jejunum, ileum, and colon, as well as over the muscularis mucosae of esophagus, fundus, and colon; binding of this radioligand is much weaker over longitudinal muscle of these regions (Burcher et al., 1984, 1986). Binding of BHNKA to a section of rat colon is shown in Fig. 1. High-affinity specific binding of INKA occurs to both circular muscle and muscularis mucosae of fundus and was characterized to NK_2 receptors (Mussap and Burcher, 1993). Binding sites for BHSP were noticeably absent over blood vessels of the rat gastrointestinal tract (Burcher et al., 1986; Mussap and Burcher, 1993).

In the guinea pig, NK_1 binding sites are visualized over both circular and longitudinal muscle (Mantyh et al., 1984; Burcher et al., 1986), as well as over myenteric ganglia *(see below)* and colon mucosa (Burcher et al., 1986). In contrast to the rat, binding sites for BHNKA were less dense in guinea pig small intestine (characterized as NK_2 sites in homogenates of longitudinal muscle-myenteric plexus) and colon (Burcher et al., 1986). Other functional studies have suggested that NK_2 receptors may be absent from guinea pig ileum, and recent evidence indicates the possible existence of unusual tachykinin receptors ("septide-preferring receptors") in this tissue (Chassaing et al., 1992; Petitet et al., 1992). No radioligand is presently available to label such receptors, and this controversial new "receptor" is the subject of ongoing research.

In the cat, there was dense BHSP binding to the ileocecal sphincter and the circular muscle of ileum and colon with weaker specific binding over the pyloric sphincter, pylorus, jejunum, and the longitudinal muscle of the distal tract (Rothstein et al., 1989, 1991). There is apparent disparity between the region-specific distribution of NK_1 receptors and the fairly uniform levels of SP-LI reported in the cat intestine (*see* Rothstein et al., 1991).

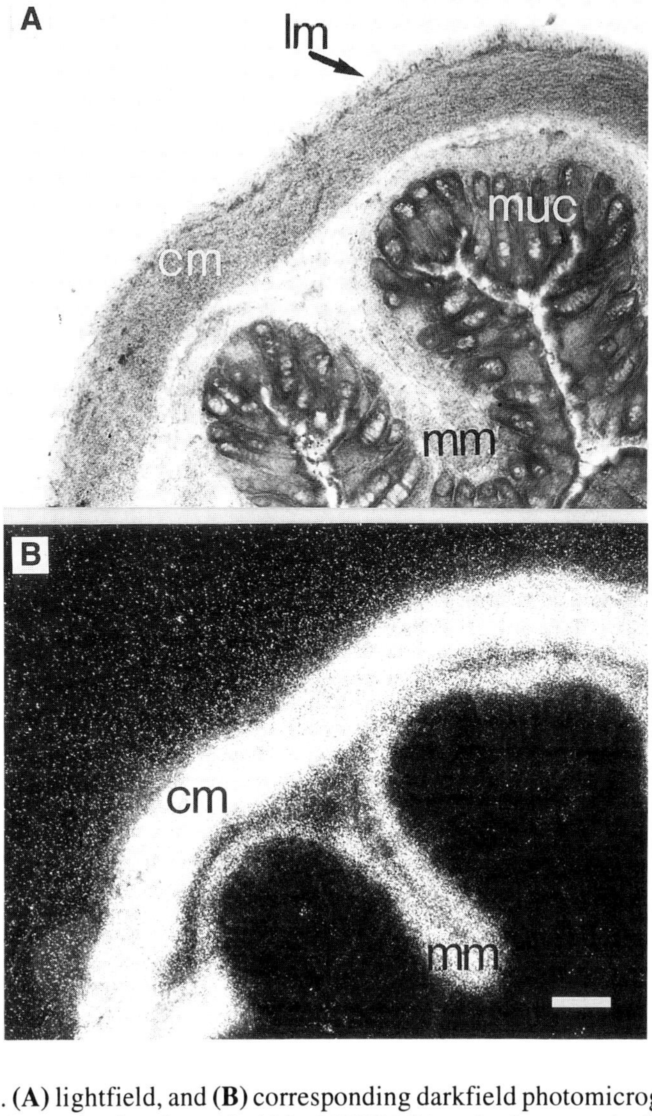

Fig. 1. **(A)** lightfield, and **(B)** corresponding darkfield photomicrograph of section of rat colon incubated in 100 p*M* BHNKA and subsequently dipped in photographic emulsion, exposed for 7 d, and stained with pyronin Y. Silver grains are densely localized to circular muscle (cm) and to muscularis mucosae (mm) only. An adjacent section, coincubated with 1 μ*M* NKA to illustrate nonspecific binding, showed no localization of binding. Other adjacent sections, incubated similarly with BHSP, showed dense binding to circular muscle and myenteric ganglia, but not to muscularis mucosae (data not shown). lm, longitudinal muscle; muc, mucosa. Bar, 200 μm.

Mantyh and coworkers have made extensive autoradiographic studies localizing binding sites for radiolabeled SP and NKA in canine and human gut. In the dog, BHSP binding sites occur over longitudinal and circular smooth muscle, as well as over myenteric ganglia, blood vessels, lymphatic tissue, and epithelial cells (Mantyh et al., 1988b). In the stomach, BHSP also labels chief cells (Vigna et al., 1989). BHNKA binding sites are found over the muscularis externa and muscularis mucosae and over blood vessels (Mantyh et al., 1988b).

In the human, NK_1 receptors are expressed by many regions, particularly circular and longitudinal smooth muscle (Gates et al., 1988; Korman et al., 1989), but also myenteric ganglia, epithelium, and the germinal centers of lymph nodules (Gates et al., 1988). In contrast, NK_2 receptors occur on muscularis mucosae, as well as circular and longitudinal muscle (Gates et al., 1988). In normal intestine, NK_1 receptors were expressed weakly by intestinal arterioles and venules, but there is a massive increase in the NK_1 sites associated with blood vessels in patients with Crohn's disease and ulcerative colitis; NK_2 sites are unaltered (Mantyh et al., 1988a; Mantyh, 1991; *see* Mantyh et al., this volume).

2.3. Neural Receptors

Tachykinin receptors, particularly NK_1 and NK_3 receptors, are widely distributed throughout the central nervous system (*see* Dam and Quirion, this volume). There is also good evidence for localization of NK_1 receptors on peripheral ganglia.

2.3.1. Sympathetic Ganglia

Immunoreactive SP is found in nerve terminals surrounding cell bodies of ganglion cells (Hokfelt et al., 1977), and SP is considered to be the excitatory transmitter mediating the slow excitatory postsynaptic potential (EPSP) in peripheral ganglia such as the celiac ganglion (*see* Bartho and Holzer, 1985). Niwa et al. (1985) described BHSP binding sites over rat sympathetic ganglia, particularly the celiac ganglion; labeling of the superior cervical ganglion was less dense (Niwa et al., 1985). BHSP binding sites have been localized over cell bodies of some (35–50%) cultured rat superior cervical ganglion neurons

(James and Burnstock, 1988). Mantyh and coworkers (1992) also demonstrated BHSP, but not BHNKA, binding sites over regions of the rat superior cervical ganglion and rabbit superior mesenteric ganglion.

2.3.2. *Enteric Ganglia*

SP-immunoreactive fibers densely innervate ganglia of the myenteric and submucous plexuses of guinea pig ileum and other regions of the gastrointestinal tract in a number of species, and SP is considered to mediate the slow EPSP in myenteric and submucous ganglia (*see* Bartho and Holzer, 1985). In accordance with this physiological evidence for neuronal receptors, binding sites for BHSP have been described on enteric ganglia of several species. Figure 2 illustrates the dense BHSP binding over myenteric ganglia in lower esophagus of the cat. High-affinity binding sites for BHSP have also been reported in the submucosa of rat esophagus, but it is unclear whether these sites represent labeling of blood vessels, neurons, and/or muscularis mucosae (Wiedermann et al., 1987). BHSP also labels the myenteric plexus in human and dog gut (Gates et al., 1988: Mantyh et al., 1988b).

Whole-mount autoradiography has been used to reveal BHSP binding sites on ganglia of the guinea pig ileum submucous plexus (Burcher and Bornstein, 1988). In this technique, the intestine is first separated into layers. Small squares of submucosa containing ganglia and blood vessels coursing within connective tissue are then arranged on subbed slides, air-dried over ice, and frozen before incubation and emulsion dipping according to standard procedures. This technique provided excellent visualization of the interconnections of the ganglia and internodal strands, as well as the branching of the blood vessels, the latter not expressing binding sites (Fig. 3). However, higher than normal nonspecific binding can occur with this technique (Burcher and Bornstein, 1988), which may reflect inadequate washing times for these comparatively thick whole mounts (as compared with cryocut sections).

Using the new NK_3 preferring radioligand, BHSCYII (Mussap and Burcher, 1990), weak specific binding was observed over gan-

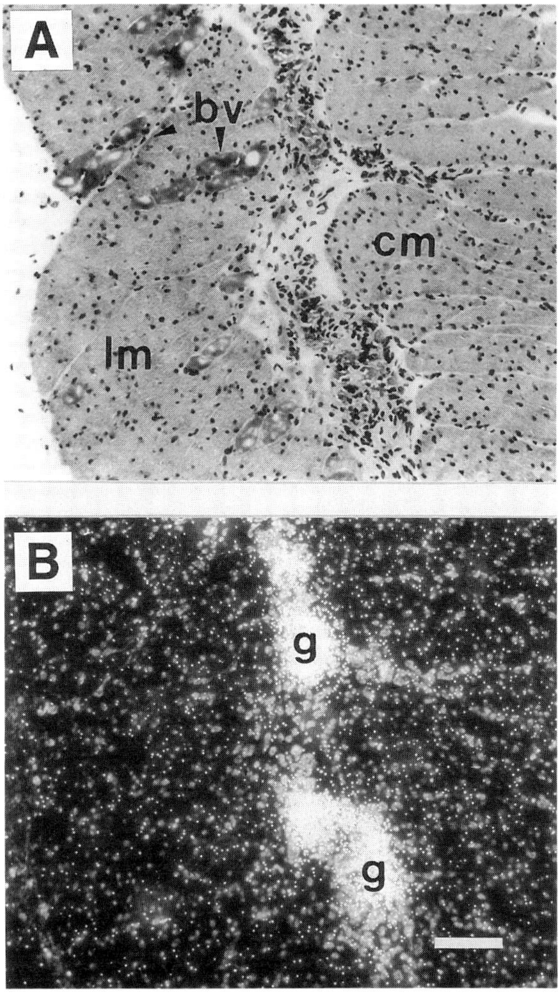

Fig. 2. (**A**) lightfield, and (**B**) corresponding darkfield photomicrograph of section of cat esophagus incubated in 100 p*M* BHSP and subsequently apposed to photographic emulsion-coated coverslips for 14 d, developed, and stained with methylene blue. Localization of silver grains occurs over neurons of the myenteric plexus ganglia (g), but not over blood vessels (bv, arrowed) or over longitudinal (lm) or circular smooth muscle (cm). Nonspecific binding, in an adjacent section coincubated with 1 μ*M* SP, was very low and showed no localization of binding (data not shown). Bar, 100 μm.

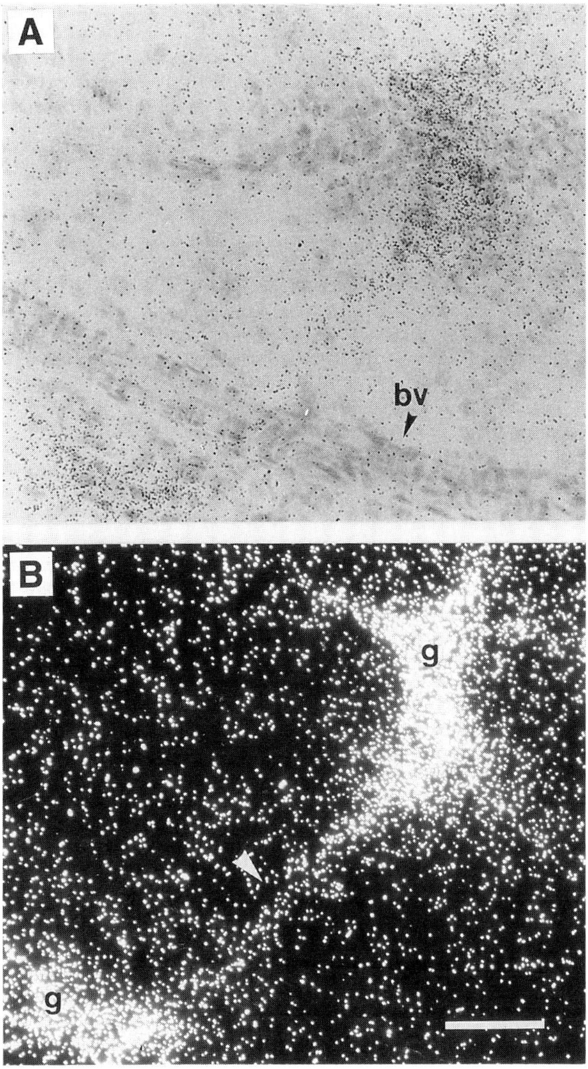

Fig. 3. **(A)** lightfield, and **(B)** corresponding darkfield high power photomicrographs of whole mount of submucosa from guinea pig ileum. Silver grains represent localization of BHSP (100 p*M*) binding sites over submucous ganglia (g), but not over blood vessel (bv). The internodal strand (arrowed in B) is weakly labeled. Bar, 50 μm.

glia of the myenteric plexus in both gut sections and myenteric plexus-longitudinal muscle whole mounts of guinea pig ileum (Mussap and Burcher, submitted). These observations are consistent with binding to neuronal NK_3 receptors thought to mediate acetylcholine release (*see* Mussap et al., 1993).

2.4. Glands

2.4.1. Salivary Glands

SP has long been established as a potent sialogogue in most species. Binding sites for BHSP were characterized comprehensively in rat submandibular and parotid glands using membrane homogenate techniques (Liang and Cascieri, 1981; Lee et al., 1983; Buck and Burcher, 1985). The submandibular gland is highly enriched with NK_1 binding sites, and this was also shown using autoradiographic techniques (Buck and Burcher, 1985), with binding sites found over secretory acini.

2.4.2. Thymus, Spleen, and Lymph Nodes

Binding sites for BHSP are found in rat spleen associated with the marginal sinus of the white pulp (Shigematsu et al., 1986). In a parallel study in rat thymus, BHSP sites occurred over vascular smooth muscle in the medulla (Shigematsu et al., 1986). Binding of BHSP was characterized in sections as binding to high-affinity NK_1 sites.

In canine lymph nodes, innervation by SP-immunoreactive fibers was described, and BHSP binding sites were present on arterioles and venules of the T-cell regions and on a small proportion of germinal centers (Popper et al., 1988). No binding sites for BHNKA were found. Taken together with reports of NK_1 receptors on T-lymphocytes (*see* Payan, 1989), a convincing role for SP in regulation of the immune response has been argued (*see* Payan, 1989).

2.5. Urinary Tract

The urinogenital system is innervated by tachykinin-containing nociceptive afferents. Both NK_1 and NK_2 receptors can be expressed on smooth muscle, whereas NK_1 receptors predominate over blood vessels. As in the cardiovascular system *(see below)*, species differences are apparent.

2.5.1. Kidney

Functional studies have described the potent diuretic and natriuretic effects of SP in the dog kidney (Gullner et al., 1979). Binding sites for BHSP were found over the endothelium of major arterial vessels including renal, arcuate, and interlobular arteries (Fig. 4), but not over veins (Stephenson et al., 1987b). Competition studies in dog kidney sections demonstrated that BHSP was binding to NK_1 receptors (Stephenson et al., 1987b). The distal and proximal tubules lacked BHSP binding sites (Fig. 5), but glomeruli were densely labeled (Fig. 5). The glomerular cell type(s) expressing binding sites is unknown, but likely candidates are endothelial or possibly mesangial cells. These data suggest that in dog, the diuretic effects of SP are probably associated with its vasodilator properties rather than to any direct effect on tubular reabsorption and secretion mechanisms. In contrast, both rat and guinea pig kidney display negligible binding of NK_1 and NK_2 preferring radioligands (Burcher, unpublished results). There have been no reported studies investigating tachykinin receptors in human kidney.

2.5.2. Ureter

The guinea pig ureter is densely innervated by capsaicin-sensitive afferents (Sann et al., 1992) and capsaicin, nerve stimulation, and SP mediate plasma extravasation in ureter (Saria et al., 1983). In autoradiographic studies, binding sites for BHSP occur over small blood vessels, predominantly venules, as well as some binding to the lower epithelium (Burcher and Watkins, unpublished data; Sann et al., 1992).

2.5.3. Urinary Bladder

Tachykinins are potent in contracting the detrusor muscle of all mammalian species. Tachykinin administration and/or antidromic nerve stimulation also result in plasma extravasation. In most cases, including human, contraction appears to be mediated by NK_2 receptors. In guinea pig and cat detrusor, NK_1 receptors predominate, whereas the rat possesses both NK_1 and NK_2 receptors (Burcher and Buck, 1986).

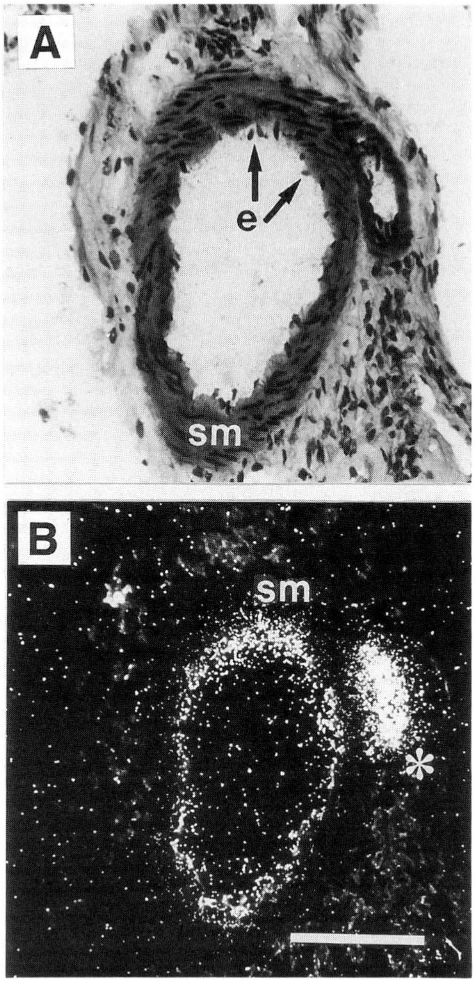

Fig. 4. **(A)** lightfield, and **(B)** corresponding darkfield high power photomicrographs of section of dog kidney incubated in 100 p*M* BHSP and subsequently apposed to photographic emulsion-coated coverslips for 14 d, developed, and stained with methylene blue. The blood vessel displayed is an artery from the cortico-medullary border with labeling over endothelial cells (e), arrowed in (A). The small branch (*) is heavily labeled. Negligible specific binding was seen over smooth muscle (sm) or adventitia (adv). Nonspecific binding, in an adjacent section coincubated with 1 μ*M* SP, was very low and showed no localization of binding (data not shown). Bar, 100 μm.

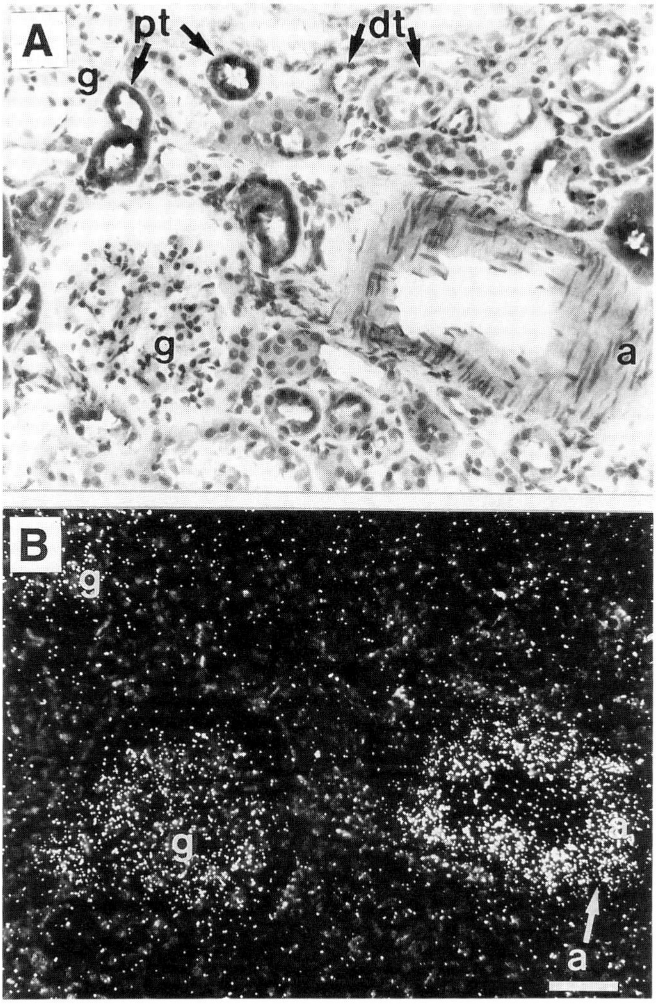

Fig. 5. **(A)** lightfield, and **(B)** corresponding darkfield photomicrographs of section of dog kidney incubated in 100 p*M* BHSP and subsequently apposed to photographic emulsion-coated coverslips for 14 d, developed, and stained with pyronin Y. The region illustrated is from the cortex and demonstrates labeling of an arcuate artery (a), with weaker labeling of two glomeruli (g). No BHSP binding was seen associated with distal convoluted tubules (dt) or with proximal convoluted tubules (pt), here (A) stained for alkaline phosphatase (naphthol AS-MX phosphate technique). Nonspecific binding, in an adjacent section coincubated with 1 μ*M* SP, was very low and showed no localization of binding (data not shown). Bar, 100 μm.

We have investigated binding of BHSP in guinea pig bladder using both membrane homogenates and slide-mounted sections (Burcher and Buck, 1986). Binding of BHSP was characterized to high-affinity NK_1 sites, with dense BHSP specific binding visualized over smooth muscle in dome, body (Fig. 6C), and trigone. In addition, very dense BHSP binding was localized to postcapillary venules in the submucosa (Fig. 6B), with weaker specific binding over submucosal arterioles (Fig. 7). In spite of some functional evidence for NK_2 receptors in guinea pig bladder (Mizrahi et al., 1985), only weak specific binding was seen using BHNKA, and this was characterized to NK_1 sites (Burcher and Buck, 1986).

In rat urinary bladder, we have observed labeling of detrusor smooth muscle with BHSP, BHSarSP, INKA, BHNKA and ^{125}I-[Lys^5,Tyr(I$_2$)^7,MeLeu^9,Nle^{10}]-NKA(4–10) (Burcher and Buck, 1986; Rau and Burcher, 1992). As in the guinea pig, there was dense labeling of NK_1 receptors over postcapillary venules, with weaker labeling of arterioles. Others have reported similar observations in the rat (Nimmo et al., 1988). These observations are consistent with the actions of SP as a spasmogen of urinary smooth muscle and mediator of plasma extravasation.

Two reports exist of labeling of the epithelium of rat bladder by NK_3 receptor recognizing radioligands. Burcher and Buck (1986) visualized BHNKB specific binding on epithelium, and more recently Morrison et al. (1989) made a similar finding using BHELE. Since both radioligands have problems of unselectivity and (for BHNKB) high nonspecific binding, such reports await confirmation using highly selective radioligands.

In dog urinary bladder, tachykinins contract the isolated detrusor muscle (Mizrahi et al., 1985), and studies with selective agonists and antagonists reveal interaction with NK_2 receptors (Burcher and Mussap, 1991; Mussap, 1992). In this species, INKA densely labels the detrusor muscle (Fig. 8), whereas specific binding of BHSarSP was confined to the endothelium of small arteries (Mussap, 1992). In contrast to the absence of INKA binding to any blood vessels in rat and guinea pig (Rau and Burcher, 1992; Burcher, unpublished data), INKA displayed dense specific binding over small arteries (Fig. 8)

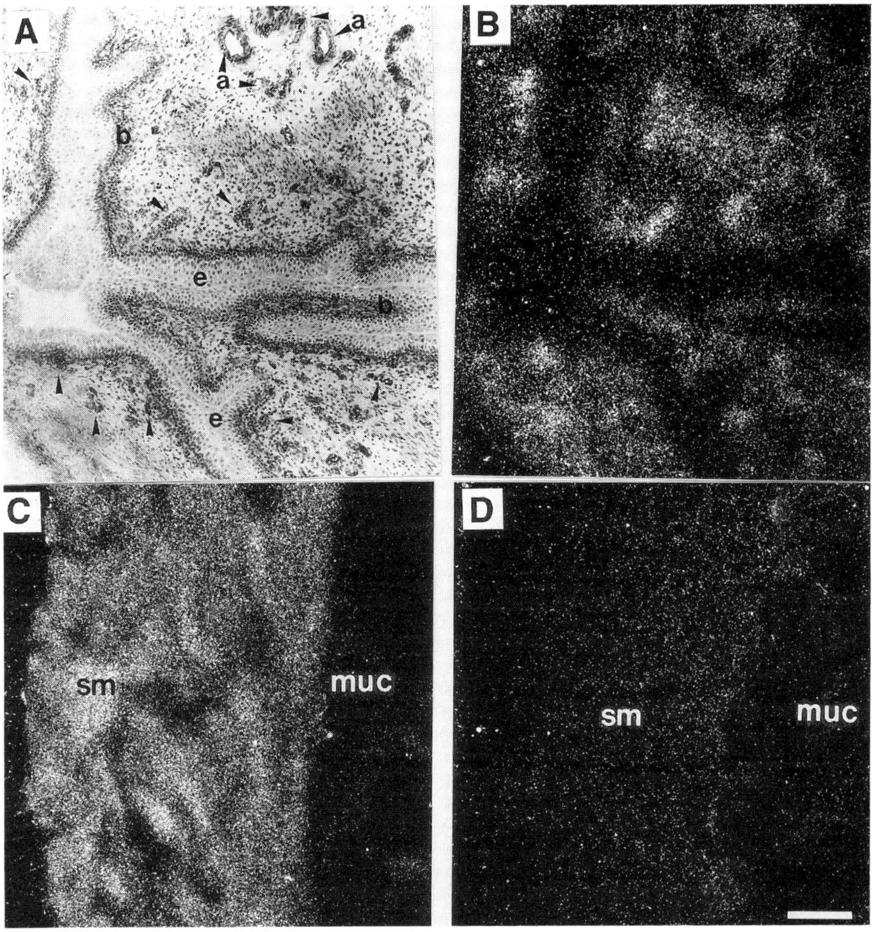

Fig. 6. Light and darkfield photomicrographs of sections of guinea pig bladder, labeled with 60 p*M* BHSP and stained with pyronin Y. **(A)** lightfield, and **(B)** corresponding darkfield photomicrographs of section showing the mucosal region of dome; **(C,D)** darkfield photomicrographs of adjacent sections displaying the muscularis of the corpus from another animal. In B, specific binding occurs over postcapillary venules (arrowed) underneath the basal epithelium (b), identified in A as a line of dark staining cells under the lighter-staining intermediate and surface transitional epithelium (e), which has negligible specific binding. Weak specific binding is seen over submucosal arterioles (a) and to the lamina propria under the basal epithelium. In C, all smooth muscle layers exhibit dense specific binding with low/negligible binding over other structures. The section shown in D (adjacent to C) was coincubated with 1 μ*M* SP and illustrates nonspecific binding that is mainly over smooth muscle. Muc, mucosa. Exposure time, 5 d. Bar, 200 μm.

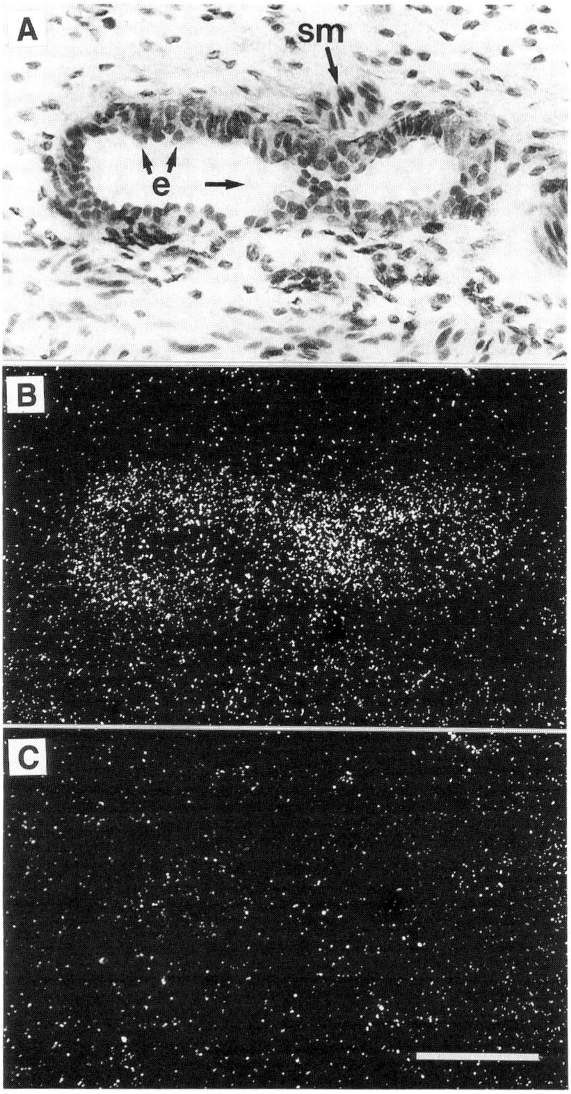

Fig. 7. High resolution photomicrographs of submucosal arterioles in sections of corpus from guinea pig bladder labeled with BHSP and stained with pyronin Y. **(A)** lightfield, **(B)** corresponding darkfield showing grain localization over the vessel wall, with binding concentrated more towards the lumen and endothelial (e) cells (round nuclei) than over the vascular smooth muscle (sm) cells (elongated nuclei). **(C)** darkfield micrograph of arterioles in adjacent sections coincubated with 1 μ*M* SP to illustrate nonspecific binding. Exposure time, 5 d. Bar, 100 μm.

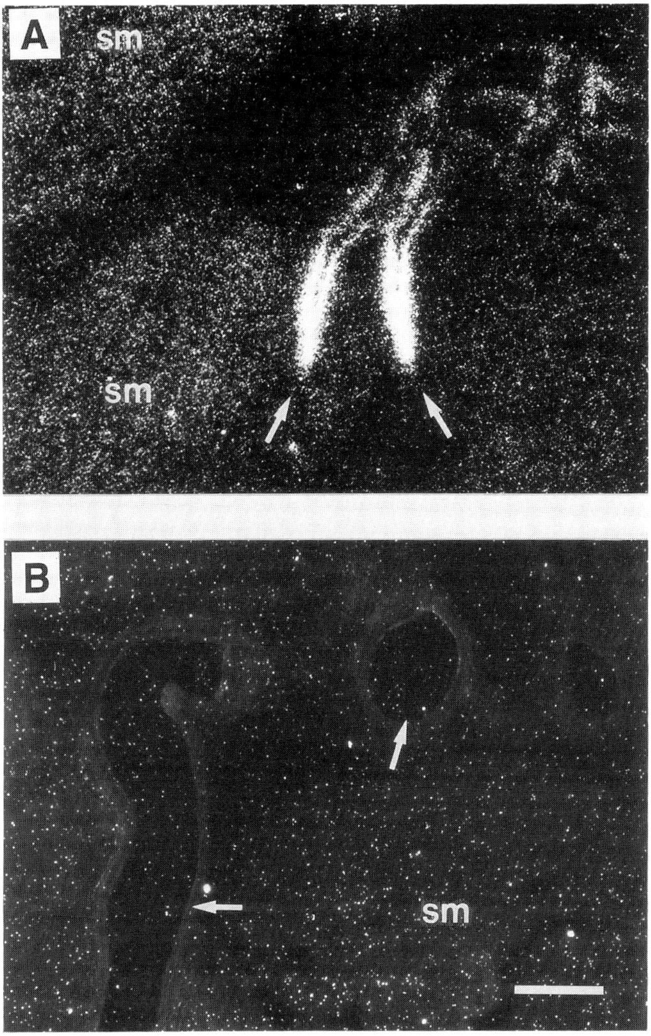

Fig. 8. Darkfield photomicrographs of sections of dog urinary bladder demonstrating total **(A)** and nonspecific binding **(B)**. Adjacent sections were incubated with 100 p*M* INKA in the absence (A) and presence (B) of 1 μ*M* NKA, and then dipped in photographic emulsion, developed after 18 d, and stained with pyronin Y. Silver grain localization in (A) occurs over smooth muscle (sm) and over arteries (arrowed) running between muscle bundles. Nonspecific binding (B) was very low and demonstrated no localization of binding. Other sections labeled with BHSarSP showed minimal specific binding over smooth muscle, but dense specific binding over endothelium of blood vessels (data not shown). Bar, 200 μm.

running between the muscle bundles in dog bladder (Burcher and Mussap, 1991; Mussap, 1992). This appears to be an unusual incidence of NK_2 receptors associated with blood vessels and raises the question of whether NKA is a vasodilator agent acting via NK_2 receptors.

Functional studies have shown that tachykinins contract the isolated human detrusor via NK_2 receptors. Using ^{125}I-[Lys^5,Tyr (I$_2$)^7,MeLeu^9,Nle^{10}]-NKA(4–10), we have found autoradiographic NK_2 binding sites on human detrusor smooth muscle, with weak specific binding over blood vessels (Burcher et al., 1993). Characterization in membranes showed unequivocal binding to high-affinity NK_2 sites.

After capsaicin pretreatment, the micturition reflex is impaired in rats (Holzer-Petsche et al., 1984), and stress incontinence in humans is alleviated (Maggi et al., 1989). Interest in tachykinins as mediators of the afferent component of the micturition reflex is therefore of considerable clinical importance (*see* Maggi, 1991).

2.6. Genital Tract

2.6.1. Vas deferens

In rat and guinea pig vas deferens, SP augments the amplitude of electrical field stimulation-induced twitches, and at high doses an increase in tone is seen (Von Euler and Hedqvist, 1974). Binding sites for BHSP were observed over the outer longitudinal muscle, but were absent over blood vessels (Mussap et al., 1989). In one of the few studies to demonstrate convincing receptor changes after capsaicin, NK_1 receptors were lost from the outer longitudinal layer and substantially increased in the inconspicuous inner longitudinal layer. These data suggest the presence of NK_1 receptors on sensory nerve endings in the outer vas, with possible upregulation of NK_1 receptors in the inner layer (Mussap et al., 1989).

2.6.2. Uterus

Tachykinins contract the rat isolated uterus via NK_2 receptors (Pennefather et al., 1993). Capsaicin-sensitive nerves are found supplying the endometrium, myometrium, and blood vessels (Papka et al., 1985). BHSP binding sites were expressed by blood vessels running through the myometrium and also by small vessels in uterine horns and vagina, and by endometrial glands (Gunasena et al., 1989; Burcher et al., 1990). No binding of BHSP is seen over uterine smooth muscle in normal or ovariectomized rats (Gunasena et al., 1989; Burcher et al., 1990). However, in estradiol-treated, ovariectomized rats, intense labeling of the myometrium by BHSP is present, and the moderate labeling seen with INKA in untreated ovariectomized rats is upregulated threefold (Gunasena et al., 1989). Functional responses to NKA are reduced in uterine horns of estradiol-treated animals, but the explanation here probably resides in the action of estradiol in enhancing the activity of the enzyme, endopeptidase 24.11 (Fisher and Pennefather, 1992).

In human fallopian tube, BHSP binding was absent over smooth muscle, but was present over blood vessels, mainly the endothelium of arterioles and capillaries (Nimmo et al., 1989).

2.7. Blood Vessels

Tachykinins are well known for their ability to cause vasodilatation in low doses. This property appears to be predominantly mediated via NK_1 receptors on the endothelium of peripheral arterial blood vessels (Furchgott, 1984), although some differences in sensitivity between species and locations may exist (e.g., dog > cat) (Burcher et al., 1977). The dog possesses a number of large blood vessels and vascular beds that are very sensitive to endothelium-dependent relaxation by SP (Burcher et al., 1977; Angus et al., 1986), and autoradiographic BHSP binding sites are prominent on the endothelium of a number of canine arteries, but not veins (Table 2; Stephenson et al., 1986, 1987a,b). BHSP sites also occur on the endothelium of other large vessels such as pig coronary artery. The binding is clearly associated with endothelial cells since denuding the vessel of endothelium by mechanical rubbing or distilled water perfusion decreases BHSP binding (Stephenson, 1988).

The function of such endothelial receptors is unclear. There is no reported innervation of endothelial cells by SP-containing nerves in these large vessels, although such nerves exist in the adventitia

Table 2
Localization of Specific BHSP Binding to Blood Vessels

Species	Vessel	Endothelium	Muscle	Adventitia
Dog	Carotid artery	+++	0	++
	Splenic artery	+	0	+
	Splenic vein	0	0	0
	Renal artery	++	0	++
	Renal vein	0	0	0
Rabbit	Pulmonary artery	0	0	+++
	Aorta	0	+	0
Rat	Aorta	0	+	0
	Renal artery	0	0	++
Cow	Aorta	0	+	
	Coronary artery	0	0	0
Pig	Coronary artery	+	0	0
Cat	Pulmonary artery	0	0	++

0 = no specific binding; + = low specific binding; ++ = moderate specific binding; +++ = dense specific binding.

Data are from Stephenson et al. (1986, 1987a, 1987b) and Stephenson (1988).

(Furness et al., 1982). Possibly, SP (or some more stable yet still active form of SP) circulating in blood is the "endogenous ligand" (*see* Pernow, 1983); other possibilities include SP reportedly present in some endothelial cells (*see* Burnstock, 1987).

The rabbit pulmonary artery responds to tachykinins by an endothelium-dependent relaxation, as well as direct smooth muscle contraction mediated by NK_2 receptors (Regoli et al., 1988). In our autoradiographic studies, this artery (Fig. 9) shows dense binding of BHSP to a single cell layer on the outer adventitia of the rabbit pulmonary artery, but negligible specific binding to endothelium or smooth muscle (Stephenson, 1988). The dense adventitial binding is thought to be associated with mesothelial cells from a continuation of the epicardium. Other examples of adventitial binding of BHSP include the dog carotid, renal, and splenic arteries, cat pulmonary

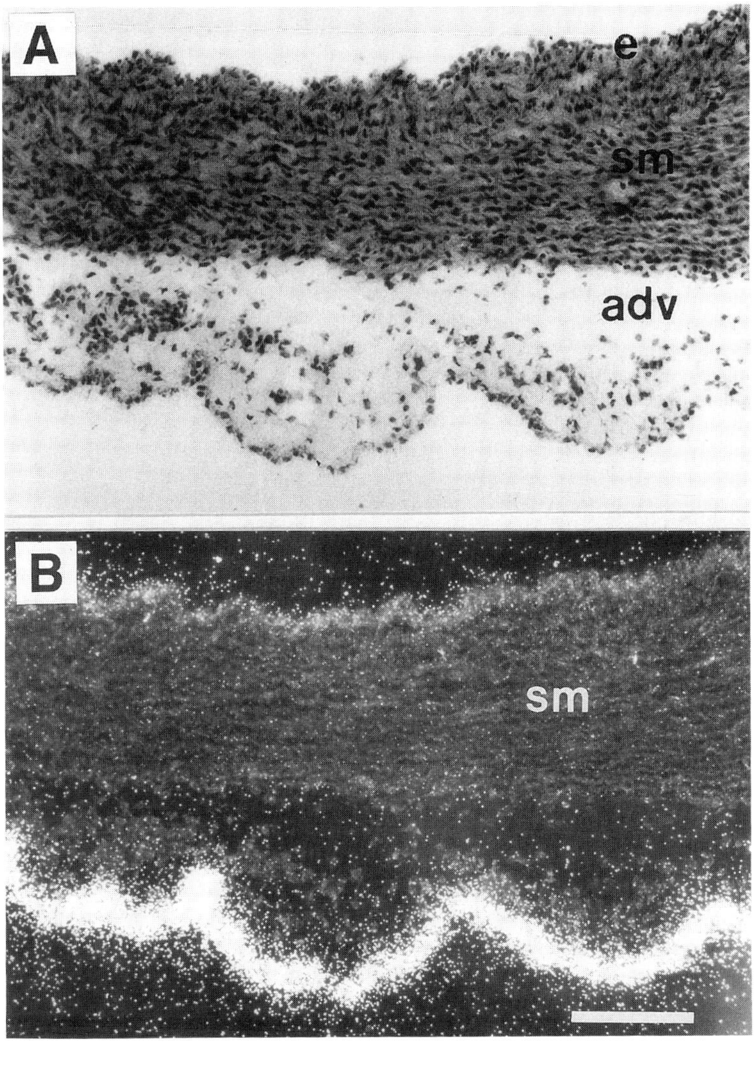

Fig. 9. (**A**) lightfield, and (**B**) corresponding darkfield high power photomicrographs of section of rabbit pulmonary artery incubated in 100 p*M* BHSP and subsequently apposed to photographic emulsion-coated coverslips for 14 d, developed, and stained with methylene blue. Silver grains are densely associated with the outer layer of the adventitia (adv). Specific binding over smooth muscle (sm) was negligible, and only weak binding was seen over the endothelium (e). Nonspecific binding, in an adjacent section coincubated with 1 μ*M* SP, was very low and showed no localization of binding (data not shown). Bar, 200 μm.

artery, and rat renal artery (Table 2). SP-immunoreactive nerves are found in the adventitia (Furness et al., 1982; Ferguson and Bell, 1985), and it is possible that these adventitial sites respond to neurally released SP. However, the functions of such sites and whether they represent functional SP receptors is not clear.

All tachykinins cause contraction of endothelium-denuded vascular smooth muscle, and the presence of all three receptors has been reported; for example, NK_1 receptors on rabbit jugular vein (Nantel et al., 1990), NK_2 receptors on rabbit pulmonary artery (Regoli et al., 1988), and NK_3 receptors on rat portal vein (Mastrangelo et al., 1986). However, we found no localization of INKA binding to any region of the rabbit pulmonary artery (Xiao et al., 1992). Other examples of "functional response/binding site mismatch" include the absence of BHSP binding sites on endothelium of rat and rabbit aorta, and bovine aorta and coronary artery, even though endothelium-dependent relaxation to SP is well documented in these vessels (Furchgott, 1984; *see* Regoli et al., 1988). It is difficult to explain the discrepancies between functional studies and the autoradiographic studies that fail to show appropriate binding sites over the expected histological structures.

Blood vessels are innervated by capsaicin-sensitive, SP-immunoreactive nerves, but veins (in guinea pigs) appear to lack such innervation (Furness et al., 1982). It is evident that there is a heterogeneous distribution of BHSP binding between arteries and veins in different arteries from the same species and in the same vessels from different species.

In addition to their actions as vasodilators and smooth muscle spasmogens, tachykinins are potent in causing plasma extravasation (Lembeck and Holzer, 1979), an action mediated via NK_1 receptors as conclusively shown recently using the selective NK_1 antagonist, CP96345 (Lembeck et al., 1992). Several tissues, including bladder and trachea, possess SP binding sites associated with postcapillary venules, and details can be found in the appropriate sections of this chapter.

2.8. *Other Tissues*

2.8.1. Skin

Tachykinins, particularly SP, are potent in causing plasma extravasation and cutaneous vasodilatation (Lembeck and Holzer, 1979), and these findings are supported by autoradiographic evidence describing NK_1 binding sites over vascular structures in rat and human skin (O'Flynn et al., 1989; Deguchi et al., 1989). Dense binding of BHSP occurs over postcapillary venules in the dermis, with moderate specific binding over arterioles (O'Flynn et al., 1989). In both rat and human, dense BHSP binding is found over dermal papillae and sweat glands, but is absent over nonvascular structures such as Pacinian corpuscles (O'Flynn et al., 1989; Deguchi et al., 1989). The identity of the cells in dermal papillae bearing binding sites is not clear—they may represent capillaries or nociceptive nerve endings. However, the study by O'Flynn et al. (1989) showed that BHSP binding over dermal papillae remains in rats treated neonatally with capsaicin. Binding was noticeably absent over mast cells, supporting other evidence that the mechanism of histamine release by SP is via non-NK_1 receptor mechanisms (*see* Mussap et al., 1993). There appears to be a good correlation between the immunohistochemical localization of SP, distribution of NK_1 receptors, and the known actions of SP in skin.

2.8.2. Joints

Joints are innervated by nociceptive, tachykinin-containing afferents present as free fibers and associated with blood vessels (Gronblad et al., 1988), and SP-LI is present in human synovial fluid (Marshall et al., 1990). High-affinity NK_1 binding sites are found over human synovial endothelial cells in patients with rheumatoid arthritis (Walsh et al., 1992). Taken together with other evidence that SP is involved in experimental arthritis (*see* Walsh et al., 1992), these findings implicate SP as an inflammatory mediator in human synovial tissue.

2.8.3. Eye

Substance P-like immunoreactivity is found in the eye of several species, mainly associated with iris smooth muscle and ciliary body vasculature (*see* Denis et al., 1991; Hall, this volume). Substance P contracts the iris sphincter muscle and has been implicated in neurogenic inflammation. In the eye of rat and rabbit, high-affinity binding sites for BHSP have been localized to the iris sphincter and posterior segment of the retina and the inner plexiform and ganglion cell layers, with weaker specific binding over choroid and scleral blood vessels (Mantyh et al., 1989; Denis et al., 1991). High-affinity binding sites for BHSP have also been characterized in homogenates of bovine retina (Osborne, 1984). The role of SP in the visual system is not well understood, but there is evidence that SP may be involved in pathophysiological phenomena such as ocular inflammation (*see* Hall, this volume).

3. Conclusions

Autoradiographic binding sites for radiolabeled tachykinins have now been described in a range of peripheral tissues. NK_1 sites are particularly widely distributed, being found in association with gastrointestinal, urinary, and respiratory smooth muscle; with blood vessels in some, but not all locations and/or species; with secretory glands; and with some, but not all, peripheral ganglia. NK_2 sites are found largely, but not exclusively on smooth muscle, particularly respiratory, gastrointestinal, and urinary, but also on uterine and vascular smooth muscle of some species. In contrast, very few reports of NK_3 binding sites exist in peripheral tissues.

In several cases, the presence of autoradiographic binding sites correlates reasonably well with the actions and functions of the corresponding peptide at that site and in many cases also with the presence of tachykinin-like immunoreactivity. In such cases where functional data and characterization of the binding supports the presence of the appropriate site, it is valid to propose that the autoradiographic sites represent locations of true receptors. Thus, autoradiographic studies carried out with appropriate selective radioligands can be not only a useful adjunct to functional studies, but can also provide a focus for further research. The concept of peptide-receptor "mismatch," discussed by Kuhar et al. (1986) in reference to the central nervous system, is much less obvious with regard to tachykinin receptors in the periphery.

One of the exceptions to the peptide-receptor "match" in the periphery is the presence of peripheral NK_3 receptors, particularly in the gastrointestinal tract. Although these receptors have been localized both functionally and autoradiographically to guinea pig myenteric ganglia, they appear to lack the appropriate "endogenous ligand," NKB, whose absence in peripheral tissues has been reported by a number of groups (e.g., Too et al., 1989). Since endogenous tachykinins are not receptor exclusive, the "endogenous ligand" for these NK_3 receptors could be SP or NKA. Possibly, low concentrations of another as yet unreported tachykinin or functional metabolite may fulfill an agonist role in vivo.

Autoradiographic and binding studies can reveal the presence of large numbers of binding sites. Does a high density of binding sites necessarily correlate with a high sensitivity or a large functional response to the peptide? In our hands, this is not always so; for example, the longitudinal muscle of guinea pig vas deferens is apparently enriched with NK_1 receptors (Mussap et al., 1989), but concentrations of SP in the micromolar range are required to produce modest increases in tension. The presence of autoradiographic sites over smooth muscle should not invariably lead to the conclusion that these represent receptors subserving smooth muscle contraction. Such binding sites may not necessarily be coupled to intracellular mechanisms or, in reality, may be associated with nerve fibers, small blood vessels, or other cells that are histologically indistinct.

Most autoradiographic studies have been carried out using tissue from laboratory animals. Human studies are sparse, but have in some cases yielded exciting data, especially in regard to changes in receptor localization and density in disease (*see* Mantyh et al., this volume. Despite occasional technical problems (tissue fragility, freezing artifacts, less than optimal resolution), in vitro receptor autoradiography remains one of the most valuable tools in identifying receptor binding site locations in heterogeneous tissues and is a useful method for delineating sites which would otherwise be difficult to resolve by conventional homogenate binding techniques.

Acknowledgments

We gratefully acknowledge the National Health and Medical Research Council of Australia, the National Heart Foundation of Australia, the Clive and Vera Ramaciotti Foundations, and the Asthma Foundations of Victoria and New South Wales for support of this work.

References

Angus, J. A., Cocks, T. M., and Satoh, K. (1986) α_2-Adrenoceptors and endothelium-dependent relaxation in canine large arteries. *Br. J. Pharmacol.* **88,** 767–777.

Badgery-Parker, T., Zeng, X.-P., Lavielle, S., and Burcher, E. (1993) A novel, selective radioligand, [^{125}I]-[Lys^5,Tyr(I$_2$)^7,MeLeu^9,Nle^{10}]-NKA(4–10), for the tachykinin N-2 receptor. *Regul. Pept.* **46,** 455–457.

Barnes P. J., Baraniuk, J. N., and Belvisi, M. G. (1991) Neuropeptides in the respiratory tract. Part 1. *Am. Rev. Respir. Dis.* **144,** 1187–1198.

Bartho, L. and Holzer, P. (1985) Search for a physiological role of substance P in gastrointestinal motility. *Neuroscience* **16,** 1–32.

Black, J. L., Diment, L. M., Alouan, L. A., Johnson, P. R. A., Armour, C. L., Badgery-Parker, T., and Burcher, E. (1992) Tachykinin receptors in rabbit airways-characterisation using functional, autoradiographic and binding studies. *Br. J. Pharmacol.* **107,** 429–436.

Black, J., Diment, L, Armour, C. L., Alouan, L., and Johnson, P., (1990) Distribution of substance P receptors in rabbit airways, functional and autoradiographic studies. *J. Pharmacol. Exp. Ther.* **253,** 381–386.

Buck, S. H. and Burcher, E. (1985) The rat submaxillary gland contains predominantly P-type tachykinin binding sites. *Peptides* **6,** 1079–1084.

Buck S. H., Helke C. J., Burcher E., Shults C. W., and O'Donohue T. L. (1986) Pharmacologic characterization and autoradiographic distribution of binding sites for iodinated tachykinins in rat central nervous system. *Peptides* **7,** 1109–1120.

Buck, S. H. and Shatzer S. A. (1988) Agonist and antagonist binding to tachykinin peptide NK-2 receptors. *Life Sci.* **42,** 2701–2708.

Burcher, E., Atterhög, J. H., Pernow, B., and Rosell, S. (1977) Cardiovascular effects of Substance P; regional differences in blood flow and effect on the heart, in *Nobel Symposium on Substance P* (Von Euler, U. S. and Pernow, B., eds.), Raven, New York, pp. 261–268.

Burcher, E., Badgery-Parker, T., Zeng, X.-P., and Lavielle, S. (1993) Characterization of a novel, selective radioligand, [^{125}I][Lys^5,Tyr(I$_2$)^7,MeLeu^9,Nle^{10}]-NKA(4–10), for the tachykinin NK2 receptor in rat fundus. *Eur. J. Pharmacol.* **233,** 201–207.
Burcher, E. and Bornstein, J. C. (1988) Localization of substance P binding sites in submucous plexus of guinea pig ileum, using whole mount autoradiography. *Synapse* **2,** 232–239.
Burcher, E. and Buck, S. H. (1986) Multiple tachykinin binding sites in hamster, rat and guinea pig urinary bladder. *Eur. J. Pharmacol.* **128,** 165–178.
Burcher, E., Buck, S. H., Lovenberg, W., and O'Donohue, T. L. (1986) Characterization and autoradiographic localization of multiple tachykinin binding sites in gastrointestinal tract and bladder. *J. Pharmacol. Exp. Ther.* **236,** 819–831.
Burcher, E. and Mussap, C. J. (1991) Neuropeptide γ-prefering subtypes of NK2 receptors in rat fundus and dog urinary bladder. *Soc. Neurosci. Abstr.* **17,** 805.
Burcher, E., Mussap, C. J., Geraghty, D. P., McClure-Sharp, J. M., and Watkins, D. J. (1991) Concepts in characterization of tachykinin receptors. *Ann. NY Acad. Sci.* **632,** 123–136.
Burcher, E., Shults, C. W., Buck, S. H., Chase, T. N., and O'Donohue, T. L. (1984) Autoradiographic distribution of substance K binding sites in rat gastrointestinal tract: a comparison with substance *P. Eur. J. Pharmacol.* **102,** 561,562.
Burcher, E., Taylor, C., and O'Flynn, N. M. (1990) Substance P binding sites associated with blood vessels in rat female reproductive tract. *Clin. Exp. Pharmacol. Physiol.* **Suppl. 17,** 90.
Burcher, E., Watkins, D. J., and O'Flynn, N. (1989) Autoradiographic localization of both substance P and neurokinin A binding sites in guinea pig lung. *Pulmonary Pharmacol.* **1,** 201–203.
Burcher, E., Zeng, X.-P. and Moore, K. H. (1993) Characterization and autoradiographic localization of tachykinin N-2 receptors in human urinary bladder. *Neuropeptides* **24,** 237.
Burnstock, G. (1987) Dual control of blood pressure by purines, in *Pharmacology* (Rand, M. J. and Raper, C., eds), Exerpta Medica Int. Cong. series 750, Elsevier, Amsterdam, pp. 245–253.
Carstairs, J. R. and Barnes, P. J. (1986) Autoradiographic mapping of substance P receptors in lung. *Eur. J. Pharmacol.* **127,** 295,296.
Chassaing, G., Lavielle, S., Brunissen, A., Carruette, A., Garret, C., Petitet, F., Saffroy, M., Beaujouan, J.-C., Torrens, Y., and Glowinski, J. (1992) [Pro^9]SP and [pGlu^6,Pro^9]SP(6–11) interact with two different receptors in the guinea pig ileum as demonstrated with new SP antagonists. *Neuropeptides* **23,** 73–79.
Deguchi, M., Niwa, M., Shigematsu, K., Fujii, T., Namba, K., and Ozaki, M. (1989) Specific [^{125}I]Bolton-Hunter substance P binding sites in human and rat skin. *Neurosci. Lett.* **99,** 287–292.
Denis, P., Fardin, V., Nordmann, J. P., Elena, P. P., Laroche, L., Saraux, H., and Rostene, W. (1991) Localization and characterization of substance P binding sites in rat and rabbit eyes. *Invest. Opthalmol. Vis. Sci.* **32,** 1894–1902.
Ferguson, M. and Bell, C. (1985) Substance P-immunoreactive nerves in the rat kidney. *Neurosci. Lett.* **60,** 183–188.

Fischer, A., Kummer, W., Courand, J. Y., Adler, D., Branscheid, D., and Heym, C. (1992) Immunohistochemical localization of receptors for vasoactive intestinal polypeptide and substance P in human trachea. *Lab. Inv.* **67,** 387–393.

Fisher, L. and Pennefather, J. N. (1992) The inactivation of neurokinin A on the rat uterus is modified by oestrogen treatment. *Regul. Pept.* **Suppl. 1,** S55.

Foster, A. C. and Tridgett, R. (1988) Comparision of the binding of radiolabelled neurokinin A and eledoisin in rat cortex synaptic membranes. *Br. J. Pharmacol.* **94,** 602–608.

Frossard, N. and Advenier, C. (1991) Tachykinin receptors and the airways. *Life Sci.* **49,** 1941–1953.

Furchgott, R. F. (1984) The role of endothelium in the responses of vascular smooth muscle to drugs. *Annu. Rev. Pharmacol. Toxicol.* **24,** 175–197.

Furness, J. B., Papka, R. E., Della, N. G., Costa, M., and Eskay, R. L. (1982) Substance P-like imnunoreactivity in nerves associated with the vascular system of guinea pigs. *Neuroscience* **7,** 447–459.

Gates, T. S., Zimmerman, R. P., Mantyh, C. R., Maggio, J. E., Walton, M. L., Passaro, E. P., Jr., and Mantyh, P. W. (1988) Substance P and substance K receptor binding sites in the human gastrointestinal tract: localization by autoradiography. *Peptides* **9,** 1207–1219.

Geraghty, D. P., Mussap, C. J., and Burcher, E. (1992) Radioiodinated substance P, neurokinin A and eledoisin bind predominantly to NK1 receptors in guinea pig lung. *Mol. Pharmacol.* **41,** 147–153.

Goldie, R. G. (1990) Receptors in asthmatic airways. *Am. Rev. Resp. Dis.* **141,** S151–156.

Gronblad, M., Konntinen, Y. T., Korkala, O., Liesi, P., Hukkanen, M., and Polak, J. (1988) Neuropeptides in the synovium of patients with rheumatoid arthritis and osteoarthritis. *J. Rheumatol.* **15,** 1807–1810.

Gullner, H. G., Campbell, W. B., and Pettinger, W. A. (1979) Effects of substance P on renin release and renal function in anaesthetized dogs. *Life Sci.* **24,** 237–246.

Gunasena, K., Morrison, J. F. B., Nimmo, A. J., and Whitaker, E.M. (1989) The effects of oestradiol and progesterone on neurokinin receptors in rat uterus. *J. Physiol.* **418,** 75P (abstract).

Herkenham, M. and Pert, C.B. (1982) Light microscopic localization of brain opiate receptors: a general autoradiographic method which preserves tissue quality. *J. Neurosci.* **2,** 1129–1149.

Hokfelt, T., Johansson, O., Kellerth, J.-O., Ljungdahl, A., Nilsson, G., Nygards, A., and Pernow, B. (1977) Immunohistochemical distribution of substance P, in *Nobel Symposium on Substance P* (Von Euler, U. S. and Pernow, B., eds.), Raven, New York, pp. 117–145.

Holzer, P. (1988) Local effector functions of capsaicin-sensitive sensory nerve endings: Involvement of tachykinins, calcitonin gene-relted peptide and other neuropeptides. *Neuroscience* **24,** 739–768.

Holzer-Petsche U. and Lembeck F. (1984) Systemic capsaicin treatment impairs the micturition reflex in the rat. *Br. J. Pharmacol.* **83,** 935–941.

Hoover, D. B. and Hancock, J. C. (1987) Autoradiographic localization of substance P binding sites in guinea pig airways. *J. Auton. Nerv. Syst.* **19,** 171–174.

James, S. and Burnstock, G. (1988) Autoradiographic localization of binding sites for ^{125}I-substance P on neurones from cultured rat superior cervical ganglion. *Brain Res.* **458,** 205–211.

Korman, L. Y., Sayadi, H., Bass, B., Moody, T. W., and Harmon, J. W. (1989) Distribution of vasoactive intestinal polypeptide and substance P receptors in human colon and small intestine. *Dig. Dis. Sci.* **34,** 1100–1108.

Kuhar, M. J. (1985) Receptor localization with the microscope, in *Neurotransmitter Receptor Binding*, 2nd ed. (Yamamura, H. I., Enna, S. J., and Kuhar, M. J., eds.), Raven, New York, pp. 153–176.

Kuhar, M. J., De Souza, E. B., and Unnerstall, J. R. (1986) Neurotransmitter receptor mapping by autoradiography and other methods. *Annu. Rev. Neurosci.* **9,** 27–59.

Lee C.-M., Javitch J. A., and Snyder S. H. (1983) ^{3}H-substance P binding to salivary gland membranes. Regulation by guanyl nucleotides and divalent cations. *Mol. Pharmacol.* **23,** 563–569.

Lembeck, F., Donnerer, J., Tsuchiya, M., and Nagahisa, A. (1992) The non-peptide tachykinin antagonist, CP-96,345, is a potent inhibitor of neurogenic inflammation. *Br. J. Pharmacol.* **105,** 527–530.

Lembeck, F. and Holzer, P. (1979) Substance P as neurogenic mediator of antidromic vasodilatation and neurogenic plasma extravasation. *Naunyn Schmiedeberg's Arch. Pharmacol.* **310,** 175–183.

Lew, R., Geraghty, D. P., Drapeau, G., Regoli, D., and Burcher, E. (1990) Binding characteristics of [[^{125}I]-Bolton-Hunter Sar^9,Met(0$_2$)^{11}]-substance P, a new selective radioligand for the NK1 receptor. *Eur. J. Pharmacol.* **184,** 97–108.

Liang, T. and Cascieri, M. A. (1981) Substance P receptor on parotid cell membranes. *J. Neurosci.* **1,** 1133–1141.

Lundberg, J. M., Hokfelt, T., Martling, C. R., Saria, A., and Cuello, C. (1984) Substance P-immunoreactive sensory nerves in the lower respiratory tract of various mammals including man. *Cell Tissue Res.* **236,** 251–261.

Maggi, C. A. (1991) The role of peptides in the regulation of the micturition reflex: an update. *Gen. Pharmacol.* **22,** 1–24.

Maggi, C. A., Barbanti, G., Santicioli, P., Beneforti, P., Misuri, D., Meli, A., and Turino, D. (1989) Cystometric evidence that capsaicin-sensitive nerves modulate the afferent branch of micturition reflex in humans. *J. Urol.* **192,** 150–154.

Mantyh, P. W. (1991) Substance P and the inflammatory and immune response. *Ann. NY Acad. Sci.* **632,** 263–271.

Mantyh, P. W., Catton, M. D., Allen, C. J., Labenski, M. E., Maggio, J. E., and Vigna, S. R. (1992) Receptor binding sites for cholecystokinin, galanin, somatostatin, substance P and vasoactive intestinal polypeptide in sympathetic ganglia. *Neuroscience* **46,** 739–754.

Mantyh, C. R., Gates, T. S., Zimmerman, R. P., Welton, M. L., Passaro, E. P., Jr., Vigna, S. R., Maggio, J. E., Kruger, L., and Mantyh, P. W. (1988a) Receptor binding sites for substance P, but not substance K or neuromedin K, are expressed in high concentrations by arterioles, venules, and lymph nodules in surgical specimens obtained from patients with ulcerative colitis and Crohn disease. *Proc. Natl. Acad. Sci.* **85,** 3235–3239.

Mantyh, P. W., Goedert, M., and Hunt, S. P. (1984) Autoradiographic visualization of receptor binding sites for substance P in the gastrointestinal tract of the guinea pig. *Eur. J. Pharmacol.* **100,** 133,134.

Mantyh, P. W., Mantyh, C. R., Gates, T., Vigna, S. R., and Maggio, J. E. (1988b) Receptor binding sites for substance P and substance K in the canine gastrointestinal tract and their possible role in inflammatory bowel disease. *Neuroscience* **25,** 817–837.

Mantyh, P. W., Gates, T., Mantyh, C. R., and Maggio, J. E. (1989) Autoradiographic localization and characterization of tachykinin receptor binding sites in the rat brain and peripheral tissues. *J. Neurosci.* **9,** 258–279.

Marshall, K. W., Chiu, B., and Inman, R. D. (1990) Substance P and arthritis: analysis of plasma and synovial fluid levels. *Arthritis Rheum.* **33,** 87–90.

Mastrangelo, D., Mathison, R., Huggel, H. G., Dion, S., D'Orleans-Juste, P., Rhaleb, N. E., Drapeau, G., Rovero, P., and Regoli, D. (1986) The rat isolated portal vein: A preparation sensitive to neurokinins, particularly neurokinin *B. Eur. J. Pharmacol.* **134,** 321–326.

Mizrahi, J., Dion, S., D'Orleans-Juste, P., Escher, E., Drapeau, G., and Regoli, D. (1985) Tachykinin receptors in smooth muscles: A study with agonists (substance P, neurokinin A) and antagonists. *Eur. J. Pharmacol.* **118,** 25–36.

Morrison, J. F. B., Nimmo, A. J., and Whitaker, E. M. (1989) The effect of neonatal capsaicin treatment on the distribution of neurokinin binding sites in rat bladder. *J. Physiol.* **423,** 79P (abstract).

Mussap, C. J. (1992) *Characterisation of tachykinin receptors: radioligand binding, functional and autoradiographic studies.* PhD thesis, University of New South Wales, Australia.

Mussap, C. J. and Burcher, E. (1990) [^{125}I]-Bolton-Hunter scyliorhinin II: a novel, selective radioligand for the tachykinin NK3 receptor in rat brain. *Peptides* **11,** 827–836.

Mussap, C. J. and Burcher, E. (1991) Multiple tachykinin receptors in rat submandibular gland. *Ann. NY Acad. Sci.* **632,** 447–452.

Mussap, C. J. and Burcher, E. (1993) Characterization and autoradiographic localization of tachykinin receptors in rat fundus. *J. Pharmacol. Exp. Ther.* **266,** 1043–1053.

Mussap, C.J., Geraghty, D.P., and Burcher, E. (1993) Tachykinin receptors: a radioligand binding perspective. *J. Neurochem.* **60,** 1987–2009.

Mussap, C. J., Lew, R., and Burcher, E. (1989) The autoradiographic distribution of substance P binding sites in guinea pig vas deferens is altered by capsaicin pretreatment. *Eur. J. Pharmacol.* **168,** 337–345.

Nantel, F., Roussi, N., Rhaleb, N.-E., Dion, S., Drapeau, G., and Regoli, D. (1990) The rabbit jugular vein is a contractile NK-1 receptor system. *Eur. J. Pharmacol.* **179,** 457–462.
Nimmo, A. J., Whitaker, E. M., Carstairs, J .R., and Morrison, J. F. B. (1989) The autoradiographic localization of calcitonin gene-related peptide and substance P receptors in human fallopian tube. *Quart. J. Exp. Physiol.* **74,** 955–958.
Nimmo, A. J., Morrison, J. F. B., and Whitaker, E. M. (1988) A comparison of the distribution of substance P and calcitonin gene-related peptide receptors in the rat bladder. *Quart. J. Exp. Physiol.* **73,** 789–792.
Niwa, M., Shigematsu, K., Plunkett, L., and Saavedra, J. (1985) High-affinity substance P binding sites in rat sympathetic ganglia. *Am. J. Physiol.* **249,** H694–H697.
Osborne, N. N. (1984) Substance P in the bovine retina: localization, identification, release, uptake and receptor analysis. *J. Physiol.* **349,** 83–93.
O'Flynn, N. M., Helme, R. D., Watkins, D. J., and Burcher, E. (1989) Autoradiographic localization of substance P binding sites in rat footpad skin. *Neurosci. Lett.* **106,** 43–48.
Payan, D. G. (1989) Neuropeptides and inflammation: The role of substance P. *Annu. Rev. Med.* **40,** 341–352.
Papka, R. E., Cotton, J. P., and Traurig, H. H. (1985) Comparative distribution of neuropeptide tyrosine-, vasoactive intestinal polypeptide-, substance P-immunoreactive, acetylcholinesterase-positive and noradrenergic nerves in the reproductive tract of the female rat. *Cell Tissue Res.* **242,** 475–490.
Pennefather, J. N., Zeng, X.-P., Gould, D., Hall, S., and Burcher, E. (1993) Mammalian tachykinins stimulate rat uterus by activating tachykinin NK-2 receptors. *Peptides* **14,** 169–174.
Pernow, B. (1983) Substance P. *Pharmacol. Rev.* **36,** 85–141.
Petitet, F., Beaujouan, J.-C., Saffroy, M., Torrens, Y., Chassaing, G., Lavielle, S., Besseyre, J., Garret, C., Carruette, A., and Glowinski, J. (1991) Further demonstration that [Pro^9]-substance P is a potent and selective ligand of NK-1 tachykinin receptors. *J. Neurochem.* **56,** 879–889.
Petitet, F., Saffroy, M., Torrens, Y., Lavielle, S., Chassaing, G., Loeuillet, D., Glowinski, J., and Beaujouan, J.-C. (1992) Possible existence of a new tachykinin receptor subtype in the guinea pig ileum. *Peptides* **13,** 383–388.
Popper, P., Mantyh, C. R., Vigna, S. R., Maggio, J. E., and Mantyh, P. W. (1988) The localization of sensory nerve fibers and receptor binding sites for sensory neuropeptides in canine mesenteric lymph nodes. *Peptides* **9,** 257–267.
Quirion, R., Dam, T. V., and Guard, S. (1991) Selective neurokinin receptor radioligands. *Ann. NY Acad. Sci.* **632,** 137–144.
Quirion, R., Shults, C. W., Moody, T. W., Pert, C. B., Chase, T. N., and O'Donohue, T. L. (1983) Autoradiographic distribution of substance P receptors in rat central nervous system. *Nature* **303,** 714–716.
Rau, D. and Burcher, E. (1992) Autoradiographic binding sites for tachykinins and calcitonin gene-related peptide in rat urinary bladder: effect of capsaicin. *Neuropeptides* **22,** 55 (abstract).

Regoli, D., Drapeau, G., Dion, S., and Couture, R. (1988) New selective agonists for neurokinin receptors: pharmacological tools for receptor characterization. *Trends Pharmacol. Sci.* **9,** 290–295.

Rothstein, R. D., Johnson, E., and Ouyang, A. (1989) Substance P: mechanism of action and receptor distribution at the feline ileocecal spincter region. *Am. J. Physiol.* **257,** G447–G453.

Rothstein, R. D., Johnson, E., and Ouyang, A. (1991) Distribution and density of substance P receptors in the feline gastrointestinal tract using autoradiography. *Gastronterology* **100,** 1676–1581.

Sann, H., Rossler, W., Hammer, K., and Pierau, F.-K. (1992) Substance P and calcitonin gene-related peptide in the ureter of chicken and guinea pig: distribution, binding sites and possible functions. *Neuroscience* **49,** 699–713.

Saria, A., Lundberg, J. M., Hua, X., and Lembeck, F. (1983) Capsaicininduced substance P release and sensory control of vascular permeability in the guinea pig ureter. *Neurosci. Lett.* **41,** 167–172.

Shigematsu, K., Saavedra, J. M., and Kurihara, M. (1986) Specific substance P binding sites in rat thymus and spleen: in vitro autoradiographic study. *Regul. Pept.* **16,** 147–156.

Stephenson, J. A. (1988) *Receptor localization in the cardiovascular system.* MSc thesis, University of Melbourne, Australia.

Stephenson, J. A., Burcher, E., and Summers, R. J. (1986) Autoradioraphic demonstration of endothelium-dependent binding of [^{125}I]-Bolton-Hunter substance P to dog carotid artery. *Eur. J. Pharmacol.* **124,** 377–378.

Stephenson, J. A., Summers, R. J., and Burcher, E. (1987a) Endothelial-dependent binding of [^{125}I]-Bolton-Hunter substance P in blood vessels, in *Substance P and Neurokinins* (Henry, J. L., Couture, R., Cuello, A.C., Pelletier, G., Quirion, R., and Regoli, D., eds.), Springer-Verlag, New York, pp. 53–55.

Stephenson, J. A., Summers, R. J., and Burcher, E. (1987b) Autoradiographic localization and characterization of substance P binding in dog kidney. *Eur. J. Pharmacol.* **142,** 391–402.

Strigas, J. and Burcher, E. (1993) Autoradiographic localization of tachykinin binding sites in guinea-pig and human airways, using selective radioligands. *Clin. Exp. Pharmacol. Physiol.* **suppl. 1,** 705 (abstract).

Too, H.-P., Cordova, J. L., and Maggio, J. E. (1989) A novel radioimmunoassay for neuromedin K. I. Absence of neuromedin K-like immunoreactivity in guinea pig ileum and bladder. II. Heterogeneity of tachykinins in guinea pig tissues. *Reg. Pept.* **26,** 93–105.

Vigna, S. R., Mantyh, C. R., Soll, A. H., Maggio, J. E., and Mantyh, P. W. (1989) Substance P receptors on canine chief cells: localization, characterization and function. *J. Neuroscience* **31,** 757–770.

Von Euler, U. and Hedqvist, P. (1974) Effect of substance P on the response of guinea pig vas deferens to transmural nerve stimulation. *Acta Physiol. Scand.* **90,** 651–653.

Walsh, D. A., Mapp, P. I., Wharton, J., Rutherford, R. A. V., Kidd, B. L., Revell, P. A., Blake, D. R., and Polak, J. M. (1992) Localization and characterization of substance P binding to human synovial tissue in rheumatoid arthritis. *Ann. Rheum. Dis.* **51,** 313–317.

Watkins, D. J. and Burcher, E. (1987) Autoradiographic localization of substance (SP) binding sites in guinea pig respiratory tract, in *Substance P and Neurokinins* (Henry, J. L., Couture, R., Cuello, A. C., Pelletier, G., Quirion, R., and Regoli, D., eds.), Springer-Verlag, New York, pp. 87–89.

Wiedermann C. J., Sertil, K., and Pert, C. B. High-affinity substance P binding sites in rat esophagus plexus submucosus. *Am. J. Physiol.* **253,** R167–171.

Xiao, X.-H., Mussap, C. J., and Burcher, E. (1992) Characterization of the tachykinin NK2 receptor subtype in the rabbit pulmonary artery. *Peptides* **13,** 281–285.

Zeng, X.-P., Lavielle, S., and Burcher, E. (1993) Evidence for tachykinin N-2 receptors in guinea pig airways from binding and functional studies, using a new selective radioligand, $[^{125}I]$-$[Lys^5, Tyr(I_2)7, MeLeu^9, Nle^{10}$-NKA(4–10). *Neuropeptides* (in press).

CHAPTER 7

Molecular Biology of Receptors

Structures, Expression, and Regulatory Mechanisms

James E. Krause, Paul Blount, and Bruce S. Sachais

1. Introduction

The tachykinin peptides comprise a family of structurally related peptides that were originally discovered on the basis of their atropine resistant, rapid stimulation of smooth muscle contraction. These peptides have been isolated from vertebrate and invertebrate species and are structurally characterized by the conserved carboxyl-terminal region of Phe-X-Gly-Leu-MetNH_2, where the X residue usually is an aromatic or aliphatic amino acid. There exist at least five mammalian tachykinin peptides including substance P (SP), neurokinin A (NKA), neurokinin B (NKB), neuropeptide K (NPK), and neuropeptide γ (NPγ) that are synthesized, processed, and secreted from tachykinin secreting neurons, as well as certain nonneural tissues and cells. NPK and NPγ are amino terminal extended forms of NKA. Each of the five

The Tachykinin Receptors Ed.: S. H. Buck © 1994 Humana Press Inc., Totowa, NJ

peptides is an agonist at each of the currently characterized tachykinin receptors, although their rank orders of potency differ. Mammalian tachykinin receptors consist of three types based upon current pharmacological, biochemical, and molecular characterization. These receptors have been named neurokinin-1 (NK_1), the receptor with highest affinity for SP; neurokinin-2 (NK_2), the receptor with highest affinity for NKA, NPK, and NPγ; and neurokinin-3 (NK_3), the receptor with highest affinity for NKB. These receptors activate their specific effector systems as a consequence of interacting with and stimulating heterotrimeric guanine nucleotide binding proteins (G-proteins) (*see* Table 1). Prior to the use of molecular biological techniques, the classification of mammalian tachykinin receptors was based predominantly on various pharmacological properties of the mammalian receptors expressed in various tissue contexts. This classification (Lee et al., 1982) was based on the potency of relatively selective agonists and peptidic antagonists and was also based on different patterns of desensitization of receptor responses in the continued presence of agonist. For example, "SP receptors" were shown in membrane preparations to display high-affinity (subnanomolar to nanomolar) for SP, were blocked by relatively weak peptide antagonists, and in tissue preparations displayed desensitization of responses. Consequently, the early classification systems were based on the availability of selective agonists and partially selective, but weak antagonists.

In the past few years, molecular biological studies coupled with expression analysis have confirmed and further clarified the existence of multiple tachykinin receptors. These studies, along with the recent discovery and use of selective, high-affinity, nonpeptidic antagonists for some of the tachykinin receptors, now conclusively document the existence of multiple tachykinin receptor types both within and between species. The extent of receptor diversity established from molecular studies to date essentially confirms that previously suggested based on pharmacological and biochemical studies performed over the past several years; however, many issues have been clarified, and now the understanding of tachykinin receptor diversity rests on a firm molecular basis. There currently exist

Table 1
Molecular Biological Characteristics of G-Protein Coupled Tachykinin Receptors

Receptor	Second messenger	Species cloned	Chromosomal location	Amino acids	Primary reference
NK_1	IP_3-Ca^{++};cAMP[a]	Human	24	407	Takeda, et al., 1991a; Girard et al.,1991; Hopkins et al., 1991; Fong et al., 1992a
	IP_3-Ca^{++};cAMP[a]	Guinea pig		407	Gorbulev et al., 1992
		Rat		407	Yokota et al., 1989; Hershey and Krause, 1990
		Mouse		407	Sundelin et al., 1992
NK_2	IP_3-Ca^{++};cAMP[a]	Human	10	398	Girard et al., 1990; Kris et al., 1991
	IP_3-Ca^{++};cAMP[a]	Cow		384	Masu et al., 1987
	IP_3-Ca^{++};cAMP[a]	Rat		390	Sasai and Nakanishi, 1989
	IP_3-Ca^{++}	Mouse		384	Sundelin et al., 1992
NK_3	human			465	Buell et al., 1992; Takahashi et al., 1992; Huang et al., 1992
	IP_3-Ca^{++};cAMP[a]	Rat		452	Shigemoto et al., 1990

[a]Ligand-stimulated cyclic AMP responses seem to be cell context dependent. *See* text for details.

three mammalian tachykinin receptor types with a single receptor having subnanomolar to nanomolar affinity for SP, for NKA, NPγ, or NPK, and for NKB. This contrasts with the multiplicity of G-protein coupled receptor subtypes displaying high-affinity for 5-hydroxytryptamine, dopamine, or acetylcholine, for example. In this chapter, we shall review the molecular characterization of mammalian tachykinin receptors based on cDNA cloning accompanied with functional expression in heterologous cell lines. The relationship of receptor structure to function will be discussed with respect to agonist and antagonist binding. The novel organization of the mammalian tachykinin receptor genes, compared to other G-protein coupled receptor genes, will be discussed in the context of gene structure, expression, and evolution. Finally, receptor gene expression will be considered in the context of factors regulating transcription and the specific transcriptional *cis* sequences that may interact with *trans*-acting transcriptional regulatory factors; the regulation of mRNA turnover will also be considered. Some of these issues have been addressed in reviews published recently (Helke et al., 1990; Nakanishi, 1991; Guard and Watson, 1991: Krause et al., 1992; Gerard et al., 1993; Krause et al., 1993a).

2. Mechanisms of Action of Tachykinin Peptides

It is well established that tachykinin receptors are activated on binding of ligand to the high-affinity receptor state brought about by G-protein coupling (MacDonald and Boyd, 1989). The classical actions of the tachykinin peptides at these receptors involve the activation of a phosphoinositide specific phospholipase C, with the subsequent breakdown of phosphatidylinositol 4,5-bisphosphate and consequent production of inositol 1,4,5-trisphosphate (IP_3) and diacylglycerol (Berridge, 1993). IP_3 has been shown to mobilize Ca^{2+} from intracellular stores, which then can play a role in the regulation of calcium-calmodulin kinases. Diacylglycerol can activate specific isoforms of protein kinase C. Receptor activation begins by the peptide binding to the G-protein coupled state of the receptor, thereby inducing a conformational change that results in GTP binding to the GDP or unliganded heterotrimeric G-protein complex, with the subsequent dissociation of this complex from the receptor and its dissociation of the α-subunit from its βγ counterpart. Due to an intrinsic GTPase, possibly along with GTPase activating proteins (Bernstein et al., 1992), the activated G_{α} subunit becomes deactivated and is subsequently able to reassociate with free $G_{\beta\gamma}$ subunits. The activated G protein subunit(s), as have been shown in other systems (Wu et al., 1993), can directly regulate ion channel function or can activate specific effector enzymes. The various receptor conformational changes that are believed to occur upon receptor stimulation have not yet been observed with any of the tachykinin receptors or with other similar G-protein coupled receptors, though they have been detected spectroscopically in studies performed with rhodopsin (Franke et al., 1990). During the conformational changes that result in G-protein activation and release, sequences within the third intracellular loop and within the cytoplasmic carboxyl-terminal tail may become exposed to the cytoplasm such that specific residues may be phosphorylated by receptor associated kinases, as well as other protein kinases. After sufficient phosphorylation, the receptor may attain a state of posttranslational modification whereby the receptor may not be able to associate with G-protein(s) and may subsequently associate with an arrestin-like molecule and become sequestered. Both cell transfection *(see below)* and reconstitution studies (MacDonald and Boyd, 1989; Kwatra et al., 1993) are consistent with this mechanism of tachykinin peptide action at the molecularly characterized G-protein coupled tachykinin receptors. The molecular mechanisms that are responsible for loss of receptor responsiveness have not been clearly elucidated at the present time, though it seems reasonable that phosphorylation, uncoupling, and sequestration may be involved in the process. The agonist occupied NK_1 receptor has recently been shown to be a substrate for β-adrenergic receptor kinases I and II (Kwatra et al., 1993). The development of appropriate cellular systems for analysis of desensitization mechanisms should facilitate studies to examine the molecular processes and subcellular compartments involved in the regulation of receptor sensitivity and turnover.

Thus, the molecular characterization of the mammalian tachykinin receptors, as well as that of the heterotrimeric GTP-binding proteins and several effector enzymes and ion channels, now provides a firm structural basis in which to further explore the molecular determinants of peptide action and intracellular signaling. Some actions of the tachykinin peptides have been suggested to be receptor independent, by directly activating G-protein complexes (Mousli et al., 1990), though studies verifying these nonclassical actions of the tachykinins have not been reported.

3. General Considerations of Tachykinin Receptor Primary Structure

Molecular cloning of tachykinin receptor cDNAs, along with their functional expression in heterologous cell systems, have unequivocally established that these molecules encode the pharmacologically characterized tachykinin receptors (Masu et al., 1987; Yokota et al., 1989; Hershey and Krause, 1990; Shigemoto et al., 1990). The primary structure of the cognate deduced proteins, when evaluated with hydropathy analysis, indicated the presence of seven putative α-helical transmembrane domains similar to the previously characterized bacteriorhodopsin, visual opsins, β-adrenergic receptor, and muscarinic receptor. To date, three types of tachykinin receptors from several species have had their primary protein structures determined by deduction after analysis of their cDNA or genomic sequences. These receptor primary structures are displayed in Figs. 1 and 2 with their conserved sequences highlighted. The primary structures of these receptors, like other G-protein coupled receptor counterparts, consist of an extracellular NH_2-terminal sequence containing one or more asparagine-linked glycosylation sites, seven putative α-helical transmembrane domains, a consensus cysteine residue approximately 15 residues beyond transmembrane domain VII, which may serve as a palmitoylation site for further membrane anchoring, and a COOH-terminal intracellular domain of variable length. Several additional features of the receptor primary structures are apparent (Figs. 1 and 2) and presumably are additionally important with regard to receptor structure–function relationships related to agonist binding, to receptor conformational changes that result in G-protein release, and to receptor desensitization. Several striking structural features, as initially pointed out by Nakanishi (1991), are discussed below:

1. In the second transmembrane domain, an acidic residue is contained in virtually all G-protein coupled receptors with the exception of the gonadotropin releasing hormone receptor (Tsutsumi et al., 1992). This residue has been implicated in the regulation by sodium of the high-affinity state of the receptor and in receptor activation (Quintana et al., 1993). In the NK_2 and NK_3 receptors cloned, all species contain an aspartic acid, whereas all NK_1 receptors cloned have a glutamic acid residue at position 78 (Fig. 1);
2. Histidine residues are present in transmembrane domains V and VI of all mammalian tachykinin receptors cloned, and this appears to be a unique feature of the tachykinin receptors;
3. The third intracellular domain of the NK_1 and NK_3 receptors are highly conserved, whereas the sequences of the NK_2 receptors have specific regions of divergence. As G-protein coupling has been shown in other receptor systems to be dependent on specific residues in this region, it would appear that this represents an important domain in signal transduction for the tachykinin receptors, and specific sequence determinants in this region may dictate coupling to distinct heterotrimeric G-protein complexes; and
4. Several serine and threonine residues exist in the third intracellular domain and in the intracellular carboxyl-terminal region that may be sites of phosphorylation. These residues differ among the three types of tachykinin receptors in both number and sequence context, with the NK_1 receptor having the most sites and the NK_2 receptors having the least. This appears to correlate with different desensitization behaviors observed on expression of the three receptors in the *Xenopus* oocyte expression system (Harada et al., 1987; Shigemoto et al., 1990).

Within the group of G-protein coupled tachykinin receptors, we determined probable evolutionary relationships between the molecularly characterized tachykinin receptor types by a phylogenetic tree

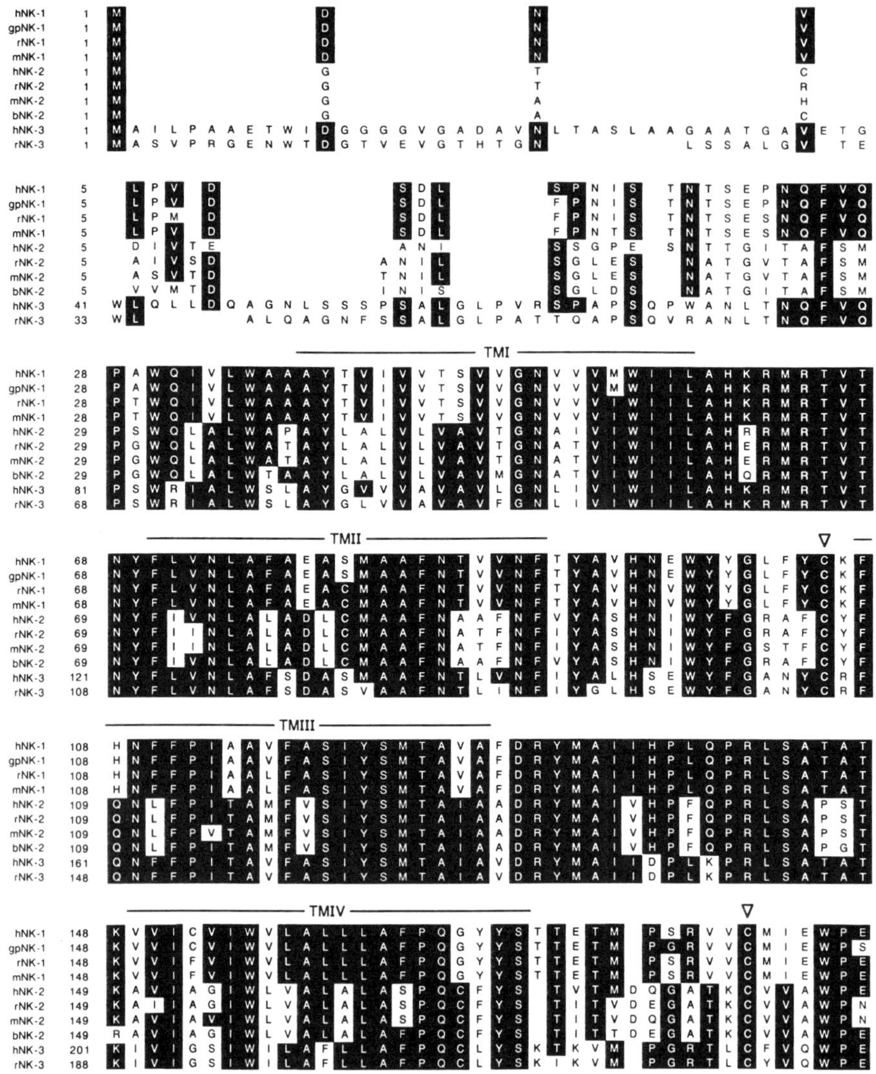

Fig. 1. Comparison of the primary structures of mammalian tachykinin receptor proteins. Residue identities present in 50% or more of the ten cloned cDNA sequences are shaded. The putative transmembrane domains (denoted TM I–VIII) are indicated by a horizontal bar above the sequences. The open

triangles denote conserved cysteine residues that may be disulfide bonded, and the closed triangle indicates cysteine residues that may be palmitoylated. The abbreviations used for the mammalian species are as follows: h, human; gp, guinea pig; r, rat; m, mouse; b, bovine.

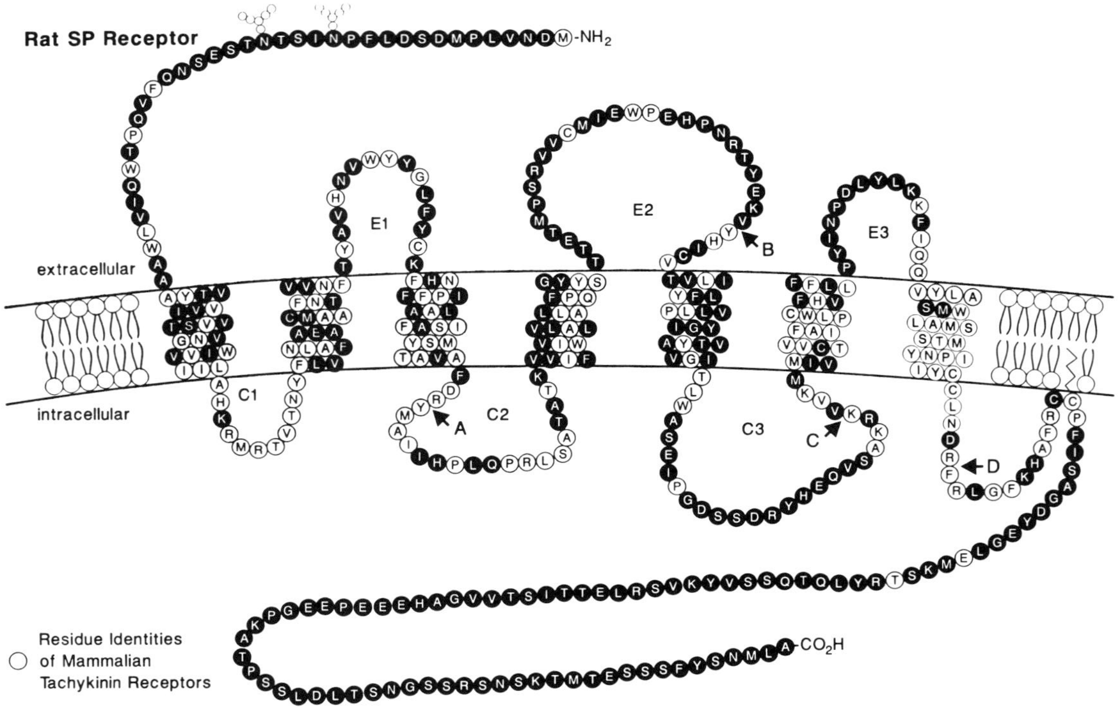

Fig. 2. Sites within tachykinin receptor open reading frames where introns of NK_1, NK_2, and NK_3 genes interrupt the receptor protein coding regions. The intron insertion sites are labeled A–D. Residue identities among the receptor types are displayed as open circles with dark letters, and divergent residues are represented by dark circles with white letters.

analysis (Nei, 1987), shown in Fig. 3. This analysis included the entire primary amino acid structures of each tachykinin receptor since there is no compelling *a priori* reason for restricting the sequence comparisons to parts of the primary sequences, and also since it appears that sequences in the amino-terminal portion of the receptors may be important for agonist binding, and those present in the carboxyl-terminal region may be important in G-protein coupling *(see below)*. It is assumed by this phylogenetic tree analysis method that the expected rate of amino acid change is constant throughout the receptor primary structure. When the distance measure is subject to errors of the stochastic type, both the topology and branch lengths may be inaccurate even if the rate of amino acid change is constant. The average distance method used (Nei, 1987), however, is often better than other distance matrix methods in determining the true species tree. The length of each branch of the phylogenetic tree correlates with the evolutionary distance between receptors. Thus, this analysis indicates that the mammalian, G-protein coupled tachykinin receptors have differentiated into three distinct major branches, each branch corresponding to a type of receptor (NK_1, NK_2, and NK_3) displaying preferential affinity for its cognate ligand. For all mammalian tachykinin receptor types identified, the interspecies variation is low such that there is a greater than 85% identity between species homologs.

Additional, related G-protein coupled receptor structures not shown in the tree (Fig. 3) have also been characterized and have been shown to display significant similarity to the mammalian tachykinin receptors. Recently, a cDNA from a human placental library was isolated by an expression cloning strategy in which a putative G-protein coupled receptor displayed 79% identity to the rat NK_3 receptor with three gaps of unrelated sequence (Xie et al., 1992). However, the expressed receptor bound opioid ligands, albeit weakly, but not tachykinin ligands. Under the same conditions, the cloned rat NK_3 receptor bound ^{3}H-eledoisin, but not opioids. These results suggest that putative receptors with relative high sequence identity do not necessarily encode a receptor with similar agonist activities. Thus, one must be cautious about the identification of receptors based on

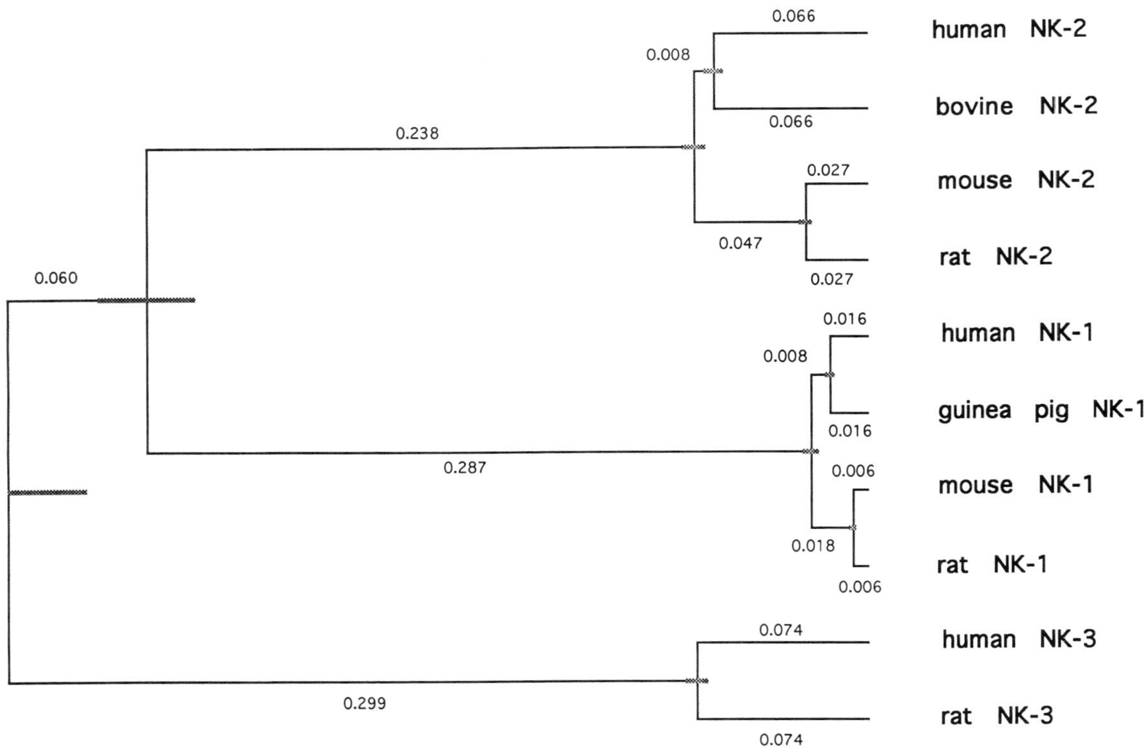

Fig. 3. Phylogenetic tree analysis of molecularly characterized mammalian tachykinin and related receptors. The deduced primary structure of the ten receptors was used to develop a distance matrix, and the tree was constructed by the unweighted pair-group method with arithmetic mean (Nei, 1987). The numbers along the horizontal bars represent the distance between operational taxonomic units, and the hatched bars represent one standard error on each side of the branching point.

homology alone with no functional expression data at hand. On the other hand, two recently described receptors from *Drosophila*, called TK_1 and TK_2, appear to be tachykinin receptors based on the fact that at micromolar concentrations some mammalian tachykinin peptides can activate responses mediated by these receptors in heterologous expression systems (Li et al., 1991; Monnier et al., 1992). However, the natural *Drosophila* agonist(s) for these receptors have not been defined, and whether there exist tachykinin peptides structurally related to their mammalian counterparts remains to be clarified. These receptors are distantly related to each other (approximately 40% identity), whereas the TK_1 receptor is more closely related to the mammalian tachykinin receptors than is the TK_2 receptor (Krause et al., 1993a).

There is an intense desire to understand the molecular basis of receptor recognition, function, and regulation from both basic science and clinical science perspectives. Some goals of the current molecular biological approaches used to determine tachykinin receptor structure–function relationships are:

1. To understand specific, receptor primary structural domains and specific residues essential for high-affinity agonist binding;
2. To determine the relationship between agonist binding to the high-affinity state of the receptor and the subsequent conformational change(s) of the receptor protein that result in its activation (i.e., release and dissociation of the G-protein heterotrimeric complex);
3. To understand specific receptor primary structural domains and specific residues essential for antagonist binding affinity and consequent blockade of the receptor;
4. Use of the information gained from structural studies to design improved synthetic agonists and antagonists for tachykinin receptors and possibly create novel agonists and antagonists for tachykinin receptors, as well other G-protein coupled receptors;
5. To understand the molecular determinants for G-protein heterotrimeric complex binding to the tachykinin receptors to help evaluate the second messenger systems activated by receptor stimulation; and
6. To define the molecular processes that result in the determination and regulation of tachykinin receptor sensitivity in normal and

pathological states in which the tachykinin peptides have been implicated.

The attainment of these research goals will greatly expand our knowledge as to the structural relevance and functions of these receptors and should aid in the chemical design and synthesis of novel receptor active molecules.

3.1. The NK_1 Receptor

The NK_1 receptor is the receptor type for which SP is the most potent, naturally occurring agonist (and, therefore, was originally named the SP receptor). Several radioligands have been used to identify and biochemically characterize the NK_1 receptor. Since SP binds to all three types of neurokinin receptors, the NK_1 receptor is identified pharmacologically by the rank order of potency of various tachykinin peptides. For the naturally occurring mammalian tachykinins, this order is SP > NPγ > NKA > NKB; for nonmammalian tachykinin peptides, this order is physalaemin > eledoisin > kassinin.

Until the recent cloning of the NK_1 receptor in several species (Figs. 1 and 2), the receptor was studied in membrane preparations from various tissues including rat brain and dog carotid artery. Early studies had suggested that this receptor was a member of the G-protein coupled superfamily because responses elicited in the oocyte expression system were slow in onset, consistent with the activation of second messengers (Parker et al., 1986), and because radioligand binding was modulated by GTP and nonhydrolyzable GTP analogs (Lee et al., 1983; MacDonald and Boyd, 1989).

All endogenous mammalian tachykinin peptides share the same carboxyl-terminal sequence of Phe-X-Gly-Leu-Met-NH_2, where X is one of several aliphatic or aromatic amino acids. In SP, this residue is Phe^8. This conserved carboxyl-terminus is recognized by the tachykinin receptors and is likely in part responsible for the cross reactivity of tachykinins among tachykinin receptors. For the NK_1 receptor, the importance of this region has been shown directly in that the carboxyl-terminal hexapeptide, Gln-Phe-Phe-Gly-Leu-Met-NH_2, is approximately equipotent to SP in tissue response assays (Lee et al., 1982; *for reviews, see* Sandberg and Iversen, 1982; Sandberg,

1985). It has been shown by alanine substitution that in SP the invariant Phe at position seven plays a dominant role in the activity of SP (Couture et al., 1979). The free acid form of the SP is on the order of 1000-fold less potent than the amidated peptide (Lee et al., 1982; Cascieri and Liang, 1983), although the methyl ester derivative of SP has been shown to be between equipotent to (in physiological assays) (Watson et al., 1983) and 50-fold less potent (radioligand competition assays) (Cascieri and Liang, 1983) than the native peptide. These studies indicate that Phe^7 and the carboxyamide moiety are important peptide regions in receptor recognition and, as discussed below, these peptidic regions may dictate either a receptor active agonist conformation or may directly interact with the receptor.

The NK_1 receptor is a member of the G-protein coupled superfamily of receptors. High-affinity binding of radiolabeled SP is sensitive to inhibition by GTP and nonhydrolyzable analogs of GTP (Lee et al., 1983; MacDonald and Boyd, 1989). This decreased binding has been attributed to the conversion of high-affinity binding sites to a lower affinity state that is minimally 20-fold lower in affinity than the high-affinity state (Luber-Narod et al., 1990). This effect is caused by an increase in the dissociation rate for the ligand and is reversible upon removal of the added guanine nucleotide (Luber-Narod et al., 1990). A loss of high-affinity, GTP-sensitive binding is also seen on alkaline treatment of NK_1 receptor containing membranes (a treatment which dissociates G-protein complexes from receptors). This binding is restored by addition of purified G_i/G_o (MacDonald and Boyd, 1989). Although G_i/G_o can act to restore the high-affinity binding of SP, these G-proteins may not be candidates for the in vivo G-proteins since G-protein modulation of SP binding is generally insensitive to treatment with pertussis toxin. Recent studies indicate that the physiological G_α-protein that couples to the NK_1 receptor as part of a heterotrimeric complex is of $G_{q/11}$ (Kwatra et al., 1993) as discussed below.

The initial design of peptide antagonists of the tachykinins was based on modifications of the primary structure of SP. Two successful approaches have been used, one using a linear peptide approach where a variety of amino acids (including natural L amino acids, unnatural L amino acids, and unnatural D amino acids) replace natural residues, and the other approach being based on the identification of key conformational requirements (i.e., constraints). The former approach has resulted in the generation of spantide (Folkers et al., 1984) and, subsequently, spantide II (Folkers et al., 1990). Spantide II, compared to spantide, has a higher pA_2 value (negative logarithm of antagonist producing a twofold shift of the agonist concentration-activity curve) of 7.7 at the NK_1 receptor and does not possess significant mast cell releasing activity. The latter approach culminated in the development of a spirolactam substituted compound called GR71251 (Ward et al., 1990), which has a pA_2 value of 7.7 at the NK_1 receptor. The peptidic structures of these and other related antagonists may be expected to limit oral bioavailability and, consequently, their testing to evaluate clinical potential.

Recently, several potent nonpeptide NK_1 receptor antagonists have been reported (Watling and Krause, 1993). The first of these antagonists was CP-96,345, a quinuclidine-based compound that has subnanomolar affinity for the NK_1 receptor of most species tested (Snider et al., 1991). CP-96,345 shows high selectivity for the NK_1 receptor over NK_2 and NK_3 receptors and is greater than 100-fold more potent than the peptide antagonists described above (Rouissi et al., 1991). Other recently discovered nonpeptide SP antagonists include WIN 51708, a steroid based molecule with a nitrogen containing heterocycle attached to the A ring (Venepalli et al., 1992) and RP 67580, a substituted perhydroisoindole (Garret et al., 1991). Finally, a fermentation product isolated from *Streptomyces violaceoniger* was catalytically hydrogenated, and this novel cyclic peptide has been shown to be an antagonist at NK_1 and NK_2 sites (Morimoto et al., 1992).

Interestingly, the nonpeptide antagonist, CP-96,345, has an affinity for rat and mouse NK_1 receptors which is 50–100-fold lower than in other species (Snider et al., 1991). This species difference is of great interest from a protein biochemical and drug discovery point of view as the NK_1 receptor from all species from which it has been cloned displays a very high identity (Fig. 2). Recently, the molecular basis of this species selectivity has been elucidated based on a mutational analysis of rat and human NK_1 receptors (Fong et al., 1992, Sachais et al., 1993; discussed further below).

The nonpeptide antagonist, WIN 51708, which has been recently described by investigators at Sterling-Winthrop (Venepalli et al., 1992), also interacts with the human NK_1 receptor differently than the rat receptor (Appell et al., 1992). In stably transfected cell lines expressing the rat or human NK_1 receptor, WIN 51708 has an IC_{50} value of 25 n*M*, whereas at the human receptor the IC_{50} is approximately 10 µ*M* (Sachais and Krause, unpublished observations). Further studies should establish the molecular basis of this dramatic species selectivity.

The NK_1 receptor has been shown to activate the hydrolysis of inositol phospholipids in many systems including central (Mantyh et al., 1984; Torrens et al., 1989) and peripheral (Merritt and Rink, 1987; Tachado et al., 1991) tissues. In the rat CNS, this activation has been shown to correlate with SP binding sites (Mantyh et al., 1984). Activation of the NK_1 receptor can also stimulate cAMP accumulation. Substance P-mediated increases in cAMP levels have been shown in some systems (Yamachita et al., 1983; Narumi and Maki, 1978; Hanley et al., 1980), but not in others (Watson, 1984; Hunter et al., 1985). SP activates either phosphatidylinositol hydrolysis or cAMP production, but not both, in iris sphincter, and this is species dependent (Tachado et al., 1991). Heterologous expression of NK_1 receptors in several cell lines has established that the NK_1 receptor is capable of coupling to both of these second messenger systems (Nakajima et al., 1992; Takeda et al., 1992; Mitsuhashi et al., 1992).

In rat parotid acinar cells, it has been shown that SP causes a rapid, but transient, increase in inositol phospholipid hydrolysis, whereas muscarinic agonists cause a more prolonged response (Sugiya et al., 1987; Merritt and Rink, 1987). Reapplication of SP does not induce a second response, eliminating peptide degradation by proteases as the mechanism underlying this transient response. Application of SP after stimulation with the muscarinic agonist does, however, induce a response similar to that seen with SP on naive cells. Application of muscarinic agonist after SP stimulation gives a prolonged response. These studies show that SP loses its ability to activate inositol phospholipid hydrolysis under conditions where muscarinic agonists are still able to do so. This phenomenon has been referred to as homologous desensitization; however, it is not entirely clear from these studies whether the origin of this phenomenon is at the level of the NK_1 receptor or some postreceptor event in the signal transduction system.

3.1.1. Human NK_1 Receptor

The human NK_1 receptor cDNA has been cloned and functionally expressed (Takeda et al., 1991; Gerard et al., 1991; Hopkins et al., 1991; Fong et al., 1992a). The primary sequence of the human NK_1 receptor is 94.5% identical to the rat NK_1 receptor *(see below)*, differing by only 22 out of 407 residues (Takeda et al., 1991) (Fig. 2). Of these differences, 6 occur in putative membrane spanning regions, 8 in extracellular regions, and 8 in intracellular regions. mRNA transcripts are present throughout the gastrointestinal system and in lung (Joslin et al., 1991). In this study, the central nervous system was not examined.

The gene for the human NK_1 receptor has also been isolated and partially characterized and found to have a similar structural organization to the rat gene (*see below* and Fig. 4; Takeda et al., 1991; Gerard et al., 1991; Takahashi et al., 1992; Krause et al., 1993). This gene has been localized to human chromosome 2 as a single copy gene based on analysis of somatic cell hybrids (Gerard et al., 1991; Hopkins et al., 1991).

3.1.2. Guinea Pig NK_1 Receptor

A cDNA for the guinea pig NK_1 receptor has been isolated from guinea pig uterus and functionally expressed in COS-7 cells (Gorbulev et al., 1992). The relative affinities of mammalian tachykinins for this receptor were shown to be SP >> NKA > NB, which is consistent with NK_1 receptors of other species.

3.1.3. Rat NK_1 Receptor

The rat NK_1 receptor cDNA was the first NK_1 receptor to be cloned and functionally expressed (Yokota et al., 1989; Hershey and Krause, 1990). A quantitative study of the tissue distribution of the NK_1 receptor mRNA in the rat has been performed (Hershey et al.,

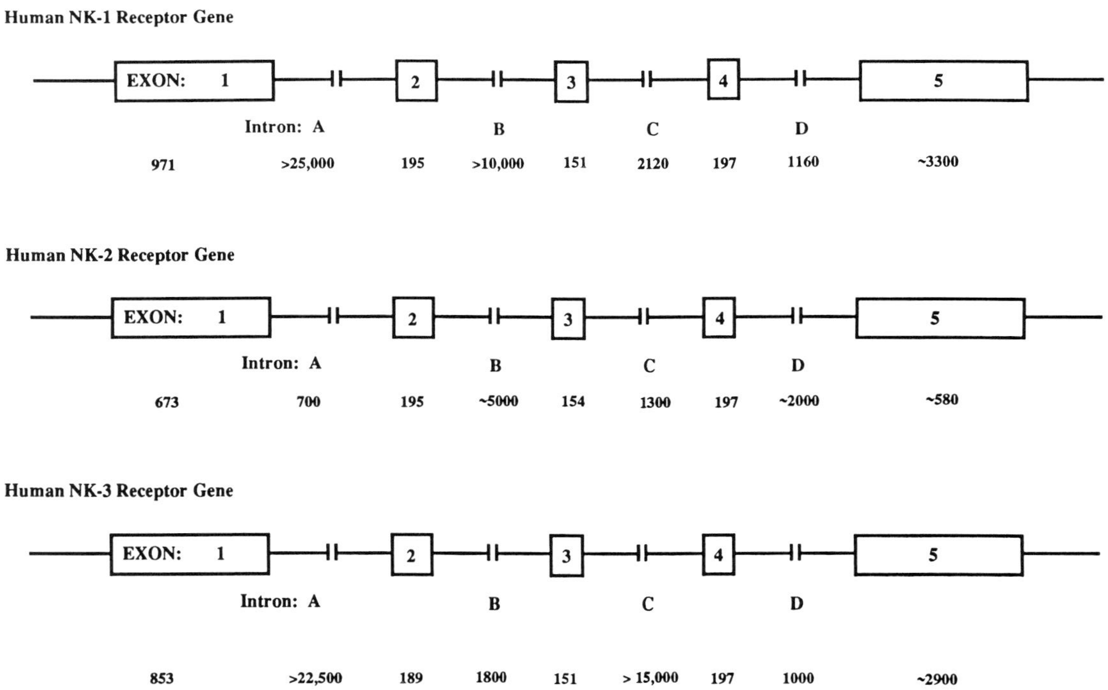

Fig. 4. Comparison of the organization and lengths of exons and introns of cloned human tachykinin receptors. For each receptor gene, the schematic displays the exons interspersed with introns. The sizes of the introns are displayed, and in the cases of NK_1 and NK_3 receptors the extents of some intron regions have not been determined. Data compiled from Hopkins et al., 1991; Takeda et al., 1991; Kris et al., 1991; Gerard et al., 1990; Takahashi et al., 1992; Buell et al., 1992; Krause et al., 1993b.

1991; Tsuchida et al., 1991). NK_1 receptor transcripts are widely distributed in the central nervous system and many peripheral tissues. Relatively high levels are present in the striatum, hypothalamus, hippocampus, olfactory bulb, and medulla, and in the gastrointestinal tract, salivary glands, and urinary bladder.

The structure of the gene encoding the rat NK_1 receptor has been reported (Hershey et al., 1991). This gene is greater than 45 kilobases in length and contains 5 exons. The splice sites of these exons occur at the borders of sequences encoding putative membrane spanning domains of the receptor, and the distinct exonic regions have been suggested to represent evolutionarily important functional units of the receptor as discussed in more detail below. The 5'-flanking region contains the consensus sequence for the cAMP response element, as well as a sequence that resembles the Ca^{2+} inducibility sequence of the c-fos gene. These and other elements are conserved between rat and human genes and may play a role in the transcriptional regulation of the NK_1 receptor (Krause et al., 1993b). Since the NK_1 receptor can activate both of these second messenger systems *(see above)*, it has been suggested that stimulation of the NK_1 receptor directly influences NK_1 receptor expression at the transcriptional level, perhaps as part of a receptor resensitization process (Hershey et al., 1991).

3.1.4. Mouse NK_1 Receptor

The murine NK_1 receptor cDNA has been cloned and functionally expressed (Sundelin et al., 1992). In an oocyte expression system, the rank order of potency for this receptor is SP > NKA > NKB, consistent with NK_1 receptors of other species. Its primary structure is 99% identical to rat and 97% identical to human NK_1 receptors. Some exons of the gene for this receptor have been isolated, and the gene appears to be greater than 30 kb in size. The gene organization appears to be similar to the rat and human NK_1 receptor genes.

3.2. NK_2 Receptor

Historically, the existence of multiple receptors for SP was proposed by Lee et al. (1982). This hypothesis was supported by work determining the rank order of potency of SP, its C-terminal fragments, analogs, and related tachykinins in systems including contracting guinea pig ileum, potentiating electrically evoked contractions of rat vas deferens preparations, and in displacement of [^{3}H]SP binding to rat brain membranes by both Lee et al. (1982) and by Rosell and coworkers (Rosell et al., 1983). Subsequent isolation of the endogenous mammalian tachykinins, NKA and NKB, supported the multiple receptor hypothesis, and further studies led to the characterization of the NK_2 receptor.

The binding properties of the NK_2 receptor are distinct from those of the NK_1 receptor. The rank order of potency of endogenous mammalian ligands is NPγ = NPK = NKA > NKB > SP, with NKB and SP having lesser affinity than the other ligands by at least one to two orders of magnitude. Similar to the NK_1 receptor, several synthesized and constrained peptide agonists and antagonists have been developed and utilized for studies to determine the requirements for receptor activation in model tissue preparations (Buck and Shatzer, 1988; Regoli et al., 1988; Dion et al., 1990; Lavielle et al., 1990; Saviano et al., 1991; Morimoto et al., 1992; Murai et al., 1992). More recently, a higher affinity nonpeptide antagonist (SR 48968) for the NK_2 receptor has been discovered (Emonds-Alt et al., 1992). Using peptide antagonists, Maggi et al. (1990) found heterogeneity between NK_2 receptors expressed in rabbit pulmonary artery and hamster trachea. In addition, two other peptidic antagonists, including a cyclic peptide, L-659,877 (McKnight et al., 1991), and a linear hexapeptide analog, R-396 (Dion et al., 1990), have been developed and these compounds also differentiate between NK_2 receptor types present in different species. The NK_2 receptors present in the rabbit pulmonary artery, when induced to contract by the NK_2 selective agonist, [β-Ala^8]-NKA(4–10), showed a rank order of antagonist potencies of MEN 10,376 > L-659,877 > R 396. By contrast, the rank order of potency of these antagonists on [β-Ala^8]-NKA(4–10)-induced contractions of hamster trachea was L-659,877 > R 396 > MEN 10,376 (Maggi et al., 1992). Heterogeneity in NK_2 receptor antagonist binding and agonist binding sensitivity to GTP analogs and divalent cations have also been observed in the hamster urinary bladder and bovine stomach (Van Giersbergen et al., 1991). The mechanisms of such heterogeneity are currently unknown, but may be attributed to species differences, to receptor subtypes, or to coupling with multiple G-proteins.

Binding studies and, more recently, analysis of RNA expression have demonstrated that NK_2 receptors have a pattern of expression distinct from the NK_1 receptor. Investigators generally agree that the NK_2 receptor is found in several peripheral tissues including duodenum, large and small intestine, urinary bladder, adrenal gland, and pulmonary artery. Several of these tissues, including rat vas deferens and rabbit pulmonary artery, have been used as model systems for the characterization of the NK_2 receptor. Several investigators have also reported the absence of detectable receptors with NK_2 binding properties in the central nervous system (Buck et al., 1986; Saffroy et al., 1988; Dietl and Palacios, 1991). However, one study by Dam et al. (1990) reported ^{125}I-NPγ binding sites in rat brain. In this study, the profile of ^{125}I-NPγ binding was similar to ^{125}I-NKA binding, but different from ^{125}I-Bolton-Hunter-SP binding supporting the hypothesis that NPγ is detecting the NK_2 receptor. RNA blot hybridization has thus far been unsuccessful at detecting NK_2 message in whole brain or brain regions in rat (Tsuchida et al., 1990); however, the more sensitive technique of nuclease protection has detected low levels in hippocampus, striatum, and spinal cord at levels 500-fold lower than urinary bladder (Takeda and Krause, 1991). Hence, the NK_2 receptor is observed in high concentrations in several peripheral tissues and is expressed in low concentrations in discrete brain regions.

The NK_2 receptor has the distinction among the tachykinin receptors of having been the first to be cloned (Masu et al., 1987). The isolation of the bovine NK_2 receptor cDNA *(below)* has allowed for the subsequent isolation and characterization of other tachykinin receptors and NK_2 receptors from other species by homologous hybridization or by PCR using oligonucleotide primers of conserved regions. The NK_2 receptors are less conserved between species than are the NK_1 receptors, often having less than 90% identity (*see* Fig. 1). Some of this divergence occurs in the carboxyl-terminal tail; the species cloned thus far have different sized "tails," with the human having a tail 14, 14, and 9 amino acids longer than the bovine, mouse, and rat predicted proteins, respectively.

3.2.1. *Human NK_2 Receptor*

The cDNA of the human NK_2 receptor has been cloned (Gerard et al., 1990; Kris et al., 1991) and expressed in 3T3 cells, where a mitogenic function has been proposed (Kris et al., 1991). Stimulation of cell growth by NKA had been previously observed in cultured arterial smooth muscle cells and human skin fibroblasts in the absence of serum (Nilsson et al., 1985). The cloning and characterization of the human gene demonstrated that, similar to the NK_1 receptor, the message is composed of five exons suggesting a common evolutionary origin for the NK_1 and NK_2 receptors. The gene encoding the NK_2 receptor has been located on the tenth chromosome (Gerard et al., 1990). Figure 3 depicts the conserved primary structure of the tachykinin receptors and the sites where introns interrupt the protein coding region. In contrast to the wealth of potential regulatory elements 5' of the rat NK_1 receptor, other than a putative TATA and GC-like box, no other established transcriptional control signals were observed within 500 bp of the presumed transcription initiation site of the human NK_2 receptor gene (Gerard et al., 1990).

3.2.2. *Bovine NK_2 Receptor*

The bovine NK_2 receptor was the first tachykinin receptor to be cloned. The oocyte expression system was used in a novel way to isolate the cDNA clone by expressing pools of cDNA and then subdividing the pools with subsequent expression to ultimately isolate the cDNA encoding this receptor (Masu et al., 1987). This expressed cDNA was most potently activated by NKA compared to NKB or SP. The analysis of the predicted protein structure of the NK_2 receptor (Masu et al., 1987) demonstrated that this receptor contained many features found in the family of G-protein coupled receptors, including 7 putative transmembrane domains.

In addition to expression in oocytes (Masu et al., 1987), the NK_2 receptor from different species has been expressed in several cell types via transfection. Such experiments have allowed for detailed binding studies and the direct monitoring of second messengers activated by receptor stimulation. In transfected CHO cells, Eistetter et al. (1991) performed a series of experiments suggesting that in this system the stimulation of bovine NK_2 receptor led to the formation of IP_3, cAMP, and output of arachidonic acid and prostaglandin E_2. In this study, a correlation of inhibition of arachidonic acid and prostaglandin E_2 release and the inhibition of cAMP production suggested a causative relationship between these responses. In contrast, expression of bovine NK_2 receptor in a rat glioma cell line, C6-2B, revealed that in this system receptor stimulation led to an increase of intracellular Ca^{2+} (presumably from IP_3 synthesis) that apparently led to a decrease in cAMP production when other receptors coupled to G_s were stimulated (DeBernardi et al., 1991). Hence, the response of the NK_2 receptor may be dependent upon the cell type within which it is expressed.

3.2.3. *Rat NK_2 Receptor*

The cDNA for the rat NK_2 receptor has also been cloned and expressed in transfected tissue culture cells. A detailed binding study of cloned rat tachykinin receptors expressed in transfected monkey kidney COS-7 cells demonstrated that the putative NK_2 receptor had many of the binding properties found in some tissues (Ingi et al., 1991). In transfected CHO cells, Nakajima et al. (1992) demonstrated that the rat NK_1 and NK_2 receptors, when stimulated, led to the stimulation in both inositol lipid hydrolysis and cAMP formation. In this system, both responses could be generated in membrane preparations suggesting that both second messengers systems were directly linked to receptor activation. In addition, a difference in the magnitude of the cAMP, but not the IP's response, has been observed between the NK_1 and NK_2 receptors (Takeda et al., 1992). More recently, the study of chimeras of rat NK_1 and NK_2 receptors expressed in stably transfected CHO cells has led to the finding that, although the NK_1 and NK_2 receptors have high homology (*see* Fig. 1), the homologous structural domains of the third cytoplasmic loop and the carboxyl tail of these receptors may not have functionally equivalent roles (Blount and Krause, 1993; *see below*). These differences in the way that NK_1 and NK_2 receptors encode functional units, combined with differences in protein expression in host cells, may explain the differences observed in duration of SP- and NKA-stimulated responses in tissues.

3.2.4. *Mouse NK_2 Receptor*

The murine NK_2 receptor genome and cDNA have been cloned, and the cDNA has been expressed in *Xenopus* oocytes (Sundelin et al., 1992). The rank order of agonist potency for this cloned mouse NK_2 receptor expressed in oocytes was consistent with what has been found in tissues, with NKA being most potent followed by NKB and SP. Similar to the NK_1 receptor and NK_2 receptors from other species, analysis of the genomic organization demonstrated that the receptor is encoded on five exons.

3.3. NK_3 Receptor

After the discovery of NKA and NKB, NK_3 receptors were initially suggested based upon ligand binding studies performed with rat central nervous system and in other tissue membrane preparations (Torrens et al., 1984; Buck et al., 1984; Cascieri et al., 1985). In these studies, the binding of radiolabeled agonists (^{125}I-Bolton-Hunter-NKA; ^{125}I-Bolton-Hunter-eledoisin) was inhibited by natural agonists, with SP most potent with the former radiolabel and NKB being most potent with the latter radiolabel. Research from Selinger's laboratory substantially clarified this issue with a combination of functional and ligand binding studies. Initially, they performed functional studies in guinea pig ileum in which SP and NKB elicit contractions both directly through activation of a muscle cell receptor and indirectly through activation of a neuronal receptor leading to acetylcholine release, which stimulates muscle contraction via muscarinic receptors (Laufer et al., 1985). In these studies, the muscle receptor was clearly distinct from the neuronal receptor, with NKB being 50 times more potent than SP or NKA at the neuronal receptor. At the muscle receptor, SP was most potent. These investigators further developed a highly selective radioligand, *NI*-([^{125}I]desamino-

3-iodotyrosyl)-[Asp^{5,6},*N*-methylPhe^{8}] substance P(5–11)heptapeptide, abbreviated ^{125}I-Bolton-Hunter-senktide (for **se**lective **NK**B receptor pep**tide**), which selectively binds to this receptor type (Laufer et al., 1986). Consequently, in rat cerebral cortex membranes, this ligand binds to the NK_3 receptor with high-affinity (K_D = 0.9 n*M*), and binding is reversible and saturable. A linear Scatchard plot and a Hill coefficient close to unity were obtained, which lends further credence to the notion that this radioligand labels a single class of binding sites. The use of this radioligand also provided further support to the suggestion that the rat brain binding site was similar or identical to the receptor site in guinea pig ileum myenteric plexus.

3.3.1. Human NK_3 Receptor

Buell et al. and Takahashi and coworkers have isolated genomic clones of a putative NK_3 receptor by cross hybridization with the cognate rat probe (Buell et al., 1992; Takahashi et al., 1992). These workers isolated and identified a genomic sequence believed to be NK_3 based upon a high degree of homology to the rat sequence. The deduced amino acid sequence indicated that the putative NK_3 receptor protein was 88.2% identical to the rat and consists of 465 residues, and the amino terminal domain is 13 amino acids longer than that of the rat NK_3 receptor. Expression studies have been performed using a PCR generated cDNA with a similar sequence used to establish that this receptor sequence does indeed encode a functional NK_3 receptor (Buell et al., 1992; Huang et al., 1992). The agonist potency profile in ligand binding studies was NKB > NKA > SP.

3.3.2. Rat NK_3 Receptor

In 1990, Nakanishi and coworkers cloned and functionally expressed the rat NK_3 receptor (Shigemoto et al., 1990). By using cDNA fragments of the bovine NK_2 receptor, reduced stringency DNA blot hybridization demonstrated multiple hybridizing cDNAs, of which one corresponded to the NK_3 receptor. This was based on electrophysiological analysis of RNA transcribed from a clone isolated from a minilibrary (which was prepared from a sucrose density fraction that contained NK_3 receptor mRNA activity based on oocyte expression). The cloned receptor was found to be activated by tachykinin peptides and most potently by NKB. The NK_3 receptor protein consists of 452 residues and, by hydropathy analysis, was shown to have seven hydrophobic domains. Electrophysiological assays of peptide-elicited responses were performed in *Xenopus* oocytes previously injected with RNA synthesized from the NK_3 receptor cDNA, and NKB was much more potent than either SP or NKA. Transient transfection studies were performed in COS cells, and ligand binding studies further documented that this cDNA encodes the NK_3 receptor. ^{125}I-Bolton-Hunter-eledoisin binding was saturable with a K_D of 0.64 nM. Inhibition studies of radiolabeled eledoisin binding were also performed and K_i values for NKB, SP, and NKA were $4.2 \times 10^{-10}M$, $2.9 \times 10^{-8}M$, and $1.9 \times 10^{-7}M$, respectively. mRNA expression studies have demonstrated that NK_3 receptor transcripts are expressed relatively abundantly in several brain regions including cerebellum, hypothalamus, and cortex, and also at lower levels in the gastrointestinal system.

4. Approaches Used to Understand the Molecular Basis of Agonist and Antagonist Binding

The molecular characterization of the three types of mammalian tachykinin receptors, coupled with previous extensive pharmacological investigations on agonist and antagonist specificity and selectivity at the three receptors, provides an interesting system with which to investigate the molecular basis for receptor activation by agonists and blockade by antagonists. Initial studies performed with the cloned rat tachykinin receptor types further documented the similarity of receptor pharmacology after recombinant receptor expression in heterologous cell systems *(see above)*. Consequently, the molecular basis of receptor specificity and selectivity must be a consequence of receptor structure. Several approaches have been used to evaluate this molecular basis and, prior to a discussion of the present results at hand, it is important to summarize the strengths and limitations of each approach. The primary questions that are being addressed include:

1. What is the chemical nature of agonist and antagonist binding pockets;
2. Which receptor residues are directly interacting with specific chemical moieties of agonists and antagonists;
3. What receptor structural changes occur upon agonist binding that result in G-protein activation; and
4. Do receptor antagonists lock the receptor in a specified conformation such that agonist is unable to bind, or does the antagonist bind to the receptor in such a way that the agonist, though able to bind, is not able to activate the receptor?

Several approaches that have been used to explore the molecular basis for agonist and antagonist binding include:

1. Construction of chimeric receptors from different receptor types in which the sequence of part or parts of one receptor type are juxtaposed using recombinant DNA methods to the part of the sequence from a different receptor type;
2. Construction of chimeric receptors using the same receptor type from different animal species;
3. Construction of site-directed mutants within a single receptor type; and
4. The use of photoaffinity agonists or antagonists for covalent attachment to the receptor with the subsequent identification of the specific residues as the covalent attachment site.

Several assumptions underly each experimental approach, and it is necessary to discuss these to determine whether these assumptions negatively impact upon the results obtained and their interpretation. A chimeric receptor approach generally implies that the specific entity under evaluation (i.e., the nature of the agonist or antagonist binding pocket) is a conserved feature of that receptor. That is to say, the same discrete regions of each receptor are involved and the nature of the interaction (either as a positive influence or negative influence on agonist binding) is similar among the different receptors used. If agonist binding domains, for example, among the three tachykinin receptors involve different regions of the primary structure, or if the same region exerts positive influences on agonist binding for one receptor type, but negative influences on agonist binding for a different receptor type, then the results from chimeric studies could be misleading in attempts to identify the relative importance of the discrete regions involved. In all mutational studies performed to examine agonist and or antagonist binding, it is important to demonstrate that cellular trafficking of the receptor is appropriate and that the general folding and cell surface assembly of the receptor occurs without gross perturbations of receptor structure. Finally, in photoaffinity labeling studies it is important that the extent of chemical reactivity of photolytically generated intermediates is understood such that receptor residue limitations do not impact on the specific conclusions drawn. Moreover, the use of short-lived photoreactive species should be sought to reduce the likelihood of unspecific labeling and crosslinking (Brunner, 1993). Thus, in the absence of X-ray crystallographic or nuclear magnetic resonance data pertaining to receptor structure, the utility of mutational approaches and photoaffinity labeling approaches, particularly when combined, should result in an enhanced understanding of how receptor primary sequence and overall receptor structure are involved in agonist and antagonist binding affinity and selectivity.

4.1. Peptide Conformation Considerations

When one considers receptor structure–activity relationships relating to agonist binding, it is important to attempt to determine or understand the agonist peptide conformation upon binding to the receptor high-affinity state. Although each of the naturally occurring mammalian tachykinin peptides has only modest selectivity, the development and use of selective analogs has presented the opportunity to use physical methods, such as nuclear magnetic resonance spectroscopy, to address whether there exist distinct predominant conformations. Two NK_1 and NK_3 receptor selective hexapeptide analogs, Ac [Arg^6, Pro^9]SP(6–11) and [$pGlu^6$, *N*-$MePhe^8$]-SP(6–11), were studied in dimethylsulfoxide-d_6 by Levian-Teitelbaum et al. (1989). The use of nuclear Overhauser enhancement spectroscopy with the analysis of coupling constants and temperature shifts resulted in the suggestion that Ac[Arg^6,Pro^9]-SP(6–11) has a *trans* conformation about the Phe^8-Pro^9 amide bond and that the preferred conformation has a type I β-turn. It was further suggested that [pGlu[6], *N*-MePhe[8]]-SP(6–11) exists as a mixture of *cis-trans*-isomers in which each isomer can adopt a different conformation (*cis* isomer adopts a type IV β-turn, the *trans*-isomer adopts a γ-turn conformation), and evidence has been provided that the two last turns are stabilized by a hydrogen bond between the syn carboxyamide proton and the Glu ring carboxyl. The conformations of the former analogs may be considered tentative, and their relationship to the receptor active conformation is not certain, but it is of interest that the suggested turn is consistent with the methyl ester for methionine carboxyamide substitution. SP methyl ester loses activity at the NK_3 site, while retaining activity at the NK_1 site, which is consistent with these notions and the tentative selective agonist conformations. Consequently, NK_3 receptor agonists may have an intramolecular hydrogen bond in which the carboxyamide participates. It is also of interest that the NK_3 selective linear analog, senktide (suc-Asp-Phe-MePhe-GIy-Leu-Met-NH_2), has a highly ordered structure in methanol and in trifluoroethanol (Sumner and Feretti, 1989) in which the conformational restriction may be attributed to both steric and electrostatic interactions produced by the *N*-methylated Phe and the negatively charged amino terminus. On the other hand, the analysis by NMR of NKA(4–10), a potent NK_2 agonist, and several analogs including the selective agonist, [β-Ala[8]]NKA(4–10), has resulted in the suggestion (Saviano et al., 1991) that these compounds have conformations incompatible with models for the NK_1 and NK_3 agonists mentioned above. NKA(4–10) contains a type I β-turn extending from Ser[5] to Gly[8], followed by a γ-turn centered on Gly[8]. With [β-Ala[8]]NKA(4–10), it was suggested that a type I β-turn extended from Ser[5] to β-Ala[8], followed by C_8 turn comprising β-Ala[8] and Leu[9], and by an additional β-turn from β-Ala[8] to the carboxyamide. Additional studies by Maggi and coworkers on NK_2 receptor active compounds with or without the carboxyamide moiety have suggested that this region of the peptide brings about agonist activity by a conformational effect on the peptide structure (Patacchini et al., 1993). Thus, taken together these structural considerations indicate an incompatibility (Saviano et al., 1991) between typical NK_1 and NK_3

agonist conformations proposed from data derived from solution studies on peptides more rigid than NKA(4–10). One likely possibility remains that the different tachykinin agonists at NK_1, NK_2, and NK_3 receptors may adopt different receptor selective conformations. These results could be considered to be consistent with the suggestion that the agonist binding domains of the cloned mammalian tachykinin receptors are not structurally conserved and that different regions of the various receptors do not subserve the same role in agonist binding *(see below)*.

4.2. Receptor Agonist Binding Domains

With the cloning of tachykinin receptors comes the ability to explore functional aspects of these receptors using mutational approaches. Ligand binding domains for both agonists and non-peptide antagonists, as well as domains involved in activation of tachykinin receptors, have begun to be elucidated by several research groups using chimeric and point-mutated rat and human tachykinin receptors.

Using chimeric rat NK_1 and NK_2 type tachykinin receptors, the Kyoto group (Yokota et al., 1992) has identified a region of the NK_1 receptor, from the carboxy terminus of the first transmembrane domain to the amino terminus of the fifth transmembrane domain, which confers an NK_1 rank order of potency (SP > NKA) to the NK_2 receptor. The amino terminus was also found to play a role in agonist selectivity since an NK_1 receptor with an NK_2 amino terminus showed a modest decrease in SP affinity, as well as an increase in affinity for NKA. The converse chimeric receptor (NK_2 receptor with an NK_1 receptor amino terminus) showed a decreased potency for NKA, which resulted in no selectivity of NKA over SP. Further, these authors concluded that sequences carboxyl-terminal to the second extracellular loop were not important in receptor type selectivity since no differences in affinity for SP or NKA were seen when domains in this region of the NK_1 receptor were replaced by their NK_2 receptor counterparts. The authors used this result to support the hypothesis that these receptor carboxy-terminal regions interact with the conserved carboxy termini of tachykinin peptides. It could also be that neither SP or NKA interact with divergent residues within these sequences. NKB affinity was not tested in these studies.

Fong and coworkers constructed chimeric NK_1/NK_3 receptors, as well as NK_1 receptor point mutations, to identify extracellular regions involved in agonist binding (Fong et al., 1992b,c). In the amino terminus of the NK_1 receptor, mutation of residues Asn^{23}, Gln^{24} (which are conserved in the NK_3 receptor), and Pro^{25} (which is conserved among all three tachykinin receptors) led to a decrease in affinity of SP, NKA, and NKB. A similar result was obtained with mutation of Asn^{96} (conserved between NK_1 and NK_2 receptors) and His^{108} (conserved as Gln in NK_2 and NK_3 receptors) located in the first extracellular loop. Since the binding of all three tachykinins was affected by these mutations, these data were taken to suggest that residues in the amino terminus and first extracellular loop may interact with the conserved carboxy terminus of the tachykinins. Since all of these residues are not conserved, the authors suggested that the peptide binding site of the tachykinin receptors may not be highly conserved. Further support for this idea of nonconserved agonist binding domains of NK_1 and NK_3 receptors has recently appeared (Gether et al., 1993c).

Substitution of residues 170–174 of the NK_1 receptor second extracellular loop by the corresponding NK_3 receptor residues resulted in a change in the rank order of potency to SP > NKB > NKA. A similar substitution of residues 271–280 in the third extracellular loop resulted in an increase in the affinities of NKA and NKB, while having no effect on the affinity of SP, such that all three tachykinins had affinities that differed less than tenfold. These data were taken as evidence that the second and third extracellular loops are important for the binding of NKB, but not SP, and consequently as support for the idea of nonconserved peptide binding sites.

These two studies lay the foundation for the molecular understanding of peptide ligand interactions with the tachykinin receptors. They do, however, offer conflicting ideas on the roles of the extracellular and transmembrane domains of these receptors. The data from these studies do not suggest a straightforward interpretation for peptide ligand binding, but rather suggest that peptide interactions at these receptors is complex. Although Fong et al. suggest that the conserved carboxy terminus of tachykinins interacts with the amino terminus and first extracellular loop regions of these receptors, Yokota et al. suggest that it is the carboxy terminal regions of the transmembrane domains of the receptors that serve this function. Since both studies present data that could support, but do not prove, their conclusions, further studies using alternative approaches are clearly needed to better understand peptide–receptor interactions. One such study has been performed by Boyd and coworkers using photoaffinity analogs of SP (Boyd et al., 1991; Boyd et al., unpublished observations). One analog, ^{125}I-Phe^{8}(pBz)-SP, has been used to photoaffinity label the NK_1 receptor from transfected CHO cells. It was shown by proteolysis, peptide isolation, and sequence analysis to label a peptide fragment (residues 178–183) from the second extracellular loop domain, placing the conserved carboxy terminus of SP in or near this loop. Future work from this group with other analogs in which the benzophenone moiety is positioned elsewhere in the agonist peptide primary structure should help to orient substance P bound to the NK_1 receptor and allow a reinterpretation of the mutational data which have been obtained.

With regard to agonist selectivity, the Merck group (Fong et al., 1992b) was able to selectively increase the affinity of NKB, but was unable to confer a preference for NKB upon the NK_1 receptor, whereas the Kyoto group (Yokota et al., 1992) was able to reverse the rank order of SP and NKA. The results from these studies suggest that ligand selectivity is not simply encoded in a single residue (or even a single domain), but is likely the result of the summation of many positive and negative interactions between the tachykinin peptides and receptors.

4.3. Receptor Antagonist Binding Domains

The discovery of the nonpeptide antagonist, CP-96,345, opened a new era in the study of tachykinin receptors. For the first time, those interested in tachykinin pharmacology had a specific, high-affinity ligand for one of the tachykinin receptors. Aside from its specificity for the NK_1 receptor, CP-96,345 has an interesting species selectivity. This antagonist is approximately 100-fold more potent at the NK_1 receptor from human as compared to rat. This fact has been exploited by two groups as a starting point for understanding the molecular interactions of this important antagonist with the NK_1 receptor.

Fong et al. (1992d) reported that, starting with the human NK_1 receptor (hNK_1), simultaneous mutation of residues 116 and 290 (in the third and seventh transmembrane domains, respectively) to their rat counterparts was able to reduce the affinity of CP-96,345 to a level similar to that for the rat NK_1 receptor (rNK_1). They also showed that the same mutations could increase the affinity of a second antagonist, RP67580, to a level identical to that for rNK_1. These authors suspected that these residues act indirectly to alter the affinity of these two antagonists.

In a separate study, Sachais et al. (1993) have shown that, starting with rNK_1, mutation of residue 290 to its human counterpart largely, but not completely, increased the affinity of CP-96,345 to that of hNK_1. They further showed that residues in the second extracellular loop (residue numbers 190,191, and 195) and in the sixth membrane spanning domain (residue 266) are also involved in imparting the human phenotype. In this study, even these residues were not sufficient to completely revert the phenotype of rNK_1 to hNK_1, although the converse receptor (hNK_1 with the above residues as rat) was able to exactly revert the affinity of CP-96,345 from hNK_1 to rNK_1. These authors hypothesized that an additional residue was required for complete expression of the human phenotype. The study of Fong et al. (1992d) suggests that this residue is residue 116

Fong et al. (1992d) state that the switch of residues 116 and 290 are "both necessary and sufficient" to impart the lower affinity of rNK_1 onto hNK_1 for CP-96,345. The study by Sachais et al. (1993) showed that residue 116 is, in fact, not necessary. Furthermore, CP-96,345 has an approximately threefold higher affinity for the h(116,290) receptor than the rNK_1 receptor in the Fong study, a difference which these authors call insignificant. For this reason, they did not identify the residues in the second extracellular loop as the Sachais study did. It is difficult to determine if this difference is significant from the data presented by the Fong group since their analysis of CP-96,345 was performed by inhibition experiments in which increasing concentrations of unlabeled antagonist were present in tenfold increments. In the Sachais study, which did call a threefold difference significant when it was seen in each repetition of side-by-side comparisons, threefold increases in CP-96,345 concentrations were used. This experimental paradigm provides more data per experiment to support smaller differences that were consistently seen. It is important for the reader of such studies to look at the data, as well as the manner in which these data are obtained, in order to carefully evaluate the conclusions presented.

A third group has exploited the NK_1 receptor specificity of CP-96345 to approach the problem of understanding the interactions of this antagonist with the NK_1 receptor. Gether et al. (1993a) have constructed chimeric receptors between the rat NK_1 and NK_3 receptors. In these studies, they have identified domains of the NK_1 receptor extracellular to transmembrane domains V and VI, which can confer high-affinity binding of CP-96,345 to the NK_3 receptor. These two domains consist of residues 183–196 in the second extracellular loop and residues 271–275 in the third extracellular loop. This study, along with the study by Sachais et al. discussed above, support a model in which CP-96,345 interacts with extracellular domains of the NK_1 receptor, at least in part. This is in contrast to models of antagonist–receptor interaction at biogenic amine neorotransmitter receptors, where binding occurs within the transmembrane domains.

Studies with an NK_2 receptor specific antagonist, SR-48,968, have also examined antagonist binding domains. Gether et al. (1993b) have shown that high-affinity binding of this antagonist can be conferred on the NK_1 receptor by replacing its sixth transmembrane domain and second extracellular loop with the corresponding sequence from the NK_2 receptor. This receptor displays a decreased affinity for CP-96,345, while retaining NK_1 receptor affinities for the peptide agonists, SP and NKA. These results (along with those of Gether et al., 1993a) are consistent with the binding domain for these two antagonists being contained entirely in the transferred domains, but it is also possible that residues conserved among tachykinin receptors also interact directly with these ligands and that these transferred regions alter the local conformation of the antagonist binding pocket.

One question that arises when interpreting data from these types of mutational experiments is whether or not the effects observed are direct or indirect. That is, is the change in affinity of a given mutant receptor for a ligand due to the loss (and/or addition) of a functionality that directly interacts with the ligand in the wild type receptor, or is it due to changes in conformation of the receptor induced by the mutation? Even when one rules out global conformational changes (for example, by showing that the mutated receptor displays wild type affinity for other ligands that may not directly depend on the mutated residue), the possibility of local conformational changes in and around the binding pocket of the relevant ligand still exists. Strong evidence for direct interactions between a receptor and its ligand in the absence of direct structural data (i.e., X-ray diffraction) must come from studies in which functionalities on both the receptor and the ligand are covaried. A good example of this comes from the beta-adrenergic receptor, where compelling data of this kind are available demonstrating the existence of two hydrogen bonds between this receptor and catecholamine agonists (Strader et al., 1989). A study has been performed along this line for the NK_1 receptor (Fong et al., 1993). Although the data are more complex than in the case of the beta-adrenergic receptor, this study is consistent with a direct interaction of His^{197} of the NK_1 receptor with the benzhydryl moiety of CP-96,345.

An issue that is central to the understanding of antagonist binding is the mechanism of the antagonism. The antagonists described here are competitive with respect to SP binding at the NK_1 receptor. They may exclude the binding of SP either by binding to a site that overlaps with the SP binding site and thereby physically exclude SP binding, or by binding to an allosteric site and locking the receptor in a conformation which disrupts the SP binding site. Several studies (Fong et al., 1992d; Sachais et al., 1993; Gether et al., 1993a) have shown that antagonist binding can be altered without affecting SP binding, which suggests that the binding sites for agonist and antagonist are different. One should exert caution not to over interpret such a result since the sites could partially overlap, even completely overlap, in physical space if the residues studied interact directly with only agonist or antagonist. Future studies, possibly combining mutational analysis with photoaffinity labeling or perhaps other physical analyses of receptor structure, should further our understanding of the physicochemical determinants of tachykinin receptor–ligand interactions.

5. Approaches to Examine Second Messenger Systems Activated by Receptor Stimulation

In order to understand the mechanisms by which SP and related peptides bring about their biological effects, it is important to determine the types of second messengers and ion channels activated by receptor stimulation, as well as their respective extents and time courses of stimulation. The mechanisms by which second messenger systems are activated can occur via direct activation of the effector system by the G_α or $G_{\beta\gamma}$ subunit complexes or by an indirect effect as a result of direct activation of a different second messenger system. The best studied action of the tachykinin peptides in terms of second messengers is that of activation of phospholipase C with the subsequent production of inositol 1,4,5-trisphosphate, which can mobilize intracellular calcium ions, and production of diacylglycerol, which can activate certain protein kinase C isoforms. As tachykinin receptor activation in many cases occurs via a pertussis toxin insensitive G-protein, the search for such G-protein heterotrimeric complexes that are able to reconstitute high-affinity agonist binding to tachykinin receptor has resulted in the identification of the $G_{q/11}$ subfamily of G_α proteins as important in this regard. Recently, Kwatra and collaborators (1993) demonstrated such a successful reconstitution study, and it was previously shown by separate groups that the subfamily of phospholipase C enzymes was activated by G_q owing to G-protein coupled receptor stimulation. Thus, it seems reasonable that SP action via its receptor in at least some systems involves a $G_{q/11}$

subunit in its mechanism of action. Since the $G_{q/11}$ subfamily has related family members that apparently activate several different phospholipases (Sternweis and Smrcka, 1992), it will be of interest to determine whether any of these pertussis toxin insensitive G-proteins are involved in the mechanisms of tachykinin peptide action. It may be that additional effectors are activated via this class of pertussis toxin insensitive G_{α} subunits as well. Additional studies (Nakajima et al., 1992; Takeda et al., 1992; Mitsuhashi et al., 1992; Blount and Krause, 1993) have demonstrated cyclic AMP responsiveness of the cloned tachykinin receptors, which seems to be a direct function of G-protein activated cyclic AMP production. Approaches that can be used for examining second messenger involvement include reconstitution with specific G-protein heterotrimeric complexes, transfection or cotransfection experiments with specific receptor types into cell lines that express the appropriate effectors, direct analysis of additional second messenger systems in animal tissues or cell lines, and, possibly, antisense RNA or gene knockout studies to evaluate the participation of various G-proteins and effector systems in the action of tachykinin peptides.

5.1. Second Messenger Response Domains

Recently, mutational studies were carried out to determine the relationship between primary structure of the NK_1 and NK_2 receptors and second messenger responses stimulated by agonists (Blount and Krause, 1993). These studies indicated that structurally homologous protein domains of the NK_1 and NK_2 receptors were not necessarily functionally equivalent. The third intracellular domain (i.e., C3 domain) of the NK_1 receptor may play an important role in G-protein and second messenger activation because NK_1 receptors containing the NK_2 receptor C3 domain were significantly dysfunctional in agonist-induced cAMP and inositol phosphate responses. The complementary chimeric receptor (NK_2 with the NK_1 C3 domain) functions at least as well as the wild type in this regard. The carboxyl-terminal domain of the NK_2 receptor may play a similar important role in G-protein and second messenger activation because NK_2 receptors containing the NK_1 receptor CT domain were significantly dysfunctional in agonist-induced cAMP responses. The complementary chimeric receptor (NK_1 with the NK_2 CT domain) showed an increase in cyclic AMP at least as great as the wild type upon agonist stimulation. It has been further suggested that the CT domain of the NK_2 receptor plays an important role in G-protein and second messenger activation, and that the same domain in the NK_1 receptor plays little if any equivalent role. The results obtained that supported this conclusion were that deletion of the CT domain of the NK_1 receptor (NK_1-Δ325) has little if any effect on agonist dependent second messenger responses. A relatively small deletion of the CT domain of the NK_2 receptor (NK_2-Δ367) has a significant effect on the magnitude of second messenger responses, especially cyclic AMP. Taken together, these results implicate both the C3 and CT domains of the tachykinin receptors in second messenger responsiveness, and it has been suggested that one reason for the diversity of tachykinin peptides may be that the receptors bring about different responses in part due to the activation of different G-proteins (Blount and Krause, 1993).

5.2. Molecular Mechanisms Regulating Receptor Sensitivity

For many G-protein coupled receptors, short term regulation of receptor sensitivity largely involves posttranslational mechanisms, of which phosphorylation, uncoupling, and sequestration have been the events most frequently examined experimentally. Less well known is the role transcriptional mechanisms play in the regulation of G-protein coupled receptor mRNA expression, both in short term (minutes to hours) and long term (hours to days) regulation. Of course, the regulation of receptor sensitivity involves mechanisms such as specific G-protein expression and coupling to the receptor, as well as expression of the effector enzymes and their appropriate localization within the cell coincident with receptor localization. The focus of this discussion is on receptor expression. For tachykinin receptor systems, tachyphylaxis of receptor responses has been known and, in fact, was a feature used to aid in the early classification of receptor types (Lee et al., 1982). As discussed above, the structural features of tachykinin receptor protein with regard to short term receptor regulation by phosphorylation are apparent and are currently under investigation by several groups. Questions concerning long term regulation and the mechanisms involved have recently surfaced due to the determination of the organization and structure of the receptor genes, particularly that of the NK_1 receptor gene.

The determination of the organization and structure of tachykinin receptor genes has resulted in the generation of several hypotheses concerning gene regulation due to the presence of discrete structural motifs that have been shown to subserve regulated expression in other systems. For the tachykinin receptors, many studies have been performed on receptor levels and affinity, but only recently have studies examined regulation of mRNA expression at the level of transcriptional activation and mRNA turnover. One common mechanism by which transcriptional regulation can be achieved is by protein binding to promoter and/or enhancer sequences with subsequent activation of gene readout. Moreover, it has become clear that combinatorial mechanisms are often utilized by a cell in transcriptional activation by transcription factors. The regulation of mRNA stability can also have important consequences for transcriptional regulation. Though this is an area of investigation for the tachykinin receptors that has only surfaced recently, it appears likely that some of the cellular mechanisms operative for the regulation of receptor sensitivity involve transcriptional and posttranscriptional processes. These are discussed in further detail below.

6. General Considerations of Receptor Gene Organization and Structure Pertaining to Regulation

The molecular cloning and functional expression of mammalian tachykinin receptors has stimulated the search for the genes encoding the receptor mRNAs. As discussed in more detail below, genes encoding NK_1, NK_2, and NK_3 receptors have recently been identified and analyzed. A schematic illustration of the organization of human NK_1, NK_2, and NK_3 receptor genes is provided in Fig. 4. The information about gene organization, structure, and patterns of expression is essential in order to understand cellular mechanisms of regulated mRNA expression, including both transcriptional mechanisms and RNA stability mechanisms. The knowledge of receptor gene promoter sequences, in the context of information regarding transcriptional activator proteins that bind to established *cis* DNA sequences that mediate transcriptional responses, may help investigators further understand and evaluate regulated tachykinin receptor gene expression. Structural information regarding the 5' and 3' untranslated sequences may help to identify mechanisms regulating mRNA translatability and stability. The organization of the tachykinin receptor genes is important to consider as the genes are split into multiple exons with introns that interrupt the protein coding region (Fig. 2). It has been hypothesized that intron insertion sites in the receptor genes may split the protein into distinct functional domains, perhaps related to agonist binding, second messenger response domains, and mechanisms of receptor desensitization (Hershey et al., 1991). Of receptors coupled to heterotrimeric G-proteins, it now is clear that different patterns of gene organization exist in which some receptor proteins are contained within a single gene exon, others are comprised of two exons, of which the second exon encodes the protein coding region, and others are expressed from genes containing three or more exons. It is of interest in this latter class of receptor genes that many of the exon splice sites are conserved with respect to their location relative to specific putative transmembrane domains of the encoded receptor protein. It has been suggested that the evolutionary relationships among G-protein coupled receptors can be further evaluated by recognition of intron locations within the gene organizations since duplication of individual exon segments may represent one mechanism by which the diversity of G-protein coupled receptors came about.

6.1. NK_1 Receptor Gene Regulation

Though much information is available regarding the locations and levels of NK_1 receptor expression in several species, only recently have the mechanisms responsible for regulating NK_1 and other tachykinin receptor gene expression been under investigation.

The availability of cDNA probes, as well as the knowledge of the organization and structure of the NK_1 receptor gene from rat and human species, has provided appropriate tools and theoretical constructs for studies designed to elucidate the mechanisms responsible for regulated gene expression. The cDNA probes, along with certain genomic probes, can be used to assess the level of mRNA expression, as well as the types of specific heteronuclear mRNA forms extant during physiological or pharmacological stimuli. The putative promoter regions of the genes have been examined by nucleotide sequence analysis methods, and specific sequence motifs, by comparison to their functional activity on transcription in other systems, have been postulated to be of importance in tachykinin receptor gene expression.

An examination of the 5' flanking region of the rat and human NK_1 receptor genes demonstrates a high degree of conservation in the putative proximal promoter region upstream from the respective TATA-dependent transcription initiation sites (Krause et al., 1993b). This region, consisting of approximately 500 base pairs, has approximately 75% identity between rat and human with several sequence motifs of 6–8 bases in length that display significant homology with previously identified sequences that mediate transcriptional regulation. These elements, which have not been shown to be functional regulators in the NK_1 gene context as yet, include AP_1, AP_2, AP_4, cyclic AMP/Ca^{2+}, Oct_2, SP_1 and NF-kB sites (Hershey et al., 1991; Takahashi et al., 1992; Krause et al., 1993b). The presence of a Ca^{2+}/cyclic AMP response element at 45–52 bases upstream from the transcription initiation site, for example, has interesting implications for regulated gene expression. Since activation of the NK_1 receptor in several tissue and cellular contexts leads to both Ca^{2+} mobilization, as well as cyclic AMP elevation, one or both of these messengers may trigger gene transcription (Hershey et al., 1991). This may be relevant during periods of intense or sustained NK_1 receptor stimulation, since the receptor presumably desensitizes and a transcriptional mechanism may exist for receptor resensitization. Recently, it was reported that forskolin treatment of U373-MG astrocytoma cells, which caused a two- to threefold elevation of intracellular cyclic AMP levels, resulted in enhanced transcription of the NK_1 receptor gene and the presence of mature mRNA 4 h after the start of pharmacological treatment (Krause et al., 1993b). On the other hand, treatment with 1,9-dideoxyforskolin, a forskolin derivative incapable of activating adenylate cyclase (Seamon and Daly, 1983), did not effect either NK_1 receptor gene transcription or steady-state mRNA levels (Krause et al., 1993b). Evidence has also been presented that indicates that phorbol ester, a molecule structurally similar to diacyglycerol, can upregulate steady-state NK_1 receptor mRNA levels after acute treatment of cultured astrocytoma cells (Krause et al., 1993b; DiMaggio and Krause, unpublished). Consequently, it appears that at least two second messenger systems can regulate NK_1 receptor gene expression, possibly as consequence of either homologous (NK_1) or heterologous receptor activation.

Ihara and Nakanishi reported (1990) that glucocorticoids dose-dependently downregulated NK_1 receptor gene expression in the rat pancreatic cell line, AR42J. The time dependence of the down-regulation was short (1-2 h onset) and was specific for glucocorticoids. In this cell line, the receptor mRNA under control situations has a short half life of approximately 30 min, and this rate was not affected by dexamethasone treatment. They also showed that the rapid turnover of NK_1 receptor mRNA was dependent on ongoing protein synthesis, and the capacity of various glucocorticoids to downregulate expression correlated with their relative potencies as glucocorticoids. Thus, it was concluded that this class of steroid hormones act at transcription initiation to reduce NK_1 receptor mRNA levels (Ihara and Nakanishi, 1990). Similar downregulatory effects of glucocorticoids have been observed on NK_1 receptor mRNA expression in U373-MG astrocytoma cells (Krause et al., 1991) and in lymphoblast IM-9 cells (Gerard et al., 1991). In the IM-9 cell line study, it was also shown that NK_1 receptor binding sites were downregulated, as well. The molecular mechanisms for this downregulation have not been established, but a glucocorticoid negative regulatory element in either the rat or human genes has not been observed to date. Consequently, this downregulation, which seems to be a general phenomenon in several NK_1 receptor expressing cells, may utilize an indirect mechanism. Since SP can act as an inflammatory agent in tissue inflammation responses, glucocorticoids may bring about their antiinflammatory actions in part by regulation of the tissue sensitivity to SP by altering receptor expression.

Recently, evidence has been provided for increased acute expression of the NK_1 receptor gene in the spinal cord dorsal horn in two models of subchronic nociceptive activation (McCarson and Krause, 1994). These animal models consist of the injection of dilute formalin or complete Freund's adjuvant into the plantar aspect of a hind paw of the rat. The formalin test (Dubuisson and Dennis, 1977; Tjolsen et al., 1992) produces a biphasic, pain-related behavioral response within the first hour after injection into the hind paw and an inflammation of the paw that is long lasting. Injection of adjuvant into one hind paw produces a longer-lasting monoarthritic inflammation that peaks after about 4 days (Iadarola et al., 1988a,b). Evidence has been presented for the involvement of NK_1 receptors in the formalin pain model (Yashpal et al., 1993), and McCarson and Krause (1994) have suggested that in these models NK_1 receptor activation may signal synthesis of new NK_1 receptors via increased mRNA levels as part of a cellular resensitization response.

6.2. NK_2 Receptor Gene Regulation

Little information is available concerning the regulation of NK_2 receptor gene expression. Two groups (Gerard et al., 1990; Graham et al., 1991) have recently isolated the NK_2 receptor gene and have described its organization. Although some inconsistencies in the reported sequence and gene organization exist, it appears that gene organization as reported by Graham et al. is consistent with that described for the rat and human NK_1 receptor gene (Gerard et al., 1991; Hershey et al., 1991; Takahashi et al., 1992; Krause et al., 1993b) and human NK_3 receptor gene (Buell et al., 1992; Takahashi et al., 1992). The NK_2 receptor gene contains a TATA sequence immediately upstream from the transcription initiation site, and the only reported recognizable 5' flanking sequence that may be involved in transcriptional regulation is a GC box-like sequence. Further studies are required to determine the mechanisms responsible for spatial, temporal, and other regulated changes in NK_2 receptor gene expression.

6.3. NK_3 Receptor Gene Regulation

Little information is available on the extent of and mechanisms responsible for regulated NK_3 receptor gene expression. The human receptor gene has been isolated (Buell et al., 1992; Takahashi et al., 1992), and the latter study determined the sequence of the putative promoter region. Potential promoter, enhancer, and other regulatory DNA elements include AP_1, AP_2, cyclic AMP responsive and SP_1 elements or sites, which are located upstream from the TATA-dependent transcription site.

To date, only one study has been reported on in vivo regulation of NK_3 receptor mRNA expression (McCarson and Krause, 1994). NK_3 receptors are located in the superficial laminae of the rat dorsal horn and have been implicated in antinociceptive processes. In the subacute pain models used for NK_1 receptor mRNA regulation studies (discussed above), dorsal horn NK_3 receptor mRNA levels also increase about twofold following nociceptive activation. Although the molecular mechanisms responsible for these changes have not been addressed, it has been suggested that nociception may trigger increased activity of intrinsic NKB antinociceptive systems that could act to limit or modify transmission of nociceptive information to higher brain centers (McCarson and Krause, 1993).

References

Appell, K. C., Fragale, B. J., Loscig, J., Singh, S., and Tomczuk, B. E. (1992) Antagonists that demonstrate species differences in neurokinin-1 receptors. *J. Pharm. Exp. Ther.* **41,** 772–778.

Bernstein, G., Blank, J. L., Jhon, D.-Y., Exton, J. H., Rhee, S. G., and Ross, E. M. (1992) Phosphlipase C-β1 is a GTPase-activating protein for $G_{q/11}$, its physiological regulator. *Cell* **70,** 411–418.

Berridge, M. J. (1993) Insoitol trisphosphate and calcium signaling. *Nature* **361,** 315–325.

Blount, P. and Krause, J. E. (1993) Functional nonequivalence of structurally homologous domains of neurokinin-1 and neurokinin-2 type tachykinin receptors. *J. Biol. Chem.* **268,** 16,388–16,395.

Boyd, N. D., White, C. F., Cerpa, R., Kaiser, E. T., and Leeman, S. E. (1991) Photoaffinity labeling the substance P receptor using a derivative of substance P containing p-benzoylphenylalanine. *Biochemistry* **30,** 336–342.

Brunner, J. (1993) New photolabelling and crosslinking methods. *Ann. Rev. Biochem.* **62,** 483–514.

Buck, S. H. and Shatzer, S. A. (1988) Agonist and antagonist binding to tachykinin peptide NK-2 receptors. *Life Sci.* **42,** 2701–2708.

Buck, S. H., Helke, C. J., Burcher, E., Shults, C. W., and O'Donohue, T. L. (1986) Pharmacologic characterization and autoradiographic distribution of binding sites for iodinated tachykinins in the rat central nervous system. *Peptides* **7,** 1109–1120.

Buck, S. H., Burcher, E., Shults, C. W., Lovenberg, W., and O'Donohue, T. L. (1984) Novel pharmacology of substance K-binding sites: a third type of tachykinin receptor. *Science* **266,** 987–989.

Buell, G., Schulz, M. F., Arkinstall, S. J., Maury, K., Missotten, M., Adami, N., Talabot, F., and Kawashima, E. (1992) Molecular characterisation, expression and localisation of human neurokinin-3 receptor. *FEBS* **299,** 90–95.

Cascieri, M.A., Chicchi, G. G., and Liang, T. (1985) Demonstration of two distinct tachykinin receptors in rat brain cortex. *J. Biol. Chem.* **260,** 1501–1507.

Cascieri, M. A. and Liang, T. (1983) Characterization of the substance P receptor in rat brain cortex membranes and the inhibition of radioligand binding by guanine nucliotides. *J. Biol. Chem.* **258,** 5158–5164.

Couture, R., Fournier, A., Magnan, J., St-Pierre, S., and Regoli, D. (1979) Structure-activity studies on substance P. *Can. J. Pysiol.* **57,** 1427–1436.

Dam, T.-V., Takeda, Y., Krause, J. E., Escher, E., and Quirion, R. (1990) γ-Preprotachykinin-(72–92)-peptide amide: An endogenous preprotachykinin I gene-derived peptide that preferentially binds to neurokinin-2 receptors. *Proc. Natl. Acad. Sci. USA* **87,** 246–250.

DeBernardi, M. A., Seki, T., and Brooker, G. (1991) Inhibition of cAMP accumulation by intracellular calcium mobilization in C6-2B cells stably transfected with substance K receptor cDNA. *Proc. Natl. Acad. Sci. USA* **88,** 9257–9261.

Dietl, M. M. and Palacios, J. M. (1991) Phylogeny of tachykinin receptor localization in the vertebrate central nervous system: Apparent absence of neurokinin-2 and neurokinin-3 binding sites in the human brain. *Brain Res.* **539,** 211–222.

Dion, S., Nantel, R. F., Jukic, D., Rhaleb, N. E., Tousignant, S., Telemasque, S., Drapeau, G., Regoli, D., Naline, C., Advenier, C., Rovero, P., and Maggi, C. A. (1990) Structure-activity study of neurokinins: Antagonists for the neurokinin-2 *Receptor. Pharm.* **41,** 184–194.

Dubuisson, D. and Dennis, S. G. (1977) The formalin test: A quantitative study of the analgenic effects of morphine, meperidine, and brain stem stimulation in rats and cats. *Pain* **4,** 161–174.

Eistetter, H. R., Church, D. J., Mills, A., et al (1991) Recombinant bovine neurokinin-2 receptor stably expressed in Chinese hamster ovary cells couples to multiple signal transduction pathways. *Cell Reg.* **2,** 767–779.

Emonds-Alt, X., Vilain, P., Goulaouic, P., Proietto, D., Van Broeck, D., Advenier, C., Naline, E., Neliat, G., LeFur, G., and Breliere, J. C. (1992) A potent and selective nonpeptide antagonist of the neurokinin A (NK-2) receptor. *Life Sci.* **50,** 101–106.

Folkers, K., Hakanson, R., Horig, J., Jie-Cheng, X., and Leander, S. (1984) Biological evaluation of substance P antagonists. *Br. J. Pharmacol.* **83,** 449–456.

Fong, T. M., Anderson, S. A., Yu, H., Huang, R.-R. C., and Strader, C. D. (1992a) Differential activation of intracellular effector by two isoforms of human neurokinin-1 receptor. *Mol. Pharmacol.* **41,** 24–30.

Fong, T. M., Huang, R. C., and Strader, C. D. (1992b) Localization of agonist and antagonist binding domains of the human neurokinin-1 receptor. *J. Biol. Chem.* **267,** 25,664–25,667.

Fong, T. M., Yu, H., Huang, R. C., and Strader, C. D. (1992c) The extracellular domain of the neurokinin-1 receptor is required for high affinity binding of peptides. *Biochemistry* **31,** 11,806–11,811.

Fong, T. M., Yu, H., and Strader, C. D. (1992d) Molecular basis for the species selectivity of the neurokinin-1 receptor antagonists CP-96,345 and RP67580. *J. Biol. Chem.* **267,** 25,668–25,671.

Fong, T. M., Cascieri, M. A., Yu, H., Bansal, A., Swain, C., and Strader, C. D. (1993) Amino-aromatic interaction between histidine 197 of the neurokinin-1 receptor and CP 96345. *Nature* **362,** 350–353.

Franke, R. R., König, B., Sakmar, T. P., Khorana, H. G., and Hofman, K. P. (1990) Rhodopsin mutants that bind but fail to activate transducin. *Science* **250,** 123–125.

Garret, C., Carruette, A., Fardin, V., Moussaoui, S., Peyronei, J., Blanchard, L., and Laduron, P. M. (1991) Pharmacological properties of a potent and selective nonpeptide substance P antagonist. *Proc. Natl. Acad. Sci. USA* **88,** 10,208–10,212.

Gerard, N. P., Bao, L., Xiao-Ping, H., and Gerard, C. (1993) Molecular aspects of the tachykinin receptors. *Regul. Pept.* **43,** 21–35.

Gerard, N. P., Garraway, L. A., Eddy, R. L., Shows, T. B., Iijima, H., Paquet, J., and Gerard, C. (1991) Human substance P receptor (NK-1): Organization of the gene, chromosome localization, and functional expression of cDNA clones. *Biochemistry* **30,** 10,640–10,646.

Gerard, N. P., Eddy, R. L., Jr., Shows, T. B., and Gerard, C. (1990) The human neurokinin A (substance K) receptor. *J. Biol. Chem.* **265,** 20,455–20,462.

Gether, U., Johansen, T. E., Snider, R. M., Lowe III, J. A., Nakanishi, S., and Schwartz, T. W. (1993a) Different binding epitopes on the NK-1 receptor for substance P and a nonpeptide antagonist. *Nature* **362,** 345–348.

Gether, U., Yokota, Y., Edmonds-Alt, X., Berliere, J.-C., Lowe, J. A., Snider, R. M., Nakanishi, S., and Schwartz, T. W. (1993b) Two nonpeptide tachykinin antagonists act through epitopes on corresponding segments of the NK-1 and NK-2 receptors. *Proc. Natl. Acad. Sci. USA* **90,** 6194–6198.

Gether, U., Johansen, T. E., and Schwartz, T. W. (1993) Chimeric NK_1 (substance P)/ NK_3 (neurokinin B) receptors: identification of domains determining the binding specificity of tachykinin agonists. *J. Biol. Chem.* **268,** 7893–7898.

Gorbulev, V., Akhundova, A., Luzius, H., and Fahrenholz, F. (1992) Molecular cloning of substance P receptor cDNA from guinea pig uterus. *Biochem. Biophys. Acta* **1131,** 99–102.

Graham, A., Hopkins, B., Powell, S. J., Danks, P. and Briggs, I. (1991) Isolation and characterization of the human lung NK-1 receptor gene using rapid amplification of cDNA ends. *Biochem. Biophys. Res. Comm.* **177,** 8–16.

Guard, S. and Watson, S. P. (1991) Tachykinin receptor types: classification and membrane signaling mechanisms. *Neurochem. Int.* **18,** 149–169.

Hanley, M. R., Lee, C.-M., Jones, L. M., and Michell, R. H. (1980) Similar effects of substance P and related peptides on salivation and phosphatidylinositol turnover in rat salivary glands. *Mol. Pharmacol.* **18,** 78–83.

Harada, Y., Takahashi, T., Kuno, M., Nakayama, K., Masu, Y., and Nakanishi, S. (1987) Expression of two different tachykinin receptors in Xenopus oocytes by exogenous mRNAs. *J. Neurosci.* **7,** 3265–3273.

Helke, C. J., Krause, J. E., Mantyh, P. W., Couture, R., and Bannon, M. J. (1990) Diversity in mammalian tachykinin peptidergic neruons: multiple peptides, receptors, and regulatory mechanisms. *FASEB J.* **4,** 1606–1615.

Hershey, A. D., Dykema, P. E., and Krause, J. E. (1991) Organization, structure, and expression of the gene encoding the rat substance P receptor. *J. Biol. Chem.* **266,** 4366–4374.

Hershey, A. D. and Krause, J. E. (1990) Molecular characterization of a functional cDNA encoding the rat substance P receptor. *Science* **247,** 958–961.

Hopkins, B., Powell, S. J., Dankal, P., Briggs, I., and Graham, A. (1991) Isolation and characterization of the human lung NK-1 receptor cDNA. *Biochem. Biophys. Res. Commun.* **180,** 1110–1117.

Huang, R.-R. C., Cheung, A. H., Mazina, K. E., Strader, C. D., and Fong, T. M. (1992) cDNA sequence and heterologous expression of the human neurokinin-3 receptor. *Biochem. Biophys. Res. Commun.* **184,** 966–972.

Hunter, J. C., Goedert, M., and Pinnock, R. D. (1985) Mammalian tachykinin-induced hydrolysis of inositol phospholipids in rat brain slices. *Biochem. Biophys. Res. Commun.* **127,** 616–622.

Iadarola, M. J., Brady, L. S., Draisci, G., and Dubner, R. (1988a) Enhancement of dynorphin gene expression in spinal cord following experimental inflammation: Stimulus specificity, behavioral parameters and opioid receptor binding. *Pain* **35,** 313–326.

Iadarola, M. J., Douglass, J., Civelli, O., and Naramjo, J. R. (1988b) Different activation of spinal cord dymorphin and enkephalin neurons during hyperalgesia: Evidence using cDNA hybridization. *Pain Res.* **455,** 205–212.

Ihara, H. and Nakanishi, S. (1990) Selective inhibition of expression of the substance P receptor mRNA in pancreatic acinar AR42J cells by glucocorticoids. *J. Biol. Chem.* **265,** 22,441–22,445.

Ingi, T., Kitaima, Y., Minamitake, Y., and Nakanishi, S. (1991) Characterization of ligand-binding properties and selectivities of three rat tachykinin receptors by transfection and functional expression of their cloned cDNAs in mammalian cells. *J. Pharmacol. Exper. Ther.* **259,** 968–975.

Joslin, G., Krause, J. E., Hershey, A. D., Adams, S. P., Fallon, R. J., and Perlmutter, D. H. (1991) Amyloid- peptide, substance P, and bombesin bind to the serpin-enzyme complex receptor. *J. Biol. Chem.* **266,** 21,897–21,902.

Krause, J. E., Takeda, J., Cremins, J. D., and Raddatz, R. (1991) Regulation of substance P receptor mRNA expression by cAMP and glucocorticoids in human and rat cell lines. *Soc. Neurosci. Abstr.* **17,** 188.

Krause, J. E., Sachais, B. S., and Blount, P. (1993a) Tachykinin receptors, in *Handbook of Receptors and Channels,* vol. 1: G protein coupled receptors (Peroutka, S. J., ed.), CRC, FL, pp. 277–298.

Krause, J. E., Bu, J.-Y., Takeda, Y., Blount, P., Raddatz, R., Sachais, B. S., Chou, K. B., Takeda, J., McCarson, K. E., and DiMaggio, D. (1993b) Structure, expression and second-messenger mediated regulation of the human and rat substance P receptors and their genes. *Regul. Pept.* **46,** 59–66.

Kris, R. M., South, V., Saltzman, A., Felder, S., Ricca, G. A., Jaye, M., Huebner, K., Kagan, J., Croce, C. M., and Schlessinger, J. (1991) Cloning and expression of the human substance K receptor and analysis of its role in mitogenesis. *Cell Growth Differen.* **2,** 15–22.

Kwatra, M. M., Schwinn, D. A., Schreurs, J., Blank, J. L., Kim, C. M., Benovic, J. L., Krause, J. E., Caron, M. G., and Lefkowitz, R. J. (1993) The substance P receptor which couples to $G_{q/11}$ is a substrate of β-adrenergic receptor kinase 1 and 2. *J. Biol. Chem.* **268,** 9161–9164.

Laufer, R., Gilon, C., Chorev, M., and Selinger, Z. (1988) Desensitization with a selective agonist discriminates between multiple tachykinin receptors. *J. Pharmacol. Exper. Ther.* **245,** 639–643.

Laufer, R., Gilon, C., Chorev, M., and Selinger, Z. (1986) Characterization of a neurokinin B receptor site in rat brain using a highly selective radioligand. *J. Biol. Chem.* **261,** 10,257–10,263.

Laufer, R., Wormser, U., Friedman, Z. Y., Gilon, C., Chorev, M., and Selinger, Z. (1985) Neurokinin B is a preferred agonist for a neuronal substance P receptor and its action is antagonized by enkephalin. *Proc. Natl. Acad. Sci. USA* **82,** 7444–7448.

Lavielle, S., Chassaing, G., Ploux, 0., Louillet, D., Besseyre, J., Julien, S., Marquet, A., Convert, O., Beauouan, J.-C., Torrens, Y., Bergstrom, L., Saffroy, M., and Glowinski, J. (1988) Analysis of tachykinin binding site interactions using constrained analogues of tachykinins. *Biochem. Pharmacol.* **37,** 41–49.

Lee, C.-M., Iversen, L. L., Hanley, M. R., and Sandberg, B. E. B. (1982) The possible existence of multiple receptors for substance P. *Naunyn Schmiedeberg's Arch. Pharmacol.* **318,** 281–287.

Lee, C.-M., Javitch, J. A., and Snyder, S. H. (1983) ^{3}H-Substance P binding to salvary gland membranes: regulation by guanyl nucleotides and divalent cations. *Mol. Pharmacol.* **23,** 563–569.

Levian-Teitelbaum, D., Kolodny, N., Chorev, M., Selinger, Z., and Gilon, C. (1989) ^{3}H-NMR Studies of receptor-selective substance P analogs reveal distinct predominant conformations in DMSO-d_6. *Biopolymers* **28,** 51–64.

Li, X. J., Wolfgang, W., Wu, Y.-N., North, R. A., and Forte, M. (1991) Cloning, heterlogous expression and developmental regulation of a *Drosophila* receptor for tachykinin-like peptides. *EMBO J.* **10,** 3221–3229.

Luber-Narod, J., Boyd, N. D., and Leeman, S. E. (1990) Guanine nucleotides decrease the affinity of substance P binding to its receptor. *Euro. J. Pharmacol.* **188,** 185–191.

MacDonald, S. G. and Boyd, N. D. (1989) Regulation of substance P receptor affinity by guanine nucliotide-binding proteins. *J. Neurochem.* **53,** 264–272.

Maggi, C. A., Eglezos, A., Quartara, R., Patacchini, R., and Giachetti, A. (1992) Heterogeneity of NK-2 tachykinin receptors in hamster and rabbit smooth muscles. *Regul. Pept.* **37,** 85–93.

Maggi, C. A., Patacchini, R., Giuliani, S., Rovero, P., Dion, S., Regoli, D., Giachetti, A., and Meli, A. (1990) Competitive antagonists discriminate between NK_2 tachykinin receptor subtypes. *Br. J. Pharmacol.* **100,** 588–592.

Mantyh, P. W., Pinnock, R. D., Downes, C. P., Goedert, M., and Hunt, S. P. (1984) Coorelation between inositol phospholipid hydrolysis and substance P receptors in rat CNS. *Nature* **309,** 795–797.

Masu, Y., Nakayama, K., Tamaki, H., Harada, Y., Kuno, M., and Nakanishi, S. (1987) cDNA cloning of bovine substance-K receptor through oocyte expression system. *Nature* **329,** 836–838.

McCarson, K. E. and Krause, J. E. (1994) NK-l and NK-3 type tachykinin receptor mRNA expression in the rat spinal cord dorsal horm is increased during aduvant or formalin-induced nociception. *J. Neurosci.*, in press.

McKnight, A. T., Maguire, J. J., Elliott, N. J., Fletcher, A. E., Foster, A. C., Tridgett, R., Williams, B. J., Longmore, J., and Iverson, L. L. (1991) Phamacological specificity of novel, synthetic, cyclic peptides as antagonists at tachykinin receptors. *Br. J. Pharm.* **104,** 355–360.

Merritt, J. E. and Rink, T. J. (1987) The effects of substance P and carbachol on inositol tris- and tetrakisphosphate formation and cytosolic free calcium in rat parotid acinar cells. *J. Biol. Chem.* **262,** 14,912–14,916.

Mitsuhashi, M., Ohashi, Y., Scichio, S., Christian, C., Sudduth-Klinger, J., Harrowe, G., and Payan, D.G. (1992) Multiple intracellular signaling pathways of the neuropeptide substance P receptor. *J. Neurosci. Res.* **32,** 437–443.

Monnier, D., Colas, J.-F., Rosay, P., Hen, R., Borrelli, E., and Maroteaux, L. (1992) NKD, a developmentally regulated tachykinin receptor in Drosophila. *J. Biol. Chem.* **267,** 1298–1302.

Morimoto, H., Murai, M., Maeda, Y., Yamaoka, M., Nishikana, M., Kiyotoh, S., and Fuii, T. (1992) FR 224, a novel cyclopeptide substance P antagonist with NK-l and NK-2 selectivity. *J. Pharmacol. Exp. Ther.* **262,** 398–402.

Mousli, M., Baeb, J.-L., Bronner, C., Rauot, B., and Landry, Y. (1990) G-protein activation: A receptor-independent mode of action for Ca^+ ionic amphiphilic neuropeptides and venom peptides. *Trends Pharmacol. Sci.* **11,** 358–362.

Murai, M., Morimoto, H., Maeda, Y., Kiyotoh, S., Nishikawa, M., and Fujii, T. (1992) Effects of FK 224, a novel compound NK-1 and NK-2 receptor antagonist, on airway constriction and airway edema induced by neurokinins and sensory nerve stimulation in guinea pigs. *J. Pharmacol. Exp. Ther.* **262,** 403–408.

Nakaima, Y., Tsuchida, K., Negishi, M., Ito, S., and Nakanishi, S. (1992) Direct linkage of three tachykinin receptors to stimulation of both phosphatidylinositol hydrolysis and cyclic AMP cascades in transfected Chinese hamster ovary cells. *J. Biol. Chem.* **267,** 2437–2442.

Nakanishi, S. (1991) Mammalian tachykinin receptors. *Annu. Rev. Neurosci.* **14,** 123–136.

Narumi, S. and Maki, Y. (1978) Stimulatory effects of substance P on neurite extension and cyclic AMP levels in cultured neuroblastoma cells. *J. Neurochem.* **30,** 1321–1326.

Nei, M. (1987) *Molecular Evolutionary Genetics.* Columbia University Press, New York.

Nilsson, J., von Euler, A. M., and Dalsgaard, C.-J. (1985) Stimulation of connective tissue cell growth by substance P and substance K. *Nature* **315,** 61–63.

Parker, I., Sumikawa, K., and Miledi, R. (1986) Neurotensin and substance P receptors expressed in *Xenopus* oocytes by messenger RNA from rat brain. *Proc. R. Soc. Lond.* **229,** 151–159.

Patacchini, R., Quartara, L., Rovero, P., Goso, C., and Maggi, C.A. (1993) Role of C-terminal amidation on the biological activity of neurokinin A derivatives with agonist and antagonist properties. *J. Pharm. Exp. Ther.* **264,** 17–21.

Quintana, J., Wong, H., and Ascoli, M. (1993) The regulation of the binding affinity of the luteinizing hormone/chorionic gonadotropin receptor by sodium ions is mediated by a highly conserved aspartate located in the second transmembrane domain of G-protein coupled receptor. *Mol. Endocrin.* **7,** 767–775.

Regoli, D., Drapeau, G., Dion, S., and Couture, R. (1988) New selective agonists for neurokinin receptors: pharmacological tools for receptor characterization. *Trends Pharmacol. Sci.* **9,** 290–295.

Regoli, D., Escher, E., Drapeau, G., D'Orleans-Juste, P., and Mizrahi, J. (1984) Receptors for substance P. III. classification by competitive antagonists. *Eur. J. Pharmacol.* **97,** 179–189.

Rosell, S., Bjorkroth, U., Xu, J., and Folkers, K. (1983) The pharmacological profile of a substance P (SP) antagonist. Evidence for the existence of subpopulations of SP receptors. *Acta Physiol. Scand.* **117,** 445–449.

Rouissi, N., Gitter, B. D., Waters, D. C., Howbert, J. J., Nixon, J. A,. and Regoli, D. (1991) Selectivity and specificity of new, nonpeptide, quinuclidine antagonists of substance P. *Biochem. Biophys. Res. Comm.* **176,** 894–901.

Sachais, B. S., Snider, R M., Lowen III, J.A., and Krause, J. E. (1993) Molecular basis for the species selectivity of the substance P antagonist CP-96,345. *J. Biol. Chem.* **268,** 2319–2323.

Saffroy, M., Beauouan, J.-C., Torrens, Y., Besseyre, J., Bergstrom, L., and Glowinski, J. (1988) Localization of tachykinin binding sites (NK1, NK2, NK3 Ligands) in the rat brain. *Peptides* **9,** 227–241.

Sandberg, B. E. B. (1985) Structure-activity relationships for substance P: a review, in *Substance P: Metabolism and Biological Actions* (Jordan, C.C. and Oehme, P., eds.), Taylor and Francis, Philadelphia, PA, pp. 65–81.

Sandberg, B. E. B. and Iversen, L. L. (1982) Substance P. *J. Med. Chem.* **25,** 1009–1015.

Sasai, Y. and Nakanishi, S. (1989) Molecular characterization of rat substance K receptor and its mRNAs. *Biochem. Biophys. Res. Commun.* **165,** 695–702.

Saviano, G., Temussi, P. A., Motta, A., Maggi, C. A., and Rovero, P. (1991) Conformation-activity relationship of tachykinin Neurokinin A (4–10) and of some [Xaa^8] analogs. *Biochemistry* **30,** 10,175–10,181.

Seamon, K. B. and Daly, J. W. (1983) Forskolin, cyclic AMP and cellular physiology. *Trends Pharmacol. Sci.* **4,** 120–123.

Shigemoto, R., Yokota, Y., Tsuchida, K., and Nakanishi, S. (1990) Cloning and expression of a rat neuromedin K receptor cDNA. *J. Biol. Chem.* **265,** 623–628.

Snider, R. M., Constantine, J. W., Lowe, J. A., et al. (1991) A potent nonpeptide antagonist of the substance P (NK1) receptor. *Science* **251,** 435–437.

Sternweis, P.C. and Smrcka, A.V. (1992) Regulation of phospholipase C by G-proteins. *Trends Biochem. Sci.* **17,** 502–506.

Strader, C. D., Candelore, M. R., Hill, W. S., Sigal, I. S., and Dixon, R. A. F. (1989) Identification of two serine residues involved in agonist activation of the beta-adrenergic receptor. *J. Biol. Chem.* **264,** 13,572–13,578.

Sugiya, H., Tennes, K. A., and Putney, J. W. Jr. (1987) Homologous desensitization of substance P induced inositol polyphosphate formation in rat parotid acinar cells. *Biochem. J.* **244,** 647–653.

Sumner, S. C. J. and Ferretti, J. A. (1989) Conformational behavior of the linear hexapeptide senktide: a receptor specific tachykinin analog. *FEBS Lett.* **253,** 117–120.

Sundelin, J. B., Provendini, D. M., Wahlestedt, C. R., Pohl, J. S., and Peterson, P. A. (1992) Molecular cloning of the murine substance K and substance P receptor genes. *Eur. J. Biochem.* **203,** 625–631.

Tachado, S. D., Akhtar, R. A., Yousufazai, S. Y. K., and Abdel-Latif, A. A. A. (1991) Species differences in the effects of substance P on inositol trisphosphate accumulation and cyclic AMP formation, and on contraction in isolated iris sphincter of the mammalian eye: Differences in receptor density. *Exp. Eye Res.* **53,** 729–739.

Takahashi, K., Tanaka, A., Hara, M., and Nakanishi, S. (1992) The primary structure and gene organization of human substance P and neuromedin K receptors. *Eur. J. Biochem.* **204,** 1025–1033.

Takeda, Y., Blount, P., Sachais, B. S., Hershey, A. D., Raddatz, R., and Krause, J. E. (1992) Ligand binding and kinetics of substance P and neurokinin A receptors stably expressed in Chinese hamster ovary cells and evidence for differential stimulation of inositol 1,4,5-trisphosphate and cyclic AMP second messenger responses. *J. Neurochem.* **59,** 740–745.

Takeda, Y. and Krause, J. E. (1991) Pharmacological and molecular biological studies on the diversity of rat tachykinin NK-2 receptor subtypes in rat CNS, duodenum, vas deferens, and urinary bladder, in *Substance P and Related Peptides* (Leeman, S. E., Krause, J. E., and Lembeck, F., eds.), *Ann. NY Acad. Sci.,* New York, **632,** 479–482.

Takeda, Y., Chou, K. B., Takeda, J., Sachais, B. S., and Krause, J. E. (1991) Molecular cloning, structural characterization and functional expression of the human substance P receptor. *Biochem. Biophys. Res. Comm.* **179,** 1232–1240.

Tjolsen, A., Berge, O.-G., Hunskaar, S., Rosland, J. H., and Hole, K. (1992) The formalin test: an evaluation of the method. *Pain* **51,** 5–17.

Torrens, Y., Daguet de Montety, M. C., El Etr, M., Beaujouan, J-C., and Glowinski, J. (1989) Tachykinin receptors of the NK-1 type (substance P) coupled positively to phospholipase C on cortical astrocytes from the newborn mouse in primary culture. *J. Neurochem.* **52,** 1913–1918.

Torrens, Y., Lavielle, S., Chassaing, G., Marquet, A., Glowinski, J., and Beaujouan, J. C. (1984) Neuromedin K, a tool to further distinguish two central tachykinin binding sites. *Eur. J. Pharmacol.* **102,** 381,382.

Tsuchida, K., Shigemoto, R., Yokota, Y., and Nakanishi, S. (1990) Tissue distribution and quantitation of the mRNA for three rat tachykinin receptors. *Eur. J. Biochem.* **193,** 751–757.

Tsutsumi, M., Zhou, W., Millar, R. P., Mellon, P. L., Roberts, J. L., Flanagan, C. A., Dong, K., Boaz, G., and Selfon, S. C. (1992) Cloning and functional expression of a mouse gonadotropin releasing hormone receptors. *Mol. Endo.* **6,** 1163–1169.

Van Giersbergen, P. L. M., Shatzer, S. A., Henderson, A. K., Lai, J., Nakanishi, S., Yamamura, H. I., and Buck, S. H. (1991) Characterization of a tachykinin peptide NK-2 receptor transfected into murine fibroblast B82 cells. *Proc. Natl. Acad. Sci. USA* **88,** 1661–1665.

Venepalli, B. R., Aimone, L. D., Appel, K. C., Bell, M. R., Dority, J. A., Goswami, R., Hall, P.L., Kumar, V., Lawrence, K. B., Logan, M. E., Scensny, P. M., Seelye, J. A., Tomczuk, B. E., and Yanni, J. M. (1992) Synthesis and substance P receptor binding activity of androstano[3,2-b]pyrymido[1,2-a]benzimidizoles. *J. Med. Chem.* **35,** 374–378.

Ward, P., Ewan, G. B., Jordan, C. C., Ireland, S. J., Hagan, R. M., and Brown, J. R. (1990) Potent and highly selective neurokinin antagonists. *J. Med. Chem.* **33,** 1848–1851.

Watling, K. J. and Krause, J. E. (1993) The rising sun shines on substance P and related peptides. *Trends Pharm. Sci.* **14,** 81–84.

Watson S. (1984) The action of substance P on contraction, inositol phospholipids, and adenylate cyclase in rat small intestine. *Biochem. Pharmacol.* **33,** 3733–3737.

Watson, S. P., Sandberg, B. E. B., Hanley, M. R., and Iversen, L. L. (1983) Tissue selectivity of substance P alkyl esters: suggestive of multiple receptors. *Eur. J. Pharmacol.* **87,** 77–84.

Wormser, U., Laufer, R., Hart, Y., Chorev, M., Gilon, C., and Selinger, Z. (1986) Highly selective agonists for substance P receptor subtypes. *EMBO J.* **5,** 2805–2808.

Wu, D., Katz, A., and Simon, M. I. (1993) Activation of phospholipase C β_2 by the α and β_γ subunits of trimeric GTP-binding protein. *Proc. Natl. Acad. Sci. USA* **90,** 5297–5301.

Xie, G.-X., Miyajima, A., and Goldstein, A. (1992) Expression cloning of cDNA encoding a seven-helix receptor from human placenta with affinity for opioid ligands. *Proc. Natl. Acad. Sci. USA* **89,** 4124–4128.

Yamachita, K., Koide, Y., and Aiyochi, Y. (1983) Effects of substance P on thyroidal cyclic AMP levels and thyroid hormone from canine thyroid slices. *Life Sci.* **32,** 2163–2166.

Yashpal, K., Radhakrishnan, V., Coderre, T. J., and Henry, J. L. (1993) CP-96,345, but not its stereoisomer, CP-96,344, blocks the nociceptive responses to intrathecally administered substance P and to noxious thermal and chemical stimuli in the rat. *Neuroscience* **52,** 1039–1047.

Yokota, Y., Sasai, Y., Tanaka, K., Fiiwana, T., Tsuchida, K., Shigemoto, R., Kakizuka, A., Ohkubo, H., and Nakanishi, S. (1989) Molecular characterization of a functional cDNA for rat substance P receptor. *J. Biol. Chem.* **264,** 17,649–17,652.

Yokota, Y., Akazawa, C., Ohkubo, H., and Nakanishi, S. (1992) Delineation of structural domains involved in the subtype specificity of tachykinin receptors through chimeric formation of substance P/substance K receptors. *EMBO J.* **11,** 3585–3591.

CHAPTER 8

Characterization of the NK_1 Receptor Using Photoaffinity Probes

Norman D. Boyd, Reinhard K. Kage, and Susan E. Leeman

1. Introduction

Although the existence of a specific receptor for substance P (SP), now termed the NK_1 receptor, was established over a decade ago, only recently has information on its structural properties become available. Receptor purification, which represents the first step in the classical approach to primary structure determination, has proved to be exceedingly difficult due to the marked decrease in SP binding affinity that occurs on detergent solubilization.

Presently, much of what has been learned about the structural characteristics of the NK_1 receptor has come through recombinant methodologies. For example, it has been possible to determine the amino acid sequence and to identify potential sites for glycosylation, phosphorylation, palmitoylation, and proteolytic modifica-

The Tachykinin Receptors Ed.: S. H. Buck © 1994 Humana Press Inc., Totowa, NJ

tion. Analysis of the amino acid sequence suggests that the NK_1 receptor, like other members of the G protein-coupled receptor family, possesses seven hydrophobic transmembrane spanning domains with an extracellular amino terminus and a cytoplasmic carboxyl terminus. A question of fundamental importance in regard to the interaction of SP with its receptor concerns the location of the ligand binding domain within this putative structural arrangement.

There are limitations in the interpretation of data provided by recombinant methodologies. Therefore, direct biochemical approaches are also important to provide a more complete definition of the structure–function relationships of these proteins. A strategy that has been employed to characterize biochemical properties of ligand-binding proteins has been to covalently attach labeled derivatives of the ligands to the proteins. By this approach, receptors and fragments derived from their binding sites can be identified during subsequent purification steps even though binding activity is lost.

Early attempts to covalently label the NK_1 receptor involved the use of homobifunctional crosslinking agents to chemically crosslink radiolabeled SP (Payan et al., 1986; Coats and Gerard, 1989) and an ^{125}I-labeled azide-containing derivative of SP for photoaffinity crosslinking (Dam et al., 1987). The usefulness of these reagents is, however, limited by the low level of covalent labeling achieved. An improvement over these approaches has been accomplished by Boyd et al. (1991b) using analogs of SP containing *p*-benzoylphenylalanine (Bpa), an amino acid with a photoreactive benzophenone side chain. In contrast to azide containing photoaffinity ligands which generate highly reactive, but relatively nonspecific, nitrene intermediates, the ketone carbonyl of the benzophenone moiety undergoes a $n \rightarrow \pi^*$ transition to give a triplet biradical that possesses a high reactivity for C–H bonds, but a very low reactivity towards water (Breslow, 1980).

This chapter will describe the development of radioiodinated derivatives of two Bpa-containing SP analogs, $[Bpa^4]SP$ and $[Bpa^8]SP$, as highly efficient photoaffinity probes to address a range of questions concerning structural and functional aspects of the NK_1 receptor.

2. Characterization of Photoaffinity Ligands

The photoreactive amino acid, *p*-benzoylphenylalanine (Bpa), is obtained in high yield by the synthetic scheme described by Kauer et al. (1986). The amino acid can be incorporated into any position of the 11 amino acid sequence of SP using the standard coupling reactions of solid-phase peptide synthesis because of the chemical stability of Bpa under conditions of peptide synthesis. The carboxyl-terminal region of SP was selected initially for introduction of Bpa to optimize the possibility that the photolabile group would be in close proximity to amino acid residues comprising the peptide binding site of the NK_1 receptor. The importance of the C-terminal region of SP for interaction with the binding site of the NK_1 receptor has been established by the demonstration that C-terminal fragments of six amino acids or larger are capable of interacting with the NK_1 receptor, whereas even shorter C-terminal fragments and all N-terminal fragments have little or no affinity (for a review, *see* Maggio, 1988). A further consideration in the design of the photoaffinity probe was the finding that, despite its location in the region critical for binding, Phe^8 can be structurally modified without causing a marked decrease in affinity or loss of biological activity (Lee et al., 1983; Viger at al., 1983). A second analog, $[Bpa^4]SP$, in which the photoreactive group is incorporated within the divergent N-terminal sequence of SP was also synthesized. This region of SP, although not essential for interaction with the NK_1 receptor, is required for binding with high-affinity (Ingi et al., 1991).

The introduction of Bpa in either position 8 or position 4 of SP is apparently well tolerated by the receptor. In competition binding studies, $[Bpa^8]SP$ inhibited radiolabeled SP binding to a rat submaxillary gland membrane preparation with an IC_{50} = 1 n*M*, a value not significantly different from that of SP itself, whereas $[Bpa^4]SP$ was only slightly less potent with an IC_{50} = 5 n*M*. Biological activity, as tested in a sialogogic assay for SP, was retained by both $[Bpa^8]SP$ and $[Bpa^4]SP$ with the same relative potencies as in the competition binding studies.

Radioiodinated derivatives of [Bpa^{8}]SP and [Bpa^{4}]SP have been prepared by conjugation with the *N*-hydroxysuccinimide ester of ^{125}I-labeled 3-(*p*-hydroxyphenyl) proprionic acid (Bolton-Hunter reagent). This reagent has been widely used for radiolabeling other peptides, including SP and related tachykinin peptides. The radiolabeled peptides were purified by HPLC to high specific activity (>2000 Ci/mmole). For some experiments involving the labeling of receptors on a preparative scale, photoaffinity labels of lower specific activity are also required. A nonradiolabeled derivative has been prepared using ^{127}I-labeled Bolton-Hunter reagent and radiolabeled peptide probes of the desired specific activity obtained by isotopic dilution with the nonradiolabeled form. Under the conditions used for conjugation with the Bolton-Hunter reagent (pH 8.6 and 4°C), the ε-amino group of Lys^{3} was found by sequence analysis to be the primary site of acylation (Fig. 1).

Under nonphotolyzing conditions, ^{125}I-[Bpa^{8}]SP binds in a saturable and reversible manner to NK_1 receptors in a membrane preparation of rat submaxillary gland with a $K_D = 0.4$ n*M* and a $B_{max} = 0.2$ pmoles/mg of membrane protein. These equilibrium parameters are the same as those characterizing the binding of the ^{125}I-labeled Bolton-Hunter conjugate of SP to the same membrane preparation. ^{125}I-[Bpa^{4}]SP also binds to the same number of sites but, as expected from the competition binding experiments described above, with a lower affinity, $K_D = 2$ n*M*.

A primary goal behind the development of a quantitative method for photoaffinity labeling the NK_1 receptor was to determine the regions of the NK_1 receptor that are in direct contact with the peptide. In addition to efficient labeling, a source of sufficient quantities of the NK_1 receptor is necessary for the isolation of radiolabeled receptor fragments and the determination of the site of covalent attachment by microsequencing. Equine salivary glands represent an abundant source of a large quantity of tissue containing a relatively high concentration of NK_1 receptors. Membranes prepared from this tissue bind ^{125}I-[Bpa^{8}]SP with high-affinity and specificity. Irradiation of the bound complex results in the formation of a covalent bond between the ligand and receptor. The photolabeled receptor protein

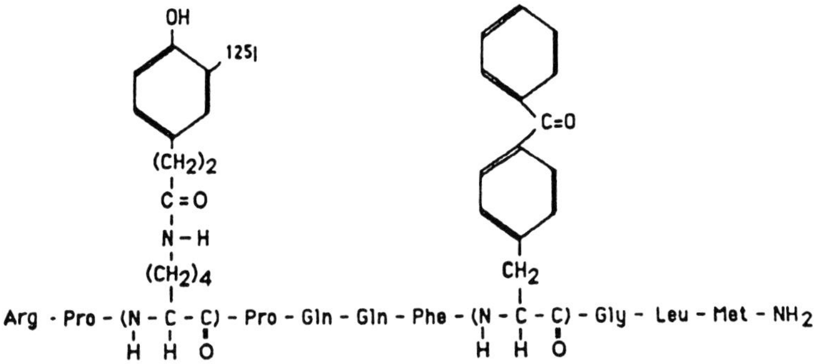

Fig. 1. Chemical structure of the ^{125}I-Bolton-Hunter conjugate of an analog of substance P in which phenylalanine in position 8 has been replaced by the photoreactive amino acid, *p*-benzoyl-L-phenylalanine (^{125}I-[Bpa^8]SP).

migrates on a SDS-PAGE gel as a broad radioactive band centered at $M_r = 53{,}000$ (Fig. 2). In the presence of 1 μ*M* SP, photolabeling is completely abolished, indicating specificity. The efficiency of photoaffinity labeling with ^{125}I-[Bpa^8]SP is unusually high; about 70% of the specifically bound peptide becomes covalently attached during the photolytic step.

Photoaffinity labeling of NK_1 receptors with ^{125}I-[Bpa^4]SP is also highly specific and efficient. However, since ^{125}I-[Bpa^8]SP displays a higher affinity for NK_1 receptor, this photoaffinity probe has been used more routinely in studies to characterize the biochemical properties of NK_1 receptor. In a later section of this chapter, the use of multiple photoaffinity probes to define the amino acid residues present at the peptide binding pocket of the NK_1 receptor is discussed together with preliminary evidence that the sites of photoincorporation of ^{125}I-[Bpa^4]SP and ^{125}I-[Bpa^8]SP are different.

3. Photolabeling NK_1 Receptors: Different Tissues

The high level of specific photoincorporation achieved using ^{125}I-[Bpa^8]SP permits the detection of NK_1 receptors in tissues and cells even where the abundance of the receptor is relatively low.

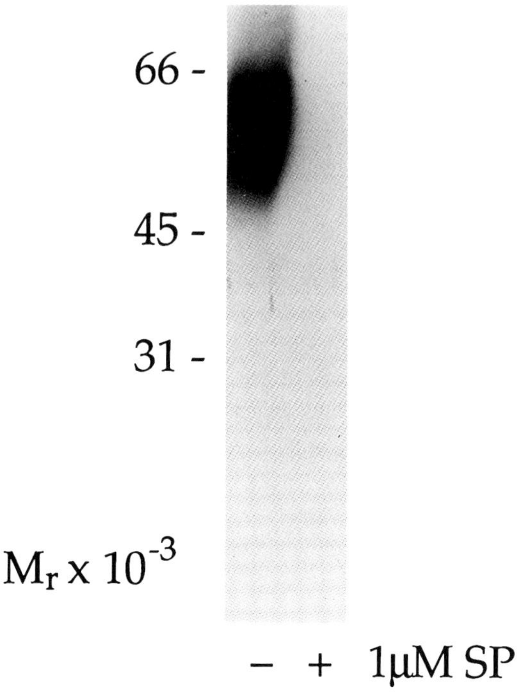

Fig. 2. Photolabeling of equine submaxillary gland membranes with ^{125}I-[Bpa^8]SP. Equine submaxillary gland membranes (1 mg of membrane protein/mL) were equilibrated at 22°C with ^{125}I-[Bpa^8]SP (0.2 n*M*) in the absence (–) and presence (+) of 1 μ*M* SP and then photolyzed at 4°C for 10 min with 350-nm light. ^{125}I-[Bpa^8]SP bound to both the photolyzed and nonphotolyzed membranes was determined by ultrafiltration. Membranes (100 μg of membrane protein) were also subjected to SDS-PAGE, and the labeled bands were visualized by autoradiography. The region of the gel containing the radiolabeled polypeptides (M_r = 43,000–55,000) was excised and the amount of radioactivity determined to assess the efficiency of photolabeling.

Photolabeling of NK_1 receptor prepared from various sources has revealed that receptors in different rat tissues differ in molecular size (Table 1). This heterogeneity is due to differences in the amount of *N*-linked carbohydrates since upon enzymatic deglycosylation with endoglycosidase F the photolabeled receptors from all sources yield

Table 1
Relative Molecular Weights
of NK-1 Receptors Photolabeled by ^{125}I-[Bpa^8]SP
in Membrane Preparations of Rat Tissues

Submaxillary gland	53,000, 46,000
Parotid gland	53,000, 46,000
Large intestine	72,000
Striatum	90,000
Olfactory bulb	90,000

a discrete radiolabeled band of M_r = 46,000. This value is in good agreement with the molecular mass of the protein core calculated from the primary structures as deduced from their cDNA sequences. In some tissues, in particular rat salivary glands, an additional receptor protein of lower molecular weight was also detected by photolabeling. An analysis of the structural relationship between these two radiolabeled proteins and a comparison of their peptide binding properties are presented later in this chapter.

The primary structures of NK_1 receptors from rat, human, mouse, and guinea pig are known from cloning studies (Nakanishi, 1991). At the NH_2-terminus of each receptor, there are two consensus sequences (Asn-X-Ser/Thr) for *N*-linked glycosylation, a property shared with other members of the G-protein coupled receptor family, including the receptors for the two other mammalian tachykinins, neurokinin A (NKA) and neurokinin B (NKB). NK_1 receptors from each of these species also have a Asn-Phe-Thr sequence starting at position 91 at the beginning of the first extracellular loop, E1, whereas rat and mouse NK_1 receptors, but not those of human and guinea pig, have a fourth potential site on the second extracellular loop, E2. It has been proposed that analogous sites on the E2 loop of the human platelet-activating factor receptor (Kunz et al., 1992) and of the rat liver angiotensin II receptor (Desarnaud et al., 1993) are glycosylated. However, it has been found that the mobility of a labeled fragment of M_r = 21,000, obtained by limited digestion of photolabeled NK_1 receptor with trypsin, is not further decreased upon treatment with endoglycosidase F (Boyd et al., 1991a). Since this tryptic fragment lacks the two amino terminal glycosylation sites, but contains the two potential internal glycosylation sites, the possibility that the latter two sites are glycosylated in the NK_1 receptor can be ruled out. In addition, the observation following short incubations with endoglycosidase F of a labeled receptor species with an M_r intermediate between the fully glycosylated and deglycosylated receptor indicates that deglycosylation is a two-step process, suggesting that both of the amino terminal sites of NK_1 receptors are glycosylated.

4. Comparison of Binding Properties of Glycosylated and Nonglycosylated Receptors

Differences in the apparent affinities of various synthetic tachykinin peptides for NK_1 receptors from rat brain and rat submaxillary gland have been interpreted as evidence that the receptors in these tissues possess different binding characteristics (Lew et al., 1990). Since only a single gene and a single mRNA for NK_1 receptor have been detected in these tissues, it has been suggested that the observed pharmacological differences may be caused by different degrees of *N*-glycosylation of these receptors (Mussap et al., 1993). The photolabeling experiments described above demonstrate that NK_1 receptors in rat brain are, indeed, more extensively glycosylated than those in rat salivary gland, but the assumption that *N*-linked sugars influence peptide binding to NK_1 receptors is not supported by the recent finding (Kage et al., 1993) that the presence of *N*-linked carbohydrates is not essential for the high-affinity and selective binding of SP to NK_1 receptors. In the latter study, Chinese hamster ovary (CHO) cells expressing NK_1 receptors were treated with the glycosylation inhibitor, tunicamycin. Following 48 h of treatment with tunicamycin, photolabeling with ^{125}I-[Bpa^8]SP revealed the presence of two forms of labeled NK_1 receptor with M_r = 80,000 and M_r = 46,000, representing glycosylated and nonglycosylated NK_1 receptors, respectively (Fig. 3). The ability of SP to prevent photolabeling of both forms of the receptor was analyzed by SDS-PAGE followed by quantitative autoradiography. These studies showed that SP bound to

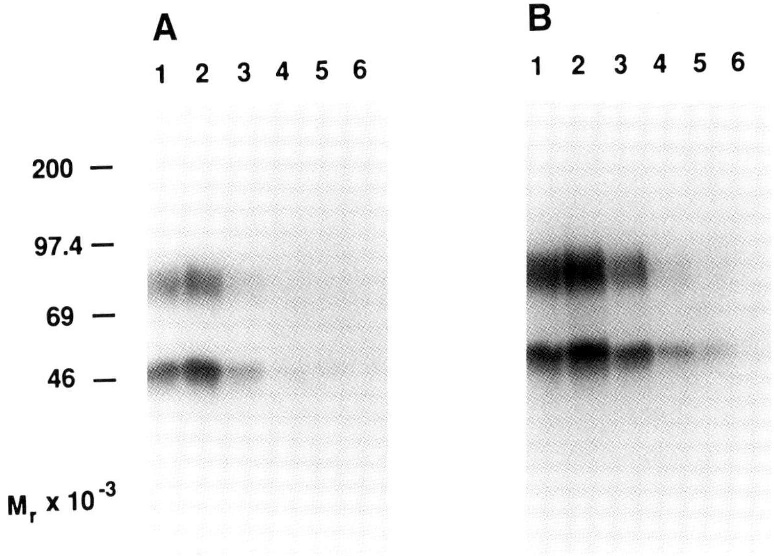

Fig. 3. Inhibition of ^{125}I-[Bpa^8]SP labeling of glycosylated and nonglycosylated NK_1 receptors by substance P and neurokinin A. CHO cells expressing the NK_1 receptor were treated with tunicamycin (5 μg/mL) for 48 h prior to photolabeling with ^{125}I-[Bpa^8]SP in the presence of increasing concentrations of substance P (**A**) and neurokinin A (**B**). Glycosylated and nonglycosylated forms of the photolabeled NK_1 receptor were resolved by SDS-PAGE and radiolabeled bands visualized by autoradiography. Regions of the gel at M_r = 80,000 and M_r = 46,000 were excised and the amount of radioactivity determined.

each form of the receptor, glycosylated and nonglycosylated, with the same affinity, K_D = 0.2 n*M*. In addition, the high selectivity of the glycosylated NK_1 receptor for SP over NKA was shown to be maintained in the nonglycosylated form; both the glycosylated and nonglycosylated forms of the receptor bound NKA with a K_D = 30 n*M*, i.e., with ~100-fold lower affinity than SP.

Support for the conclusion that *N*-linked glycosylation is not required for high-affinity binding of SP is provided by the observation of Fong et al. (1992b) showing that binding of SP to a human NK_1 receptor mutant lacking both potential glycosylation sites at the amino terminus was not different from that of the native receptor.

5. Characterization of Two Different Receptor Forms in Rat Submaxillary Gland

When NK_1 receptors in rat submaxillary glands were first photolabeled by Boyd et al. (1991b) using ^{125}I-[Bpa^8]SP, two proteins with M_r = 53,000 and 46,000 were detected. This was surprising since saturation analysis of the equilibrium binding under nonphotolyzing conditions showed binding of ^{125}I-[Bpa^8]SP was to an apparent single class of binding sites. Although other studies using peptide analogs and nonpeptide antagonists have been interpreted as providing evidence for heterogeneity in the binding properties of NK_1 receptors, the basis for such heterogeneity is currently unknown, but may be due to the existence of receptor subtypes, species differences, or coupling to different G-proteins (*see* other chapters in this volume). The close to quantitative incorporation of ^{125}I-[Bpa$^{8]}$SP provides a sensitive and convenient method for comparing directly the binding properties of forms of the receptor that differ in molecular weight. The ability of SP to protect against photolabeling of both forms of the receptor was assessed in the presence of increasing concentrations of SP and other tachykinins followed by resolution of the labeled receptors on SDS-PAGE and quantitation by densitometric scanning of the resulting autoradiograms. The labeling of receptors with M_r = 53,000 and M_r = 46,000 was equally sensitive to inhibition by SP, indicating that SP interacts with both forms of the receptor with the same affinity (Fig. 4). Further studies using highly selective analogs of SP and other tachykinins showed that both forms of the receptor displayed the same peptide specificity, demonstrating that, by pharmacological criteria, the two receptor proteins, although differing in their molecular size, do not represent subtypes of the NK_1 receptor (Boyd et al., 1991b). The two most likely reasons for a difference in molecular weight of this magnitude are differential glycosylation or a difference in the lengths of the protein backbones.

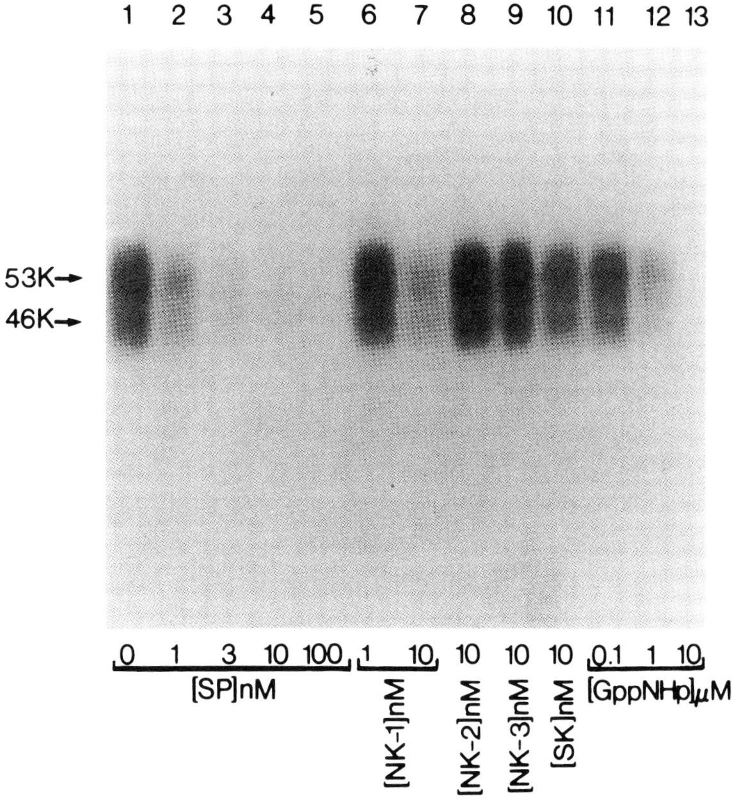

Fig. 4. Peptide specificity and guanine nucleotide sensitivity of ^{125}I-[Bpa^8]SP photoincorporation into the two forms of the NK_1 receptor present in rat submaxillary gland membranes. Membranes were photoaffinity labeled with ^{125}I-[Bpa^8]SP alone (lane 1) or in the presence of the specified concentrations of competing tachykinin peptides (lanes 2–10) or GppNHp (lanes 11–13). Samples were subjected to SDS-PAGE and autoradiography. The amount of labeling at M_r = 53,000 and M_r = 46,000 was assessed by densitometric scanning of the individual lanes of the autoradiogram. Terms: NK_1, [Sar^9,Met(O_2)^{11}]SP; NK_2, [Nle^{10}]NKA(4-10); NK_3, [MePhe^7]NKB; SK, substance K (neurokinin A).

Because it was possible to specifically photolabel these receptors in crude membrane preparations, the biochemical basis for the differences in molecular weight could be analyzed without needing to isolate these NK_1 receptor proteins, currently a formidable and still not accomplished task.

A series of experiments was conducted to distinguish between these two possibilities (Kage et al., 1993b). When the two forms of the receptor were treated individually with endoglycosidase F, both decreased in molecular weight by the same amount (M_r = 10,000), establishing that both forms of the receptor contain the same amount of *N*-linked carbohydrates (Fig. 5). It was therefore concluded that the two forms most likely differ in the lengths of their protein backbone. Experiments in which photolabeled membranes were treated with carboxypeptidase Y to remove the portions of the carboxy termini that extend beyond the membrane provided evidence that the two forms differ in the lengths of their carboxy termini.

How these two forms of the receptor for SP arise is unknown. As stated previously, only one gene that encodes the NK_1 receptor has been described. Whether the receptor with the truncated carboxy terminus arises by alternate splicing of a single gene (Fong et al., 1992a) or by the action of a proteolytic enzyme that removes most of the carboxy terminus after the protein is synthesized is not known. Nor is it known whether the two forms of the receptor have different physiological properties. It is known that both forms can couple to G proteins (Boyd et al., 1991b). An interesting possibility that merits further study is that the two forms differ in their ability to undergo desensitization. Phosphorylation of the β_2-adrenergic receptor that occurs at multiple serine and threonine residues located at the carboxy terminus has been linked to desensitization of this receptor (Hausdorff et al. 1989). The carboxy terminus of the NK_1 receptor also contains many serines and threonines that could serve as enzymatic phosphorylation sites. Absence of the C-terminal phosphorylation sites could result in an impaired ability of the NK_1 receptor to desensitize. In fact, Putney et al. (1991) found that after continuous stimulation of parotid cells with SP, an elevated level of inositol 1,4,5-trisphosphate still remained, indicating an incomplete desensitization process. Indeed, photolabeling of membranes from rat parotid glands with ^{125}I-[Bpa^8]-SP also reveals the presence of the two forms of the NK_1

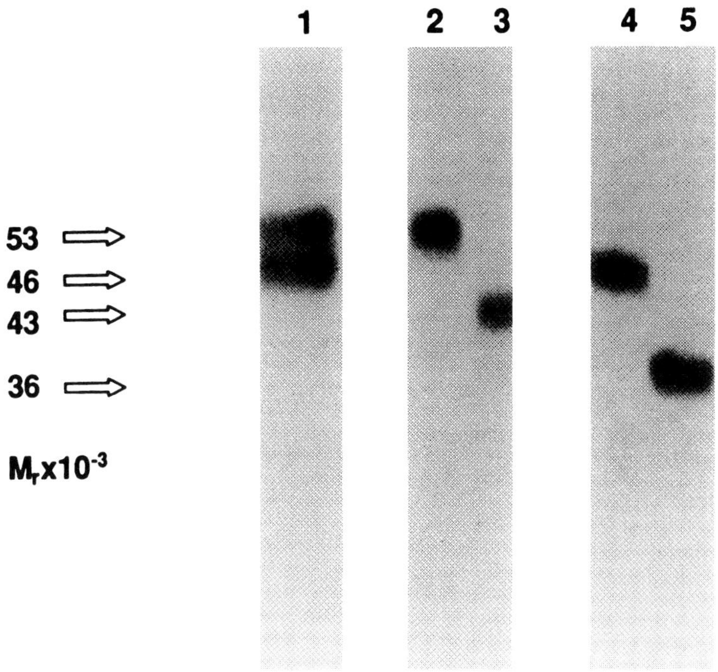

Fig. 5. Endoglycosidase F treatment of the resolved 53,000 and 46,000 photolabeled NK_1 polypeptides. The 53,000 and 46,000 photolabeled NK_1 polypeptides were resolved by semipreparative SDS-PAGE and electroelution. Photolabeled polypeptides were incubated at 37°C for 12 h in 50 m*M* sodium phosphate buffer (pH 6.0) containing 1% (wt/vol) Nonidet P-40 in the presence (+) and absence (-) of endoglycosidase F (5 U/mL). Following the incubation period, samples were analyzed by SDS-PAGE and autoradiography. Lanes 2 and 3: the 53,000 photolabeled polypeptide incubated without and with endoglycosidase F, respectively. Lanes 4 and 5: the 46,000 photolabeled polypeptide incubated without and with endoglycosidase F, respectively.

receptor. The sustained secretion of saliva observed after prolonged infusion of SP into rats may be explained by the presence of a form of the NK_1 receptor that does not desensitize. Further studies will be necessary to examine the potential physiological role of the two forms of the NK_1 receptor in rat salivary gland tissue.

6. Mapping the Peptide Binding Pocket

Current knowledge of the structure of agonist binding sites of G-protein coupled receptors is based primarily on studies of receptors whose endogenous agonists are small molecules such as catecholamines and acetylcholine. Two approaches have proved particularly useful. One involves an analysis of the binding properties of mutated receptors either by substitution of individual amino acids or by construction of chimeric receptors. The second involves the use of affinity ligands containing either chemically reactive or photoreactive functional groups that covalently label amino acid residues present in the binding site, which can then be identified by fragmentation and sequence analysis. Application of these two approaches to the β-adrenergic and muscarinic receptors has led to a model for the ligand binding site of these receptors in which the agonist is bound within a pocket formed by the seven hydrophobic transmembrane α-helices that characterize this family of receptors (Dohlman et al., 1991).

Recently, site mutated NK_1 receptors and tachykinin receptor chimeras have been generated in an attempt to delineate the peptide binding site. From these studies, a useful, although as yet somewhat broad, definition of the regions involved in SP binding has emerged, with both extracellular and transmembrane domains being implicated. Since the binding of SP involves a much larger number of receptor contacts than small nonpeptide agonists, a considerably larger number of mutant receptors will have to be analyzed to more closely define the interaction of SP with its receptor. A caution in the use of this approach is that mutation of an amino acid may affect ligand binding owing to changes in conformation at the binding pocket without that specific residue necessarily being directly involved in interactions with the ligand. It is therefore difficult to establish with certainty whether or not a particular amino acid is directly involved with binding.

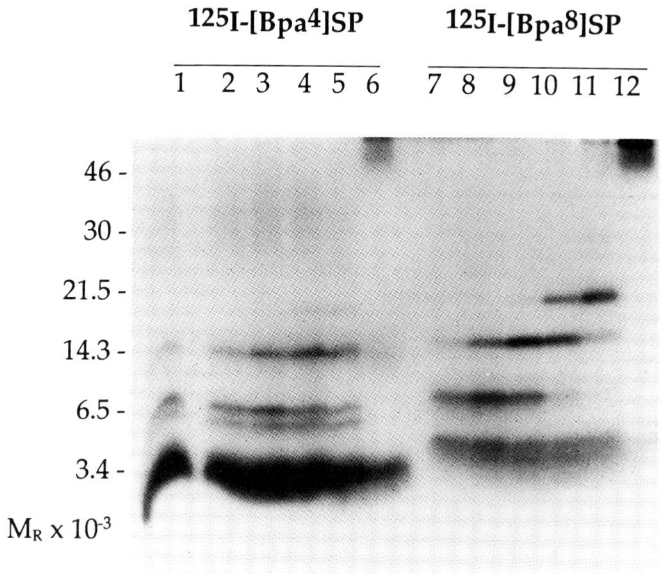

Fig. 6. Autoradiograph tryptic fragments of the NK_1 receptor photoaffinity labeled with ^{125}I-[Bpa^4]SP and ^{125}I-[Bpa^8]SP. Chinese hamster ovary cells expressing the NK_1 receptor were photoaffinity labeled with either ^{125}I-[Bpa^4]SP or ^{125}I-[Bpa^8]SP. Membranes containing labeled NK_1 receptor were prepared from the cells and were incubated for 60 min at 22°C with trypsin at the following concentrations (mg/mL): 2, 0.6, 0.2, 0.06, 0.02, and 0; lanes 1–6: ^{125}I-[Bpa^4]SP labeled; and lanes 7–12: ^{125}I-[Bpa^8]SP labeled. The tryptic fragments were resolved on a Tricine-SDS-PAGE gel and visualized by autoradiography.

In contrast, determination of the site of incorporation of photoaffinity probes such as ^{125}I-[Bpa^8]SP and ^{125}I-[Bpa^4]SP identifies directly the region of the receptor that is in direct contact with a specified position of the peptide. The most straightforward approach to determining the site of insertion of a photoaffinity probe is by fragmentation and sequencing. Prior to preparative scale isolation and sequencing of receptor binding domains, analyses of enzymatic and/or chemical fragmentation procedures conducted on an analytical scale are important for optimization of fragment yield and evaluation of purification steps. Since the photoaffinity probes are themselves peptides, a key consideration in the evaluating potential fragmentation schemes is the requirement that the covalently attached probe remains intact under receptor fragmentation conditions. The results of this type of experiment are shown in Fig. 6 in which NK_1 receptors expressed in CHO cells were photolabeled with either ^{125}I-[Bpa^8]SP or ^{125}I-[Bpa^4]SP and subjected to digestion with increasing concentrations of trypsin and the resulting radiolabeled fragments analyzed by Tricine-SDS-PAGE.

Tryptic digestion of receptor photolabeled with ^{125}I-[Bpa^8]SP presented a different fragmentation pattern than tryptic digestion of receptor photolabeled with ^{125}I-[Bpa^4]SP. These results suggest that different amino acid residues serve as the site of covalent attachment for each photoaffinity probe. These experiments further suggest that, by this approach, sufficient quantities of labeled fragments can be generated and amino acid attachment sites identified by purification and microsequencing. Extending this approach to additional SP derivatives should provide detailed information on the amino acid residues that comprise the peptide binding pocket and thus provide an understanding at the molecular level of peptide-receptor interaction.

References

Boyd, N. D., Macdonald, S. G., Kage, R., Luber-Narod, J., and Leeman, S. E. (1991a) Substance P receptor, biochemical characterization and interactions with G proteins. *Ann. N.Y. Acad. Sci.* **632,** 79–93.

Boyd, N. D., White, C. F., Cerpa, R., Kaiser, E. T., and Leeman, S. E. (1991b) Photoaffinity labeling the substance P receptor using a derivative of substance P containing *p*-benzoylphenylalanine. *Biochemistry* **30,** 336–342.

Breslow, R. (1980) *Acc. Chem. Res.* **13,** 170–177.

Cascieri, M. A., Huang, R.-R. C., Fong, T. M., Cheung, A. H., Sadowski, S., Ber, E., and Strader, C. D. (1992) Determination of the amino acid residues in substance P conferring selectivity and specificity for the rat neurokinin receptors. *Mol. Pharmacol.* **41,** 1096–1099.

Coats, S. R. and Gerard, N. P. (1989) Characterization of the substance P receptor in guinea pig lung tissues. *Am. J. Respir. Cell Mol. Biol.* **1,** 269–275.

Dam, T.-V., Escher, E., and Quirion R. (1987) Apparent molecular weight of substance P/Neurokinin-1 receptors determined using a photoaffinity labelled probe [(3'-^{125}I) D-Tyr^o, (4'-N_3)Phe^8, Nle^{11}]-Substance P. *Biochem. Biophys. Res. Commun.* **149,** 297–303.

Desarnaud, F., Marie, J., Lombard, C., Larguier R., Seyer, R., Lorca, T., Jard S., and Bonnafous, J.-C. (1993) Deglycosylation and fragmentation of purified rat liver angiotensin II receptor: application to the mapping of hormone-binding domains. *Biochem. J.* **289,** 289–297.

Dohlman, H. G., Thorner, J., Caron, M. G., and Lefkowitz, R. J. (1991) Model systems for the study of seven-transmembrane-segment receptors. *Annu. Rev. Biochem.* **60,** 653–688.

Fong, T. M. and Strader, C. D. (1992) The extracellular domain of substance P (NK_1) receptor comprises part of the ligand binding site. *Biophys. J.* **62,** 59–60.

Fong, T. M., Anderson, S. A., Yu, H., Huang R. -R. C., and Strader, C. D. (1992a) Differential activation of intracellular effector by two isoforms of human neurokinin-1 receptor. *Mol. Pharmacol.* **41,** 24–30.

Fong, T. M., Yu, H., Hunang, R. R. C., and Strader, C. D. (1992b) The extracellular domain of the neurokinin-1 receptor is required for high-affinity binding of peptides. *Biochemistry* **31,** 11,806–11,811.

Gether, U., Teit J., and Schwartz, T. W. (1993) Chimeric NK_1 (Substance P)/NK_3 (Neurokinin B) Receptors. *J. Biol. Chem.* **268,** 7893–7898.

Hausdorff, W. P., Bouvier, M., O'Dowd, B. F., Irons, G. P., Caron, M. G., and Lefkowitz, R. J. (1989) Phosphorylation sites of the β_2-adrenergic receptor are involved in distinct pathways of receptor desensitization. *J. Biol. Chem.* **37,** 12,657–12,665.

Hershey, A. D., Dykema, P. E., and Krause, J. E. (1991) Organization, structure, and expression of the gene encoding the rat substance P receptor. *J. Biol. Chem.* **266,** 4366–4374.

Hershey, A. D., Polenzani, L., Woodward, R. M., Miledi, R., and Krause, J. E. (1991) Molecular and genetic characterization, functional expression, and mRNA expression patterns of a rat substance P receptor. *Ann. NY Acad. Sci.* **632,** 63–78.

Ingi, T., Kitajima, Y., Minamitake, Y., and Nakanishi, S. (1991) Characterizationn of ligand-binding properties and selectivities of three rat tachykinin receptors by transfection and functional expression of their cloned cDNAs in mammalian cells. *J. Pharm. Exper. Ther.* **259,** 969–975.

Kage, R., Hershey, A. D., Krause, J. E., Boyd, N. D., and Leeman, S. E. (1993a) Characterization of the substance P (NK-1) receptor in tunicamycin-treated transfected cells using a photoaffinity analog of substance P. *J. Neurochem.* in press.

Kage, R., Leeman, S. E., and Boyd, N. D. (1993b) Biochemical characterization of two different forms of the substance P receptorin rat submaxillary gland. *J. Neurochem.* **60,** 347–351.

Kauer J. C., Erickson-Viitanen S., Wolfe Jr. H. R., and DeGrado W. R. (1986) p-Benzoyl-L-phenylalanine, a new photoreactive amino acid. *J. Biol. Chem.* **261,** 10,695–10,700.

Kunz, D., Gerard, N. P., and Gerard, C. (1992) The human leukocyte platelet-activating factor receptor: cDNA cloning, cell surface expression and construction of a novel epitope-bearing analog. *J. Biol. Chem.* **267,** 9101–9106.

Lee C.-M., Javitch J. A., and Snyder S. H. (1983) ^{3}H-SP binding to salivary gland membranes. Regulation by guanyl nucleotides and divalent cations. *Mol. Pharmacol.* **23,** 563–569.

Lew R., Geraghty D. P., Drapeau G., Regoli D., and Burcher E. (1990) Binding characteristics of [^{125}I] Bolton-Hunter [Sar^9.-Met(O$_2$)^{11}]substance P a new selective radiolagand for the NK$_1$ receptor. *Eur. J. Pharmacol.* **184,** 97–108.

Maggio, J. E. (1988) Tachykinins. *Annu. Rev. Neurosci.* 1**1,** 13–28.

Mussap, C. J., Geraghty, D. P., and Burcher E. (1993) Tachykinin receptors: A radioligand binding prespective. *J. Neurochem.* **60,** 1987–2009.

Nakanishi, S. (1991) Mammalian tachykinin receptor. *Annu. Rev. Neurosci.* **14,** 123–136.

Payan, D. G., McGillis, J. P., and Organist, M. L. (1986) Binding characteristics and affinity labeling of protein constitutents of the human IM9 lymphoblast receptor for substance P. *J. Biol. Chem.* **261,** 14,321–14,329.

Putney, J. W., Bird, G. St. J., Horstman, D. A., Hughes, A. R., Menniti, F. S., Nogimori, K., Obie, J., Oliver. K. G., Sugiya, H., and Takemura H. (1991) Role of inositol phosphates in the actions of substance P on NK$_1$ receptors in exocrine gland cells. *Ann. NY Acad. Sci.* **632,** 94–102.

Viger A., Beaujouan C.-C., Torrens Y., and Glowinski J. (1983) Specific binding of a ^{125}I-SP derivative to rat brain synaptosomes. *J. Neurochem.* **40,** 1030–1039.

Yokota, Y., Akazawa, C., Ohkubo, H., and Nakanishi, S. (1992) Delineation of strucural domains involved in the subtype specificity of tachykinin receptors through chimeric formation of SP/substance K receptor. *EMBO J.* **11,** 3585–3591.

CHAPTER 9

Modeling of G-Protein-Coupled Receptors

Application to the NK_1 Tachykinin Receptor

Susanne Trumpp-Kallmeyer, Jan Hoflack, and Marcel Hibert

1. Background

Substance P (SP), an undecapeptide with the sequence Arg-Pro-Lys-Pro-Gln-Gln-Phe-Phe-Gly-Leu-Met-NH_2, acts as a neurotransmitter/neuromodulator in the central and peripheral nervous systems. SP belongs to the family of so-called tachykinins. All of these peptides share the common C-terminal sequence Phe-X-Gly-Leu-Met-NH_2, whereas they show great variation in their N-terminal region. The biological actions of tachykinins are mediated by at least three different transmembrane G-protein-coupled receptors (GPCR). Based on extensive pharmacological studies (Buck and Shatzer, 1988; Guard and Watson, 1991), they have been classified into the neurokinin-1 (NK_1), neurokinin-2 (NK_2), and neurokinin-3

The Tachykinin Receptors Ed.: S. H. Buck © 1994 Humana Press Inc., Totowa, NJ

(NK_3) receptor types, which demonstrate preferential binding of SP, neurokinin A, and neurokinin B, respectively. Recently, these receptors have been cloned and sequenced (Masu et al., 1987; Yokota et al., 1989; Hershey and Krause, 1990; Shigemoto et al., 1990). The deduced peptide sequences revealed that they are single polypeptide chains consisting of seven hydrophobic transmembrane-spanning domains with an extracellular amino terminus and a cytoplasmic carboxy terminus, structural features that are characteristic of this family of membrane receptors. Extensive studies have implicated SP in a variety of physiological functions, such as transmission of pain stimuli, exocrine gland secretion, intestinal motility, vasodilation, and the neuronally mediated inflammatory skin reaction (Hansch, 1990). Antagonists of the NK_1 receptor are potential drugs for the treatment of pain, neurogenic inflammation, and asthma. Therefore, a detailed understanding of the interactions between the receptor and its ligand could be highly valuable in the design of substances that modulate receptor function. A graphics-computer-generated, three-dimensional model of the NK_1 receptor was constructed using extensive primary structure comparisons, secondary structure predictions, and three-dimensional homology building with bacteriorhodopsin (Henderson et al., 1990) as a template for the packing of the α-helices. Focusing on the tachykinin and other related peptide receptors allowed us to analyze the putative neuropeptide binding site in the transmembrane domain. Residues likely to be responsible for the receptor affinity, selectivity, and activation were identified. By analogy with cationic neurotransmitter receptors (Hibert et al., 1991; Trumpp-Kallmeyer et al., 1992), the neuropeptide binding site may be located in a hydrophobic pocket near the extracellular surface and consisting of helices 2, 3, 4, 5, 6, and 7. The C-terminal part of SP is proposed to bind in an open turn conformation to this site, forming hydrogen bonds and hydrophobic interactions with conserved amino acids. The model represents an important working hypothesis to direct future experimental exploration of SP binding to the NK_1 receptor and its subsequent activation. This information could be used to design peptidic analogs and nonpeptidic agonists and antagonists.

2. Introduction

A major goal of medicinal chemistry and pharmacology is to understand the molecular basis for interactions of drugs with their targets. Many drugs act by binding to cell surface receptors either as agonists, which stimulate receptor function, or as antagonists, which block the action of endogenous agonists. One such class of receptors couples through guanine nucleotide binding proteins (G-proteins) to modulate the intracellular levels of a variety of second messengers. On binding of agonists, receptors of this class undergo a conformational change that induces the exchange of GDP for GTP in the G-protein that is coupled at the cytoplasmic side to the receptor. The activated GTP-G-protein complex then dissociates from the receptor and its subunits and interacts with a specific effector protein to modulate the level of an intracellular second messenger. Members of this class include cationic neurotransmitter receptors and peptide receptors such as tachykinin, endorphin, and endothelin receptors. Other GPCR, such as the FSH and the LSH receptors, bind large glycoproteins. Furthermore, sensory receptors such as the opsin and odorant receptors also belong to this receptor superfamily.

Given their key role in a variety of major physiological processes, many GPCR have become important pharmacological targets in the treatment of a variety of disease states. Therefore, a detailed understanding of the three-dimensional (3D) structure of these receptor proteins and their interactions with ligands and with G-protein partners is highly valuable. Unfortunately, as for most membrane proteins, the nondegenerative purification and crystallization of GPCR still remain a problem. Given this, molecular modeling is the only methodology to address the 3D structure of GPCR.

More than 70 different GPCR have now been cloned and identified. Various methods, including hydrophobicity analyses, bindings of antibodies raised against putative extracellular domains, proteolysis studies, and affinity labeling (Findlay and Pappin, 1986; Wang et al., 1989) have indicated a seven transmembrane α-helical domain arrangement. Moreover, site directed mutagenesis, chemical labeling, and biophysical studies allowed a more detailed understanding of the receptor structure–function relationships. Thus, it appears that small ligands like cationic neurotransmitters bind to the transmembrane domain of their respective receptor, whereas in the case of peptide ligands the extracellular loops are also involved in binding (Strader et al., 1987; Tung Ming Fong et al., 1992). Furthermore, it has been shown that the cytoplasmic loops are involved in the selective coupling to the G-proteins (O'Dowd et al., 1988). In some cases, an extra functional module is found at the extended N-terminal part that can play "catalytic" (thrombin; Vu et al., 1991), "catching and driving" (gonadothyrotropin; Hoflack et al., 1993), or ligand binding (glutamate; O'Hara et al., 1993) roles. Glycosylation sites are found on the N-terminal fragments or the extracellular loops, whereas phosphorylation sites are found at different intracellular locations and play a key role in GPCR desensitization (Hausdorff et al., 1990).

The predicted structure of the GPCR, containing an extracellular amino terminus and seven hydrophobic membrane spanning α-helices connected by extracellular and intracellular loops, has become the hallmark of this receptor superfamily. The proposed structure is remarkably similar to that of bacteriorhodopsin, identified by electron diffraction and sequence analysis (Henderson and Unwin, 1975). Bacteriorhodopsin is a proton pump, isolated from *Halobacterium halobium*, which is not linked to a G-protein. As it has been predicted for GPCR, bacteriorhodopsin consists of seven transmembrane spanning α-helices that are connected by extracellular and intracellular loops. Furthermore, the mammalian opsins that belong to the GPCR superfamily bind the same ligand (retinal) and are activated in the same way as bacteriorhodopsin by the isomerization of retinal following irradiation. Recently, a refined 3D model of bacteriorhodopsin was disclosed that allowed the study of intra- and intermolecular interactions at the atomic level (Henderson et al., 1990).

Taken together, these pieces of information reasonably allowed a 3D modeling attempt of GPCR. In this chapter, we describe the 3D modeling of the NK_1 receptor and its proposed interaction with the C-terminal part of SP. Owing to the potent actions of SP on many organs and systems, tachykinin receptors have become a subject of intense investigation in the past decade. The 3D model of the NK_1 receptor presented here can be a useful tool in the design of new peptidic and nonpeptidic agonists and antagonists.

3. Modeling of the NK_1 Receptor

As a first step in the modeling of GPCR, a comparison of the primary sequences of these receptors is essential. This led to the identification of the seven putative transmembrane regions and of conserved amino acids that might play a role in the receptor folding and function and in ligand binding.

3.1. Primary Structure Comparison

The primary sequences of the following six tachykinin receptors were taken into consideration: the human and rat NK_1 (Hershey and Krause, 1990; Takeda et al., 1991), the human NK_2 (Gerard et al., 1991), bovine NK_2 (Masu et al., 1987) and rat NK_2 (Sasai et al., 1989), and the rat NK_3 (Shigemoto et al., 1990) receptors. These sequences were compared to the primary sequences of 35 other GPCR taken from published data (Trumpp-Kallmeyer et al., 1992; and references cited therein). The sequence alignments were performed using the method of Needleman and Wunsch (Needleman and Wunsch, 1970) as implemented in the University of Wisconsin Genetic's Computer Group Software package (Devereux et al., 1984) using the Dayhoff similarity table for amino acids (Schwarz et al., 1978). The resulting alignment was refined manually in order to minimize the number of insertions and deletions. The final alignment of the transmembrane regions and the corresponding residue numbering are given in Fig. 1.

The primary structure comparison of the tachykinin receptors leads to high homology scores within this receptor family, with the highest conservation in the seven putative transmembrane helices (62–67%). As discussed in our previous papers (Hibert et al., 1991; Trumpp-Kallmeyer et al., 1992), several residues are conserved in

	101
Nk1r$Human	V L **W** A A A **Y** T V I V V T S V V G N V V V M **W**
Nk1r$Rat	V L **W** A A A **Y** T V I V V T S V V G N V V V I **W**
Nk2r$Human	A L **W** A P A **Y** L A L V L V A V T G N A I V I **W**
Nk2r$Bovin	A L **W** T A A **Y** L A L V L V A V M G N A T V I **W**
Nk2r$Rat	A L **W** A T A **Y** L A L V L V A V T G N A T V I **W**
Nk3r$Rat	A L **W** S L A **Y** G L V V A V A V F G N L I V I **W**
Bacr$Halh	W I **W** L A L G T A L M G L G T L Y F L V K G M

TM1

	201
Nk1r$Human	V N L A F A E A S M A A F **N** T V V **N** F T **Y** A V
Nk1r$Rat	V N L A F A E A C M A A F **N** T V V **N** F T **Y** A V
Nk2r$Human	V N L A L A D L C M A A F **N** A A F **N** F V **Y** A S
Nk2r$Bovin	V N L A L A D L C M A A F **N** A A F **N** F V **Y** A S
Nk2r$Rat	I N L A L A D L C M A A F **N** A T F **N** F I **Y** A S
Nk3r$Rat	V N L A F S D A S V A A F **N** T L I **N** F I **Y** G L
Bacr$Halh	K K F Y A I T T L V P A I A F T M Y L S M L L

TM2

	301
Nk1r$Human	L F Y C K F H **N** F F **P** I A A V F A S I **Y** **S**
Nk1r$Rat	L F Y C K F H **N** F F **P** I A A L F A S I **Y** **S**
Nk2r$Human	R A F C Y F Q **N** L F **P** I T A M F V S I **Y** **S**
Nk2r$Bovin	R A F C Y F Q **N** L F **P** I T A M F V S I **Y** **S**
Nk2r$Rat	R A F C Y F Q **N** L F **P** I T A M F V S I **Y** **S**
Nk3r$Rat	A N Y C R F Q **N** F F **P** I T A V F A S I **Y** **S**
Bacr$Halh	W A R Y A D W L F T T P L L L L D L A L L

TM3

	401
Nk1r$Human	V I W V L A L L L A F P **Q** G Y **Y**
Nk1r$Rat	V I W V L A L L L A F P **Q** G Y **Y**
Nk2r$Human	G I W L V A L A L A S P **Q** C F **Y**
Nk2r$Bovin	G I W L V A L A L A F P **Q** C F **Y**
Nk2r$Rat	G I W L V A L A L A S P **Q** C F **Y**
Nk3r$Rat	S I W I L A F L L A F P **Q** C L **Y**
Bacr$Halh	G T I L A I V G A D G I M I G T

TM4

Fig. 1. Alignment of the seven selected transmembrane regions (labeled TM1 to TM7). The receptors considered are: 1. NK_1 human; 2. NK_1 rat; 3. NK_2 human; 4. NK_2 bovine; 5. NK_2 rat; 6. NK_3 rat; 7. bacrhalh, bacteriorhodopsin from *Halobacterium halobium*. The residues are numbered 101, 102, *(continued)*

```
                501
Nk1r$Human  H I C V T V L I Y F L P L L V I G Y A Y T
  Nk1r$Rat  H I C V T V L I Y F L P L L V I G Y A Y T
Nk2r$Human  H L V V I A L I Y F L P L A V M F V A Y S
Nk2r$Bovin  H L I V I A L I Y F L P L V V M F V A Y S
  Nk2r$Rat  H L V V F V L I Y F L P L L V M F G A Y S
  Nk3r$Rat  H I I V I I L V Y C F P L L I M G V T Y T
 Bacr$Halh  W W A I S T A A M L Y I L Y V L F F G F T
                              TM5

                601
Nk1r$Human  M M I V V V C T E A I C W L P F H I F F L L
  Nk1r$Rat  M M I V V V C T E A I C W L P F H V F F L L
Nk2r$Human  T M V L V V L T E A I C W L P Y H L Y F I L
Nk2r$Bovin  T M V L V V V T E A I C W L P Y H L Y F I L
  Nk2r$Rat  A M V L V V L T E A I C W L P Y H L Y F I L
  Nk3r$Rat  M M I I V V V T E A I C W L P Y H V Y F I L
 Bacr$Halh  T F K V L R N V T V V L W S A Y P V V W L I
                              TM6

                701
Nk1r$Human  K F I Q Q V Y L A I M W L A M S S T M Y N P I I
  Nk1r$Rat  K F I Q Q V Y L A S M W L A M S S T M Y N P I I
Nk2r$Human  K F I Q Q V Y L A L E W L A M S S T M Y N P I I
Nk2r$Bovin  K F I Q Q V Y L A L E W L A M S S T M Y N P I I
  Nk2r$Rat  K F I Q Q V Y L A L E W L A M S S T M Y N P I I
  Nk3r$Rat  K Y I Q Q V Y L A S E W L A M S S T M Y N P I I
 Bacr$Halh  N I E T L L F M V L D V S A K V G F G L I L L R
                              TM7
```

(Fig. 1 continued) and so on in helix 1, 201, 202, and so on in helix 2, and so forth. Residues conserved in most of the GPCR are underlined, and functional residues conserved in the NK receptors are in bold. The alignment of bacteriorhodopsin results from the *a posteriori* comparison of the three dimensional structures.

most GPCR, and these are also found in the neurokinin receptors. (The numbering of the amino acid residues is taken from the sequence alignment shown in Fig. 1 and does not correspond to the actual numbering of the receptor sequences). They include the Gly 117, Asn 118 sequence on helix 1, Leu 203 and Asp 207 (or Glu) on helix 2, Cys 304 and Ser 318 on helix 3, Trp 403 and Pro 412 on helix 4, Pro 512 and Tyr 520 on helix 5, Phe 609, Trp 613, and Pro 615 on helix 6, and Asn 721, Pro 722 on helix 7. As proposed for the cationic neurotransmitter and mammalian opsin receptors, these residues could form a network of aromatic–aromatic and hydrogen bonding interactions and could play a crucial role in triggering the proposed conformational change of the receptor during ligand binding.

Several residues are conserved specifically in the family of tachykinin receptors and thus might play a specific role in ligand binding: Asn 214 and Asn 218 on helix 2, Asn 308 on helix 3, Gln 413 on helix 4, Phe 616 (or Tyr 616) and His 617 on helix 6, and Gln 704 and 705 on helix 7. It is interesting to note that the transmembrane domain of the tachykinin receptors contains a relatively high number of aromatic and hydrogen bonding residues. This is in striking contrast to other peptide receptors, where more basic and acidic amino acids are present. In addition to overall conserved amino acids and those invariant within all tachykinin receptors, a large number of residues are conserved only in the NK_1 receptor variants (*see* Fig. 1). Some of these might be responsible for ligand specificity, and this will be addressed in the 3D model.

3.2. Three-Dimensional Modeling of the NK_1 Receptor

The construction of the NK_1 receptor was performed using methods already described (Hibert et al., 1991; Trumpp-Kallmeyer et al., 1992). The transmembrane regions were selected based on primary structure comparison and hydropathicity profiles and were constructed in an α-helical conformation. The single helices were geometry optimized with the Kollman "all atom" force field parameter set implemented in Sybyl 5.3 (Weiner and Kollman, 1986; Sybyl by Tripos E&S, St. Louis, MO). In the minimization procedure, a cutoff of 8 Å was used. To account to some extent for the membrane environment, a dielectric constant of 5 was chosen. The minimized helices were packed according to the model of bacteriorhodopsin and oriented in such a way that a maximum number of charged, hydrophilic, and conserved residues faced inward and hydrophobic residues faced outward. The adjustment of side chain torsion angles was accomplished using interactive graphics techniques. Despite the fact that the seven transmembrane regions in the GPCR can be localized easily using extensive primary structure comparisons and hydropathicity analyses, none of these methods is accurate enough to identify the precise beginning and ending of the helical segments. This leads to some ambiguity in the relative position of each helix in the bundle. In our 3D models, the different helices were positioned in such a way that full agreement was obtained with data from site directed mutagenesis experiments on bovine rhodopsin and cationic neurotransmitter receptors (Strader et al., 1987; Zhukovsky and Oprian, 1989; Tomoko and Gorbind Korana, 1990, 1991). The construction of 3D models for most of GPCR sequenced until now, followed by analysis of the amino acid environment in the transmembrane regions and their agreement with structure activity relationship studies, further helped to refine the assembly of the helix bundle. The 3 extracellular and 3 intracellular loops were constructed with the loop search procedure implemented in Sybyl 5.3. The second extracellular loop was adjusted in order to obtain the disulfide bridge between Cys 105 at the beginning of helix 3 and Cys 180 in the second extracellular loop with an optimal C-S-S-C angle of 90°. The receptor model thus obtained was again energy minimized to improve helix–helix interactions and to relieve steric conflicts. In the next step, the C-terminal part of SP (last 6 residues) was docked into the proposed ligand binding site. The receptor–ligand complex was again energy minimized and analyzed in terms of interaction energies and conformational properties.

3.3. Docking of SP in the NK_1 Receptor

Structure–activity relationship studies have shown that the C-terminal end of SP is important for biological activity. C-terminal fragments as small as hexapeptides were equipotent to SP as agonists in several tests (Cascieri et al., 1992). The N-terminal fragments, SP(1–10)-NH_2, SP(1–9)-NH_2, and SP(1–8)-NH_2 were inactive on the guinea pig ileum, but were weak agonists in the rat salivation model and in several CNS tests (Dion et al. 1987).

Studies of SP, based on conformational energy calculations, have pointed out several nearly equivalent low energy conformers (Cotrait and Hospital, 1982), which differ not only in the position of the amino acid side chains but also in the peptide backbone conformation. NMR and circular dichroism techniques also did not show any evidence for ordered structures in aqueous solution. However, preferred conformers with a turn or α-helical conformation for the C-terminal part of SP (Chassaing et al., 1986; Levian-Teitelbaum et al., 1989; Saulitis et al., 1992) seem to be present in various other solvents. In our docking study of the 6 C-terminal residues of SP, a set of low energy conformations as determined by Monte Carlo simulations (Benmbarek, 1991) was used. These conformers were positioned in several different orientations in the ligand binding site of the NK_1 receptor in order to obtain the best possible interaction complexes as determined by energy calculations. The conformer with the most favorable receptor–ligand interactions possesses an open turn conformation in the last four residues (Phe^8-Gly^9-Leu^{10}-Met^{11}-NH_2). Figure 2 shows a stereoplot of the proposed receptor bound conformation of the C-terminal end of SP.

4. Discussion of the NK_1 Model

The 3D model of the NK_1 receptor has many features in common with those for receptors already published (Hibert et al., 1991; Trumpp-Kallmeyer et al., 1992), consisting of seven tightly packed α-helices defining a narrow cleft. Amino acids conserved in all GPCR are distributed across the seven helices with the aromatic residues concentrated on helices 3, 4, 5, 6, and 7 near the extracellular side and the hydrogen bonding and charged residues near the intracellular side of the receptor. Amino acids conserved only in tachykinin receptors, which could play a role in ligand binding, are located in a pocket near the extracellular entrance surrounded by helices 2, 3, 4, 5, 6, and 7. A comparison of the amino acid composition in NK_1, NK_2, and NK_3 receptors shows that only 3 amino acids are different at the inside of the transmembrane cleft; in the NK_2 and NK_3 receptor types, a Gln is found at the position of His 307 on helix 3, Thr 505 is changed

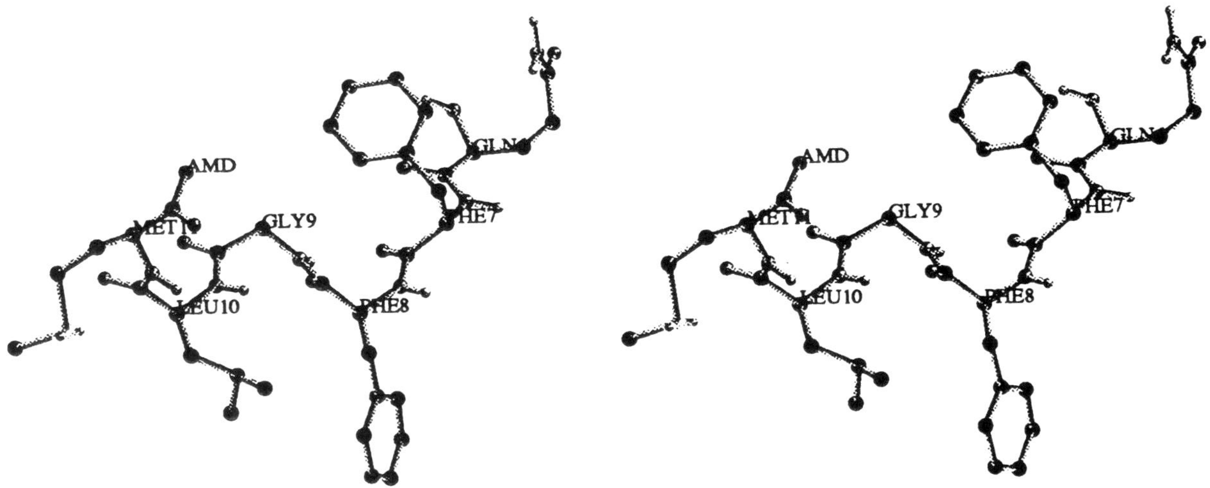

Fig. 2. Stereoplot of the proposed receptor bound conformation of SP. The phi/psi angles are: Phe^7 (50.3/–177.3); Phe^8 (-99.2/118.6); Gly^9 (-62/32.5); Leu^{10} (-60.5/–60); Met^{11} (-177.4).

to Ile 505 on helix 5, and helix 7 contains Phe instead of Met 711. In contrast, many more residues are different at the outside of the receptors facing the lipid bilayer.

Consistent with structure–activity relationships, the three-dimensional modeling study suggests that the last 6 C-terminal residues of SP bind to the transmembrane domain of the NK_1 receptor and that the N-terminus binds to residues located at the entrance of the binding cleft. Noteworthy, the binding domain of the last 4 residues of SP corresponds to the agonist binding pocket proposed for the cationic neurotransmitter and mammalian opsin receptors (Hibert et al., 1991; Trumpp-Kallmeyer et al., 1992). The border of this hydrophobic binding pocket is defined by Tyr 509, Trp 613, and Phe 616, as in cationic neurotransmitter receptors. In addition, a number of tachykinin receptor specific amino acids are present in this pocket: Asn 308 and Pro 311 on helix 3; Ala 406, Leu 409, Ala 410 and Gln 413 on helix 4; Phe 616 (or Tyr) and His 617 on helix 6; and Gln 704 and 705 on helix 7.

In our proposed receptor–ligand complex, the C-terminal end of SP with Met^{11}-NH_2 is located about 15 Å below the entrance of the transmembrane region in a hydrophobic pocket consisting of helices 4, 5, and 6, whereas Gln^6 of SP is located at the entrance of the transmembrane region between helices 2, 3, and 7. Within this hydrophobic environment, several amino acid residues are appropriately positioned for interaction with the C-terminus of SP or SP analogs.

Asn 308 is conserved in the tachykinin receptors only and can form a hydrogen bond with the Phe^8-Gly^9 and/or Phe^7-Phe^8 peptide bond of SP. His 617 and Gln 413, also invariant in the tachykinin receptors, can make hydrogen bonds to the C-terminal carboxamide group. Gln 413 can form an additional hydrogen bond with the Gly^9-Leu^{10} peptide bond in SP. SP itself, in its proposed receptor bound conformation, possesses an intramolecular hydrogen bond from the carbonyl oxygen of Phe^8 to the amide nitrogen of Leu^{10} (Fig. 2). The amino acid side chains of Leu^{10} and Met^{11} of SP are located between helices 4, 5, and 6. Met^{11} can interact with Tyr 509 forming π–σ interactions. Trp 613 is located at the bottom of the ligand binding pocket and is part of the hydrophobic environment of Leu^{10}. This binding pocket corresponds to the site that binds the aromatic part of the aromatic cationic neurotransmitter ligands in their respective receptors. The side chain of Phe^8 in SP is located in a second hydrophobic pocket consisting of helices 3 and 7 and makes interactions with His 307 and Met 711. Both residues are conserved in the NK_1 receptors, but not in NK_2 or NK_3 receptor types, and could be responsible for the selectivity of C-terminal hexapeptides that possess a Phe-Gly-Leu-Met-NH_2 terminus. In contrast, NK_2 and NK_3 receptors possess a Gln at position 307 and a Phe at position 711 in the ligand binding site and simultaneously, instead of a Phe, the NKA and NKB peptides contain a Val residue. The binding site proposed for the Phe^8 side chain corresponds to the region where the positively charged head group of cationic neurotransmitters bind to their receptors. The binding pocket of Phe^7 and Gln^6 is located between helices 2, 3, and 7 at the entrance of the transmembrane region. In the model of the receptor–ligand interaction complex, the side chain of Gln 704 is able to interact with the Gln^6-Phe^7 amide bond and/or with the side chain of Gln^6. The side chain of Phe^7 is positioned between helices 6 and 3 and can form aromatic–aromatic interactions with Phe 620. Figure 3 shows a stereoplot of the C-terminal portion of SP in its proposed binding site.

5. Signal Transduction

As proposed for the cationic neurotransmitter receptors, several conserved aromatic and Pro residues might be involved in the conformational change of receptors during ligand binding (Hibert et al., 1991; Trumpp-Kallmeyer et al., 1992). The overall conserved residues (Phe 609, Trp 613, Pro 615, Pro 512, Tyr 520, Trp 403, and Pro 412), together with residues characteristic for the tachykinin receptors (Asn 308, Pro 311, Gln 413, Phe 616, His 617, and Gln 704), are probably involved in binding of the C-terminal end of SP and might play an essential function in triggering the conformational change of the receptor. More particularly, Trp on helix 6 appears to have a key role since it seems from our modeling studies that this amino acid represents the bottom of the ligand binding pocket in most

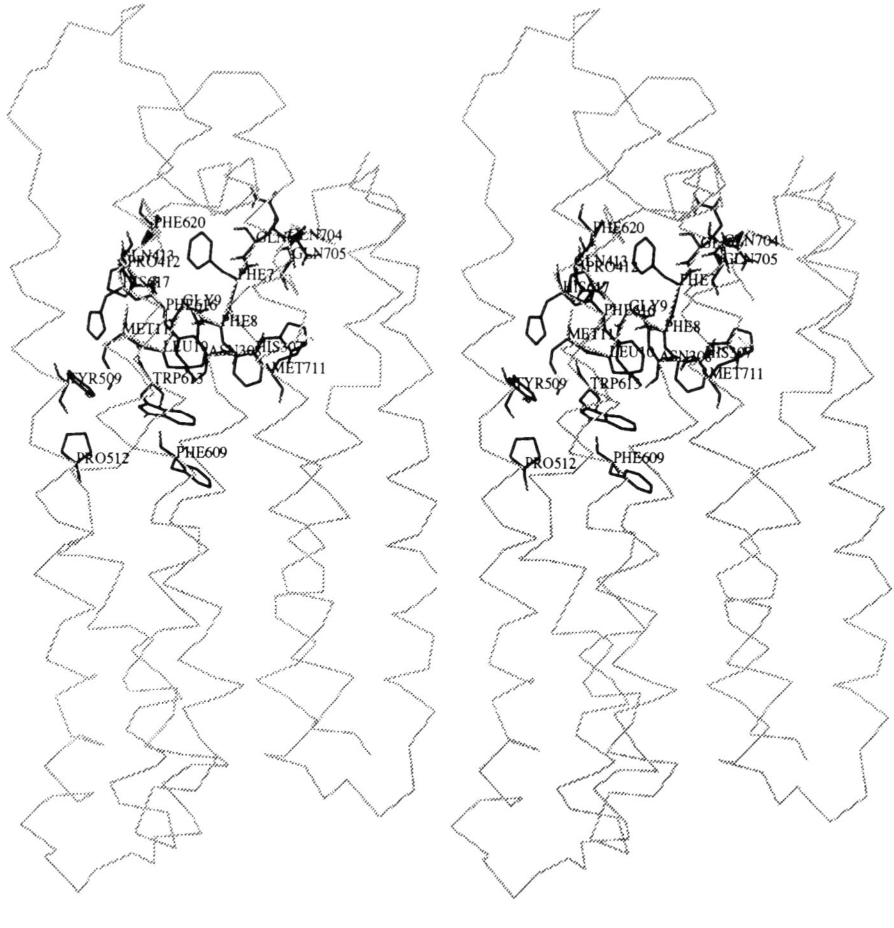

Fig. 3. Stereoplot of a sideview of the NK$_1$ receptor. The seven helices and the extra- and intracellular loops are indicated by a C_α-carbon chain trace. Only residues in the binding region and the C-terminal residues of SP are displayed. The last four residues are proposed to bind in an open turn conformation to the receptor. Asn 308 can form a hydrogen bond with the Phe^8-Gly^9 and/or Phe^7-Phe^8 amide bond; Gln 413 can form a hydrogen bond with the Gly^9-Leu^{10} peptide bond and can simultaneously interact with the carboxamide of SP. His 617 might form a second hydrogen bond with the carboxamide. Trp 613 interacts with the side chain of Leu^{10} and builds the bonom of the ligand binding site. Phe^8 is located between helices 3 and 7 in a hydrophobic pocket and can interact with His 307 and Met 711.

GPCR. The importance of this residue is supported by site directed mutagenesis studies on the corresponding Trp in bovine rhodopsin. Mutation to Phe or Ala produced mutant receptors with abnormal bleaching behavior and a reduced G-protein activation (Tomoko and Gobind Khorana, 1991). A rearrangement of some of these residues during agonist binding could produce a conformational change in the ligand binding site and propagate to the intracellular part of the receptor along other conserved aromatic, proline, and hydrogen bonding residues in the neighboring helices. At the level of the intracellular loops, conserved charged residues appear to be involved in crucial electrostatic interactions, which can influence the receptor conformation. Thus, it seems from our models that Arg 329 in the conserved Asp-Arg-Tyr sequence at the bottom of helix 3 can interact with Tyr 725 on helix 7 and may serve to constrain different receptor conformations. This can affect the accessibility of conserved basic residues in the cytoplasmic loops located at the interface between the phospholipid bilayer and the cytoplasmic medium. In particular, secondary structure predictions and the study of short peptide fragments suggest the presence of two short amphiphilic α-helices at the intracellular loop segment between helix pair 5 and 6 (Sukumar et al., 1992). These amphiphilic α-helices are crucial for the interaction with the G-protein as implicated by site directed mutagenesis (Strader et al., 1987). Especially, the helical segment near the amino terminus of transmembrane region 6 possesses a number of basic Arg and Lys residues which are present in all GPCR. In our 3D model, these residues are located on one side of helix 6 at the interface between the lipid bilayer and the cytoplasm. This could enable electrostatic interactions between these conserved basic residues and the C-terminal part of G-proteins in the G-protein activated state. Studies with peptide fragments corresponding to this helical segment support this hypothesis (Hamm et al., 1988; Cheung et al., 1991). In contrast, in the G-protein inactivated state, the basic residues near the amino terminus of transmembrane region 6 could interact with the phospholipid bilayer. It thus appears from our 3D models that ligand binding can induce a cascade of events involving conserved aromatic, hydrogen bonding, charged and proline residues.

6. Conclusions

Although only a limited amount of experimental data on tachykinin receptors is available at the moment, we have been able to generate a reasonable model for the NK_1 receptor using the close homology to the cationic neurotransmitter and mammalian opsin receptors. A possible binding site and active conformation for the C-terminal part of SP have been identified and enabled a detailed study of the interaction of SP with its receptor. Our model provides hypotheses concerning the residues likely to be responsible for SP binding and receptor activation. In addition, a hypothesis has been formulated that can explain the selectivity of short C-terminal SP fragments (e.g., SP[6–11]-NH_2) for the NK_1 receptor. A comparison of the amino acid composition at the inside of the transmembrane region in all tachykinin receptors reveals that they differ only in 3 amino acids corresponding to His 307, Thr 505, and Met 711 in NK_1. In our receptor–ligand interaction model, Phe^8 of SP (which is Val^7 in NKA and NKB) is located in a hydrophobic pocket between helix 3 and helix 7, which contains two of the amino acids different among the tachykinin receptor types. Thus, a difference between NK_1 specific and NK_2 and NK_3 specific C-terminal fragments may be in their preferred interactions with His 307/Met 711 or Gln 307/Phe 711, respectively.

It is clear that the 3D model of the NK_1 receptor and its interactions with SP are very hypothetical because of the small amount of experimental data available for these receptor types. Nevertheless, since a large part of this modeling study is based on comparisons with other GPCR for which much more experimental data are available (cationic neurotransmitter receptors and mammalian opsins), we believe that at least some of the predictions we have made will prove to be useful for a better understanding of the mechanisms involved in tachykinin receptor activation. In any event, the 3D model presented here, despite the obvious limitations, can be a useful tool for orienting site directed mutagenesis experiments and in the design of new peptidic SP analogs and nonpeptidic antagonists.

References

Benmbarek, A. (1991) Etude conformationelle de molecules peptidiques à l'interet Pharmacologique: Enkephalines, Substance P et Cholecystokinine. Ph.D. Thesis, University Louis Pasteur, Institute of Theoretical Chemistry, Strasbourg, France.

Buck, S. H. and Shatzer, S. A. (1988) Agonist and antagonist binding to tachykinin peptide NK-2 receptors. *Life Sci.* **42,** 2701–2708.

Cascieri, M. A., Huang, C., Ming Fong, T., Cheung, A. H., Sadowski, S., Ber, E., and Strader, C. D. (1992) Determination of the amino acid residues in substance P conferring selectivity and specifity for the rat neurokinin receptors. *Mol. Pharmacol.* **41,** 1096–1099.

Chassaing, G., Convert, O., and Lavielle, S. (1986) Preferential conformation of substance P in solution. *Eur. J. Biochem.* **154,** 77–85.

Cheung, A. H., Huang, R. R. C., Graziano, M. P., and Strader, C. D. (1991) Specific activation of G_s by synthetic peptides corresponding to an intracellular loop of the adrenergic receptor. *Fed. Eur. Biochem. Soc.* **279,** 277–280.

Cotrait, M. and Hospital M. (1982) Conformational studies on SP C-terminal pentapeptide pGlu-Phe-Phe-Gly-Leu-Met-NH_2. *Biochem. Biophys. Res. Commun.* **109,** 1123–1128.

Devereux, J., Haeberli, P., and Smithies, O. (1984) A comprehensive set of sequence analysis programs for the vax. *Nucleic Acid Res.* **12,** 387–395 as implemented in the GCG software at the University of Wisconsin, Biotechnology Center, 1710 University Ave., Madison, WI 53705.

Dion, S., Drapeau, G., Rhaleb, N. E., D'Orléans-Juste, P., and Regoli, D. (1987) Receptors for substance P and neurokinins, correlation between binding and biological activities. *Eur. J. Pharmacol.* **138,** 125–128.

Findlay, J. B. C. and Pappin, D. J. (1986) The opsin family of proteins. *Biochem. J.* **238,** 625–642.

Gerard, N. P., Eddy, R. L., Shows, T. B., and Gerard, C. (1991) The human neurokinin A receptor. Molecular cloning of the gene, chromosome localization and isolation of cDNA from tracheal and gastric tissues. *J. Biol. Chem.* **266,** 1354,1355.

Guard, S. and Watson, S. P. (1991) Tachykinin receptor-types: Classification and membrane signalling mechanisms. *Neurochem. Int.* **18,** 149–165.

Hamm, H. E., Deretic, D., Arendt, A., Hargrave, P. A., Koenig, B., and Hofmann, K. P. (1988) Site of G protein binding to rhodopsin mapped with synthetic peptides to the α-subunit. *Science* **241,** 832–835.

Hansch, C. (1990) *Comprehensive Medicinal Chemistry,* vol. 3, Pergamon Press, Oxford, UK.

Hausdorff, W. P., Lohse, M. J., Bouvier, M., Ligget, S. B., Caron, M. G., and Lefkowitz, R. J. (1990) Two kinases mediate agonist dependent phosphorylation and desensitization of the β_2-adrenergic receptor. *Symp. Soc. Exp. Biol.* **44,** 225–240.

Henderson, R. and Unwin, P. N. T. (1975) Three dimensional model of the purple membrane obtained by electron microscopy. *Nature* **257,** 28–32.

Henderson, R., Baldwin, J., Ceska, T. H., Zemlin, F., Beckamn, E., and Downing, K. (1990) Model of the structure of bacteriorhodopsin based on high resolution electron cryo-microscopy. *J. Mol. Biol.* **213,** 899–929

Hershey, A. D. and Krause, J. E. (1990) Molecular characterization of a functional cDNA encoding the rat substance P receptor. *Science* **247,** 958–962.

Hibert, M., Trumpp-Kallmeyer, S., Bruinvels, and Hoflack, J. (1991) 3D models of cationic neurotransmitter G-protein-coupled receptors. *Mol. Pharmacol.* **40,** 8–15.

Hoflack, J., Hibert, F. M., and Trumpp-Kallmeyer, S. (1993) Three-dimensional models of gonado-thyrotropin hormone receptor transmembrane domain. *Drug Design Dis.* **10,** 157–171.

Levian-Teitelbaum, D., Kolodny, N., Chorev, M., Selinger, Z., and Gilon, C. (1989) 1H-NMR studies of receptor selective substance P analogs reveal distinct predominant conformations in DMSO-d6. *Biopolymers* **28,** 51–64.

Ming Fong, T., Hong Y., and Strader, C. (1992) The extracellular domain of substance P (NK1) receptor comprises part of the ligand binding site. *Biophys. J.* **62,** 59–60.

Needleman, S. B. and Wunsch, C. D. (1970) A general method applicable to the search for similarities in the amino acid sequence of two proteins. *J. Mol. Biol.* **48,** 443–453.

O'Dowd, B. F., Huntowick, M., Regan, J. W., Leader, W. M., Caron, M. G., and Lefkowitz, R. J. (1988) Site directed mutagenesis of the cytoplasmic domains of the human β_2-adrenergic receptor. Localization of regions involved in G protein receptor coupling. *J. Biol. Chem.* **263,** 15,985–15,992.

O'Hara, P. J., Sheppard, P. O., Thøgeren, H., Venezia, D., Haldeman, B. A., McCrane, V., Houamed, K. M., Thomsen, C., Gilbert, T. L., and Mulvihill, E. R. (1993) The ligand binding domain in metabotropic glutamate receptors is related to bacterial periplasmic binding proteins. *Neuron* **11,** 41–52.

Sasai, Y. and Nakanishi, S. (1989) Molecular cloning and characterization of the rat substance K receptor. *Biochem. Biophys. Res. Com.* **165,** 695–702.

Saulitis, J., Mierke, D. F., Byk, G., Gilon, C., and Kessler, H. (1992) Conformation of cyclic analogues of SP: NMR and molecular dynamics in dimethyl sulfoxide. *J. Am. Chem. Soc.* **114,** 4818–4827.

Schwarz, R. M., Dayhoff, M. O., and Orcutt, B. C. (1978) in *Atlas of Protein Sequence and Strucuture* **5** (Supp. 3), 353–358. National Biomedical Research Foundation, Washington, DC.

Shigemoto, R., Yokota, Y., Tsuchida, K., and Nakanishi, S. (1990) Cloning and expression of a rat neuromedin K receptor DNA. *J. Biol. Chem.* **265,** 623–628.

Strader, C., Sigal, I. S., Register, R. B., Candelore, M. R., Rands, E., and Dixon, R. A. F. (1987) Identification of residues required for ligand binding to the β-adrenergic receptor. *Proc. Natl. Acad. Sci. USA* **84,** 4384–4388.

Strader, C. D., Dixon, R. A. F., Cheung, A. H., Candelore, M. H., Blake, A. D., and Sigal, I. S. (1987) Mutations that uncouple the adrenergic receptor from G_s and increase agonist affinity. *J. Biol. Chem.* **262,** 16,439–16,443.

Sukumar, M. and Higashijima, T. (1992) G protein-bound conformation of mastoparan-X, a receptor-mimetic peptide. *J. Biol. Chem.* **267,** 21,421–21,424.

Takeda, Y., Chou, K. B., Takeda, J., Sachais, B. S., and Krause, J. E. (1991) Molecular cloning, structural characterization and functional expression of the human substance P receptor. *Biochem. Biophys. Res. Commun.* **179,** 1232–1240.

Vu, T.-K. H., Wheaton, V. I., Hung, D. T., Charo, I., and Coughlin, S. R. (1991) Domains specifying thrombin-receptor interaction. *Nature* **353,** 674–677.

Tomoko, A. and Gobind Korana, H. (1990) Orientation of retinal in bovine rhodopsin determined by crosslinking studies using a photoactivatable analog of 11-cis-retinal. *J. Biol. Chem.* **265,** 15,762–15,769.

Tomoko, A. and Gobind Khorana, H. (1991) Mapping of the amino acids in membrane embbeded helices that interact with the retinal chromophore in bovine rhodopsin. *J. Biol. Chem.* **266,** 4269–4275.

Trumpp-Kallmeyer, S., Hoflack, J., Bruinvels, A., and Hibert, M. (1992) Modeling of G-protein-coupled receptors: Application to dopamine, adrenaline, serotonin, acelcholine and mammalian opsin receptors. *J. Med. Chem.* **35,** 3448–3462.

Yokota, Y., Sasai, Y., Tanaka, K., Fujiwara, T., Tsuchida, K., Shigemoto, R., Kazizuka, A., Ohkubo, H., and Nakanishi, S. (1989) Molecular characterization of a functional cDNA for rat substance P receptor. *J. Biol. Chem.* **264,** 17,649–17,652.

Wang, H.-Y., Lipfert, L., Malbon, C. C., and Bahouth, S. (1989) Site directed antipeptide antibodies define the topography of the β-adrenergic receptor. *J. Biol. Chem.* **264,** 14,424–14,431.

Weiner, S. J. and Kollman, P. A. (1986) An all atom force field for simulations of proteins and nucleic acids. *J. Comput. Chem.* **7,** 230–252.

Zhukovsky, E. A. and Oprian, D. (1989) Effect of carboxylic acid side chains on the absorption maximum of visual pigments. *Science* **248,** 928–930.

Part III

Mechanisms of Tachykinin Receptor Action

CHAPTER 10

Receptors and the Inositol Phosphate–Calcium Signaling System

James W. Putney, Jr.

1. Introduction

1.1. Historical Aspects of the Inositol Phosphate–Calcium Signaling System

Extracellular stimuli modify $[Ca^{2+}]_i$ through a variety of signaling systems. The most widespread of these is a system that signals Ca^{2+} mobilization through the formation of Ca^{2+}-mobilizing messengers, the inositol polyphosphates. This inositol polyphosphate–Ca^{2+} signaling system is associated with many growth factor, hormone, and neurotransmitter receptor mechanisms and is the focus of this chapter. A schematic diagram of the components of this system is presented in Fig. 1.

Our current understanding of the inositol polyphosphate–Ca^{2+} system originates from a review by Michell (Michell, 1975), who recognized the link between the "phosphoinositide effect" first described by the Hokins in 1954 (Hokin and Hokin, 1954) and Ca^{2+} mobilization (*see* Michell, 1986 for an extensive historical review).

The Tachykinin Receptors Ed.: S. H. Buck © 1994 Humana Press Inc., Totowa, NJ

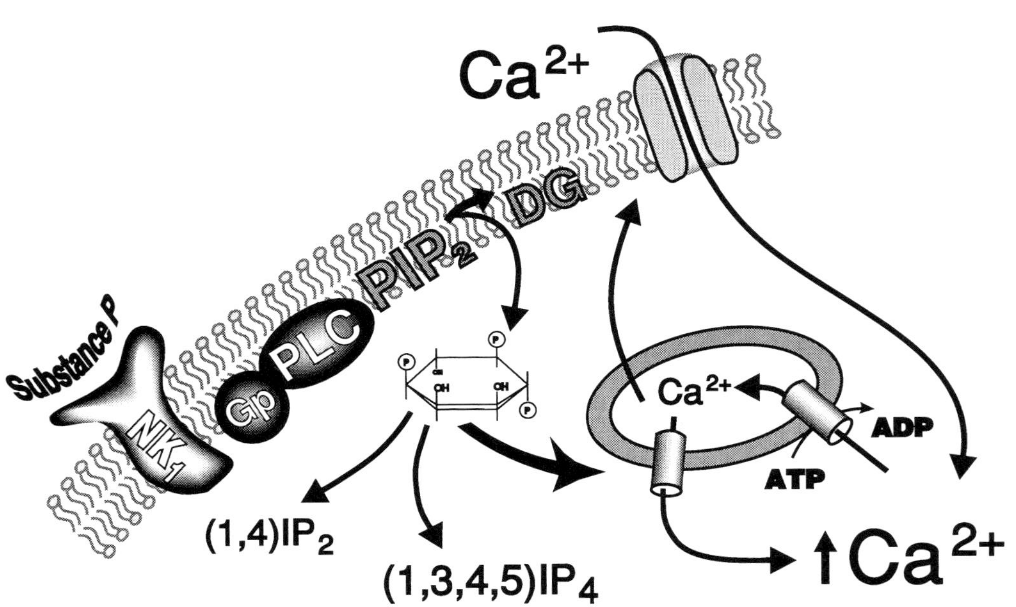

Fig. 1. The NK_1 receptor and the inositol phosphate–Ca^{2+} signaling system. The SP receptor coupled to this signaling system is primarily the NK_1 type, which activates a polyphosphoinositide-specific phospholipase C (PLC) via an intermediary guanine nucleotide-dependent regulatory protein (Gp). This phospholipase C cleaves phosphatidylinositol 4,5-*bis*phosphate (PIP_2) into diacylglycerol (DG) and inositol 1,4,5-trisphosphate (structure shown). The latter is metabolized either by dephosphorylation to inositol 1,4-bisphosphate ($(1,4)IP_2$) or by phosphorylation to inositol 1,3,4,5-tetrakisphosphate ($(1,3,4,5)IP_4$). Inositol 1,4,5-trisphosphate acts to cause the discharge of Ca^{2+} from an intracellular organelle, which accumulates Ca^{2+} by means of an ATP hydrolyzing Ca^{2+} pump. The depletion of Ca^{2+} from this organelle by the action of inositol 1,4,5-trisphosphate may secondarily signal Ca^{2+} entry across the plasma membrane.

However, the molecular basis for this link was not clearly established until the discovery that the initial reaction in receptor-stimulated phosphoinositide turnover involved not the major inositol lipid, phosphatidylinositol, but rather a minor phosphorylated derivative, phosphatidylinositol 4,5-bisphosphate (Abdel-Latif et al., 1977; Kirket al., 1981). Thus, receptor activation, through stimulation of phospholipase C, causes the breakdown of this minor membrane lipid, and Berridge (1983) proposed that the water soluble product of this reaction, inositol 1,4,5-trisphosphate [(1,4,5)IP_3], is the intracellular messenger that activates Ca^{2+} mobilization. With collaborators in England and Germany, he demonstrated that micromolar concentrations of (1,4,5)IP_3 rapidly released Ca^{2+} from a nonmitochondrial store in permeabilized pancreatic acinar cells (Streb et al., 1983). This result was quickly confirmed in several different laboratories working with preparations from a variety of cells, and it was soon established that (1,4,5)IP_3 is the intracellular messenger that mediates receptor stimulated Ca^{2+} mobilization.

In the years since the discovery of (1,4,5)IP_3 and its action, much about the inositol polyphosphate–Ca^{2+} signaling system has been learned. Here is presented a brief discussion of some of these many developments, emphasizing the role of this system in the signal transduction mechanism of NK_1 receptors. This topic was reviewed in 1991 by Putney et al., and this chapter will emphasize progress since that time. The reader will also be directed to selected references and more detailed reviews on specific aspects of this body of research (for example, Putney, 1992b).

1.2. Evidence that Tachykinins Activate the Inositol Phosphate-Calcium Signaling System

Nonexcitable exocrine gland cells have served as excellent models for investigating the roles of inositol phosphates and Ca^{2+} signaling in the actions of substance P (SP). Indeed, it was the potent sialagogic action of SP that provided the basis for its isolation by Chang and Leeman (1970). In rat parotid gland, SP was found to stimulate breakdown (Jones and Michell, 1978) and turnover (Hanley et al., 1980) of inositol lipids, and this was subsequently shown to involve degradation of PIP_2 (Weiss et al., 1982) and generation of (1,4,5)IP_3 (Aub and Putney, 1985; Sugiya et al., 1987). SP methyl ester activates (1,4,5)IP_3 formation in parotid cells with a potency and efficacy approaching that of SP (Sugiya and Putney, 1988b), indicating that the receptor type in these cells is the SP-P type, or in the modern nomenclature, NK_1 (Watson and Abbott, 1990). The activation of phospholipase C in parotid gland by SP appears to involve a pertussis toxin-insensitive, guanine nucleotide-dependent regulatory protein (Taylor et al., 1986), as has been shown for a variety of phospholipase C-linked receptors (Taylor and Merritt, 1986). Associated with this hydrolysis of polyphosphoinositides, substance P causes a rapid increase in cytosolic Ca^{2+} ($[Ca^{2+}]_i$) in parotid acinar cells, attributable to a release of intracellular Ca^{2+}, followed by an influx of Ca^{2+} from the extracellular space (Rudich and Butcher, 1976; Marier et al., 1978; Merritt and Rink, 1987a; Merritt and Rink, 1987b). (1,4,5)IP_3 rapidly releases Ca^{2+} from an intracellular store in permeabilized parotid acinar cells (Schulz et al., 1985; Putney et al., 1986; Thevenod and Schulz, 1988), and thus it is likely that this action of (1,4,5)IP_3 is responsible for the initial intracellular release of Ca^{2+} seen in intact cells stimulated with SP. The mechanism by which SP and other agonists of this class regulate Ca^{2+} entry is less well understood and is discussed below in more detail.

AR4-2J is a cell line derived from a rat exocrine pancreatoma (Womack et al., 1985). Unlike the nontransformed rat exocrine pancreatic acinar cells, AR4-2J cells express functional receptors for SP, and Womack et al. (1985) have pointed out their potential utility for investigations into SP receptor mechanisms. These cells respond to SP methyl ester with increases in $[Ca^{2+}]_i$, as well as increases in (1,4,5)IP_3, indicating the presence of an NK_1 type receptor (Putney et al., 1991). In AR4-2J cells, it was possible to devise experimental conditions for labeling cellular inositol lipids with [^{3}H]inositol to demonstrated isotopic equilibrium (Horstman et al., 1988). This has permitted estimation of the mass of the inositol phosphates formed on activation of SP receptors in these cells without the necessity of assumptions regarding homogeneity of labeling of cellular lipid pools. Thus, Horstman et al. (1988) demonstrated that in AR4-2J cells the basal level of (1,4,5)IP_3 appeared to be around 2 μ*M* and was increased by SP activation to around 20 μ*M* (Horstman et al., 1988). This finding was surprising in light of the observation that, in permeable AR4-2J cells, (1,4,5)IP_3 released intracellular Ca^{2+} with an apparent K_D of around 100 n*M* (Bird et al., 1991a). Although the identity of the material as (1,4,5)IP_3 in these studies was based solely on the chromatographic mobility of the ^{3}H-labeled material, in a subsequent report qualitative analysis of the apparent (1,4,5)IP_3 formed in these cells revealed that this material, in both basal and stimulated cells, was essentially pure (1,4,5)IP_3 (Nogimori et al., 1990). Furthermore, subcellular fractionation and cell permeabilization studies failed to detect substantial compartmentalization of (1,4,5)IP_3 in these cells (Bird et al., 1991a). Thus, it is possible that the sensitivity of (1,4,5)IP_3 receptors is regulated by an unknown mechanism in AR4-2J cells, as well as in other cell types.

2. Inositol Phosphate Formation and Metabolism

2.1. *Inositol Phosphates Involved in Calcium Signaling*

Initially, the metabolism of inositol polyphosphates generated in response to the activation of surface membrane receptors was thought to be a rather simple process. It was believed that (1,4,5)IP_3 was dephosphorylated to (1,4)IP_2, to (1)IP, and finally to free inositol by a lithium-sensitive inositol 1-phosphatase. More recently, the pathways of inositol phosphate metabolism have become much more complex (Putney, 1987; Majerus et al., 1988; Menniti et al., 1993) (Fig. 2). This is attributable largely to the development of HPLC analytical procedures that can resolve inositol phosphates with only subtle differences in structure. (1,4,5)IP_3 is dephosphorylated by a 5-phosphatase to (1,4)IP_2 (Downes et al., 1982), and (1,4)IP_2 is believed to be dephosphorylated almost exclusively to (4)IP by (1,3,4)IP_3/(1,4)IP_2-1-phosphatase (Inhorn et al., 1987). In addition to the dephosphorylation of (1,4,5)IP_3 by the 5-phosphatase, most tissues contain a 3-kinase that transfers a phosphate from ATP to the 3 position of (1,4,5)IP_3 to form (1,3,4,5)IP_4 (Irvine et al., 1986a). This

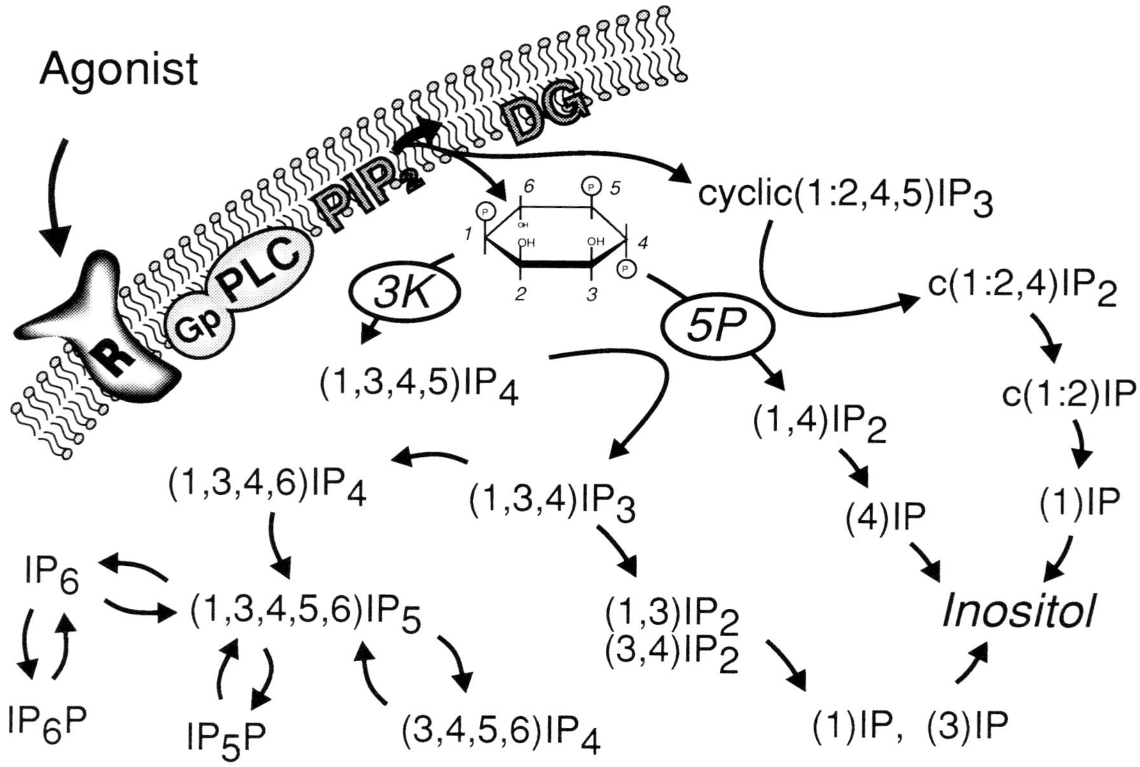

Fig. 2. Pathways of inositol phosphate metabolism in activated cells. Agonist activation of a surface receptor (R) activates, usually through a G-protein (Gp), a polyphosphoinositide-specific phospholipase C (PLC) which catalyzes the breakdown of phosphatidylinositol 4,5-bisphosphate (PIP_2) into the Ca^{2+}-signaling messenger, $(1,4,5)IP_3$ (structure shown), and the protein kinase C activator, diacylglycerol (DG). A minor product of PLC action on PIP_2 is a cyclic derivative of $(1,4,5)IP_3$, cyclic$(1:2,4,5)IP_3$ (Connolly et al., 1986). $(1,4,5)IP_3$ is metabolized by two enzymes, a 5-phosphatase and a 3-kinase, initiating complex degradative and synthetic pathways of inositol phosphate metabolism (*see text* for details).

molecule is then dephosphorylated by the same 5-phosphatase which degrades (1,4,5)IP_3 to form an isomeric inositol trisphosphate, (1,3,4)IP_3. (1,3,4)IP_3 is then dephosphorylated by the (1,3,4)IP_3/(1,4)IP_2-1-phosphatase to (3,4)IP_2 (Inhorn et al., 1987) and to (1,3)IP_2 (Bansal et al., 1987) by an enzyme that is not well characterized. These bisphophates are then dephosphorylated primarily to a mixture of (1)IP and (3)IP, which are stereoisomers not resolved by conventional HPLC.

The complexity of the metabolic pathway for metabolism of (1,4,5)IP_3 has suggested to some that, in addition to the Ca^{2+}-mobilizing actions of (1,4,5)IP_3, other biological functions may be regulated by one or more of these metabolites. For example, Irvine and his collaborators have presented evidence for a role of (1,3,4,5)IP_4 in either potentiating the action of (1,4,5)IP_3 or in modulating Ca^{2+} entry in a variety of cell types (Irvine and Moor, 1986; Morris et al., 1987; Changya et al., 1989a,b). The evidence for and against this idea will be discussed in more detail in a subsequent section.

An additional complication in the inositol polyphosphate metabolic scheme, discovered by Wilson et al. (1985a), is the formation of inositol cyclic 1:2,4,5-trisphosphate ((c1:2,4,5)IP_3). This compound accumulates as a proportion of the product of phospholipase C action on phosphatidylinositol 4,5-bisphosphate. In some (Wilson et al., 1985b; Irvine et al., 1986b), but not all (Lee and Hokin, 1989; Willcocks et al., 1989), studies (c1:2,4,5)IP_3 mobilized intracellular Ca^{2+} at concentrations similar to those required for (1,4,5)IP_3. However, kinetic studies in intact cells (Hughes et al., 1988; Wong et al., 1988) indicate that this inositol phosphate is unlikely to function in $[Ca^{2+}]_i$ signaling, at least in a direct way. This conclusion is derived from the observation that the cellular degradation rate of (c1:2,4,5)IP_3 is very slow (half-time, about 10 min) relative to the time required to terminate $[Ca^{2+}]_i$ signals (a few seconds). In fact, it is this relatively slow rate of metabolism that allows accumulation and detection of (c1:2,4,5)IP_3 following prolonged stimulation of cells. It was estimated that in parotid acinar cells, only about 1% of the product of phospholipase C-mediated hydrolysis of phosphatidylinositol 4,5-bisphosphate is (c1:2,4,5)IP_3, the balance being the potent Ca^{2+}-

mobilizing signal, $(1,4,5)IP_3$. It is possible that the apparent biological actions of $(c1:2,4,5)IP_3$ in vitro result from nonenzymatic hydrolysis of the cyclic phosphodiester bond, thereby generating $(1,4,5)IP_3$ (Willcocks et al., 1989).

2.2. Higher Phosphorylated Inositol Phosphates

As shown in Fig. 2, in the pathway of metabolism of $(1,3,4,5)IP_4$, a significant branch point occurs. The majority of $(1,3,4)IP_3$, the dephosphorylation metabolite of $(1,3,4,5)IP_4$, is further dephosphorylated eventually to inositol, but some $(1,3,4)IP_3$ is phosphorylated by a 6-kinase to form $(1,3,4,6)IP_4$ (Shears et al., 1987). This compound turns over rather slowly in comparison to $(1,4,5)IP_3$, arguing against a direct role in Ca^{2+} signaling, although Ca^{2+} mobilization has been observed following its injection into *Xenopus* oocytes (Ivorra et al., 1991) and neuroblastoma cells (Gawler et al., 1991). Perhaps of more general significance, phosphorylation of $(1,3,4,6)IP_4$ provides the major, if not exclusive, pathway for the formation of $(1,3,4,5,6)IP_5$ (Menniti et al., 1990) and possibly also IP_6, assuming that $(1,3,4,5,6)IP_5$ is its major *de novo* precursor. The phosphorylation of $(1,3,4,6)IP_4$ appears to be rate-limiting and proceeds at near maximal rates even in cells whose phospholipase C has not been activated by a surface receptor agonist (Menniti et al., 1990). As a consequence, the rate of $(1,3,4,5,6)IP_5$ synthesis is not usually increased when phospholipase C is activated and thus may not be regulated by phospholipase C-linked agonists.

However, the metabolism of $(1,3,4,5,6)IP_5$ is subject to receptor regulation in another way. Menniti et al. (1990) demonstrated that another inositol tetrakisphosphate isomer, $(3,4,5,6)IP_4$, is derived from $(1,3,4,5,6)IP_5$, and its formation is markedly increased by phospholipase C-linked agonists. Agents that bypass regulation of phospholipase C, such as the Ca^{2+}-mobilizing agent, thapsigargin, and the protein kinase C-activator, phorbol myristate acetate, did not mimic agonist activation by increasing cellular levels of $(3,4,5,6)IP_4$. Thus, this reaction is apparently regulated through some mechanism other than the protein kinase C–Ca^{2+} signaling pathway. Regardless of its mode of regulation, it appears likely that this reaction provides some function in cell regulation, although at present the nature of this function is a mystery. Levels of (1,3,4,5,6)IP_6 are rather high in most cell types (Menniti et al., 1993) and are thus reduced only slightly by agonists, suggesting that it is the increased level of (3,4,5,6)IP_4 (which can be severalfold) that affects some cellular function.

The recent discovery of inositol phosphates more polar than IP_6 has added a new level of complexity and mystery to the field of inositol phosphate biochemistry (Menniti et al., 1992; Stephens et al., 1992). These highly polar compounds appear to be pyrophosphorylated derivatives of (1,3,4,5,6)IP_5 and IP_6, and their presence has been documented in mammalian cells (Menniti et al., 1992) and in *Dictyostelium* (Stephens et al., 1992). They apparently arise from relatively rapid cycles of phosphorylation and dephosphorylation of (1,3,4,5,6)IP_5 and IP_6. There is no evidence as yet that the metabolism of these unusual inositol phosphates is under control of hormones or neurotransmitters, and, as for (3,4,5,6)IP_4, their function remains a mystery.

3. Phases of Calcium Signaling

3.1. Receptors for (1,4,5)IP_3 and Intracellular Calcium Release

In contrast to the case for the surface membrane receptors, the intracellular receptors for (1,4,5)IP_3 initiate the $[Ca^{2+}]_i$ signal in a relatively simple and straightforward manner. Supattapone et al. (1988) first purified the (1,4,5)IP_3 receptor and demonstrated by reconstitution studies that this single molecule (or molecular complex) was capable of binding (1,4,5)IP_3, as well as acting as a Ca^{2+} channel regulated by (1,4,5)IP_3. The functional (1,4,5)IP_3 receptor/channel appears to be a homotetramer containing four binding sites for (1,4,5)IP_3. More than one form of the receptor has been purified, and it appears that these arise from alternate splicing of mRNA in the monomer subunits (Ferris and Snyder, 1992).

Although the initiation of Ca^{2+} release by (1,4,5)IP_3 involves a relatively simple channel gating process, it has now become clear that the interaction of (1,4,5)IP_3 with its receptor also involves complex and poorly understood regulatory interactions among the receptor, (1,4,5)IP_3, and Ca^{2+}. The Ca^{2+} exerts its influence from both the cytoplasmic and lumenal aspects of the receptor. Jean and Klee (1986) first demonstrated that elevation of Ca^{2+} on the cytoplasmic surface of the receptor appears to inhibit the ability of (1,4,5)IP_3 to activate the channel. This effect may play an important role in the generation of Ca^{2+} oscillations discussed later. Also, regulation of the (1,4,5)IP_3 receptor by Ca^{2+} may play a role in the "quantal" release of Ca^{2+}. This latter phenomenon was first suggested from studies of Muallem et al. (1989) whose data indicated that submaximal concentrations of (1,4,5)IP_3 did not induce a partially activated permeability to Ca^{2+} throughout the intracellular Ca^{2+} pool, but rather appeared to release Ca^{2+} fully from a small subset of the total Ca^{2+} pool (*see also,* Meyer and Stryer, 1990; Taylor and Potter, 1990). As outlined in a recent review by Taylor (1992), two basic interpretations may be placed on these observations. Either (1) some feedback mechanism, presumably initiated by the fall in Ca^{2+} within the (1,4,5)IP_3-sensitive organelle, reduces the sensitivity of the Ca^{2+}-gating receptor to (1,4,5)IP_3, thereby abruptly terminating release at some set level (Irvine, 1990), or (2) the intracellular (1,4,5)IP_3-sensitive Ca^{2+}-releasing organelles release virtually all of their stored Ca^{2+} in response to activation by threshold concentrations of (1,4,5)IP_3, but they are markedly heterogeneous in their sensitivity to (1,4,5)IP_3; i.e., there are different thresholds for an "all-or-none" release of Ca^{2+} (Oldershaw et al., 1991). Although not entirely conclusive, at the present the preponderance of evidence favors the former mechanism—a loss of (1,4,5)IP_3 receptor sensitivity for its ligand as the lumenal concentration of Ca^{2+} decreases (Missiaen et al., 1992; Nunn and Taylor, 1992) (*see also* Combettes et al., 1992; Shuttleworth, 1992). Some caution should be taken in interpreting studies addressing Ca^{2+} pool sizes in permeable cells, since there is evidence that permeabilization may artifactually cause functional fragmentation of intracellular Ca^{2+} pools (Menniti et al., 1991; Bird et al., 1992a). It will be important to distinguish among possible explanations for this phenomenon, since mechanisms controlling receptor sensitivity to (1,4,5)IP_3 are likely to play important roles in determining the temporal and spatial aspect of Ca^{2+} signaling that give rise to $[Ca^{2+}]_i$ oscillations and waves (discussed in detail in Berridge, 1990; Cuthbertson and Cobbold, 1991; Meyer and Stryer, 1991; Berridge, 1992; Thomas et al., 1992).

3.2. *Regulation of Calcium Entry*

The Ca^{2+} mobilizing action of $(1,4,5)IP_3$ is capable of inducing rapid and precipitous increases in $[Ca^{2+}]_i$, but because the capacity of the intracellular Ca^{2+} stores is finite, the maintenance of sustained $[Ca^{2+}]_i$ signals requires an increased rate of entry of Ca^{2+} across the plasma membrane. This generally results in a signal that is biphasic in nature; a transient rise in $[Ca^{2+}]$ occurs that is largely independent of the presence of extracellular Ca^{2+}, and this is followed by a sustained phase of elevated $[Ca^{2+}]_i$ that is absolutely dependent on the presence of extracellular Ca^{2+} (Fig. 3). In theory, the regulation of the second, Ca^{2+} entry phase of $[Ca^{2+}]_i$ signaling could be rather simple since the initial event—receptor binding—and the functional endpoint—increased permeability to Ca^{2+}—both occur in the plasma membrane. However, it has become quite clear that, as shown for the intracellular release phase of the $[Ca^{2+}]_i$ response, inositol phosphates can also provide the signal for activating Ca^{2+} entry across the plasma membrane. The evidence that this is so is straightforward: Injection of cells with $(1,4,5)IP_3$ or its analogues activates both phases of the $[Ca^{2+}]_i$ signal (Irvine and Moor, 1986; Slack et al., 1986; Irvine and Moor, 1987; Llano et al., 1987; Morris et al., 1987; Penner et al., 1988; Changya et al., 1989b), sometimes appearing to require some $(1,3,4,5)IP_4$ *(discussed below)*, and injection of the $(1,4,5)IP_3$ receptor antagonist, heparin, completely blocks both phases of the response (Changya et al., 1989a; Bird et al., 1992b).

On the other hand, the mechanism by which inositol phosphates regulate entry of Ca^{2+} is not only unclear, but controversial. On the surface, the simplest scenario would be for $(1,4,5)IP_3$ to bind to a receptor on the plasma membrane with properties similar to the internal receptor and thus activate an associated Ca^{2+} channel. There is evidence for this idea. Kuno and Gardner (1987) demonstrated that when $(1,4,5)IP_3$ was applied to the cytoplasmic surface of patches excised from lymphocytes, the activity of Ca^{2+} channels in the

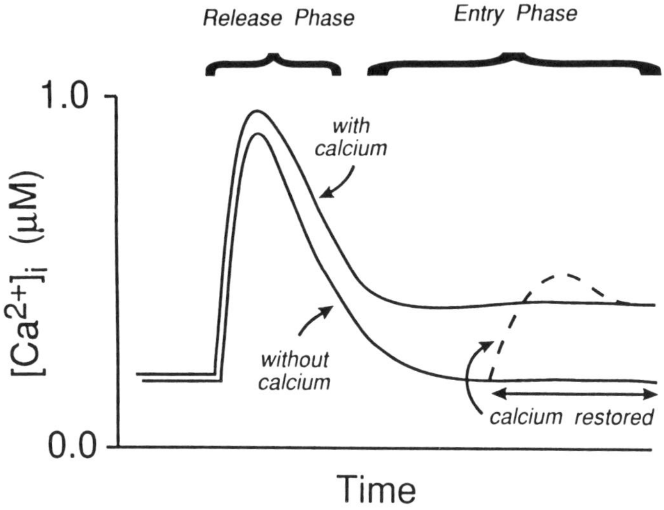

Fig. 3. Stylized kinetics of $(Ca^{2+})_i$ signaling owing to activation of NK_1 and other phospholipase C-linked receptors. Calcium signaling occurs in two phases. An initial Release Phase, which reflects release of Ca^{2+} from intracellular stores, is followed by an Entry Phase, which is due to Ca^{2+} entry across the plasma membrane. The two phases are clearly delimited by their dependence on extracellular Ca^{2+}; note that in medium lacking Ca^{2+}, the Release Phase persists, whereas the Entry Phase is blocked. In the latter instance, entry can be restored if Ca^{2+} is restored extracellularly (dashed line).

patches increased. (1,4,5)IP_3 increased the permeability of platelet plasma membrane vesicles (Rengasamy and Feinberg, 1988), but several negative reports regarding actions of (1,4,5)IP_3 on plasma membrane vesicles have appeared (Delfert et al., 1986; Ueda et al., 1986; Dargemont et al., 1988). In addition, Khan et al. (1992), using fluorescent antibodies against the (1,4,5)IP_3 receptor, have demonstrated the presence of this receptor, or something antigenically closely related to it, in the plasma membrane of T-cells.

This rather simple view of (1,4,5)IP_3 regulating the permeability of the plasma membrane to Ca^{2+} in a direct way is at odds with another body of evidence that suggests a much more indirect action of (1,4,5)IP_3. According to the *capacitative model* for Ca^{2+} entry, it is the Ca^{2+} content of the intracellular Ca^{2+} pool which determines the rate of Ca^{2+} entry (Fig. 1). This theory was originally proposed (Casteels and Droogmans, 1981; Putney, 1986) on the basis of circumstantial evidence, which has largely been supplanted by more direct proofs. These proofs are derived from experiments employing a variety of strategies to deplete the agonist- and (1,4,5)IP_3-sensitive intracellular Ca^{2+} pools, which in each case resulted in activation of Ca^{2+} entry (for recent reviews, *see* Irvine, 1990; Putney, 1990; Meldolesi et al., 1991; Putney, 1992a). The most popular strategy has been the application of drugs such as thapsigargin (Jackson et al., 1988; Takemura et al., 1989; Thastrup, 1990), which specifically inhibit microsomal (and not plasmalemmal) Ca^{2+}-ATPase, thereby causing depletion of intracellular Ca^{2+} pools. In many instances, such drugs activate Ca^{2+} entry to the same or greater extent as phospholipase C-linked agonists. For example, thapsigargin induces sustained Ca^{2+} entry in parotid acinar cells. Subsequent addition of a muscarinic agonist does not further increase the steady-state entry, despite its ability to increase cellular levels of inositol phosphates (Takemura et al., 1989). Other intracellular Ca^{2+}-ATPase inhibitors that have been used in this way are cyclopiazonic acid (Deng and Kwan, 1991; Mason et al., 1991; Demaurex et al., 1992) and 2,5-di-(*tert*-butyl)-1,4-hydroquinone (Llopis et al., 1991; Mason et al., 1991; Zhang et al., 1991). Because these compounds are structurally unrelated, but all share the ability to inhibit intracellular Ca^{2+}-ATPase as well as activate Ca^{2+} entry, it is unlikely that the latter effect can be ascribed to a coincidental side effect unrelated to the depletion of intracellular pools. The capacitative model and its historical basis have been discussed in some detail recently, and the interested reader is referred to these more detailed treatments (Putney, 1986, 1990).

Irvine and his collaborators have proposed that (1,3,4,5)IP_4, the phosphorylated product of (1,4,5)IP_3 metabolism, plays a significant role in the regulation of Ca^{2+} entry (Irvine et al., 1988; Irvine, 1989, 1990). Irvine and Moor (Irvine and Moor, 1986, 1987), from experiments with sea urchin eggs, provided the first evidence for a biological role for (1,3,4,5)IP_4, reporting that a full fertilization response to injected inositol polyphosphates depended on the presence of this product of (1,4,5)IP_3 metabolism. In these and in subsequent studies, (1,3,4,5)IP_4 did not induce any apparent $[Ca^{2+}]_i$ signal on its own, but always depended on the simultaneous presence of (1,4,5)IP_3 or one of its more stable analogs (i.e., (2,4,5)IP_3). In whole-cell, patch-perfusion studies of lacrimal acinar cells by Petersen, Irvine, and their collaborators (Morris et al., 1987; Changya et al., 1989a,b), it appeared that a maximal, sustained $[Ca^{2+}]_i$ response could not be obtained when the patch pipet was perfused with (1,4,5)IP_3 alone, but required the simultaneous presence of (1,3,4,5)IP_4 in the pipet. The results obtained in this system led to the model proposed by Irvine (Irvine, 1990) according to which IP_3 receptors on the intracellular Ca^{2+}-releasing organelle and (1,3,4,5)IP_4 receptors on the plasma membrane interact to regulate Ca^{2+} entry. Paradoxically, however, in this same preparation acetylcholine, which activates endogenous production of (1,4,5)IP_3, was fully capable of activating robust and sustained $[Ca^{2+}]_i$ signals. Furthermore, in mouse lacrimal cells (Kwan et al., 1990), the same type of cell used by Petersen and Irvine in the patch-clamp studies, thapsigargin is capable of quantitatively reproducing or even exceeding the level of Ca^{2+} entry obtained with phospholipase C-activating agonists, but thapsigargin does not increase the cellular levels of any inositol phosphates. Apparently, this entry occurs solely by virtue of the ability of thapsigargin to deplete critical intracellular stores of their Ca^{2+}. In addition, there are cell types that produce very little (1,3,4,5)IP_4 and are still capable of demonstrating an activated Ca^{2+} entry. For example, when AR4-2J pancreatoma cells were stimulated with SP, there was a substantial increase in the cellular level of (1,4,5)IP_3, but (1,3,4,5)IP_4 could not be detected (Horstman et al., 1988). However, SP caused a substantial mobilization of cellular Ca^{2+}. This rise in cytosolic $[Ca^{2+}]_i$ was partially dependent on external Ca^{2+}, and was partially, but not completely, inhibited by the Ca^{2+} antagonist, La^{3+}. Thus, these cells appear capable of responding to agonist activation by release of intracellular Ca^{2+} and by entry of Ca^{2+} across the plasma membrane, despite the inability to produce substantial quantities of (1,3,4,5)IP_4.

To gain further insight into the roles played by (1,4,5)IP_3 and (1,3,4,5)IP_4 in $[Ca^{2+}]_i$ signaling in lacrimal acinar cells, Bird et al.

(1992b) employed two techniques to introduce inositol phosphates into the cells and measure their effects on Ca^{2+} entry: Microinjection into fura-2-loaded cells with $[Ca^{2+}]_i$ monitored with microspectrofluorimetry, and, as used in the experiments of Petersen and Irvine and others (Morris et al., 1987; Changga et al., 1989a,b), perfusion of patch clamp pipets in the whole cell configuration while monitoring Ca^{2+}-activated K^+ conductance as an indicator of intracellular Ca^{2+}. This study differed in one aspect from the earlier work on (1,3,4,5)IP_4 in utilizing primarily the (2,4,5)IP_3 isomer, which was shown not to be phosphorylated by lacrimal cell 3-kinase. In addition, studies in permeable lacrimal cells demonstrated that substantially higher concentrations of (1,4,5)IP_3 (and its analogues, such as (2,4,5)IP_3) are required for Ca^{2+} release than expected from previous studies in other cell types (Bird et al., 1992b). The reason for this apparent lower sensitivity of the IP_3 receptor is not known, but it was clear that in the earlier patch clamp studies (Morris et al., 1987; Changya et al., 1989b) the concentrations of (1,4,5)IP_3 and its analogs were below those required for maximal activation of intracellular release. Thus, when higher concentrations of (1,4,5)IP_3 or (2,4,5)IP_3 were introduced into cells by patch pipet perfusion, sustained $[Ca^{2+}]_i$ signals were observed. When microinjected into lacrimal acinar cells, (2,4,5)IP_3 was shown to mimic fully the actions of muscarinic receptor activation in stimulating both intracellular release of Ca^{2+}, as well as entry of Ca^{2+} across the plasma membrane (Bird et al., 1992b). Similar actions of (2,4,5)IP_3 were demonstrated in hepatocytes under somewhat different experimental conditions (Burgess et al., 1991). Thus, these findings indicate that (1,4,5)IP_3 serves as a signal that is both necessary and sufficient for receptor activation of Ca^{2+} entry across the plasma membrane in lacrimal acinar cells, and there does not appear to be a physiological role for (1,3,4,5)IP_4, in this process. It is not clear why such definitive effects of (1,3,4,5)IP_4 were obtained in the experiments by Irvine, Petersen and collaborators (Morris et al., 1987; Changya et al., 1989a,b). However, two recent reports have indicated that under some experimental conditions (1,3,4,5)IP_4 may have direct, Ca^{2+}-independent effects on K^+ channels (Molleman et al., 1991; Wu et al., 1991). In addition, it is noteworthy that effects of (1,3,4,5)IP_4 were only observed when inositol phosphates were introduced in patch pipet perfusion experiments; even low concentrations of (2,4,5)IP_3 activated Ca^{2+} entry when microinjected into cells (Bird et al., 1992b), and (1,3,4,5)IP_4 did not potentiate these responses (Bird and Putney, unpublished). Thus, it may be that (1,3,4,5)IP_4 only affects Ca^{2+} signaling when soluble cellular contents have been dialyzed; however, a more specific explanation for these discrepant findings will no doubt require additional experiments, perhaps employing novel approaches to this important problem.

4. Regulation of Tachykinin Receptors Linked to Inositol Phosphate Formation

The rapid and substantial increase in cellular (1,4,5)IP_3 in parotid acinar cells, as well as in AR4-2J cells, following application of a maximal concentration of SP is a transient phenomenon, lasting on average from one to two minutes (Sugiya et al., 1987; Bird et al., 1991b). Thus, as for many peptide receptors, the SP receptor on exocrine gland cells undergoes desensitization. Although the onset of this desensitization is rapid, occurring over an interval of from five to thirty seconds after agonist addition, the desensitization is persistent, requiring one or two hours after removal of agonist to restore full responsiveness (Sugiya et al., 1987). Interestingly, in parotid cells in the continued presence of SP, a residual steady-state increased level of (1,4,5)IP_3 and $[Ca^{2+}]_i$ is maintained for long periods after the initial transient response, whereas in AR4-2J cells the response returns to baseline. Thus, although desensitization is equally rapid in onset and equally persistent in the two cell types, it appears that the desensitization process is a complete one in AR4-2J cells and a partial one in parotid cells. The simplest explanation for this discrepancy is that the residual response in parotid cells results from a population of SP receptors that are not susceptible to desensitization and that this subpopulation of receptors is missing from the AR4-2J cells. However, there is as yet no direct evidence for such heterogeneity of SP receptors in parotid cells.

The desensitization of the SP response appears to be homologous. Activation of other phospholipase C-linked receptors does not produce persistent desensitization of the SP response, and following the induction of persistent desensitization of the SP response, subsequent responses to agonists acting on other, phospholipase C-linked receptors are not attenuated (Sugiya et al., 1987). Associated with the development of desensitization in parotid cells is a loss of surface binding sites for SP, indicating that desensitization may reflect sequestration and/or downregulation of the SP receptors.

There are numerous documented instances of desensitization of phospholipase C-coupled receptor that are believed to be mediated through activation of protein kinase C (Labarca et al., 1984; Johnson et al., 1986; Liles et al., 1986; Muldoon et al., 1987; Orellana et al., 1987; Osugi et al., 1987; Willems et al., 1987; Hepler et al., 1988; Crouch and Lapetina, 1989). The primary evidence for such a mechanism comes from the observation that phorbol esters, potent activators of protein kinase C, inhibit cellular responses to agonists of this class. This mechanism is believed to represent a negative feedback mechanism since activation of phospholipase C would increase cellular levels of diacylglycerol, the physiological activator of protein kinase C. However, the findings with the NK_1 receptor in exocrine gland cells suggest that it may be incorrect, or at least premature, to conclude that such a mechanism contributes to regulation of these signaling pathways during agonist stimulation. Whereas in parotid acinar cells, phorbol esters inhibit SP-induced (1,4,5)IP_3 formation (Sugiya et al., 1988), the activation of protein kinase C does not appear to be involved in the persistent desensitization seen in these cells for the following reasons:

1. The desensitization is homologous in nature (Sugiya et al., 1987), while inhibitory mechanisms attributed to second messengers should be equally activated by other phospholipase C-linked agonists;
2. It is possible to dissociate activation of phospholipase C from desensitization by addition of antimycin, which completely blocks SP-induced (1,4,5)IP_3 formation without inhibiting the development of persistent desensitization (Sugiya and Putney, 1988b);

3. Unlike the agonist-induced persistent desensitization, the inhibitory effects of phorbol esters are rapidly reversed upon removal of the drug, and phorbol esters do not cause a loss of surface membrane binding sites for SP (Sugiya et al., 1988); and
4. Drugs that inhibit protein kinase C completely prevent the inhibitory effects of phorbol esters on the SP response, but do not affect the time course of $(1,4,5)IP_3$ generation owing to SP receptor activation nor the development of persistent desensitization (Sugiya and Putney, 1988a).

For the case of homologous desensitization of NK_1, as well as other phospholipase C-linked receptors, a mechanism may operate that is similar to that for the more extensively studied adenylyl cyclase-coupled, beta-adrenergic receptor (Benovic et al., 1988). For this receptor, homologous desensitization also results in receptor sequestration and downregulation and is believed to be signaled by the phosphorylation of the receptor by a specific beta-adrenergic receptor kinase (Strasser et al., 1986). This mechanism provides a reasonable working hypothesis for future studies on regulation of the phospholipase C-linked NK_1 receptor in exocrine gland cells and other SP-responsive systems.

5. Summary and Conclusions

This review has attempted to summarize some of our current knowledge about the mechanisms by which cellular Ca^{2+} is regulated by phospholipase C-linked receptors such as the NK_1 receptor. It is well established that the initial, agonist-induced $[Ca^{2+}]_i$ signal is derived from the Ca^{2+}-releasing actions of $(1,4,5)IP_3$ on at least one specific intracellular Ca^{2+}-sequestering organelle, and there is substantial evidence that this mechanism applies to the NK_1 receptor as well. The metabolism of the $(1,4,5)IP_3$ formed in activated cells is exceedingly complex, and the significance of this complexity to cellular signaling mechanisms is not yet fully understood. In virtually all cells that utilize the phosphoinositide-linked Ca^{2+} signaling system, the initial release of intracellular Ca^{2+} caused by agonist activation is followed or accompanied by an accelerated entry of Ca^{2+}

into the cytoplasm across the plasma membrane. The mechanism by which this process is regulated has been somewhat more elusive and may vary in different cellular systems. Yet, there is strong evidence that in many cell types depletion of agonist-sensitive intracellular Ca^{2+} stores by (1,4,5)IP_3 generates a secondary signal of unknown nature that activates Ca^{2+} entry. There appear to be rapid mechanisms for regulating the SP signaling system in cells at the level of the receptor. A rapid loss of surface membrane binding sites is observed in SP-stimulated cells, representing a mechanism for homologous desensitization of this pathway. The continued investigation of these and other interesting aspects of the phospholipase C-linked NK_1 receptor pathway should provide us with many additional insights into the complex control mechanisms underlying this intriguing peptide signaling system.

References

Abdel-Latif, A. A., Akhtar, R., and Hawthorne, J. N. (1977) Acetylcholine increases the breakdown of triphosphoinositide of rabbit iris muscle prelabeled with [^{32}P]phosphate. *Biochem. J.* **162,** 61–73.

Aub, D. L. and Putney, J. W., Jr. (1985) Properties of receptor-controlled inositol trisphosphate formation in parotid acinar cells. *Biochem. J.* **225**, 263–266.

Bansal, V. S., Inhorn, R. C., and MaJerus, P. W. (1987) The metabolism of inositol 1,3,4-trisphosphate to inositol 1,3-bisphosphate. *J. Biol. Chem.* **262,** 9444–9447.

Benovic, J. L., Bouvier, M., Caron, M. G., and Lefkowitz, R. J. (1988) Regulation of adenylyl cyclase-coupled-adrenergic receptors. *Annu. Rev. Cell Biol.* **4,** 405–428.

Berridge, M. J. (1983) Rapid accumulation of inositol trisphosphate reveals that agonists hydrolyse polyphosphoinositides instead of phosphatidylinositol. *Biochem. J.* **212,** 849–858.

Berridge, M. J. (1990) Calcium oscillations. *J. Biol. Chem.* **266,** 9583–9586.

Berridge, M. J. (1992) Inositol trisphosphate and calcium oscillations. *Adv. Second Mess. Phosphoprotein Res.* **26,** 211–223.

Bird, G. St. J., Oliver, K. G., Horstman, D. A., Obie, J. F., and Putney, J. W., Jr. (1991a) Relation between the calcium-mobilizing action of inositol 1,4,5-trisphosphate in permeable AR4-2J cells and the estimated levels of inositol 1,4,5-trisphosphate in intact AR4-2J cells. *Biochem. J.* **273,** 541–546.

Bird, G. St. J., Takemura, H., Thastrup, O., Putney, J. W., Jr., and Menniti, F. S. (1991b) Mechanism of activated Ca^{2+} entry in the rat pancreatoma cell line, AR4-2J. *Cell Calcium* **13,** 49–58.

Bird, G. St. J., Obie, J. F., and Putney, J. W., Jr. (1992a) Functional homogeneity of the nonmitochondrial Ca^{2+}-pool in intact mouse lacrimal acinar cells. *J. Biol. Cem.* **267,** 18,382–18,386.

Bird, G. St. J., Rossier, M. F., Hughes, A. R., Shears, S. B., Armstrong, D. L., and Putney, J. W., Jr. (1992b) Activation of Ca^{2+} entry into acinar cells by a nonphosphorylatable inositol trisphosphate. *Nature* **362,** 162–165.

Burgess, G. M., Bird, G. St. J., Obie, J. F., and Putney, J. W., Jr. (1991) The mechanism for synergism between phospholipase C- and adenylyl cyclase-linked hormones in liver. Cyclic AMP-dependent kinase augments inositol trisphosphate-mediated Ca^{2+} mobilization without increasing the cellular levels of inositol polyphosphates. *J. Biol. Chem.* **266,** 4772–4781.

Casteels, R. and Droogmans, G. (1981) Exchange characteristics of the noradrenaline-sensitive calcium store in vascular smooth muscle cells of rabbit ear artery. *J. Physiol. (Lond.)* **317,** 263–279.

Chang, M. M. and Leeman, S. E. (1970) Isolation of a sialogogic peptide from bovine hypothalamic tissue and its characterization as SP. *J. Biol. Chem.* **245,** 4784–4790.

Changya, L., Gallacher, D. V., Irvine, R. F., and Petersen, O. H. (1989a) Inositol 1,3,4,5-tetrakisphosphate and inositol 1,4,5-trisphosphate act by different mechanisms when controlling Ca^{2+} in mouse lacnmal cells. *FEBS Lett.* **251,** 43–48.

Changya, L., Gallacher, D. V., Irvine, R. F., Potter, B. V. L., and Petersen, O. H. (1989b) Inositol 1,3,4,5-trisphosphate is essential for sustained activation of the Ca^{2+}-dependent K^+ current in single internally perfused lacrimal cells. *J. Membrane Biol.* **109,** 85–93.

Combettes, L., Claret, M., and Champeil, P. (1992) Do submaximal $InsP_3$ concentrations only induce the partial discharge of permeabilized hepatocyte pools because of the concomitant reduction of intraluminal Ca^{2+} concentration? *FEBS Lett.* **301,** 287–290.

Connolly, T. M., Wilson, D. B., Bross, T. E., and Majerus, P. W. (1986) Isolation and characterization of the inositol cyclic phosphate products of phosphoinositide cleavage by phospholipase C. Metabolism in cell-free extracts. *J. Biol. Chem.* **261,** 122–126.

Crouch, M. F. and Lapetina, E. G. (1989) Dual mechanisms of platelet hormone receptor desensitization. Differential importance between agonists of protein kinase C-dependent and -independent pathways. *J. Biol. Chem.* **264,** 584–588.

Cuthbertson, K. S. R. and Cobbold, P. H. (eds.) (1991) Oscillations in cell calcium (Collected papers and reviews). *Cell Calcium* **12,** 61–268.

Dargemont, C., Hilly, M., Claret, M., and Mauger, J.-P. (1988) Characterization of Ca^{2+} fluxes in rat liver plasma-membrane vesicles. *Biochem. J.* **256,** 117–124.

Delfert, D. M., Hill, S., Pershadsingh, H. A., and Sherman, W. R. (1986) *myo*-Inositol 1,4,5-trisphosphate mobilizes Ca^{2+} from isolated adipocyte endoplasmic reticulum but not from plasma membranes. *Biochem. J.* **236,** 37–44.

Demaurex, N., Lew, D. P., and Krause, K-H. (1992) Cyclopiazonic acid depletes intracellular Ca^{2+} stores and activates an influx pathway for divalent cations in HL-60 cells. *J. Biol. Chem.* **267,** 2318–2324.

Deng, H.-W. and Kwan, C.-Y. (1991) Cyclopiazonic acid is a sarcoplasmic reticulum Ca^{2+}-pump inhibitor of rat aortic smooth muscle. *Acta Pharmacol. Sinica* **12,** 1–6.

Downes, C. P., Mussat, M. C., and Michell, R. H. (1982) The inositol trisphosphate phosphomonoesterase of the human erythrocyte membrane. *Biochem. J.* **203,** 169–177.

Ferris, C. D. and Snyder, S. H. (1992) Inositol 1,4,5-trisphosphate-activated calcium channels. *Annu. Rev. Physiol.* **64,** 469–488.

Gawler, D. J., Potter, B. V. L., Gigg, R., and Nahorski, S. R. (1991) Interactions between inositol tris- and tetrakis-phosphates. Effects on intracellular Ca^{2+} mobilization in SH-SY5Y cells. *Biochem. J.* **276,** 163–167.

Hanley, M. R., Lee, C. M., Jones, L. M., and Michell, R. H. (1980) Similar effects of SP and related peptides on salivation and on phosphatidylinositol turnover in rat salivary glands. *Mol. Pharmacol.* **18,** 78–83.

Hepler, J. R., Earp, H. S., and Harden, T. K. (1988) Long-term phorbol ester treatment down-regulates protein kinase C and sensitizes the phosphoinositide signaling pathway to hormone and growth factor stimulation. Evidence for a role of protein kinase C in agonist-induced desensitization. *J. Biol. Chem.* **263,** 7610–7619.

Hokin, M. R. and Hokin, L. E. (1954) Effects of acetylcholine on phospholipides in the pancreas. *J. Biol. Chem.* **209,** 549–558.

Horstman, D. A., Takemura, H., and Putney, J. W., Jr. (1988) Formation and metabolism of [^{3}H]inositol phosphates in AR42J pancreatoma cells: SP-induced Ca^{2+} mobilization in the apparent absence of inositol 1,4,5-trisphosphate 3-kinase activity. *J. Biol. Chem.* **263,** 15,297–15,303.

Hughes, A. R., Takemura, H., and Putney, J. W., Jr. (1988) Kinetics of inositol 1,4,5-trisphosphate and inositol cyclic 1:2,4,5-tIisphosphate metabolism in intact rat parotid acinar cells: Relationship to calcium signaling. *J. Biol. Chem.* **263,** 10,314–10,319.

Inhorn, R. C., Bansal, V. S., and Majerus, P. W. (1987) Pathway for 1,3,4-trisphosphate and 1,4-bisphosphate metabolism. *Proc. Natl. Acad Sci. USA* **84,** 2170–2174.

Irvine, R. F., Letcher, A. J., Heslop, J. P., and Berridge, M. J. (1986a) The inositol tris/tetrakisphosphate pathway—demonstration of Ins(1,4,5)P3 3-kinase activity in animal tissues. *Nature* **320,** 631–634.

Irvine, R. F., Letcher, A. J., Lander, D. J., and Berridge, M. J. (1986b) Specificity of inositol phosphate-stimulated Ca^{2+} mobilization from Swiss-mouse 3T3 cells. *Biochem. J.* **240,** 301–304.

Irvine, R. F., Moor, R. M., Pollock, W. K., Smith, P. M., and Wreggett, K. A. (1988) Inositol phosphates: proliferation, metabolism and function. *Phil. Trans. R. Soc. Lond. B* **320,** 281–298.

Irvine, R. F. (1989) How do inositol 1,4,5-trisphosphate and inositol 1,3,4,5-tetrakisphosphate regulate intracellular Ca^{2+}? *Biochem. Soc. Trans.* **17,** 6–9.

Irvine, R. F. (1990) "Quantal" Ca^{2+} release and the control of Ca^{2+} entry by inositol phosphates—a possible mechanism. *FEBS Lett.* **263,** 5–9.

Irvine, R. F. and Moor, R. M. (1986) Micro-injection of inositol 13,4,5-tetrakisphosphate activates sea urchin eggs by a mechanism dependent on external Ca^{2+}. *Biochem. J.* **240,** 917–920.

Irvine, R. F. and Moor, R. M. (1987) Inositol(1,3,4,5)tetrakisphosphate-induced activation of sea urchin eggs requires the presence of inositol trisphosphate. *Biochem. Biophys. Res. Comm.* **146,** 284–290.

Ivorra, I., Gigg, R., Irvine, R. F., and Parker, I. (1991) Inositol 1,3,4,6-tetrakisphosphate mobilizes calcium in *Xenopus* oocytes with high potency. *Biochem. J.* **273,** 317–321.

Jackson, T. R., Patterson, S. I., Thastrup, O., and Hanley, M. R. (1988) A novel tumour promoter, thapsigargin, transiently increases cytoplasmic free Ca^{2+} without generation of inositol phosphates in NG115-401L neuronal cells. *Biochem. J.* **263,** 81–86.

Jean, T. and Klee, C. B. (1986) Calcium modulation of inositol 1,4,5-trisphosphate-induced calcium release from neuroblastoma x glioma hybrid (NG108-15) microsomes. *J. Biol. Chem.* **261,** 16,414–16,420.

Johnson, R. M., Connellly, P. A., Sisk, R. B., Pobiner, B. F., Hewlett, E. L., and Garrison, J. C. (1986) Pertussis toxin or phorbol 12-myristate 13-acetate can distinguish between epidermal growth factor- and angiotensin-stimulated signals in hepatocytes. *Proc. Nat. Acad. Sci. USA* **83,** 2032–2036.

Jones, L. M. and Michell, R. H. (1978) Enhanced phosphatidylinositol breakdown as a calcium-independent response of rat parotid fragments to substance P. *Biochem. Soc. Trans.* **6,** 1035–1037.

Khan, A. A., Steiner, J. P., Klein, M. G., Schneider, M. F., and Snyder, S. H. (1992) IP_3 receptor: Localization to plasma membrane of T cells and cocapping with the T cell receptor. *Science* **257,** 815–818.

Kirk, C. J., Creba, J. A., Downes, C. P., and Michell, R. H. (1981) Hormone-stimulated metabolism of inositol lipids and its relationship to hepatic receptor function. *Biochem. Soc. Trans.* **9,** 377–379.

Kuno, M. and Gardner, P. (1987) Ion channels activated by inositol 1,4,5-trisphosphate in plasma membrane of human T-lymphocytes. *Nature* **326,** 301–304.

Kwan, C. Y., Takemura, H., Obie, J. F., Thastrup, O., and Putney, J. W., Jr. (1990) Effects of methacholine, thapsigargin and La^{3+} on plasmalemmal and intracellular Ca^{2+} transport in lacrimal acinar cells. *Am. J. Physiol.* **258,** C1006–C1015.

Labarca, R., Janowsky, A., Patel, J., and Paul, S. M. (1984) Phorbol esters inhibit agonist-induced [^{3}H] inositol-1-phosphate accumulation in rat hippocampal slices. *Biochem. Biophys. Res. Commun.* **123,** 703–709.

Lee, C. H. and Hokin, L. E. (1989) Inositol 1,2-cyclic 4,5-trisphosphate is an order of magnitude less potent than inositol 1,4,5-trisphosphate in mobilizing intracellular stores of calcium in mouse pancreatic acinar cells. *Biochem. Biophys. Res. Commun.* **159,** 561–565.

Liles, W. C., Hunter, D. D., Meier, K. E., and Nathanson, N. M. (1986) Activation of protein kinase C induces rapid internalization and subsequent degradation of muscarinic acetylcholine receptors in neuroblastoma cells. *J. Biol. Chem.* **261,** 5307–5313.

Llano, I., Marty, A., and Tanguy, J. (1987) Dependence of intracellular effects of GTP(γ)S and inositoltrisphosphate on cell membrane potential and on external Ca ions. *Pflüger's Arch.* **409,** 499–506.

Llopis, J., Chow, S. B., Kass, G. E. N., Gahm, A., and Orrenius, S. (1991) Comparision between the effects of the microsomal Ca^{2+}-translocase inhibitors thapsigargin and 2,5-di-(*t*-butyl)-1,4-benzohydroquinone on cellular calcium fluxes. *Biochem. J.* **277,** 553–556.

Majerus, P. W., Connolly, T. M., Bansal, V. S., Inhorn, R. C., Ross, T. S., and Lips, D. L. (1988) Inositol phosphates: Synthesis and degradation. *J. Biol. Chem.* **263,** 3051–3054.

Marier, S. H., Putney, J. W., Jr., and Van de Walle, C. M. (1978) Control of calcium channels by membrane receptors in the rat parotid gland. *J. Physiol. (Lond.)* **279,** 141–151.

Mason, M. J., Garcia-Rodriquez, C., and Grinstein, S. (1991) Coupling between intracellular Ca^{2+} stores and the Ca^{2+} permeability of the plasma membrane. Comparison of the effects of thapsigargin, 2,5-di-(*tert*-butyl)-1,4-hydroquinone, and cyclopiazonic acid in rat thymic lymphocytes. *J. Biol. Chem.* **266,** 20,856–20,862.

Meldolesi, J., Clementi, E., Fasolato, C., Zacchetti, D., and Pozzan, T. (1991) Ca^{2+} influx following receptor activation. *Trends Pharmacol. Sci.* **12,** 289–292.

Menniti, F. S., Oliver, K. G., Nogimori, K., Obie, J. F., Shears, S. B., and Putney, J. W., Jr. (1990) Origins of myoinositol tetrakisphosphates in agonist-stimulated rat pancreatoma cells: Stimulation by bombesin of myoinositol (1,3,4,5,6) pentakisphosphate breakdown to myoinositol (3,4,5,6)tetrakisphosphate. *J. Biol. Chem.* **265,** 11,167–11,176.

Menniti, F. S., Bird, G. St. J., Takemura, H., Thastrup, O., Potter, B. V. L., and Putney, J. W., Jr. (1991) Mobilization of calcium by inositol trisphosphate from permeabilized rat parotid acinar cells. Evidence for translocation of calcium from uptake to release sites within the inositol 1,4,5-trisphosphate- and thapsigargin-sensitive calcium pool. *J. Biol. Chem.* **266,** 13,646–13,653.

Menniti, F. S., Miller, R. N., Putney, J. W., Jr., and Shears, S. B. (1993) Rapid turnover of inositol polyphosphate pyrophosphates in pancreatoma cells. *J. Biol. Chem.* **268,** 3850–3856.

Menniti, F. S., Oliver, K. G., Putney, J. W., Jr., and Shears, S. B. (1993) Inositol phosphates and cell signaling: New views of $InsP_6$ and $InsP_6$. *Trends Biochem. Sci.* **18,** 53–56.

Merritt, J. E. and Rink, T. J. (1987a) The effects of SP and carbachol on inositol tris- and tetrakisphosphate formation and cytosolic free calcium in rat parotid acinar cells. A correlation between inositol phosphate levels and calcium entry. *J. Biol. Chem.* **262,** 14,912–14,916.

Merritt, J. E. and Rink, T. J. (1987b) Regulation of cytosolic free calcium in fura-2-loaded rat parotid acinar cells. *J. Biol. Chem.* **262,** 17,362–17,369.

Meyer, T. and Stryer, L. (1990) Transient calcium release induced by successive increments of inositol 1,4,5-trisphosphate. *Proc. Natl. Acad. Sci. USA* **87,** 3841–3845.

Meyer, T. and Stryer, L. (1991) Calcium spiking. *Annu. Rev. Biophys. Biophys. Chem.* **20,** 153–174.

Michell, R. H. (1986) Inositol Lipids and their role in receptor function: History and general principles, in *Phosphoinositides and Receptor Mechanisms* (Putney, J. W., Jr., ed.), Alan Liss, New York, pp. 1–24.

Michell, R. H. (1975) Inositol phospholipids and cell surface receptor function. *Biochim. Biophys. Acta* **416,** 81–147.

Missiaen, L., DeSmedt, J., Droogmans, G., and Casteels, R. (1992) Ca^{2+} release induced by inositol 1,4,5-trisphosphate is a steady-state phenomenon controlled by luminal Ca^{2+} in permeabilized cells. *Nature* **367,** 599-602.

Molleman, A., Hoiting, B., Duin, M., van den Akker, J., Nelemans, A., and Den Hertog, A. (1991) Potassium channels regulated by inositol 1,3,4,5-tetrakisphosphate and internal calcium in DDT 2 smooth muscle cells. *J. Biol. Chem.* **266,** 5658–5663.

Morris, A. P., Gallacher, D. V., Irvine, R. F., and Petersen, O. H. (1987) Synergism of inositol trisphosphate and tetrakisphosphate in activating Ca^{2+}-dependent K^+ channels. *Nature* **330,** 653–655.

Muallem, S., Pandol, S. J., and Beeker, T. G. (1989) Hormone-evoked calcium release from intracellular stores is a quantal process. *J. Biol. Chem.* **264,** 205–212.

Muldoon, L. L., Jamieson, G. A., Jr., and Villereal, M. L. (1987) Calcium mobilization in permeabilized fibroblasts: Effects of inositol trisphosphate, orthovanadate, mitogens, phorbol ester, and guanosine triphosphate. *J. Cell. Physiol.* **130,** 29–36.

Nogimori, K, Menniti, F. S., and Putney, J. W., Jr. (1990) Identification in extracts from AR4-2J cells of inositol (1,4,5)trisphosphate by its susceptibility to inositol (1,4,5)trisphosphate 3-kinase and 5-phosphatase. *Biochem. J.* **269,** 195–200.

Nunn, D. L. and Taylor, C. W. (1992) Luminal Ca^{2+} increases the sensitivity of the Ca^{2+} stores to inositol 1,4,5-trisphosphate. *Mol. Pharmacol.* **41,** 115–119.

Oldershaw, K. A., Nunn, D. L., and Taylor, C. W. (1991) Quantal Ca^{2+} mobilization stimulated by inositol 1,4,5-trisphosphate in permeabilized hepatocytes. *Biochem. J.* **278,** 704–708.

Orellana, S., Solski, P. A., and Brown, J. H. (1987) Guanosine 5′-O-(thiophosphate)-dependent inositol trisphosphate formation in membranes is inhibited by phorbol ester and protein kinase C. *J. Biol. Chem.* **262,** 1638–1643.

Osugi, T., Imaizumi, T., Mizushim, A., Uchida, S., and Yoshida, H. (1987) Phorbol ester inhibits bradykinin-stimulated inositol trisphosphate formation and calcium mobilization in neuroblastoma x glioma hybrid NG108-15 cells. *J. Pharmacol. Exp. Ther.* **240,** 617–622.

Penner, R., Matthews, G., and Neher, E. (1988) Regulation of calcium influx by second messengers in rat mast cells. *Nature* **334,** 499–504.

Putney, J. W., Jr. (1986) A model for receptor-regulated calcium entry. *Cell Calcium* **7,** 1–12.

Putney, J. W., Jr., Aub, D. L., Taylor, C. W., and Merritt, J. E. (1986) Formation and biological action of inositol 1,4,5-trisphosphate. *Fed. Proc.* **45,** 2634–2638.

Putney, J. W., Jr. (1987) Calcium-mobilizing receptors. *Trends Pharmacol. Sci.* **8,** 481–486.

Putney, J. W., Jr. (1990) Capacitative calcium entry revisited. *Cell Calcium* **11,** 611–624.

Putney, J. W., Jr., Bird, G. St. J., Horstman, D. A., Hughes, A. R., Menniti, F. S., Nogimori, K., Obie, J. F., Oliver, K. G., Sugiya, H., and Takemura, H. (1991) Role of inositol phosphates in the actions of SP on NK_1 receptors in exocrine gland cells. *Ann. NY. Acad. Sci.* **632,** 94–103.

Putney, J. W., Jr. (1992a) Inositol phosphates and calcium entry. *Adv. Second Messenger Phosphoprotein Res.* **26,** 143–160.

Putney, J. W., Jr. (ed.) (1992b) *Inositol Phosphates and Calcium Signaling,* vol. 26: *Advances in Second Messenger and Phosphprotein Research,* Raven, New York.

Rengasamy, A. and Feinberg, H. (1988) Inositol 1,4,5-trisphosphate-induced calcium release from platelet plasma membrane vesicles. *Biochem. Biophys. Res. Commun.* **150,** 1021–1026.

Rudich, L. and Butcher, F. R. (1976) Effect of substance P and eledoisin on K efflux, amylase release, and cyclic nucleotides in slices of rat parotid gland. *Biochim. Biophys. Acta* **444,** 704–711.

Schulz, I., Streb, H., Bayerdorffer, E., and Thevenod, F. (1985) Stimulus-secretion coupling in exocrine glands: The role of inositol-1,4,5-trisphosphate, calcium and cAMP. *Current Eye Res.* **4,** 467–473.

Shears, S. B., Parry, J. B., Tang, E. K. Y., Irvine, R. F., Michell, R. H., and Kirk, C. J. (1987) Metabolism of D-myo-inositol 1,3,4,5-tetrakisphosphate by rat liver, including the synthesis of a novel isomer of myo-inositol tetrakisphosphate. *Biochem. J.* **246,** 139–147.

Shuttleworth, T. J. (1992) Ca^{2+} release from inositol trisphosphate-sensitive stores is not modulated by intraluminal [Ca^{2+}]. *J. Biol. Chem.* **267,** 3573–3576.

Slack, B. E., Bell, J. E., and Benos, D. J. (1986) Inositol 1,4,5-trisphosphate injection mimics fertilization potentials in sea urchin eggs. *Am. J. Physiol.* **250,** C340–C344.

Stephens, L., Radenberg, T., Thiel, U., Vogel, G., Khoo, K.-H., Dell, A., Jackson, T. R., and Hawkins, P. T. (1992) Pyrophosphorylated inositol hexakisphosphates: A new form of high energy phosphate in cells. *J. Biol. Chem.* submitted.

Strasser, R. H., Benovic, J. L., Caron, M. G., and Lefkowitz, R. J. (1986) β-agonist- and prostaglandin E_1-induced translocation of the β-adrenergic receptor kinase: Evidence that the kinase may act on multiple adenylate cyclase-coupled receptors. *Proc. Natl. Acad. Sci. USA* **83,** 6362–6366.

Streb, H., Irvine, R. F., Berridge, M. J., and Schulz, I. (1983) Release of Ca^{2+} from a nonmitochondrial store in pancreatic cells by inositol-1,4,5-trisphosphate. *Nature* **306,** 67,68.

Sugiya, H., Tennes, K. A., and Putney, J. W., Jr. (1987) Homologous desensitization of substance-P-induced inositol polyphosphate formation in rat parotid acinar cells. *Biochem. J.* **244,** 647–653.

Sugiya, H., Obie, J. F., and Putney, J. W., Jr. (1988) Two modes of regulation of the phospholipase C-linked substance P receptor in rat parotid acinar cells. *Biochem. J.* **253,** 459–446.

Sugiya, H. and Putney, J. W., Jr. (1988a) Protein kinase C-dependent and -independent mechanisms regulating the parotid substance P receptor as revealed by differential effects of protein kinase C inhibitors. *Biochem. J.* **266,** 677–680.

Sugiya, H. and Putney, J. W., Jr. (1988b) Substance P receptor desensitization requires activation of receptor, but not phospholipase *C. Am. J. Physiol.* **255,** C149–C154.

Supattapone, S., Worley, P. F., Baraban, J. M., and Snyder, S. H. (1988) Solubilization, purification, and characterization of an inositol trisphosphate receptor. *J. Biol. Chem.* **263,** 1530–1534.

Takemura, H., Hughes, A. R., Thastrup, O., and Putney, J. W., Jr. (1989) Activation of calcium entry by the tumor promoter, thapsigargin, in parotid acinar cells. Evidence that an intracellular calcium pool, and not an inositol phosphate, regulates calcium fluxes at the plasma membrane. *J. Biol. Chem.* **264,** 12,266–12,271.

Taylor, C. W., Merritt, J. E., Putney, J. W., Jr., and Rubin, R. P. (1986) A guanine nucleotide-dependent regulatory protein couples substance P receptors to phospholipase C in rat parotid gland. *Biochem. Biophys. Res. Commun.* **136,** 362–368.

Taylor, C. W. (1992) Kinetics of inositol 1,4,5-trisphosphate-stimulated Ca^{2+} mobilization, in *Inositol Polyphosphates and Calcium Signalling.* vol. 26: *Advances in Second Messenger and Phosphoprotein Research* (Putney, J. W., Jr., ed.), Raven, New York, pp. 109–142.

Taylor, C. W. and Merritt, J. E. (1986) Receptor coupling to polyphosphoinositide turnover: A parallel with the adenylate cyclase system. *Trends Pharmacol. Sci.* **7,** 238–242.

Taylor, C. W. and Potter, B. V. L. (1990) The size of inositol 1,4,5-trisphosphate-sensitive Ca^{2+} stores depends on inositol 1,4,5-trisphosphate concentration. *Biochem. J.* **266,** 189–194.

Thastrup, O. (1990) Role of Ca^{2+}-ATPases in regulation of cellular Ca^{2+} signaling, as studied with the selective microsomal Ca^{2+}-ATPase inhibitor, thapsigargin. *Agents and Actions* **29,** 8–15.

Thevenod, F. and Schulz, I. (1988) H^+-dependent calcium uptake into an IP_3-sensitive calcium pool from rat parotid gland. *Am. J. Physiol.* **255,** G429–G440.

Thomas, A. P., Renard, D. C., and Rooney, T. A. (1992) Spatial organization of Ca^{2+} signaling and Ins(1,4,5)*P*3 action. *Adv. Second Mess. Phosphoprotein Res.* **26,** 225–263.

Ueda, T., Church, S. H., Noel, M. W., and Gill, D. L. (1986) Influence of inositol 1,4,5-trisphosphate and guanine nucleotides on intracellular calcium release within the N1E-115 neuronal cell line. *J. Biol. Chem.* **261,** 3184–3192.

Watson, S. P. and Abbott, A. (1990) TIPS receptor nomenclature supplement: Tachykinin receptors. *Trends Pharmacol. Sci.* **11**(Supp.), 25.

Weiss, S. J., McKinney, J. S., and Putney, J. W., Jr. (1982) Receptor-mediated net breakdown of phosphatidylinositol 4,5-bisphosphate in parotid acinar cells. *Biochem. J.* **206,** 555–560.

Willcocks, A. L., Strupish, J., Irvine, R. F., and Nahorski, S. R. (1989) Inositol 1:2-cyclic,4,5-trisphosphate is only a weak agonist at inositol 1,4,5-trisphosphate receptors. *Biochem. J.* **267,** 297–300.

Willems, P. H. G. M., van Nooij, I. G. P., Haenen, H. E. M. G., and de Pont, J. J. H. H. M. (1987) Phorbol ester inhibits cholecystokinin octapeptide-induced amylase secretion and calcium mobilization, but is without effect on secretagogue-induced hydrolysis of phosphatidylinositol 4,5-bisphosphate in rabbit pancreatic acini. *Biochim. Biophys. Acta* **930,** 230–236.

Wilson, D. B., Bross, T. E., Sherman, W. R., Berger, R. A., and Majerus, P. W. (1985a) Inositol cyclic phosphates are produced by cleavage of phosphatidylphosphoinositols (polyphosphoinositides) with purified sheep seminal vesicle phospholipase C enzymes. *Proc. Natl. Acad. Sci. USA* **82,** 4013–4017.

Wilson, D. B., Connolly, T. M., Bross, T. E., Majerus, P. W., Sherman, W. R., Tyler, A. N., Rubin, L. J., and Brown, J. E. (1985b) Isolation and characterization of the inositol cyclic phosphate products of polyphosphoinositide cleavage by phospholipase C. Physiological effects in permeabilized platelets and Limulus photoreceptor cells. *J. Biol. Chem.* **260,** 13,496–13,501.

Womack, M. D., Hanley, M. R., and Jessel, T. M. (1985) Functional substance P receptors on a rat pancreatic acinar cell line. *J. Neurosci.* **5,** 3370–3378.

Wong, N. S., Barker, C. J., Shears, S. B., Kirk, C. J., and Michell, R. H. (1988) Inositol 1:2(cyclic),4,5-trisphosphate is not a major product of inositol phospholipid metabolism in vasopressin-stimulated WRK1 cells. *Biochem. J.* **252,** 1–5.

Wu, J. T., Gong, Q., Chou, R. H., and Wieland, S. J. (1991) Ca^{2+}-insensitive modulation of a K^+ conductance by inositol phosphates. *J. Biol. Chem.* **266,** 14,893–14,895.

Zhang, Y., Duszynski, J., Hreniuk, S., Waybill, M. M., and LaNoue, K. F. (1991) Regulation of plasma membrane permeability to calcium in primary cultures of rat hepatocytes. *Cell Calcium* **12,** 559–575.

CHAPTER 11

Signal Transduction Mechanisms of Tachykinin Effects on Ion Channels

Yasuko Nakajima and Shigehiro Nakajima

1. Introduction

In 1971, Konishi and Otsuka found that application of physalaemin produced a depolarization in spinal motoneurons. Physalaemin is one of the tachykinins (named by Erspamer and Anastasi, 1966) extracted from frog skin (Erspamer et al., 1964). This was the first electrophysiological experiment on tachykinin effects. Subsequently, Otsuka et al. (1972) applied substance P (SP), which had just been identified and synthesized by Leeman's group (Chang and Leeman, 1970; Chang et al., 1971; Tregear et al., 1971), to spinal motoneurons and obtained similar depolarizations. Since then, many papers have appeared dealing with the effects of tachykinins on the electrical activities of various kinds of cells, and in almost all cases tachykinins produced excitation (Konishi and Otsuka, 1974; Krnjevic and Morris, 1974; Henry et al., 1975; Henry, 1976; Guyenet and Aghajanian, 1977; Krnjevic, 1977; Randic and Miletic, 1977; Otsuka

The Tachykinin Receptors Ed.: S. H. Buck © 1994 Humana Press Inc., Totowa, NJ

and Takahashi, 1977; Katayama and North, 1978; Nicoll, 1978; Katayama et al., 1979; Dun and Karczmar, 1979; Konishi et al., 1979; Vincent and Barker, 1979; Nicoll et al., 1980; Dun and Minota, 1981; Jan and Jan, 1982; Nowak and Macdonald, 1982; Adams et al., 1983; Cheeseman et al., 1983; Nakajima et al., 1985; Stanfield et al., 1985; Murase et al., 1986, 1989; Yamaguchi et al., 1990; Koyano et al., 1991, 1993; Konishi et al., 1992; Shen and North, 1992).

It is only recently that electrophysiologists have started asking questions about the intracellular signal transduction process. We now know that, unlike the effect of a *fast* transmitter such as L-glutamate, electrical responses produced by tachykinins are relatively *slow*. The slowness of the SP response was emphasized by Krnjevic and Morris (1974), Henry et al. (1975), Krnjevic (1977), and Nicoll (1978) (*but see* Vincent and Barker, 1979, for a very rapid response). The slowness originates from the fact that tachykinins produce their effects via a cascade of events that involves several transduction molecules. This cascade, which takes place between the binding of a tachykinin and the eventual changes in channel activity, is the signal transduction process. The signal transduction cascades consist of several mediators (transducers or messengers). Among various mediators, interest has been centered on G-proteins, phosphoinositide turnover, and arachidonic acid metabolism. The main effector channels are inwardly rectifying K channels, M-channels, nonselective channels, and voltage-gated Ca channels.

This article reviews the signal transduction mechanisms of tachykinin effects. Many papers cited here are concerned with SP rather than other tachykinins; the precise ionic and signal transduction mechanisms of other tachykinins have not been studied extensively.

2. Tachykinin Effects on K Channels

In some neurons, an obvious effect of SP application is a slow depolarization near the resting potential. Two different types of depolarizations have been observed in various preparations. One is a depolarization accompanied by an increase in membrane conductance (Otsuka and Konishi, 1977; Nicoll, 1978), and the other is a depolarization accompanied by a decrease in conductance (Krnjevic, 1977; Katayama and North, 1978; Katayama et al., 1979; Nowak and Macdonald, 1982; Adams et al., 1983; Stanfield et al., 1985; Yamaguchi et al., 1990).

Dun and Karczmar (1979) and Konishi et al. (1979) noticed that in sympathetic ganglion cells of the guinea pig either a conductance increase or a conductance decrease, depending on the cell, was induced by SP. Dun and Minota (1981), by manually clamping the membrane potential, further analyzed the ionic mechanisms of these two responses and concluded that SP produces depolarizations by either increasing Na conductance or by inhibiting K conductance or both. These dual mechanisms have been confirmed in various neurons using different experimental techniques (Murase et al., 1986; Koyano et al., 1991, 1993; Konishi et al., 1992; Shen and North, 1992).

First, we shall deal with the tachykinin-induced inhibition of K conductance. There are many types of K channels (Hille, 1992a). Among them, the inwardly rectifying K channel and M-channel have been shown to be effectors for SP action.

3. Tachykinin Effects on the Inwardly Rectifying K Channel

3.1. Properties of the SP-Suppressed Current

In the basal forebrain, there are cholinergic nuclei that supply acetylcholine to wide areas of the cerebral cortex and the hippocampus. These cholinergic neurons play an essential role in cognition and memory, and it has been proposed that degeneration of these cholinergic neurons is responsible for Alzheimer's disease (Coyle et al., 1983; Terry and Katzman, 1983). These neurons are richly innervated by SP-containing nerve fibers (Bolam et al., 1986; Beach et al., 1987). Nakajima et al. (1985) succeeded in making dissociated neuronal cell cultures from rat basal forebrain nuclei (such as the nucleus basalis of Meynert) and observed that SP evoked a long-lasting depolarization near the resting level. Stanfield et al. (1985) and Yamaguchi et al. (1990) investigated this SP effect in cultured nucleus basalis neurons by using the whole-cell patch-clamp technique. SP produced an inward current accompanied by a decrease in membrane conductance (Fig. 1A). The SP-suppressed current, obtained by digitally subtracting the current during SP action from the control current, did not show a marked relaxation (slow increase or decrease of the current) upon sudden hyperpolarization (Fig. 1B,C). The current–voltage relation of this SP-suppressed current revealed inward rectification with a reversal potential approximately equal to the potassium equilibrium potential (E_K) (Fig. 1D). External Cs^+ (0.1 m*M*) blocked the SP-suppressed current in a voltage-dependent manner with an equivalent valence for Cs^+ of 1.9, suggesting that the blockage occurs in a multi-ion pore (Yamaguchi et al., 1990). External Ba^{2+} (0.05 m*M*) blocked the SP-suppressed current, but this blockage was less voltage-dependent (Yamaguchi et al., 1990). Rb^+ also blocked the SP-suppressed current (Stanfield et al., 1985). Thus, the *channel properties* of the K channel responsible for the SP-suppressed current are very similar to those of the inward rectifiers described in skeletal muscle or oocytes (Katz, 1949; Stanfield et al., 1981; Hagiwara, 1983).

The experiments on the SP-suppressed current were conducted on cultured cells. However, it is unlikely that the modulation of the inward rectifier was due to culture artifacts, since work from North's laboratory has shown that in neurons of brain slice preparations the modulation of inwardly rectifying K currents is responsible for de- or hyperpolarizations produced by various slow transmitters (North et al., 1987; Williams et al., 1988; North, 1989; North and Uchimura, 1989). (The modulation of inwardly rectifying K currents by slow transmitters has also been investigated in various peripheral and CNS neurons: *see* Trussel and Jackson, 1985, 1987; Mihara et al., 1987; Yakel et al., 1988; Inoue et al., 1988; Surprenant and North, 1988; Kim et al., 1990; Farkas et al., 1992. *See also* Benson and Levitan, 1983, for an early paper on *Aplysia* neurons.)

3.2. SP Suppresses an Inwardly Rectifying K Channel Through a Pertussis Toxin-Insensitive G-Protein

Nakajima et al. (1988) investigated the role of G-proteins in the SP effect described above. One of the tests for the involvement of G-

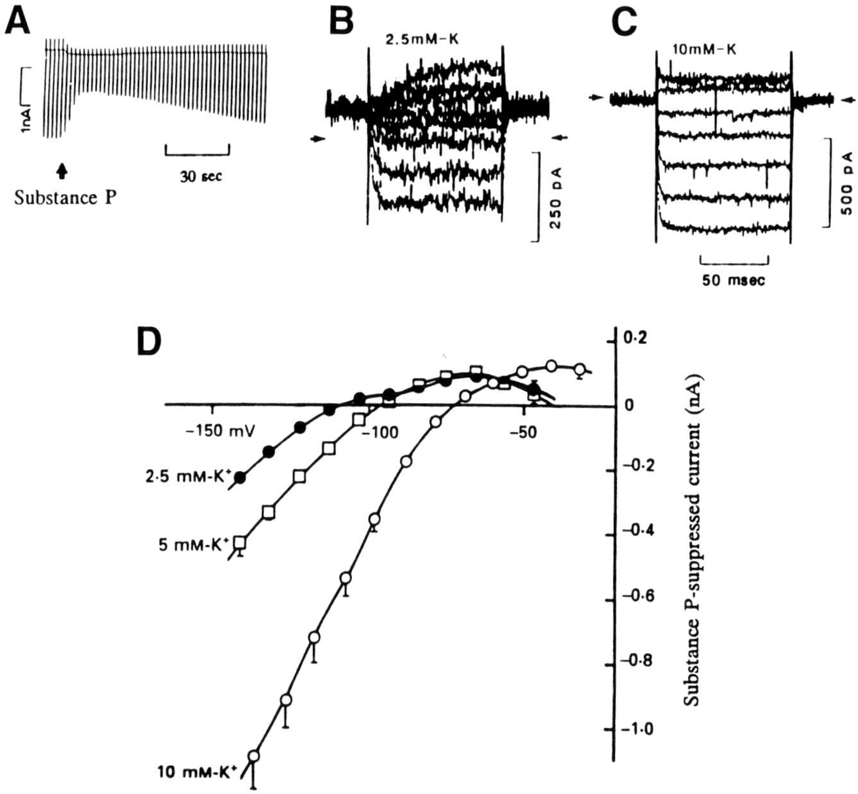

Fig. 1. Substance P excites cholinergic neurons cultured from the rat nucleus basalis by suppressing the inward rectifier K channel. **A:** Application of 3 μ*M* SP produced an inward current accompanied by a reduction in conductance under voltage-clamp in 5 m*M*-K^+ external solution. Intermittent voltage pulses (20 mV and –60 mV) were applied. The holding potential was –76 mV. **B:** SP-suppressed currents in 2.5 m*M*-K^+ external solution were obtained by subtracting the record during SP action from the control record. The holding potential was –84 mV. The potential level for each record was (from the uppermost): –56, –66, –75, –94, –103, –113, –122, and –132 mV. **C:** SP-suppressed currents in 10 m*M*-K^+ external solution with the holding potential at –69 mV. The potential level for each record was (from the uppermost): –41, –51, –60, –79, –88, –98, –107, and –117 mV. **D:** Mean values of SP-suppressed currents plotted against potential for three different $[K^+]_o$ solutions. Modified from Stanfield et al. (1985) (record A) and from Yamaguchi et al. (1990) (records B, C, D).

proteins is to apply intracellularly a hydrolysis-resistant GTP analog, such as GTPγS (guanosine 5'-[γ-thio] triphosphate). This would cause an irreversible activation of the G-protein either spontaneously or upon applying the agonist, resulting in sustained activation of the effector (Gilman, 1987) (For reviews on G-protein regulation of ion channels, *see* Brown, 1990; Brown and Birnbaumer, 1990; Dolphin, 1990; Schultz et al., 1990; Hille, 1992b).

Application of SP to a nucleus basalis neuron loaded with GTPγS resulted in an almost irreversible decrease in the inwardly rectifying K conductance (Fig. 2). Another hydrolysis-resistant GTP analog, Gpp (NH)p (5'-guanylyl imidodiphosphate), had a similar but weaker effect than that of GTPγS. These results strongly suggest that the SP effect on the K channel is mediated through a G-protein (Nakajima et al., 1988). This conclusion agrees with the results of studies on the molecular structure of the NK_1 receptor (SP receptor), which demonstrate structural features common to G-protein-coupled receptors (Yokota et al., 1989; Hershey and Krause, 1990).

Many α-subunits of G-proteins have been cloned (Kaziro et al., 1991; Simon et al., 1991), and it is important to know which one of these G-proteins mediates the SP effect on the inward rectifier. Certain G-proteins, such as Gi and Go, are inactivated by pertussis toxin through ADP-ribosylation (Katada and Ui, 1982; West et al., 1985; Iyengar et al., 1987; Katada et al., 1987). However, pertussis toxin treatment of cultured nucleus basalis neurons did not alter the SP effect on the inward rectifier (Nakajima et al., 1988), whereas the same pertussis toxin treatment completely inactivated the somatostatin-induced response in locus ceruleus neurons (Inoue et al., 1988; Nakajima et al., 1988). These results indicate that the G-protein mediating the SP effect is pertussis toxin-insensitive. Further, studies of biochemical and binding properties of the NK_1 receptor (Boyd et al., 1991) and tachykinin-induced stimulation of phospholipase C (Laniyonu et al., 1988) have revealed that a pertussis toxin-insensitive G-protein is coupled to the receptor. The substrate for the pertussis toxin-induced ADP-ribosylation is the cysteine residue fourth from the carboxyl terminus (West et al., 1985). Thus, the G-protein mediating the SP effect would lack the cysteine residue at this

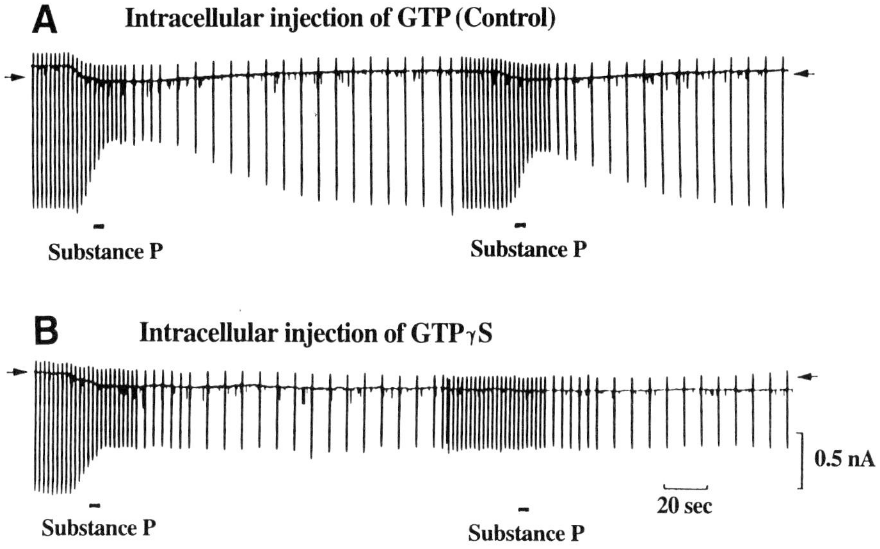

Fig. 2. Responses produced by SP (300 n*M*) persisted when nucleus basalis neurons were loaded with GTPγS. The neurons were recorded 5 min after rupture of the patch with the holding potential at –74 mV. Hyperpolarizing square-wave pulses (50 mV, 100 ms) were intermittently imposed. **A:** Control neuron. The patch pipet contained 100 μ*M* GTP. **B:** The patch pipet contained 100 μ*M* GTPγS. Modified from Nakajima et al. (1988).

position. Several kinds of pertussis toxin-insensitive G-proteins have been reported: Gs, G(z or x) (Gz and Gx are the same G-protein, named differently) (Matsuoka et al., 1988; Fong et al., 1988), G-proteins of the Gq family (Gq, G11, G14, G15, and G16) and G12 family (G12 and G13) (Amatruda et al., 1991; Simon et al., 1991; Wilkie et al., 1991), and low molecular weight GTP-binding proteins.

At present, we suspect that the G-proteins involved in the SP effect on ionic channels are members of the Gq family (particularly, Gq or G11). These G-proteins are pertussis toxin-insensitive (Strathmann and Simon, 1990), and their mRNAs are located in the brain. Tachykinins, including SP, stimulate phosphoinositide turnover by activating phospholipase C (Hanley et al., 1980; Taylor et al., 1986; Pfaffinger et al., 1988; Putney et al., 1991). Mau and Saermark

(1991) reported that this SP action on phosphoinositide turnover involves a G-protein insensitive to both pertussis toxin and cholera toxin. Gqα (α-subunit of Gq) and G11α similarly activate phospholipase C (particularly phospholipase Cβ1) (Gutowski et al., 1991; Smrcka et al., 1991; Taylor, 1991; Lee et al., 1992; Wu et al., 1992). Therefore, it is likely that Gq or G11 couples SP receptors, and that SP actions on the K channel, as well as on phospholipase C, are mediated via these G-proteins.

3.3. SP and Somatostatin (or Met-Enkephalin) Modulate the Same Inward Rectifier K Channel in Opposite Ways

Noradrenergic neurons of the locus ceruleus are innervated by several types of nerve fibers and respond to various transmitters. In locus ceruleus neurons, SP produces excitation (Guyenet and Aghajanian, 1977; Cheeseman et al., 1983; Koyano et al., 1991, 1993; Shen and North, 1992) partly by suppressing an inwardly rectifying K current (Koyano et al., 1991, 1993; Shen and North, 1992), whereas somatostatin or a μ- or δ-opioid receptor agonist causes inhibition by enhancing the K conductance (North et al., 1987; Inoue et al., 1988).

Inoue et al. (1988) studied somatostatin effects on cultured locus ceruleus neurons. The application of somatostatin produced an outward current accompanied by an increase in membrane conductance (Fig. 3A). Somatostatin-induced current, obtained by subtracting the control current from the current during the somatostatin effect, is shown in Fig. 3B,C. The somatostatin-induced current exhibits properties very similar to those of the SP-suppressed current (compare Fig. 1 with Fig. 3), except, of course, that the directions of the effects are opposite; namely, SP suppresses the current, whereas somatostatin increases it.

As for the signal transduction process, Mihara et al. (1987) in peripheral autonomic neurons, and Inoue et al. (1988) in locus ceruleus neurons, showed that intracellular application of GTPγS made the somatostatin-induced response almost irreversible. Further, the pretreatment of locus ceruleus neurons with pertussis toxin

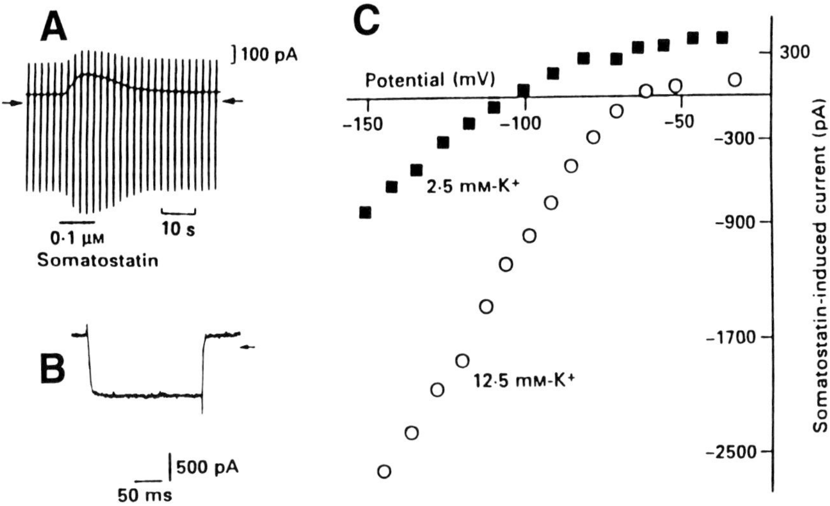

Fig. 3. Somatostatin inhibits cultured locus ceruleus neurons by inducing inwardly rectifying currents. **A**: Somatostatin (0.1 μ*M*) elicited an outward current accompanied by an increase in membrane conductance under voltage clamp. Square-wave hyperpolarizing pulses (50 mV, 100 ms) were applied once every 2 s. The holding potential was –81 mV. **B**: Somatostatin-induced current computed by subtracting the control current from the current during application of somatostatin. The current does not show relaxation. The holding potential was –81 mV and a hyperpolarizing square-wave pulse from –81 to –151 mV was applied. **C:** Current-voltage relation of the somatostatin-induced currents in 2.5 m*M* K^+ (■) and in 12.5 m*M* K^+ (○). Modified from Inoue et al. (1988).

completely abolished the somatostatin response (Inoue et al., 1988). Thus, the somatostatin effect on the inward rectifier is mediated through a pertussis toxin-sensitive G-protein (Gi or Go).

The marked similarity of the somatostatin-induced inward rectifier (in locus ceruleus neurons) and the SP-suppressed inward rectifier (in nucleus basalis neurons) has led to a hypothesis that SP and somatostatin modulate the same K channel in opposite ways (Nakajima et al., 1991a,b). Recently, Velimirovic et al. (1991) obtained strong evidence that this hypothesis is indeed correct. They used cultured locus ceruleus neurons, which respond not only to somatostatin, as in Fig. 3, but also to SP. According to Shen and North (1992) and Koyano et al. (1991, 1993), SP suppresses the inward rectifier in locus ceruleus neurons in the same way as in nucleus basalis neurons. In addition, SP increases a nonselective (cation) current *(see later)*. For now, however, we focus on the SP effect on the inward rectifier.

The most crucial experiment done by Velimirovic et al. (1991) was as follows. First, somatostatin was applied to a GTPγS-loaded locus ceruleus neuron. This produced an almost irreversible increase in inwardly rectifying K current (somatostatin-induced current). At this point, SP was applied; this produced not only complete neutralization of the somatostatin-induced current, but in addition suppressed the resting conductance that existed before somatostatin was applied. Obviously, SP eliminated the extra-conductance created by somatostatin, indicating that the same K channel is regulated by both somatostatin and SP. The same kind of interaction was observed between actions of met-enkephalin and SP. Figure 4 shows a scheme based on the effects of SP and somatostatin on the inward rectifier in locus ceruleus neurons.

3.4. Single-Channel Properties of the G-Protein-Coupled Inward Rectifier

What are the unitary properties of the inwardly rectifying K channel that is suppressed by SP? There are no published single-channel records of this K channel. However, as discussed in Section 3.3., in locus ceruleus neurons, the same channel is regulated by both SP and somatostatin (Velimirovic et al., 1991). Hence, the properties of the somatostatin-activated K channel would correspond to those of the SP-suppressed K channel.

Miyake et al. (1989) were the first to describe the single-channel properties of the K channel induced by a μ-opioid receptor agonist, somatostatin, or an α_2-adrenoceptor agonist. They worked on acutely dissociated locus ceruleus neurons by using the cell-attached patch clamp. They could record K channel activity only if the patch pipet contained the agonist. This channel had a unitary conductance of 45 pS at resting potential with $[K]_o$ = 150 m*M* at 31 °C. An important

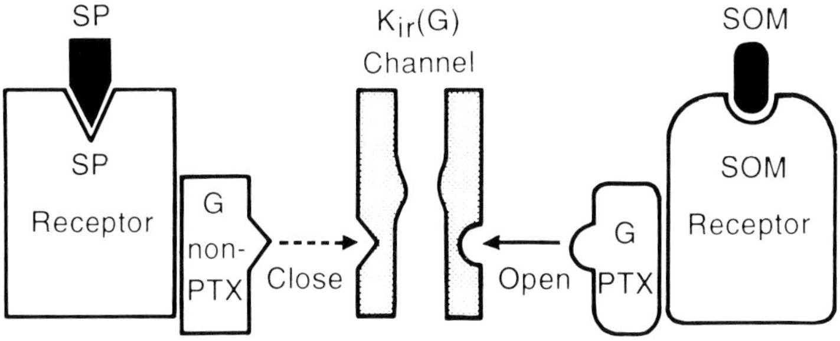

Fig. 4. A scheme based on results of SP (SP) and somatostatin (SOM) experiments showing that SP and SOM have opposite effects on the same inward rectifier K-channel (K_{ir}(G) channel). The scheme shows that SP binds to the SP receptor. This activates a pertussis toxin-*insensitive* G-protein ($G_{non\text{-}PTX}$) (which could be Gq or G11). When GDP is released and exchanged with GTP, G-proteins are separated into α-subunits and βγ-subunits. The channel could be closed directly and/or indirectly by Gqα or G11α. The SOM part of the scheme is almost a mirror image of the SP part. However, the SOM effect is the opening of the channel and is mediated by a pertussis toxin-*sensitive* G-protein (G_{PTX}), probably Gi1, Gi2, Gi3, Go-1, or Go-2. In SOM effects, the βγ-subunits or α-subunits could be involved.

finding is that when the μ-opioid agonist was applied to the area of membrane outside the patch, no channel activity was recorded in the patch, indicating that no diffusible messengers carry the signal to channels in the patch.

Grigg et al. (1991b, 1992, data to be published) used a different approach. They used an inside-out patch. The patch pipet contained an agonist (either somatostatin or met-enkephalin). When medium on the cytoplasmic side was switched from a GDP-containing solution to a GTP-containing solution, a large increase in K channel activity occurred (*see* Kurachi et al., 1986; and Yatani et al., 1987, for experiments on atrial cells). The single-channel conductance was ≈33 pS at –95 mV with $[K]_o$ = 156 m*M* and $[K]_i$ = 124 m*M* at 23°C. The unitary current indicated a reversal at approximately E_K and showed substantial rectification in the inward direction. The properties of this GTP-dependent inwardly rectifying K channel approximately correspond to those of the K channel that Miyake et al. (1989) observed. Again, no diffusible messenger is involved in the activation, since the channel was active using the inside-out patch-clamp method. Importantly, Grigg et al. observed that the opening frequency of this 33 pS channel was almost negligible in the GDP-containing solution compared to the GTP-containing solution. This suggests that this channel hardly opens without the activated G-protein. In other words, the channel is G-protein-coupled (corresponding to K(G) in Hille, 1992a). The channel may not be voltage-gated; the rectification may simply be caused by channel blockage (Matsuda et al., 1987; Vandenberg, 1987). Thus, this 35 pS channel will be called the G-protein-coupled inwardly rectifying K channel, K_{ir}(G). This channel is analogous to the muscarine-induced K channel in atrial cells (Sakmann et al., 1983; Soejima and Noma, 1984; Breitwieser and Szabo, 1985; Pfaffinger et al., 1985; Kurachi et al., 1986; Logothetis et al., 1987; Yatani et al., 1987). However, there is an important difference; in neurons, the same K channel is regulated in both directions (opening and closing) by different G-proteins (Velimirovic, 1991), whereas in atrial cells the channel is regulated in one direction only (opening).

3.5. Second Messenger Systems

This section discusses whether the G-protein responsible for the SP effect on the inward rectifier works directly on the channel or indirectly via second messengers.

3.5.1. Cyclic AMP and Cyclic GMP

Among the pertussis toxin-insensitive G-proteins described above, Gs mediates the β-adrenergic effect on Ca channels in cardiac cells. A β-adrenergic receptor activates Gs, which stimulates adenylate cyclase, leading to an increase in the cyclic AMP level. Cyclic AMP dependent phosphorylation stimulates the activity of the L-type Ca channel (reviewed by Dolphin, 1990).

To test whether the SP effect on the K channel is also mediated by the adenylate cyclase–cyclic AMP system, an experiment to increase intracellular cyclic AMP was carried out. Intracellular application of cyclic AMP (100 μ*M*) and IBMX (3-isobutyl-1-methylxanthine) (1 m*M*) through the patch pipet did not affect the SP response (Nakajima et al., 1988). Since the K_m of cyclic AMP to protein kinases is 0.01–0.15 μ*M* (Nestler and Greengard, 1984), 100 μ*M* is a very high concentration. Thus, our result suggests that cyclic AMP is not a mediator of the SP effect. However, this conclusion does not exclude the possibility that cyclic AMP could still act as a "modulator" of the effectiveness of the signal transduction.

The role of cyclic GMP in the signal transduction has not been investigated extensively. *See* Konishi and Okamoto (1985) and Yamaguchi et al. (1990) for the effects of nitroprusside on SP effects.

3.5.2. Role of Arachidonic Acid Metabolism in the SP Effect on the Inward Rectifier

Piomelli et al. (1987) demonstrated that the 12-lipoxygenase pathway of arachidonic acid metabolism produces a second messenger for FMRFamide effects in *Aplysia* neurons. However, in vertebrates the role of arachidonic acid metabolism in transmitter-induced effects is not so simple. Kim et al. (1989) observed that in cardiac cells arachidonic acid and metabolites of the 5-lipoxygenase pathway stimulated the muscarinic K channel (the muscarine induced K channel). After analyzing this phenomenon, they suggested the following scheme: βγ-subunits of the G-protein that couples to the muscarinic receptor activate phospholipase A2, which initiates the arachidonate cascade, and some metabolites of the 5-lipoxygenase pathway activate the muscarinic K channel (Kim et al., 1989). Further, Schweitzer et al. (1990) presented data suggesting that metabolites of the 5-lipoxygenase pathway mediate the somatostatin-induced increase in the M-current in hippocampal neurons.

There are, however, results that contradict the idea that arachidonate metabolites mediate transmitter effects on K channels. According to the work by Kurachi's group (Kurachi et al., 1989; Nakajima et al., 1991; Ito et al., 1992), some arachidonate metabolites activate the muscarinic K channel in atrial cells. However, after further investigating this effect, they concluded that arachidonate metabolites do not *mediate* the muscarinic effect on the K channel, but instead *modulate* the turnover of the G-protein that couples to the muscarinic receptor (Nakajima et al., 1991; Ito et al., 1992). Further, Scherer and Breitwieser (1990), also using atrial cells, analyzed the effect of a leukotriene on the kinetics of G-protein activation and concluded that the leukotriene increases the rate of the interaction of the G-protein with the K channel. This work, in agreement with the conclusion reached by the Kurachi group, emphasizes the modulatory action of arachidonate metabolites on the function of G-proteins.

Recently, Koyano et al. (1990) investigated the role of arachidonate metabolism in the somatostatin-induced enhancement and the SP-induced suppression of the K channel in nucleus basalis and locus ceruleus neurons. They showed that application of 5-lipoxygenase inhibitors (but not 12-lipoxygenase inhibitors or cyclooxygenase inhibitors) inhibited the SP-induced suppression of the K conductance, as well as the somatostatin-induced increase in the K conductance. As described above, there is strong evidence indicating that the same inwardly rectifying K channel is regulated in opposite ways by SP and somatostatin. Thus, the results that 5-lipoxygenase inhibitors suppress both SP and somatostatin effects do not agree with the idea that the lipoxygenase metabolites are a mediator of either one of the signal transduction pathways. Rather, the results of Koyano et al. (1990) suggest that some 5-lipoxygenase metabolites *modulate*, but do not *mediate*, the transmitter-induced signal transductions, in general agreement with the idea put forward by the groups of Kurachi (Kurachi et al., 1989; Nakajima et al., 1991; Ito et al., 1992) and Breitwieser (Scherer and Breitwieser, 1990).

In summary, at present it seems unlikely that cyclic AMP or arachidonate metabolites mediate the SP effect on the inwardly rectifying K channel. However, it is likely that some arachidonate metabolites modulate the signal transduction pathway. The role of cyclic GMP or phosphoinositide turnover in the tachykinin effects on the inward rectifier remains an open question.

Referring to the schematic diagram of Fig. 4, there are no diffusible messengers on the somatostatin side as explained above; thus, the transduction is membrane-delimited (the term coined by Brown and Birnbaumer, 1990). On the other hand, there is no definitive answer as to whether or not the SP side of the transduction process involves a diffusible messenger.

4. Tachykinin Effects on the M-Current

4.1. Tachykinins Inhibit the M-Current

The M-current is a time- and voltage-dependent K current that is inhibited by muscarine and other agonists (Brown and Adams, 1980; Adams et al., 1982). The M-current is activated over the voltage range more positive than –60 mV. The unitary conductance of the M-channel is very small and has not easily been recognized as a channel event. Noise analysis of patch recordings yielded values of 3.1 pS (Neher et al., 1988), 1–2 pS (Owen et al., 1990), or 2.8–5.0 pS (Marrion et al., 1992) (but Selyanko et al., 1992, gave larger values).

Nowak and Macdonald (1982), working on cultured spinal cord neurons, observed that tachykinins (SP and eledoisin-related peptide) inhibited a K current which they supposed to be the M-current. More direct evidence was given by Adams et al. (1983), who, by using the single electrode voltage clamp technique, showed that SP suppressed the M-current in frog sympathetic neurons (*see also* Ramirez and Chiappinelli, 1987). Jones (1985) further investigated the effects of tachykinins on the M-current in detail. The M-current was suppressed by SP, neurokinin A, eledoisin-related peptide, eledoisin, kassinin, and physalaemin. Putative antagonists of the SP receptor, [D-Pro^{2},D-Trp^{7,9}] SP and [D-Arg^{1},D-Pro^{2},D-Trp^{7,9},Leu^{11}] SP (Bjökroth et al., 1982; Konishi et al., 1983), did not exhibit antagonism to the SP effect on the M-current. Jones (1985) concluded that the most important functional role of the SP inhibition of the M-current is to enhance "the number of spikes produced by a given depolarizing current," and not to produce a depolarization near the resting potential (also, Jan and Jan, 1982, indicate that the suppression of the M-current cannot account for the LHRH-induced depolarization near the resting potential in sympathetic neurons).

Suppression of the M-current by SP was also reported in smooth muscle cells (Sims et al., 1986).

4.2. SP Modulates the M-Current Through a Pertussis Toxin Insensitive G-Protein

In frog sympathetic ganglion cells, loading the cell with GTPγS produced a spontaneous suppression of the M-current. Also, when SP, LHRH, muscarine, or UTP (uridine triphosphate) was applied to a GTPγS-loaded cell, M-current inhibition occurred, but the inhibition became partially irreversible (Lopez and Adams, 1989). These results suggest that the SP effect is mediated by a G-protein. (*See* Simmons and Mather, 1991, for differential actions of GDPβS on different agonist effects.)

Only recently was it reported that the G-protein that mediates the SP effect on the M-current is pertussis toxin-insensitive (Tokimasa et al., 1992). Previously, pertussis toxin-insensitivity had been reported in the muscarine- and LHRH-induced suppression of the M-current (Pfaffinger, 1988; Lopez and Adams, 1989).

4.3. Second Messengers

4.3.1. Protein Kinase C

SP activates phospholipase C, and consequently stimulates phosphoinositide turnover (Hanley et al., 1980; Pfaffinger et al., 1988). This action is mediated by a G-protein (Taylor et al., 1986), which is insensitive to both pertussis toxin and cholera toxin (Mau and Saermark, 1991). Thus, a question arises whether phosphoinositide turnover mediates the SP effect on the M-current.

Phorbol esters or diacylglycerol analogs inhibit the M-current in a neuroblastoma-glioma hybrid cell line (NG108-15) (Higashida and Brown, 1986; Brown and Higashida, 1988) in frog sympathetic neurons (Brown and Adams, 1987; Tsuji et al., 1987; Pfaffinger et al., 1988), in cultured rat sympathetic neurons (Brown et al., 1989), and in gastric smooth muscle cells (Clapp et al., 1988). This effect resembles the inhibitory action of bradykinin, muscarine, and LHRH on the M-current. Further, the effect of phorbol esters occluded the effects of bradykinin, muscarine, or LHRH (Higashida and Brown, 1986; Brown and Adams, 1987). Thus, it was suggested that protein kinase C may be a messenger of the effect of these agonists on the M-current (Higashida and Brown, 1986; Brown and Adams, 1987; Brown and Higashida, 1988; Brown et al., 1989). In contrast, Malenka et al. (1986) and Dutar and Nicoll (1988) obtained results that suggest that IP_3 (inositol 1,4,5-triphosphate), but not protein kinase C, is involved in the muscarinic suppression of the M-current in hippocampal neurons in slice. Even though tachykinins were not used in these investigations, the results are indirectly relevant to our main topic.

Pfaffinger et al. (1988) observed in frog sympathetic neurons that agonists such as SP, muscarine, and LHRH stimulated phosphoinositide turnover, elevated intracellular [Ca^{2+} (presumably via IP_3), and suppressed the M-current. Yet, they concluded that internal [Ca^{2+}] is not a mediator of these agonist effects on the M-current because the injection of IP_3 or a Ca^{2+}-buffer did not affect the inhibition of the M-current by either phorbol esters or by the agonists. They also doubted that protein kinase C is a messenger of these agonist effects because the suppressing effect of phorbol esters was only partial.

Bosma and Hille (1989) carried out experiments to critically evaluate the role of protein kinase C in the SP- or LHRH-induced suppression of the M-current in frog sympathetic ganglion cells. They confirmed that the agonists (SP and LHRH), as well as a phorbol ester or a diacylglyerol analog, suppressed the M-current. They also observed that the SP effect is occluded by the phorbol ester. However, protein kinase C inhibitors, such as staurosporine (Tamaoki et al., 1986), PKC(19-36) (House and Kemp, 1987), or H-7 (Kawamoto and Hidaka, 1984), suppressed the effect on the M-current of the phorbol ester or the diacylglycerol, as was expected, whereas the suppressing effect by SP and LHRH was not affected. This result convincingly negated the hypothesis that protein kinase C mediates the SP suppression of the M-current, at least in frog sympathetic neurons. (We naturally expect that *occlusion* is a necessary but not a sufficient criterion for identifying a substance as a mediator of the signal transduction pathway. Bosma and Hille's (1989) experiment is a good example to show that occlusion alone is not a sufficient criterion.)

4.3.2. Other Messengers

Beech et al. (1991) observed in rat sympathetic neurons that intracellular application of a high concentration (20 m*M*) of BAPTA, a fast-acting Ca chelator, interfered with the suppressing effect of *muscarine* on the M-current (as well as on the voltage-gated Ca currents, *see later*). They proposed that some unknown diffusible messenger, which is sensitive to intracellular [Ca^{2+}], mediates the muscarine effect on the M-current (*also see* Hille, 1992b). However, whether this is the case for the *SP* effect on the M-current has not been tested, leaving unanswered the question about a diffusible messenger in the SP effect.

4.3.3. Comments on the Modulation of the M-Current

In summary, even though SP definitely enhances phosphoinositide turnover, there is no decisive evidence that it mediates suppression of the M-current. Nevertheless, there is a possibility that an unknown diffusible messenger that is BAPTA sensitive carries the tachykinin message to the M-channel. It may well be that the G-protein, which couples to the tachykinin receptors, suppresses the M-type K channel directly (or via the membrane-delimited pathway). The functional significance of the M-current inhibition will be discussed in Section 8. of this chapter.

5. Tachykinin Effects on Nonselective Ion Channels

5.1. SP Induces Two Ionic Mechanisms Near the Resting Potential

As stated in Section 2., two types of SP-induced depolarization have been described: one accompanied by a conductance increase and the other by a conductance decrease. The ionic mechanism of the depolarization was first analyzed by Dun and Minota (1981) in guinea pig mesenteric ganglion cells. They concluded that the SP-induced depolarization was caused by a combination of two mechanisms: an increase in Na^+ permeability and a decrease in K^+ permeability.

Randic's group (Murase et al., 1986, 1989) reported dual ionic mechanisms for SP-induced depolarizations of dorsal horn neurons in the spinal cord in slice; one is caused by an increase in a "Ca^{2+}-sensitive, voltage-insensitive conductance" (Murase et al., 1989), and the other is caused by an inhibition of the M-current. It is not clear, however, whether this Ca^{2+}-sensitivity is caused by a rise in intracellular [Ca^{2+}] or by a mechanism in which Ca ions act as a charge carrier for the inward current.

More recently, Konishi et al. (1992) reported fast and slow phases in the depolarization produced by SP in sympathetic neurons. They interpreted the fast phase as having originated from the activation of a Ca-dependent cation channel (thus, it will be treated in the section on Ca-dependent channels, Section 7.1.). Also, *see* Hanani and Burnstock, 1984, and Willard and Nishi, 1985, for reports on SP-induced fast and slow depolarizations.

In cultured locus ceruleus neurons, dual ionic mechanisms for SP induced depolarizations were reported: an initial activation of a nonselective ionic current followed by suppression of the G-protein-coupled inward rectifier (Koyano et al., 1991; 1993). Also, in locus ceruleus neurons in slice, Shen and North (1992) reported similar dual ionic mechanisms induced by tachykinins (SP, neurokinin A, and neurokinin B); one mechanism is the activation of a nonselective cation conductance and the other inhibition of an inward rectifier. The nonselective current shows an almost linear current–voltage relationship with a reversal about 0 to +10 mV (Shen and North, 1992; Koyano et al., 1993).

Dual ionic mechanisms for slow depolarizations are not confined to the SP effect. There is strong evidence for the dual mechanisms in the slow depolarizations induced by muscarine or LHRH in sympathetic neurons and in hippocampal neurons (Kuba and Koketsu, 1976; Jan and Jan, 1982; Akasu et al., 1984; Kuffler and Seinowski, 1983; Benson et al., 1988).

In summary, it is likely that in all types of neurons dual ionic mechanisms, one a suppression of the K conductance and the other an activation of the nonselective current, are initiated by tachykinins. Both mechanisms produce slow depolarizations. Depending on the preparation, expression of these two components may be quantitatively different; in some cells, one may dominate the other, whereas in other cells both substantially contribute to the depolarization. The physiological significance of the nonselective channel will be discussed further in the last section.

5.2. Signal Transduction Mechanisms of SP Effects on Nonselective Ion Channels

Not much is known about the signal transduction pathway of the SP-induced increase of the nonselective ionic conductance. As explained above, Konishi et al. (1992) proposed a Ca^{2+}-dependent mechanism activated by internal Ca^{2+} (*see* Section 7.1.). In cultured locus ceruleus neurons, Koyano et al. (1991, 1993) reported that in GTPγS-loaded cells SP evoked an irreversible suppression of the inward rectifier, whereas SP did not produce an irreversible activation of the nonselective ionic current. This suggests that the latter effect is not mediated by a G-protein. Further, they observed that the nonselective conductance is not affected by intracellular application of high concentrations of cyclic AMP or cyclic GMP.

6. Tachykinin Effects on Voltage-Gated Ca Channels

6.1. SP Suppresses Voltage-Gated Ca Channels

SP suppresses voltage-gated Ca currents in frog and rat sympathetic neurons (Bley and Tsien, 1990; Elmslie, 1992; Shapiro and Hille, 1993) and in cultured cholinergic neurons from the nucleus basalis (Grigg et al., 1991a). In frog sympathetic neurons, SP, as well as LHRH, suppresses the N-type Ca channels (Bley and Tsien, 1990). Desensitation of the effect of SP on the Ca current developed more quickly when using the perforated patch technique than when using conventional whole-cell recording, suggesting that some essential ingredients for the development of desensitization were lost in the whole-cell configuration (Bley and Tsien, 1990).

Shapiro and Hille (1993) investigated the effects of SP and somatostatin on the N-type Ca current in rat sympathetic neurons by using the whole-cell patch-clamp technique. They noticed that the

SP-induced suppressions, measured on the tail current after a conditioning depolarization to either 0 mV or to +120 mV, were not different; thus, the SP-induced suppression of the Ca channel was *not* voltage-dependent. (But the records by Bley and Tsien, 1990, and Elmslie, 1992, suggest that the SP effect is voltage-dependent in *frog* sympathetic neurons.) On the other hand, the somatostatin-induced inhibition of the Ca channel was strongly voltage-dependent, in agreement with the mode of Ca channel inhibition by many other transmitters (Luini et al., 1986; Marchetti et al., 1986; Tsunoo et al., 1986; Bean, 1989; Ikeda and Schofield, 1989; Elmslie et al., 1990; Beech et al., 1992; Kasai, 1992).

In spinal dorsal horn neurons in slice preparations, Ryu and Randic (1990) reported that SP enhanced both low- and high-voltage activated Ca currents. Apart from tachykinin effects, there are several instances in which a slow transmitter enhances a Ca channel (Schultz et al., 1990). The mechanism of this enhancement is not well understood except for a case of cyclic AMP-induced stimulation of the cardiac L-type Ca channel *(see later).*

It is known that various inhibitory and excitatory neurotransmitters suppress voltage-gated calcium channels (Dunlap and Fischbach, 1978; Dolphin, 1990; Schultz et al., 1990; Anwyl, 1991). On the surface, suppression of the Ca channel by excitatory transmitters such as SP appears puzzling. We shall speculate on the physiological significance of this phenomenon later (*see* Section 8.).

6.2. Inhibition of Ca Channels by SP Is Mediated Through a Pertussis Toxin-Insensitive G-Protein

Shapiro and Hille (1993) reported that in rat sympathetic neurons loading the cell with 2 m*M* GDPβS reduced the SP inhibition of the N-type Ca current by about one-half, suggesting that the SP effect is mediated through a G-protein. This SP effect was insensitive to pertussis toxin treatment, in contrast to the pertussis toxin sensitivity of the somatostatin effect on the Ca channel. As discussed in Section 3.2. (the inward rectifier), we suspect that Gq or G11 is the strongest candidate for the G-protein mediating the SP effect on the N-type Ca channel. Another pertussis toxin-insensitive G-protein is G(x or z)

(Fong et al., 1988; Matsuoka et al., 1988). However, according to Bley and Tsien (1990), intracellular application of G(x or z) into frog sympathetic neurons had no effect on the Ca currents.

6.3. Second Messengers

Bley and Tsien (1990) examined effects of various second messengers on Ca currents in frog sympathetic neurons. Activators of protein kinase C, phorbol esters, or a diacylglycerol analog did not mimic the SP- or LHRH-induced suppression of the Ca current. Instead, such activators of protein kinase C enhanced the Ca current (Bley and Tsien, 1990; Yang and Tsien, 1992). Intracellular application of IP_3, external application of 8-Br-cAMP or 8-Br-cGMP, or external application of arachidonic acid all did not affect the Ca current.

Shapiro and Hille (1993) have investigated the effect of the intracellular application of BAPTA on the SP effect on the Ca current in rat sympathetic neurons. Previously, Bernheim et al. (1991) demonstrated that the degree of Ca current suppression caused by muscarinic agonists depends on BAPTA concentrations; with a high BAPTA concentration (20 m*M*) the muscarinic suppression of Ca current is inhibited. They call it BAPTA sensitive inhibition of the Ca channel. Their interpretation of the BAPTA sensitivity is that there exists an unknown diffusible second messenger that is disrupted by a high concentration of BAPTA and/or a low concentration of Ca ions. Shapiro and Hille (1993) showed that, unlike the case of muscarinic inhibition, the SP-induced suppression of the Ca channel was independent of the BAPTA concentration (BAPTA-insensitive), indicating that the BAPTA-sensitive diffusible messenger does not mediate the SP effect. Further, Shapiro and Hille (1993) observed that the application of SP or somatostatin to the extra-patch region of the cell did not affect calcium channels inside the patched membrane, indicating an absence of any diffusible second messenger for the SP or somatostatin effect. Both Bley and Tsien (1990) and Shapiro and Hille (1993) suggested that the G-protein that couples to the NK_1 receptor acts in a membrane-delimited way, possibly directly on the Ca channel.

Although these results from SP application on sympathetic ganglion cells (both in rats and frogs) are relatively straightforward, this situation may not apply to other cell types. In fact, there exist variations among preparations in the ways Ca channels are modulated by neurotransmitters. We have already mentioned that SP *enhances* Ca channel activity in spinal dorsal horn neurons (Ryu and Randic, 1990). There are additional examples that transmitters enhance voltage-gated Ca currents without involving cyclic AMP (reviewed by Schultz et al., 1990). Furthermore, the role of protein kinase C in the transmitter-induced suppression of Ca channels is different between chick dorsal root ganglion cells and sympathetic neurons. In the former case, there is strong evidence for protein kinase C being a mediator (Rane et al., 1989), whereas in the latter case, as already described, there is strong evidence against the participation of protein kinase C as a mediator (Wanke et al., 1987; Bley and Tsien; 1990; Bernheim et al., 1991).

We hope that this complex situation can soon be clarified as more data accumulate. Part of the complexity may originate from the possibility that there may be many types of Ca channels with different degrees of sensitivity for receiving different modulating signals. Also, dual or multiple signals may be initiated by a single transmitter. For example, in sympathetic neurons tachykinins send a suppressing signal to the N-type Ca channel without involving diffusible messengers (Bley and Tsien, 1990; Shapiro and Hille, 1993), while at the same time tachykinins stimulate phosphoinositide turnover and enhance protein kinase C activity. This increase in protein kinase C activity potentially enhances the N-type Ca current as shown by Bley and Tsien (1990) and Yang and Tsien (1992). Ca currents would be either suppressed or enhanced by tachykinins depending on which signal is stronger (a membrane delimited signal transducer or a diffusible messenger).

7. Tachykinin Effects on Other Channels

7.1. Tachykinin Effects on Ca-Dependent Channels

SP stimulates the turnover of phosphoinositides. This would lead to an elevation of intracellular $[Ca^{2+}]$ via IP_3. Womac et al. (1988) observed a SP-induced rise in intracellular $[Ca^{2+}]$ measured by using a Ca^{2+}-indicator dye in dissociated dorsal horn cells of the spinal cord. This SP-induced Ca^{2+} came from intracellular storage sites. Pfaffinger et al. (1988) observed a modest rise in the intracellular $[Ca^{2+}]$ induced by SP, LHRH, and muscarine in sympathetic neurons.

In certain cases, elevation of the intracellular $[Ca^{2+}]$ activates Ca-dependent channels. In the pancreatic acinar cell line AR4-2J, the application of SP elevated intracellular $[Ca^{2+}]$ via IP_3 formation; this in turn produced depolarization by activating a Ca-dependent cation conductance (Song et al., 1988). Intracellular application of IP_3 or GTPγS mimicked and occluded this SP effect (Song et al., 1988). More recently, Konishi et al. (1992) reported that in rat sympathetic neurons there are fast and slow phases in the depolarization produced by SP, neurokinin A, and neurokinin B. They presented some evidence indicating that the fast phase is caused by a Ca^{2+}-dependent cationic conductance, namely, a cationic conductance activated by a rise in internal $[Ca^{2+}]$, a mechanism similar to that found in secretory cells (Petersen and Maruyama, 1984).

In addition to Ca mobilization from the intracellular Ca reservoir, SP enhances Ca entry from the external solution in certain cells, and under certain conditions the elevated internal $[Ca^{2+}]$ activates Ca-dependent K channels. Thus, according to Mayer et al. (1990), application of a very low concentration of SP (1–100 p*M*) on longitudinal colonic muscle cells increased the opening probability of a large conductance Ca-dependent K channel. This response was suppressed by nifedipine, suggesting that Ca ions came from the external solution through the L-type Ca channel.

Matthews et al. (1989a) reported results of simultaneous measurements of whole-cell currents and the internal $[Ca^{2+}]$ in rat mast cells by using the Ca^{2+}-indicator dye, fura-2. They showed that secretagogues, such as SP or compound 48/80, caused the release of Ca^{2+} from the internal calcium reservoir, probably via IP_3. In addition, the secretagogues induced Ca-influx through two voltage-independent pathways; one is a Ca-influx activated by IP_3, associated with a very small electric current, and the other is a Ca-influx through large conductance (≈50 pS) cation channels. This second type of Ca-influx is independent of IP_3 and is G-protein-mediated (Matthews et al., 1989a).

However, these SP effects on mast cells may be independent of the NK_1 receptor. In fact, Devillier et al. (1989) showed that SP or bradykinin related peptides released histamine from mast cells in a nonspecific way, unrelated to their action on the NK_1 or the bradykinin receptor. Further, it has been shown that SP, compound 48/80, or mastoparan stimulates a G-protein directly without activating the receptor (Mousli et al., 1990). This G-protein is pertussis toxin-sensitive. Thus, it is likely that the SP effect described here is totally different from the SP effect on neurons, since in neurons SP acts on the NK_1 receptor and activates a pertussis toxin-insensitive G-protein.

7.2. Tachykinin Effects on Cl Channels

Activation of Cl currents by SP was reported in mast cells (Matthews et al., 1989b), in astrocytes (Backus et al., 1991), and in colonic smooth muscle (Rangachari et al., 1990; Sun et al., 1992). Matthews et al. (1989b) in mast cells and Sun et al. (1992) in colon smooth muscle described single channel properties of SP-induced Cl channels. The properties of the Cl channels in these two materials are quite different. In mast cells, SP induced a slowly developed Cl current. This Cl current was reduced by a Cl channel inhibitor, DIDS (4,4'-diisothiocyano-2,2'-stilbenedisulfonate) (Matthews et al., 1989b). Single-channel currents were too small to be recognized as individual events. The unitary conductance calculated from noise analysis was 1.6 pS. Intracellular application of GTPγS activated this Cl current, suggesting the involvement of a G-protein. High intracellular $[Ca^{2+}]$ or intracellular application of cyclic AMP also activated this channel. They speculate that the Cl current would keep the membrane potential inside more negative after the application of secretagogues and thus facilitate the calcium influx necessary for secretion.

In contrast, the unitary conductance of the SP-induced Cl channel in colonic smooth muscle cells is large, ≈300 pS with $[Cl^-]$ ≈130 m*M* on both sides (Sun et al., 1992). This Cl current is induced through the NK_1 receptor and suppressed by NPPB (5 nitro-2(3-phenylpropylamino)-benzoate) and DIDS. Again, a G-protein is involved in the activation of the Cl channel. Sun et al. (1992) speculate that activation of the Cl channel causes the tachykinin-induced depolarization since the internal $[Cl^-]$ in smooth muscle cells is quite high.

7.3. Tachykinin Effects on Nicotinic Acetylcholine Receptor/Channels

SP downregulates the activity of nicotinic receptors in various types of cells: Renshaw cells (Belcher and Ryall, 1977; Krnjevic and Lekic, 1977), PC12 cells (Stallcup and Patrick, 1980; Boyd and Leeman, 1987; Simasko et al., 1987), ciliary ganglion cells (Role, 1984; Margiotta and Berg, 1986), sympathetic ganglion neurons (Simmons et al., 1990), and chromaffin cells (Livett et al., 1979; Role et al., 1981; Clapham and Neher, 1984). This modulation of nicotinic receptor activity is caused by an acceleration of receptor desensitization by SP. In bovine chromaffin cells, Clapham and Neher (1984) noticed that SP alone did not produce any effect. Yet, SP shortened the mean open time of the nicotinic receptor/channel without affecting the amplitude of the single channel current. Further, SP increased the interburst intervals, shortened the burst duration, and reduced the number of openings per burst, indicating that SP enhanced the rate of desensitization or produced channel blockage of the nicotinic receptor/channel.

Using cell-attached single channel recording in sympathetic neurons, Simmons et al. (1990) observed that SP decreased the mean open time and opening frequency of the nicotinic receptor/channel. They concluded that this effect is mediated through a diffusible messenger since SP application on the cell membrane outside the patched region produced the effect. This is an interesting mode of interaction between two transmitter effects. Activation of one receptor influences the function of another receptor thorough a diffusible second messenger.

8. Conclusions and Speculations on Functional Significance

8.1. The Inwardly Rectifying K Channel, M-Channel, and Nonselective Channel

The inwardly rectifying K channel is one of the effector channels of SP and other transmitters in many types of CNS and peripheral neurons. This inward rectifier is G-protein-coupled. As summarized by Fig. 4, this inward rectifier molecule in neurons, unlike that in atrial cells, is regulated in both ways, having receiving sites for opening signals from inhibitory transmitters and for closing signals from excitatory transmitters. The opening signal is mediated by a pertussis toxin-sensitive G-protein, whereas the closing signal from SP is transmitted by a pertussis toxin-insensitive G-protein.

The inwardly rectifying K channel operates over the potential range more negative than –60 mV, which is subthreshold for the action potential. As shown in Fig. 5 (I_{ir}), over the membrane potential below the threshold of action potentials (–90 to –60 mV) neurons in the CNS are constantly producing slow inhibitory and excitatory synaptic potentials. These synaptic potentials arise partly from the regulation of the G-protein-coupled inward rectifier. Below the threshold range, the interplay of slow inhibitory synaptic potentials and slow excitatory synaptic potentials sets the membrane potential and thus is an important determinant of *excitability* (i.e., the reciprocal of the threshold current). One might ask whether regulating the inward rectifier is a very effective way to achieve changes in the membrane potential. When SP depolarizes the membrane to some degree (say, halfway between the resting potential and the threshold of action potential), the unitary conductance of the inward rectifier decreases (inward rectification); therefore, further decrease in channel activity (by a higher SP concentration) appears not to be very effective in producing more depolarization. But this is not the case. Take a simple case of a neuron. Assume that the conductance of the inward rectifier at rest (–80 mV) is 10 nS (100 MΩ) and the depolarizing current (mainly supplied by the constant current source

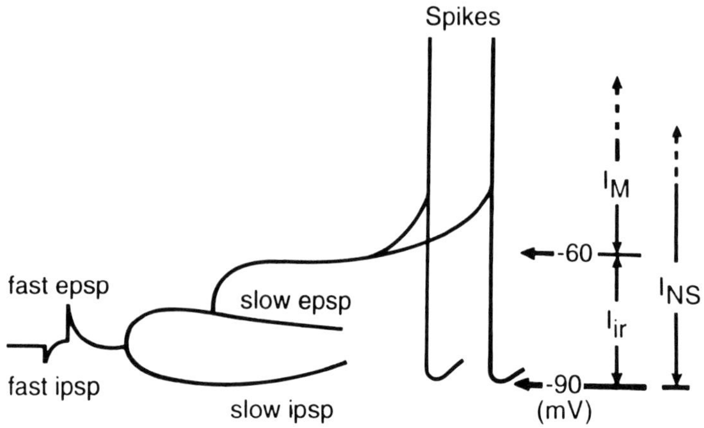

Fig. 5. Scheme showing the operating ranges of the inwardly rectifying K channel (I_{ir}), the M-current (I_M), and the nonselective channel (I_{NS}) in relation to fast and slow excitatory postsynaptic potentials (epsps) and to fast and slow inhibitory postsynaptic potentials (ipsps).

of the Na battery) is assumed to be 100 pA. If SP decreases the inward rectifier conductance by 25% (from 10 to 7.5 nS; namely, the channel activity is suppressed by 25%), it will produce a depolarization of *3.3* mV. Next, we repeat the same calculation starting from a depolarized level of –70 mV (instead of –80 mV). At –70 mV, the inward rectifier conductance is only 5 nS. If SP again produces a 25% decrease in conductance (conductance changes from 5 to 3.75 nS), it will depolarize the cell by *6.7 mV*. This is twice the depolarization produced by the same relative change in channel activity at –80 mV. Put in a more generalized way, because the depolarization is caused by an almost constant current source provided by the Na battery, as the conductance of the inward rectifier decreases due to depolarization, a given change of channel activity will cause larger changes of membrane potential (a similar argument was presented by Stanfield et al., 1985).

The real situation, however, is more complicated because of the presence of a leak conductance (owing to background K and Cl channel activities unrelated to G-protein), in parallel with the inward rectifier. Depending on the value assumed to this leak conductance,

SP effect could become less or more at a depolarized level than that at the resting level. In addition to the suppression of the inward rectifier, tachykinins induce activation of a nonselective current. This mechanism also contributes to the tachykinin-induced depolarization. Because the current-voltage relation of the nonselective current is almost linear with the reversal near 0 mV, the channel would contribute to depolarization over quite a large range, from –90 mV to beyond the threshold of the action potential (Fig. 5, I_{NS}). In most cells, both the nonselective channel and the G-protein-coupled K channel contribute to the depolarization induced by slow excitatory transmitters, the relative influence of these two mechanisms differing greatly from cell type to cell type. Assuming everything else to be equal, the depolarization produced by the nonselective channel would consume more cellular energy than the depolarization resulting from the suppression of the K channel.

Tachykinins suppress another K channel, the M-current. M-current is activated over the potential more positive than –60 mV (Fig. 5, I_M). M-current is time-dependent in the sense that after sudden changes in potential it takes some time to reach a new value. The unitary conductance is very small (1.6–3 pS). Again, the suppression of the M-current by SP is mediated by a pertussis toxin-insensitive G-protein.

Since the M-current is activated at potentials more positive than –60 mV, which is near the threshold of action potentials, even a small change in the M-current would greatly influence the threshold and firing frequency of action potentials. On the other hand, in the subthreshold range of –60 to –90 mV, the M-channel is largely silent, thus not participating in slow excitatory or inhibitory synaptic potentials.

8.2. Ca Channels

Slow inhibitory transmitters, such as somatostatin or opioids, inhibit the N-type Ca channel via a pertussis toxin-sensitive G-protein, while these transmitters increase the K channel activity, again via a pertussis toxin-sensitive G-protein. The functional significance of these two processes is easily explained. Both the enhancement of K channel activity and the suppression of Ca channel activity synergistically cause inhibition of neuronal functions; neuronal excitability is inhibited, action potentials are shortened, and transmitter release is suppressed.

On the other hand, the action of excitatory transmitters such as SP is difficult to understand. SP inhibits the N-type Ca channels via a pertussis toxin-insensitive G-protein. At the same time, SP excites the neurons both by activating the nonselective channel and by suppressing the K channel activity. On the surface, the simultaneous suppression of the Ca channel and the enhancement of excitability is paradoxical. It has been known that there are prejunctional alpha-2 adrenergic receptors in sympathetic neurons, and released norepinephrine acts on these receptors to suppress further release of norepinephrine (autoinhibition). Lipscombe et al. (1989) proposed that the suppression of Ca channels by norepinephrine is responsible for this autoinhibition. According to Shapiro and Hille (1993), Ca current suppression by SP seemed to occur in SP-containing neurons in rat sympathetic ganglion cells. They also suggested that the suppression of Ca currents by SP serves as an autoinhibitory mechanism. Bley and Tsien (1990) proposed an additional functional significance. If an excitatory transmitter inhibits the Ca-channel, this will lead to less activation of the Ca-dependent K channel, and this will ensure a sustained excitation with less adaptation.

In the authors' opinion, this paradox may be important for the survival of the neuron. The suppression of the Ca channel by SP or other slow excitatory transmitters would prevent the neuronal soma from being overloaded with Ca ions during prolonged depolarizations. The excitotoxic effect of NMDA is well documented and is known to be caused by Ca inflow through NMDA receptors. Excitotoxic effects of a prolonged depolarization and a long train of spike potentials caused by SP would be quite intense were it not for this protective mechanism of suppressing calcium channels.

8.3. Second Messengers

Phosphoinositide turnover is enhanced by SP and other tachykinins. However, it is not clear how the increased phosphoinositide turnover is related to neuronal electrical events, an important functional expression of neurons. After all, phosphoinositide turnover may not be a mediator of the SP-induced suppression of the K channels or of the Ca channels. Instead, it could be a mediator of very long-term signals (longer than the duration of slow synaptic potentials), which are essential for such cell functions as long-term potentiation, gene regulation, replenishment of each element of the transduction cascade, growth, plasticity, structural changes, and so forth. Hille (1992b) emphasizes multiple consequences of exposing the cells to a G-protein-coupled slow transmitter. Creating slow synaptic potentials that last for a few minutes may be a relatively simple task compared to the complexity of maintaining and reshaping the entire cellular machinery of the neuron.

Acknowledgments

We thank Bertil Hille and Peter R. Stanfield for reading the manuscript and giving valuable suggestions. We are indebted to John J. Grigg, Tohru Kozasa, Konomi Koyano, Bratislav M. Velimirovic, and Ronald H. Farkas for stimulating discussion and for reading the manuscript. We also thank Linda Johnston for editorial work. The work was supported by PHS grants AG06093 and NS24711.

References

Adams, P. R., Brown, D. A., and Constanti, A. (1982) Pharmacological inhibition of the M-current. *J. Physiol.* **332,** 223–262.

Adams, P. R., Brown, D. A., and Jones, S. W. (1983) Substance P inhibits the M-current in bullfrog sympathetic neurones. *Br. J. Pharmacol.* **79,** 330–333.

Akasu, T., Gallagher, J. P., Koketsu, K., and Shinnick-Gallagher, P. (1984) Slow excitatory post-synaptic currents in bull-frog sympathetic neurones. *J. Physiol.* **351,** 583–593.

Amatruda III, T. T., Steele, D. A., Slepak, V. Z., and Simon, M. I. (1991) Gα16, a G protein α subunit specifically expressed in hematopoietic cells. *Proc. Natl. Acad. Sci. USA* **88,** 5587–5591.

Anwyl, R. (1991) Modulation of vertebrate neuronal calcium channels by transmitters. *Brain Res. Rev.* **16,** 265–281.

Backus, K. H., Berger, T., and Kettenmann, H. (1991) Activation of neurokinin receptors modulates K^+ and Cl^- channel activity in cultured astrocytes from rat cortex. *Brain Res.* **541,** 103–109.

Beach, T. G., Tago, H., and McGeer, E. G. (1987) Light microscopic evidence for a substance P-containing innervation of the human nucleus basalis of Meynert. *Brain Res.* **408,** 251–257.

Bean, B. P. (1989) Neurotransmitter inhibition of neuronal calcium currents by changes in channel voltage dependence. *Nature* **340,** 153–156.

Beech, D. J., Bernheim, L., and Hille, B. (1992) Pertussis toxin and voltage dependence distinguish multiple pathways modulating calicum channels of rat sympathetic neurons. *Neuron* **8,** 97–106.

Beech, D. J., Bernheim, L., Mathie, A., and Hille, B. (1991) Intracellular Ca^{2+} buffers disrupt muscarinic suppression of Ca^{2+} current and M current in rat sympathetic neurons. *Proc. Natl. Acad. Sci. USA* **88,** 652–656.

Belcher, G. and Ryall, R. W. (1977) Substance P and renshaw cells: a new concept of inhibitory synaptic interactions. *J. Physiol.* **272,** 105–119.

Benson, D. M., Blitzer, R. D., and Landau, E. M. (1988) An analysis of the depolarization produced in guinea-pig hippocampus by cholinergic receptor stimulation. *J. Physiol.* **404,** 479–496.

Benson, J. A. and Levitan, I. B. (1983) Serotonin increases an anomalously rectifying K^+ current in the *Aplysia* neuron R15. *Proc. Natl. Acad. Sci. USA* **80,** 3522–3525.

Bernheim, L., Beech, D. J., and Hille, B. (1991) A diffusible second messenger mediates one of the pathways coupling receptors to calcium channels in rat sympathetic neurons. *Neuron* **6,** 859–867.

Bjökroth, U., Rosell, S., Xu, J.-C., and Folkers, K. (1982) Pharmacological characterization of four related substance P antagonist. *Acta Physiol. Scand.* **116,** 167–173.

Bley, K. R. and Tsien, R. W. (1990) Inhibition of Ca^{2+} and K^+ channels in sympathetic neurons by neuropeptides and other ganglionic transmitters. *Neuron* **4,** 379–391.

Bolam, J. P., Ingham, C. A., Izzo, P. N., Levey, A. I., Rye, D. B., Smith, A. D., and Wainer, B. H. (1986) Substance P-containing terminals in synaptic contact with cholinergic neurons in the neostriatum and basal forebrain: A double immunocytochemical study in the rat. *Brain Res.* **397,** 279–289.

Bosma, M. M. and Hille, B. (1989) Protein kinase C is not necessary for peptide-induced suppression of M current or for desensitization of the peptide receptors. *Proc. Natl. Acad. Sci. USA* **86,** 2943–2947.

Boyd, N. D., MacDonald, S. G., Kage, R., Luber-Narod, J., and Leeman, S. E. (1991) Substance P receptor biochemical characterization and interactions with G proteins, in *Substance P and Related Peptides: Cellular and Molecular Physiology. The Annals of the New York Academy of Sciences* **632** (Leeman, S. E., Krause, J. E., and Lembeck, F., eds.), NY Academy of Sciences, New York, pp. 79–93.

Boyd, N. D. and Leeman, S. E. (1987) Multiple actions of substance P that regulate the functional properties of acetylcholine receptors of clonal rat PC12 cells. *J. Physiol.* **389,** 69–97.

Breitwieser, G. E. and Szabo, G. (1985) Uncoupling of cardiac muscarinic and β-adrenergic receptors from ion channels by a guanine nucleotide analogue. *Nature* **317,** 538–540.

Brown, A. M. and Birnbaumer, L. (1990) Ionic channels and their regulation by G protein subunits. *Annu. Rev. Physiol.* **52,** 197–213.

Brown, D. A. (1990) G-proteins and potassium currents in neurons. *Annu. Rev. Physiol.* **52,** 215–242.

Brown, D. A. and Adams, P. R. (1980) Muscarinic suppression of a novel voltage-sensitive K^+ current in a vertebrate neurone. *Nature* **283,** 673–676.

Brown, D. A. and Adams, P. R. (1987) Effects of phorbol dibutyrate on M currents and M current inhibition in bullfrog sympathetic neurons. *Cell. Mol. Neurobiol.* **7,** 255–269.

Brown, D. A. and Higashida, H. (1988) Inositol 1,4,5-Trisphosphate and diacylglycerol mimic bradykinin effects on mouse neuroblastoma x rat glioma hybrid cells. *J. Physiol.* **397,** 185–207.

Brown, D. A., Marrion, N. V., and Smart, T. G. (1989) On the transduction mechanism for muscarine-induced inhibition of M-current in cultured rat sympathetic neurones. *J. Physiol.* **413,** 469–488.

Chang, M. M. and Leeman, S. E. (1970) Isolation of a sialogogic peptide from bovine hypothalamic tissue and its characterization as substance P. *J. Biol. Chem.* **245,** 4784–4790.

Chang, M. M., Leeman, S. E., and Niall, H. D. (1971) Amino-acid sequence of substance P. *Nature New Biology* **232,** 86,87.

Cheeseman, H. J., Pinnock, R. D., and Henderson, G. (1983) Substance P excitation of rat locus coeruleus neurones. *Eur. J. Pharmacol.* **94,** 93–99.

Clapham, D. E. and Neher, E. (1984) Substance P reduces acetylcholine-induced currents in isolated bovine chromaffin cells. *J. Physiol.* **347,** 255–277.

Clapp, L. H., Sims, S. M., Singer, J. J., and Walsh, J. V., Jr. (1988) Cholinergic suppression of both endogenous and isoproterenol-induced M-current in isolated smooth muscle cells is mimicked by a diacylglycerol analog. *Soc. Neurosci. Abstr.* **14,** 1088.

Coyle, J. T., Price, D. L., and DeLong, M. R. (1983) Alzheimer's disease: A disorder of cortical cholinergic innervation. *Science* **219,** 1184–1190.

Devillier, P., Drapeau, G., Renoux, M., and Regoli, D. (1989) Role of the N-terminal arginine in the histamine-releasing activity of substance P, bradykinin and related peptides. *Eur. J. Pharm.* **168,** 53–60.

Dolphin, A. C. (1990) G-protein modulation of calcium currents in neurons. *Annu. Rev. Physiol.* **52,** 243–255.

Dun, N. J. and Karczmar, A. G. (1979) Actions of substance P on sympathetic neurons. *Neuropharmacology* **18,** 215–218.

Dun, N. J. and Minota, S. (1981) Effects of substance P on neurones of the inferior mesenteric ganglia of the guinea-pig. *J. Physiol.* **321,** 259–271.

Dunlap, K. and Fischbach, G. D. (1978) Neurotransmitters decrease the calcium component of sensory neurone action potentials. *Nature* **276,** 837–839.

Dutar P. and Nicoll, R. A. (1988) Classification of muscarinic responses in hippocampus in terms of receptor subtypes and second-messenger systems: electrophysiological studies *in vitro*. *J. Neurosci.* **8,** 4214–4224.

Elmslie, K. S. (1992) Calcium current modulation in frog sympathetic neurones: Multiple neurotransmitters and G proteins. *J. Physiol.* **451,** 229–246.

Elmslie, K. S., Zhou, W., and Jones, S. W. (1990) LHRH and GTP-γ-S modify calcium current activation in bullfrog sympathetic neurons. *Neuron* **5,** 75–80.

Erspamer, V. and Anastasi, A. (1966) Polypeptides active on plain muscle in the amphibian skin, in *Hypotensive Peptides* (Erdös, E.G., Back, N., and Sicuteri, F., eds.), Springer-Verlag, New York, pp. 63–75.

Erspamer, V., Anastasi, A., Bertaccini, G., and Cei, J. M. (1964) Structure and pharmacological actions of physalaemin, the main active polypeptide of the skin of *Physalaemus fuscumaculatus. Experientia.* **20,** 489,490.

Farkas, R., Grigg, J. J., Nakajima, S., and Nakajima, Y. (1992) Effects of neurotensin on magnocellular cholinergic neurons from the nucleus basalis. *Soc. Neurosci. Abstr.* **18,** 1500.

Fong, H. K. W., Yoshimoto, K. K., Eversole-Cire, P., and Simon, M. I. (1988) Identification of a GTP-binding protein α subunit that lacks an apparent ADP-ribosylation site for pertussis toxin. *Proc. Natl. Acad. Sci. USA* **85,** 3066–3070.

Gilman, A.G. (1987) G proteins: transducers of receptor-generated signals. *Annu. Rev. Biochem.* **56,** 615–649.

Grigg, J. J., Koyano, K., Nakajima, S., and Nakajima, Y. (1991a) Peptide-induced reduction in calcium currents in rat nucleus basalis and locus ceruleus neurons. *Biophys. J.* **59,** 83a.

Grigg, J. J., Kozasa, T., Nakajima, S., and Nakajima, Y. (1991b) Single channel properties of somatostatin- or met-enkephalin-activated, GTP-dependent inwardly rectifying K-conductance in cultured rat brain neurons. *Soc. Neurosci. Abstr.* **17,** 1097.

Grigg, J. J., Kozasa, T., Nakajima, Y., and Nakajima, S. (1992) Single channel properties of the inwardly rectifying K-current activated by somatostatin or met-enkephalin in cultured rat brain neurons. *Biophys. J.* **61,** A14.

Gutowski, S., Smrcka, A., Nowak, L., Wu, D., Simon, M., and Sternweis, P. C. (1991) Antibodies to the α_q subfamily of guanine nucleotide-binding regulatory protein α subunits attenuate activation of phosphatidylinositol 4,5-bisphosphate hydrolysis by hormones. *J. Biol. Chem.* **266,** 20,519–20,524.

Guyenet, P. G. and Aghajanian, G. K. (1977) Excitation of neurons in the nucleus locus coeruleus by substance P and related-peptides. *Brain Res.* **136,** 178–184.

Hagiwara, S. (1983) *Membrane Potential-Dependent Ion Channels in Cell Membrane. Phylogenetic and Developmental Approaches,* Raven Press, New York.

Hanani, M. and Burnstock, G. (1984) Substance P evokes slow and fast responses in cultured myenteric neurons of the guinea pig. *Neurosci. Lett.* **48,** 19–23.

Hanley, M. R., Lee, C. M., Jones, L. M., and Michell, R. H. (1980) Similar effects of substance P and related peptides on salivation and on phosphatidylinositol turnover in rat salivary glands. *Mol. Pharmacol.* **18,** 78–83.

Henry, J. L. (1976) Effects of substance P on functionally identified units in cat spinal cord. *Brain Res.* **114,** 439–451.

Henry, J. L., Krnjevic, K., and Morris, M. E. (1975) Substance P and spinal neurones. *Can. J. Physiol. Pharmacol.* **53,** 423–432.

Hershey, A. D. and Krause, J. E. (1990) Molecular characterization of a functional cDNA encoding the rat substance P receptor. *Science* **247,** 958–962.

Higashida, H. and Brown, D. A. (1986) Two polyphosphatidylinositide metabolites control two K^+ currents in a neuronal cell. *Nature* **323,** 333–335.

Hille, B. (1992a) *Ionic Channels of Excitable Membranes,* Sinauer Assoc., Sunderland, MA.

Hille, B. (1992b) G protein-coupled mechanisms and nervous signaling. *Neuron* **9,** 187–195.

House, C. and Kemp, B. E. (1987) Protein kinase C contains a pseudosubstrate prototope in its regulatory domain. *Science* **238**, 1726–1728.

Ikeda, S. R. and Schofield, G. G. (1989) Somatostatin blocks a calcium current in rat sympathetic ganglion neurones. *J. Physiol.* **409,** 221–240.

Inoue, M., Nakajima, S., and Nakajima, Y. (1988) Somatostatin induces an inward rectification in rat locus coeruleus neurones through a pertussis toxin-sensitive mechanism. *J. Physiol.* **407,** 177–198.

Ito, H., Tung, R. T., Sugimoto, T., Kobayashi, I., Takahashi, K., Katada, T., Ui, M., and Kurachi, Y. (1992) On the mechanism of G protein βγ subunit activation of the muscarinic K+ channel in guinea pig atrial cell membrane. *J. Gen. Physiol.* **99,** 961–983.

Iyengar, R., Rich, K. A., Herberg, J. T., Grenet, D., Mumby, S., and Codina, J. (1987) Identification of a new GTP-binding protein. *J. Biol. Chem.* **262,** 9239–9245.

Jan, L. Y. and Jan, Y. N. (1982) Peptidergic transmission in sympathetic ganglia of the frog. *J. Physiol.* **327,** 219–246.

Jones, S. W. (1985) Muscarinic and peptidergic excitation of bull-frog sympathetic neurones. *J. Physiol.* **366,** 63–87.

Kasai, H. (1992) Voltage- and time-dependent inhibition of neuronal calcium channels by a GTP-binding protein in a mammalian cell line. *J. Physiol.* **448,** 189–209.

Katada, T., Oinuma, M., Kusakabe, K., and Ui, M. (1987) A new GTP-binding protein in brain tissues serving as the specific substrate of islet-activating protein, pertussis toxin. *FEBS Lett.* **213,** 353–358.

Katada, T. and Ui, M. (1982) Direct modification of the membrane adenylate cyclase system by islet-activating protein due to ADP-ribosylation of a membrane protein. *Proc. Natl. Acad. Sci. USA* **79,** 3129–3133.

Katayama, Y. and North, R. A. (1978) Does substance P mediate slow synaptic excitation within the myenteric plexus? *Nature* **274,** 387,388.

Katayama, Y., North, R. A., and Williams, J. T. (1979) The action of substance P on neurons of the myenteric plexus of the guinea-pig small intestine. *Proc. R. Soc. Lond. B* **226,** 191–208.

Katz, B. (1949) Les constantes électriques de la membrane du muscle. *Arch. Sci. Physiol.* **3,** 285–299.

Kawamoto, S. and Hidaka, H. (1984) 1-(5-Isoquinolinesulfonyl)-2-methylpiperazine (H-7) is a selective inhibitor of protein kinase C in rabbit platelets. *Biochem. Biophys. Res. Comm.* **125,** 258–264.

Kaziro, Y., Itoh, H., Kozasa, T., Nakafuku, M., and Satoh, T. (1991) Structure and function of signal-transducing GTP-binding proteins. *Annu. Rev. Biochem.* **60,** 349–400.

Kim, D., Lewis, D. L., Graziadei, L., Neer, E. J., Bar-Sagi, D., and Clapham, D. E. (1989) G-protein βγ-subunits activate the cardiac muscarinic K^+-channel via phospholipase A_2. *Nature* **337,** 557–560.

Kim, K-M., Nakajima, Y., and Nakajima, S. (1990) Effects of dopaminergic agonists on cultured substantia nigra neurons. *Soc. Neurosci. Abstr.* **16,** 81.

Konishi, S. and Okamoto, T. (1985) The relationship between substance P-mediated synaptic response and intracellular cyclic nucleotide formation in the guinea-pig inferior mesenteric ganglion. *Neurosci. Res. Suppl.* **1,** S102.

Konishi, S. and Otsuka, M. (1971) Actions of certain polypeptides on frog spinal neurons. *Jpn.. J. Pharmacol.* **21,** 685–687.

Konishi, S. and Otsuka, M. (1974) Excitatory action of hypothalamic substance P on spinal motoneurones of newborn rats. *Nature* **252,** 734–735.

Konishi, S., Otsuka, M., Folkers, K., and Rosell, S. (1983) A substance P antagonist blocks non-cholinergic slow excitatory postsynaptic potential in guinea-pig sympathetic ganglia. *Acta Physiol. Scand.* **117,** 157–160.

Konishi, S., Song, S.-Y., Ogawa, T., and Kanazawa, I. (1992) Fast and slow depolarizations produced by substance P and other tachykinins in sympathetic neurons of rat prevertebral ganglia. *Neurosci. Res.* **14,** 81–95.

Konishi, S., Tsunoo, A., and Otsuka, M. (1979) Substance P and noncholinergic excitatory synaptic transmission in guinea pig sympathetic ganglia. *Proc. Jpn. Acad.* **55,** 525–530.

Koyano, K., Grigg, J. J., Nakajima, S., and Nakajima, Y. (1990) Role of arachidonic acid metabolism in modulating the inward rectifier of brain neurons. *Biophys. J.* **57,** 314a.

Koyano, K., Velimirovic, B. M., Grigg, J. J., Nakajima, S., and Nakajima, Y. (1993) Two signal transduction mechanisms of substance P-induced depolarization in locus coeruleus neurons. *Eur. J. Neurosci.* **5,** 1189–1197.

Koyano, K., Velimirovic, B., Grigg, J. J., Nakajima, Y., and Nakajima, S. (1991) Substance P modulates two ionic currents through different transduction pathways in locus coeruleus neurons. *Biophys. J.* **59,** 391a.

Krnjevic, K. (1977) Effects of substance P on central neurons in cats, in *Substance P* (von Euler, U.S. and Pernow, B., eds.), Raven Press, New York, pp. 217–230.

Krnjevic, K. and Lekic, D. (1977) Substance P selectively blocks excitation of Renshaw cell by acetylcholine. *Can. J. Physiol. Pharmacol.* **55,** 958–961.

Krnjevic, K. and Morris, M. E. (1974) An excitatory action of substance P on cuneate neurones. *Can. J. Physiol. Pharmacol.* **52,** 736–744.

Kuba, K. and Koketsu, K. (1976) Analysis of the slow excitatory postsynaptic potential in bullfrog sympathetic ganglion cells. *Jpn. J. Physiol.* **26,** 651–669.

Kuffler, S. W. and Sejnowski, T. J. (1983) Peptidergic and muscarinic excitation at amphibian sympathetic synapses. *J. Physiol.* **341,** 257–278.

Kurachi, Y., Ito, H., Sugimoto, T., Shimizu, T., Miki, I., and Ui, M. (1989) Arachidonic acid metabolites as intracellular modulators of the G protein-gated cardiac K^+ channel. *Nature* **337,** 555–557.

Kurachi, Y., Nakajima, T., and Sugimoto, T. (1986) On the mechanism of activation of muscarinic K^+ channels by adenosine in isolated atrial cells: Involvement of GTP- binding proteins. *Pflüger's Arch.* **407,** 264–274.

Laniyonu, A., Sliwinski-Lis, E., and Fleming, N. (1988) Different tachykinin receptor subtypes are coupled to the phosphoinositide or cyclic AMP signal transduction pathways in rat submandibular cells. *FEBS Lett.* **240,** 186–190.

Lee, C. H., Park, D., Wu, D., Rhee, S. G., and Simon, M. I. (1992) Members of the G_q α subunit gene family activate phospholipase C β isoenzymes. *J. Biol. Chem.* **267,** 16,044–16,047.

Lipscombe, D., Kongsamut, S., and R. W. Tsien (1989) α-adrenergic inhibition of sympathetic neurotransmitter release mediated by modulation of N-type calcium-channel gating. *Nature* **340,** 639–642.

Livett, B. G., Kozousek, V., Mizobe, F., and Dean, D. M. (1979) Substance P inhibits nicotinic activation of chromaffin cells. *Nature* **278,** 256,257.

Logothetis, D. E., Kurachi, Y., Galper, J., Neer, E. J., and Clapham, D. E. (1987) The $\beta\gamma$ subunits of GTP-binding proteins activate the muscarinic K^+ channel in heart. *Nature* **325,** 321–326.

Lopez, H. S. and Adams, P. R. (1989) A G protein mediates the inhibition of the voltage-dependent potassium M current by muscarine, LHRH, substance P and UTP in bullfrog sympathetic neurons. *Eur. J. Neurosci.* **1,** 529–542.

Luini, A., Lewis, D., Guild, S., Schofield, G., and Weight, F. (1986) Somatostatin, an inhibitor of ACTH secretion, decreases cytosolic free calcium and voltage-dependent calcium current in a pituitary cell line. *J. Neurosci.* **6,** 3128–3132.

Malenka, R. C., Madison, D. V., Andrade, R., and Nicoll, R. A. (1986) Phorbol esters mimic some cholinergic actions in hippocampal pyramidal neurons. *J. Neurosci.* **6,** 475–480.

Marchetti, C. Carbone, E., and Lux, H. D. (1986) Effects of dopamine and noradrenaline on Ca channels of cultured sensory and sympathetic neurons of chick. *Pflüger's Arch.* **406,** 104–111.

Margiotta, J. F. and Berg, D. K. (1986) Enkephalin and substance P modulate synaptic properties of chick ciliary ganglion neurons in cell culture. *Neuroscience* **18,** 175–182.

Marrion, N. V., Adams, P. R., and Gruner, W. (1992) Multiple kinetic states underlying macroscopic M-currents in bullfrog sympathetic neurons. *Proc. R. Soc. Lond. B* **248,** 207–214.

Matsuda, H., Saigusa, A., and Irisawa, H. (1987) Ohmic conductance through the inwardly rectifying K channel and blocking by internal Mg^{2+}. *Nature* **325,** 156–159.

Matsuoka, M., Itoh, H., Kozasa, T., and Kaziro, Y. (1988) Sequence analysis of cDNA and genomic DNA for a putative pertussis toxin-insensitive guanine nucleotide-binding regulatory protein subunit. *Proc. Natl. Acad. Sci. USA* **85,** 5384–5388.

Matthews, G., Neher, E., and Penner, R. (1989a) Second messenger-activated calcium influx in rat peritoneal mast cells. *J. Physiol.* **418,** 105–130.

Matthews, G., Neher, E., and Penner, R. (1989b) Chloride conductance activated by external agonists and internal messengers in rat peritoneal mast cells. *J. Physiol.* **418,** 131–144.

Mau, S. E. and Saermark, T. (1991) Substance P stimulation of polyphosphoinositide hydrolysis in rat anterior pituitary membranes involves a GTP-dependent mechanism. *J. Endocrin.* **130,** 63-70.

Mayer, E. A., Loo, D. D. F., Snape, W. J., Jr., and Sachs, G. (1990) The activation of calcium and calcium-activated potassium channels in mammalian colonic smooth muscle by substance P. *J. Physiol.* **420,** 47–71.

Mihara, S., North, R. A., and Surprenant, A. (1987) Somatostatin increases an inwardly rectifying potassium conductance in guinea-pig submucous plexus neurones. *J. Physiol.* **390,** 335–355.

Miyake, M., Christie, M. J., and North, R. A. (1989) Single potassium channels opened by opioids in rat locus ceruleus neurons. *Proc. Natl. Acad. Sci. USA* **86,** 3419–3422.

Mousli, M., Bronner, C., Bockaert, J., Rouot, B., and Landry, Y. (1990) Interaction of SP, compound 48/80 and mastoparan with the α-subunit C-terminus of G protein. *Immunology Lett.* **25,** 355–358.

Murase, K., Ryu, P. D., and Randic, M. (1986) Substance P augments a persistent slow inward calcium-sensitive current in voltage-clamped spinal dorsal horn neurons of the rat. *Brain Res.* **365,** 369–376.

Murase, K., Ryu, P. D., and Randic, M. (1989) Tachykinins modulate multiple ionic conductances in voltage-clamped rat spinal dorsal horn neurons. *J. Neurophysiol.* **61,** 854–865.

Nakajima, T., Sugimoto, T., and Kurachi, Y. (1991) Platelet-activating factor activates cardiac G_K via arachidonic acid metabolites. *FEBS Lett.* **289,** 239–243.

Nakajima, Y., Nakajima, S., and Inoue, M. (1988) Pertussis toxin-insensitive G protein mediates substance P-induced inhibition of potassium channels in brain neurons. *Proc. Natl. Acad. Sci. USA* **85,** 3643–3647.

Nakajima, Y., Nakajima, S., and Inoue, M. (1991a) Substance P induced inhibition of potassium channels via a pertussis toxin-insensitive G protein, in *Substance P and Related Peptides: Cellular and Molecular Physiology. The Annals of the New York Academy of Sciences* **632** (Leeman, S. E., Krause, J. E., and Lembeck, F., eds.), NY Academy of Sciences, New York, pp. 103–111.

Nakajima, Y., Nakajima, S., Obata, K., Carlson, C. G., and Yamaguchi, K. (1985) Dissociated cell culture of cholinergic neurons from nucleus basalis of Meynert and other basal forebrain nuclei. *Proc. Natl. Acad. Sci. USA* **82,** 6325–6329.

Nakajima, Y., Stanfield, P.R., Yamaguchi, K. and Nakajima, S. (1991b) Substance P excites cultured cholinergic neurons in the basal forebrain, in *The Basal Forebrain: Anatomy to Function* (Napier, T.C., Kalivas, P.W., and Hanin, I., eds.), Plenum, New York, pp. 157–167.

Neher, E., Marty, A., Fukuda, K., Kubo, T., and Numa, S. (1988) Intracellular calcium release mediated by two muscarinic receptor subtypes. *FEBS Lett.* **240,** 88–94.

Nestler, E. J. and Greengard, P. (1984) *Protein Phosphorylation in the Nervous System,* Wiley, New York.

Nicoll, R. A. (1978) The action of thyrotropin-releasing hormone, substance P and related peptides on frog spinal motoneurons. *J. Pharmacol. Exp. Ther.* **207,** 817–824.

Nicoll, R. A., Schenker, C., and Leeman, S. E. (1980) Substance P as a transmitter candidate. *Annu. Rev. Neurosci.* **3,** 227–268.

North, R. A. (1989) Drug receptors and the inhibition of nerve cells. *Br. J. Pharmacol.* **98,** 13–28.

North, R. A. and Uchimura, N. (1989) 5-Hydroxytryptamine acts at 5-HT_2 receptors to decrease potassium conductance in rat nucleus accumbens neurones. *J. Physiol.* **417,** 1–12.

North, R. A., Williams, J. T., Surprenant, A., and Christie, M. J. (1987) μ and δ receptors belong to a family of receptors that are coupled to potassium channels. *Proc. Natl. Acad. Sci. USA* **84,** 5487–5491.

Nowak, L. M. and Macdonald, R. L. (1982) Substance P: Ionic basis for depolarizing responses of mouse spinal cord neurons in cell culture. *J. Neurosci.* **2,** 1119–1128.

Otsuka, M. and Konishi, S. (1977) Electrophysiological and neurochemical evidence for substance P as a transmitter of primary sensory neurons, in *Substance P* (von Euler, U.S. and Pernow, B., eds.), Raven Press, New York, pp. 207–216.

Otsuka, M., Konishi, S., and Takahashi, T. (1972) A further study of the motoneuron-depolarizing peptide extracted from dorsal roots of bovine spinal nerves. *Proc. Jpn. Acad.* **48,** 747–752.

Otsuka, M. and Takahashi, T. (1977) Putative peptide neurotransmitters. *Annu. Rev. Pharmacol. Toxicol.* **17,** 425–439.

Owen, D. G., Marsh, S. J., and Brown, D. A. (1990) M-current noise and putative M-channels in cultured rat sympathetic ganglion cells. *J. Physiol.* **431,** 269–290.

Petersen, O. H. and Maruyama, Y. (1984) Calcium-activated potassium channels and their role in secretion. *Nature* **307,** 693–696.

Pfaffinger, P. (1988) Muscarine and t-LHRH suppress M-current by activating an IAP-insensitive G-protein. *J. Neurosci.* **8,** 3343–3353.

Pfaffinger, P. J., Leibowitz, M. D., Subers, E. M., Nathanson, N. M., Almers, W., and Hille, B. (1988) Agonists that suppress M-current elicit phosphoinositide turnover and Ca^{2+} transients, but these events do not explain M-current suppression. *Neuron* **1,** 477–484.

Pfaffinger, P. J., Martin, J. M., Hunter, D. D., Nathanson, N. M., and Hille, B. (1985) GTP-binding proteins couple cardiac muscarinic receptors to a K channel. *Nature* **317,** 536–538.

Piomelli, D., Volterra, A., Dale, N., Siegelbaum, S. A., Kandel, E. R., Schwartz, J. H., and Belardetti, F. (1987) Lipoxygenase metabolites of arachidonic acid as second messengers for presynaptic inhibition of *Aplysia* sensory cells. *Nature* **328,** 38–43.

Putney, J. W. Jr., Bird, G. St. J., Horstman, D. A., Hughes, A. R., Menniti, F. S., Nogimori, K., Obie, J., Oliver, K. G., Sugiya, H., and Takemura, H. (1991) Role of inositol phosphates in the actions of substance P on NK_1 receptors in exocrine gland cells, in *Substance P and Related Peptides: Cellular and Molecular Physiology. The Annals of The New York Academy of Sciences* **632** (Leeman, S. E., Krause, J. E., and Lembeck, F., eds.), NY Academy of Sciences, New York, pp. 94–102.

Ramirez, O. A. Chiappinelli, V. A. (1987) Properties of tachykinin receptors examined by intracellular recording from chick sympathetic ganglia. *Brain Res.* **414,** 228–238.

Randic, M. and Miletic, V. (1977) Effect of substance P in cat dorsal horn neurones activated by noxious stimuli. *Brain Res.* **128,** 164–169.

Rane, S. G., Walsh, M. P., McDonald, J. R., and Dunlap, K. (1989) Specific inhibitors of protein kinase C block transmitter-induced modulation of sensory neuron calcium current. *Neuron* **3,** 239–245.
Rangachari, P. K., Prior, T., and McWade, D. (1990) Epithelial and mucosal preparations from canine colon: responses to substance P. *J. Pharmacol. Exp. Ther.* **254,** 1076–1083.
Role, L. W. (1984) Substance P modulation of acetylcholine-induced currents in embryonic chicken sympathetic and ciliary ganglion neurons. *Proc. Natl. Acad. Sci. USA* **81,** 2924–2928.
Role, L. W., Leeman, S. E., and Perlman, R. L. (1981) Somatostatin and substance P inhibit catecholamine secretion from isolated cells of guinea-pig adrenal medulla. *Neuroscience* **6,** 1813–1821.
Ryu, P. D. and Randic, M. (1990) Low- and high-voltage-activated calcium currents in rat spinal dorsal horn neurons. *J. Neurophysiol.* **63,** 273–275.
Sakmann, B., Noma, A., and Trautwein, W. (1983) Acetylcholine activation of single muscarinic K^+ channels in isolated pacemaker cells of the mammalian heart. *Nature* **303,** 250–253.
Scherer, R. W. and Breitwieser, G. E. (1990) Arachidonic acid metabolites alter G protein-mediated signal transduction in heart. Effects on muscarinic K^+ channels. *J. Gen. Physiol.* **96,** 735–755.
Schultz, G., Rosenthal, W., Hescheler, J., and Trautwein, W. (1990) Role of G proteins in calcium channel modulation. *Annu. Rev. Physiol.* **52,** 275–292.
Schumann, M. A. and Gardner, P. (1989) Modulation of membrane K^+ conductance in T-lymphocytes by substance P via a GTP-binding protein. *J. Membrane Biol.* **111,** 133–139.
Schweitzer, P., Madamba, S., and Siggins, G.R. (1990) Arachidonic acid metabolites as mediators of somatostatin-induced increase of neuronal M-current. *Nature* **346,** 464–467.
Selyanko, A. A., Stansfeld, C. E., and Brown, D. A. (1992) Closure of potassium M-channels by muscarinic acetylcholine-receptor stimulants requires a diffusible messenger. *Proc. R. Soc. Lond. B* **250,** 119–125.
Shapiro, M. S. and Hille, B. (1993) Substance P and somatostatin inhibit calcium channels in rat sympathetic neurons via different G protein pathways. *Neuron* **10,** 11–20.
Shen, K.-Z. and North, R.A. (1992) Substance P opens cation channels and closes potassium channels in rat locus coeruleus neurons. *Neuroscience* **50,** 345–353.
Simasko S. M., Durkin, J. A., and Weiland, G. A. (1987) Effects of substance P on nicotinic acetylcholine receptor function in PC12 cells. *J. Neurochem.* **49,** 253–260.
Simmons, L. K., Schuetze, S. M., and Role, L. W. (1990) Substance P modulates single-channel properties of neuronal nicotinic acetylcholine receptors. *Neuron* **2,** 393–403.
Simmons, M. A. and Mather, R. J. (1991) Selectivity of the effects of guanosine-5'-0-(2-thiodiphosphate) on agonist inhibition of the M-current in amphibian sympathetic neurons. *J. Neurosci.* **11,** 2130–2134.

Simon, M. I., Strathmann, M. P., and Gautam, N. (1991) Diversity of G proteins in signal transduction. *Science* **252,** 802–808.

Sims, S. M., Walsh, J. V., Jr., and Singer, J. J. (1986) Substance P and acetylcholine both suppress the same K^+ current in dissociated smooth muscle cells. *Amer. J. Physiol.* **251,** C580–C587.

Smrcka, A. V., Hepler, J. R., Brown, K. O., and Sternweis, P. C. (1991) Regulation of polyphosphoinositide-specific phospholipase C activity by purified G_q. *Science* **251,** 804–807.

Soejima, M. and Noma, A. (1984) Mode of regulation of the ACh-sensitive K-channel by the muscarinic receptor in rabbit atrial cells. *Pflüger's Arch.* **400,** 424–431.

Song, S.-Y., Iwashita, S., Noguchi, K., and Konishi, S. (1988) Inositol triphosphate-linked calcium mobilization couples substance P receptors to conductance increase in a rat pancreatic acinar cell line. *Neurosci. Lett.* **95,** 143–148.

Stallcup, W. B. and Patrick, J. (1980) Substance P enhances cholinergic receptor desensitization in a clonal nerve cell line. *Proc. Natl. Acad. Sci. USA* **77,** 634–638.

Stanfield, P. R., Nakajima, Y., and Yamaguchi, K. (1985) Substance P raises neuronal membrane excitability by reducing inward rectification. *Nature* **315,** 498–501.

Stanfield, P. R., Standen, N. B., Leech, C. A., and Ashcroft, F. M. (1981) Inward rectification in skeletal muscle fibres, in *Adv. Physiol. Sci.*. vol. 5: *Molecular and Cellular Aspects of Muscle Function* (Varger, E., Kövér, A., Kovács, T., and Kovács, L., eds.), Pergamon Press, New York, pp. 247–262.

Strathmann, M. and Simon, M. I. (1990) G-protein diversity: A distinct class of α subunits is present in vertebrates and invertebrates. *Proc. Natl. Acad. Sci. USA* **87,** 9113–9117.

Sun, X. P., Supplisson, S., Torres, R., Sachs, G., and Mayer, E. (1992) Characterization of large-conductance chloride channels in rabbit colonic smooth muscle. *J. Physiol.* **448,** 355–382.

Surprenant, A. and North, R. A. (1988) Mechanism of synaptic inhibition by noradrenaline acting at α_2-adrenoceptors. *Proc. R. Soc. Lond. B.* **234,** 85–114.

Tamaoki, T., Nomoto, H., Takahashi, I., Kato, Y., Morimoto, M., and Tomita, F. (1986) Staurosporine, a potent inhibitor of phospholipid/Ca^{++} dependent protein kinase. *Biochem. Biophys. Res. Comm.* **135,** 397–402.

Taylor, C. W., Merritt, J. E., Putney, J. W., Jr., and Rubin, R. P. (1986) A guanine nucleotide-dependent regulatory protein couples substance P receptors to phospholipase C in rat parotid gland. *Biochem. Biophys. Res. Comm.* **136,** 362–368.

Taylor, S. J., Chae, H. Z., Rhee, S. G., and Exton, J. H. (1991) Activation of the β1 isozyme of phospholipase C by α subunits of the G_q class of G proteins. *Nature* **350,** 516–518.

Terry, R. D. and Katzman, R. (1983) Senile dementia of the Alzheimer type. *Ann. Neurol.* **14,** 497–506.

Tokimasa, T., Ishimatsu, M., and Akasu, T. (1992) Substance-P modulates separate set of potassium currents in bullfrog primary afferent neurons. *Soc. Neurosci. Abstr.* **18,** 796.

Tregear, G. W., Niall, H. D., Potts, J. T. Jr., Leeman, S. E., and Chang, M. M. (1971) Synthesis of substance P. *Nature New Biol.* **232,** 87–89.

Trussell, L. O. and Jackson, M. B. (1985) Adenosine-activated potassium conductance in cultured striatal neurons. *Proc. Natl. Acad. Sci. USA* **82,** 4857–4861.

Trussell, L. O. and Jackson, M. B. (1987) Dependence of an adenosine-activated potassium current on a GTP-binding protein in mammalian central neurons. *J. Neurosci.* **7,** 3306–3316.

Tsuji, S., Minota, S., and Kuba, K. (1987) Regulation of two ion channels by a common muscarinic receptor-transduction system in a vertebrate neuron. *Neurosci. Lett.* **81,** 139–145.

Tsunoo, A., Yoshii, M., and Narahashi, T. (1986) Block of calcium channels by enkephalin and somatostatin in neuroblastoma-glioma hybrid NG108-15 cells. *Proc. Natl. Acad. Sci. USA* **83,** 9832–9836.

Vandenberg, C. A. (1987) Inward rectification of a potassium channel in cardiac ventricular cells depends on internal magnesium ions. *Proc. Natl. Acad. Sci. USA* **84,** 2560–2564.

Velimirovic, B., Koyano, K., Nakajima, S., and Nakajima, Y. (1991) Substance P, somatostatin and met-enkephalin regulate the same K-channel in cultured noradrenergic neurons from the locus ceruleus. *Soc. Neurosci. Abstr.* **17,** 1474.

Vincent, J.-D. and Barker, J. L. (1979) Substance P: evidence for diverse roles in neuronal function from cultured mouse spinal neurons. *Science* **205,** 1409–1412.

Wanke, E., Ferroni, A., Malgaroli, A., Ambrosini, A., Pozzan, T., and Meldolesi, J. (1987) Activation of a muscarinic receptor selectively inhibits a rapidly inactivated Ca^{2+} current in rat sympathetic neurons. *Proc. Natl. Acad. Sci. USA* **84,** 4313–4317.

West, R. E. Jr., Moss, J., Vaughan, M., Liu, T., and Liu, T.-Y. (1985) Pertussis toxin-catalyzed ADP-ribosylation of transducin. *J. Biol. Chem.* **260,** 14,428–14,430.

Wilkie, T. M., Scherle, P. A., Strathmann, M. P., Slepak, V. Z., and Simon, M. I. (1991) Characterization of G protein α subunits in the G_q class: Expression in murine tissues and in stromal and hematopoietic cell lines. *Proc. Natl. Acad. Sci. USA* **88,** 10,049–10,053.

Willard, A. L. and Nishi, R. (1985) Neurons dissociated from rat myenteric plexus retain differentiated properties when grown in cell culture. *Neuroscience* **16,** 201–211.

Williams, J. T., North, R. A., and Tokimasa, T. (1988) Inward rectification of resting and opiate-activated potassium currents in rat locus coeruleus neurons. *J. Neurosci.* **8,** 4299–4306.

Womack, M. D., MacDermott, A. B., and Jessell, T. M. (1988) Sensory transmitters regulate intracellular calcium in dorsal horn neurons. *Nature* **334,** 351–353.

Wu, D., Lee, C. H., Rhee, S. G., and Simon, M. I. (1992) Activation of phospholipase C by the α subunits of the G_q and G_{11} proteins in transfected Cos-7 cells. *J. Biol. Chem.* **267,** 1811–1817.

Yakel, J. L., Trussell, L. O., and Jackson, M. B. (1988) Three serotonin responses in cultured mouse hippocampal and striatal neurons. *J. Neurosci.* **8,** 1273–1285.

Yamaguchi, K., Nakajima, Y., Nakajima, S., and Stanfield, P. R. (1990) Modulation of inwardly rectifying channels by substance P in cholinergic neurones from rat brain in culture. *J. Physiol.* **426,** 499–520.

Yang, J. and Tsien, R. W. (1992) Enhancement of neuronal N- and L-type Ca^{2+} channel activity by protein kinase C. *Soc. Neurosci. Abstr.* **18,** 1271.

Yatani, A., Codina, J., Brown, A. M., and Birnbaumer, L. (1987) Direct activation of mammalian atrial muscarinic potassium channels by GTP regulatory protein G_K. *Science* **235,** 207–211.

Yokota, Y., Sasai, Y., Tanaka, K., Fujiwara, T., Tsuchida, K., Shigemoto, R., Kakizuka, A., Ohkubo, H. and Nakanishi, S. (1989) Molecular characterization of a functional CDNA for rat substance P receptor. *J. Biol. Chem.* **264,** 17,649–17,652.

PART IV

TACHYKININ RECEPTOR FUNCTION

CHAPTER 12

Structure–Activity Relationships of Agonist and Antagonist Ligands

Scott L. Harbeson and Paolo Rovero

1. NK_1 Receptor Ligands

1.1. Receptor Agonists

1.1.1. Structure–Activity Relationships (SAR)

1.1.1.1. PEPTIDE LENGTH. The biological activity of substance P (SP) C-terminal partial sequences has been previously reported (Bury and Mashford, 1976). The minimum sequence that maintains significant biological activity in the isolated guinea pig ileum is the pentapeptide, SP_{7-11}. This concept has been recently confirmed by binding experiments to cloned rat NK_1, NK_2, and NK_3 receptors heterologously expressed in Chinese hamster ovary cells (Cascieri et al., 1992). C-terminal partial sequences of SP, neurokinin A (NKA), and neurokinin B (NKB) have also been tested on monoreceptorial preparations selective for each of the three tachykinin receptor types (Dion et al., 1987b). The results obtained with SP and its fragments on the isolated dog carotid artery clearly indicate that, at the NK_1 receptor, SP is more potent than its fragments, although these fragments main-

The Tachykinin Receptors Ed.: S. H. Buck © 1994 Humana Press Inc., Totowa, NJ

tain full agonist activity. Since some of the C-terminal fragments of NKA are more active than NKA itself at the NK_2 receptor (*see* Section 2.1.1.), the activity of the fragments has been used as a second criterion for the classification of the tachykinin receptors, in addition to the rank order of potency of natural ligands (Regoli et al., 1987). Thus, the three tachykinin receptor types are defined as follows:

NK_1: SP > NKA and > NKB.
The tachykinins are more active than their respective fragments.
NK_2: NKA > NKB > SP.
Some NKA fragments are more selective than their precursor.
NK_3: NKB > NKA > SP.
Fragments of NKB are less active than their precursor.

Several biological actions have been ascribed to N-terminal fragments of SP (*see,* for review, Maggi et al., 1993). The structure–activity relationships of these peptides will not be reported here. The tachykinin receptors do not bind N-terminal fragments of tachykinins; therefore, by definition, these biological actions do not involve activation of NK_1, NK_2, and NK_3 receptors (Cascieri et al., 1992).

1.1.1.2. Systematic Amino Acid Substitutions. A complete series of SP analogs was synthesized in 1979 in which each residue was systematically replaced by Ala (Couture et al., 1979). The reported biological activity (guinea pig ileum assay) indicates that the first five residues and Gly^9 can be replaced by Ala without change of activity, but the substitution of the residues in the C-terminal half of the peptide (namely, Phe^7, Phe^8, Leu^{10}, and Met^{11}) brings about variable losses of potency, the relative affinity being 0.5, 23, 26, and 10%, respectively, as compared to SP. All active analogs are full agonists.

More recently, the systematic replacement of each residue in the sequence of SP with the respective D-enantiomer has been performed (Duplaa et al., 1991). Binding affinities and biological activities have been measured in preparations representative of each of the three receptor types. For the NK_1 receptor, the results obtained with the Ala-analogs of SP were confirmed. Changing to the D-enantiomer in the N-terminal part of the peptide does not modify the pharmacological pattern of the analogs. Introduction of a D-residue in the 6–11 positions drastically decreases potency (20- to 40-fold as compared to SP in the guinea pig ileum), with the exception of [D-Leu^{11}]SP, which is only three times less potent than SP. The D-residue analogs have also been analyzed from a conformational point of view (Convert et al., 1991). The results of NMR experiments and energy calculations have been discussed with respect to the bioactive conformation of SP previously postulated by the same authors (Lavielle et al., 1988b). This model involves five anchoring points between SP and the NK_1 receptor:

1. The guanidinium of Arg^1;
2. The aromatic rings of Phe^7 and Phe^8; and
3. The sulfur atom and the carboxamide function of the C-terminal Met.

Replacement of one of the residues involved in the binding (1,7,8 or 11) by its enantiomer should therefore decrease the binding potency of the corresponding analog, as actually observed for [D-Phe^7]SP, [D-Phe^8]SP, and [D-Met^{11}]SP. On the other hand, inverting the chirality of the other residues (2–6, 9, and 10) will affect the backbone conformation; this may hamper to a lesser extent the binding of the anchoring points.

1.1.1.3. Amino Acid Substitutions That Enhance NK_1 Selectivity. As soon as the first evidence of tachykinin receptor heterogeneity began to accumulate, the search for a selective synthetic agonist was undertaken. A selective agonist, as compared to the natural tachykinins, would aid in the confirmation of the three receptor hypothesis. The first SP analog reported to display some tissue selectivity was its C-terminal methyl ester, SP-OMe (compound **1**, Fig. 1), which is as potent as SP on the guinea pig ileum and other NK_1 preparations, while being virtually inactive at the two other receptors (Cascieri et al., 1981; Watson et al., 1983). Subsequently, the work of several groups focused on the role of position 9 (Gly) in the C-terminal hexapeptide, SP_{6-11}. In fact, the chirality of a Pro residue introduced in place of Gly was found to strongly affect selectivity

Arg-Pro-Lys-Pro-Gln-Gln-Phe-Phe-Gly-Leu-Met-OMe
SP-OMe (1)

pGlu-Phe-Phe-Pro-Leu-Met-NH_2
Septide (2)

$H_2N-(CH_2)_4$-CO-Phe-Phe-Pro-(N-Me-Leu)-Met-NH_2
GR73632 (3)

Orn-Phe-Phe-Gly-Leu-Glu(δ–O-*t*-Bu)-NH_2
[Orn^6,Glu(δ-*t*-Bu)^{11}] $SP_{6\text{-}11}$ (4)

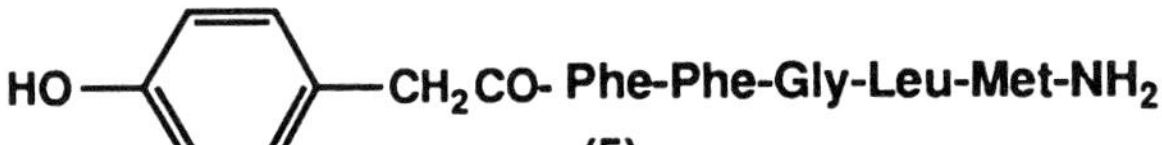

(5)

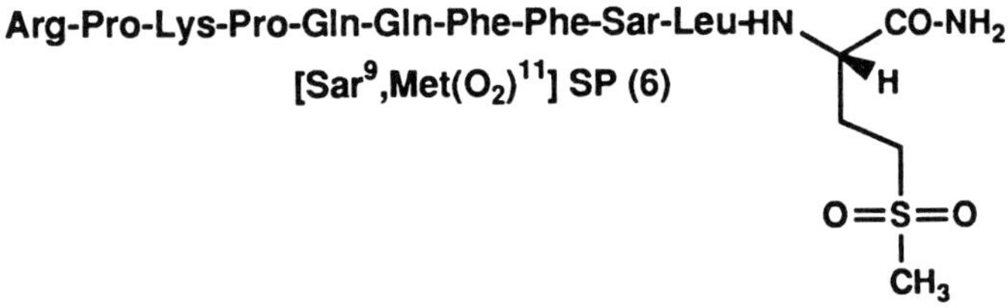

[Sar^9,$Met(O_2)^{11}$] SP (6)

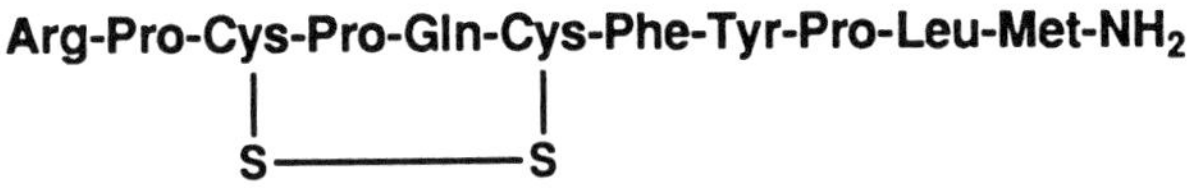

[$Cys^{3,6}$, Tyr^8, Pro^9] SP (7)

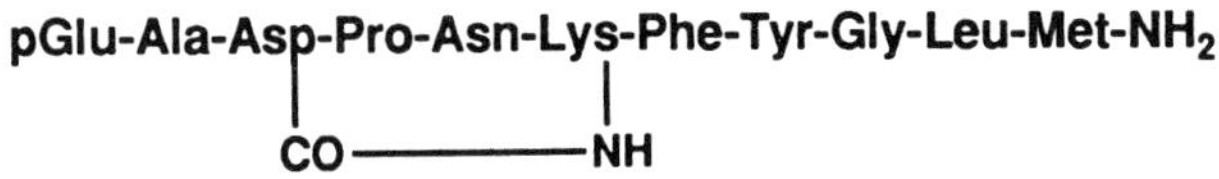

(8)

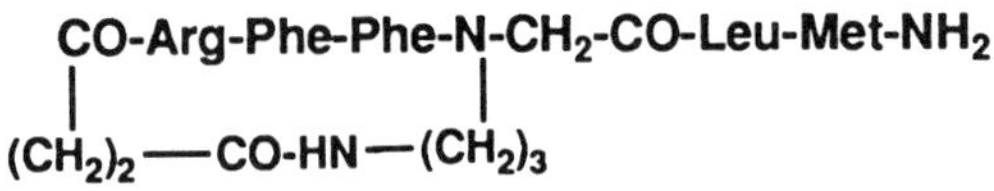

Cycloseptide (9)

Fig. 1. NK_1 selective agonists.

(Piercey et al., 1985). Indeed, both functional assays (Laufer et al., 1986) and binding data (Lee et al., 1986) indicate that [pGlu^6, Pro^9]SP$_{6–11}$ (**2**) displays good selectivity for the NK$_1$ receptor. This compound was designated "septide" (selective SP receptor peptide); water solubility was achieved by replacing the pGlu^6 residue by *N*-Ac-Arg, a modification that does not affect the biological activity. Recently, replacement of pGlu^6 by γ-aminovaleric acid (Ava) has been proposed in order to increase the metabolic stability of the synthetic analog. Additional *N*-methylation of Leu^{10} gives rise to a peptide, [Ava^6,Pro^9,*N*-Me-Leu^{10}]SP$_{6–11}$ (GR73632, compound **3**), that is endowed with good selectivity and enhanced enzymatic stability (Hagan et al., 1989). Still working on hexapeptidic sequences, Poulos et al. (1986) reported that [Orn^6]SP$_{6-11}$ is as active as the parent hexapeptide in the guinea pig ileum, being more soluble and not prone to contamination with the corresponding pyroglutamyl deriviative, as compared to SP$_{6-11}$. Subsequent work by the same group (Poulos et al., 1987; Karagiannis et al., 1991) on a class of [Orn^6]SP$_{6–11}$ analogs in which the Met^{11} residue is replaced by glutamate as its γ-alkylester indicates that when the thiomethyl ether of Met^{11} is replaced by a *t*-butyl ester (peptide **4**), the potency of the hexapeptide is enhanced at all three tachykinin receptors. Particularly, compound **4** is as active as SP-OMe at the NK$_1$ receptor. N-terminal acylated pentapeptide analogs have also been reported (Michelot et al., 1988; Tschirhart et al., 1989). One of the analogs, namely [*N*-(p-hydroxyphenylacetyl)]SP$_{7–11}$ (**5**), was reported to be more potent than SP at the NK$_1$ receptor (guinea pig ileum). Interestingly, most of the modified pentapeptides are quite potent at the NK$_2$ receptor, as tested in the guinea pig trachea (GPT), thus confirming the observation that shorter peptides interact better with the NK$_2$ receptor (Regoli et al., 1987). Working on both positions 9 and 11 of undecapeptidic sequences, Regoli and coworkers have reported that replacement of Gly^9 by Sar with concomitant oxidation of the thioether function of Met^{11} to the corresponding sulfone gives rise to an analog, namely [Sar^9, Met(O$_2$)^{11}]SP (**6**), that is as active as SP at the NK$_1$ receptor, while being inactive at the two other receptors (Drapeau et al., 1987a).

In conclusion, the most selective analogs of SP obtained by amino acid substitutions are characterized by modifications of either Gly^9 (with introduction of limited conformational constraint) or the C-terminal Met-NH_2 or both. These results strenghthen the proposed critical role of the C-terminus (common to all of the tachykinins) for receptor activation. The key residues for recognition and binding to the NK_1 receptor that are therefore required for agonist activity appear to be the aromatic residues, namely Phe^7 and Phe^8.

1.1.1.4. Conformationally Constrained Peptide Agonists. The observation that the Gly residue in position 9 of SP plays a crucial conformational role as a hinge between the two important aromatic residues (Phe^7-Phe^8) and the C-terminal dipeptide (Leu^{10}-Met^{11}-NH_2) prompted the development of the first conformationally constrained SP analogs. Indeed, the replacement of Gly^9 by Pro (as in septide and analogs) or by Sar (as in compound **6**) introduces limited conformational constraints in the peptide backbone, resulting in increased selectivity. The introduction of more rigid constraints, i.e., cyclization, has been attempted. Thus, on the basis of the bioactive conformation deduced by NMR and energy calculations (*see* Section 1.1.1.2.), Lavielle and coworkers reported a series of cyclic analogs of SP designed with the aim of stabilizing the three-dimensional feature observed, i.e., the α-helical structure of the Pro^4 to Phe^8 moiety (Lavielle et al., 1988b). Cyclization was achieved by introduction of a disulfide bridge; the best results were obtained using a Cys^3-Cys^6 bridge. Selectivity was improved by further replacement of Phe^8 by Tyr and Gly^9 by Pro, yielding [$Cys^{3,6}$,Tyr^8,Pro^9]SP (**7**), but the affinity for the NK_1 receptor was reduced fivefold as compared to SP. A similar peptide has been recently described which was obtained by cyclization of physalaemin (the amphibian counterpart of SP). Cyclization is achieved by formation of an amide bond between the side chains of Asp^3 and Lys^6 (compound **8**, Holzemann et al., 1991). The pharmacological evaluation, as performed on monoreceptorial preparations, indicates that biological activity is substantially maintained at the three receptor types, thus indicating that this kind of conformational constraint does not enhance the selectivity of physalaemin. A new approach to the synthesis of cyclic peptides, termed "backbone cyclization," and its application to the C-terminal hexapeptide of SP has been recently published (Gilon et al., 1991a,b). In order to preserve the side chains and the C-terminal carboxamide, which appear to be necessary for receptor activation, the constraint is achieved by linking the *N*-terminal amino group of position 6 of SP (where Gln^6 has been replaced by Arg in order to increase the water solubility) to the α-N of Gly^9 through an alkyl chain (compound **9**). The resulting cyclic peptide (termed "cycloseptide") maintains the same activity at the NK_1 receptor as its linear precursor and a selectivity comparable to that of the original septide, but appears to be more stable to metabolic degradation.

1.2. Receptor Antagonists

1.2.1. Peptide Antagonists

1.2.1.1. ANALOGS OF SUBSTANCE P. The most notable early SP antagonist that showed significant receptor affinity and some selectivity for the NK_1 receptor is spantide (compound **10**, Fig. 2) (Folkers et al., 1984). This peptide was reported to have a pA_2 value of 7.1 in guinea pig ileum. The activity of this peptide has been confirmed in numerous studies; however, the selectivity is poor in that a recent study has shown spantide to have a pA_2 value of 6.7 in guinea pig ileum, but also a pA_2 of 6.2 in rat portal vein, a tissue in which the NK_3 receptor predominates (Jukic et al., 1991). A more recent analog, spantide II (**11**), is a more potent antagonist ($pA_2 = 7.7$ vs SP in guinea pig taenia coli), although this antagonist still lacks NK_1 specificity (Folkers et al., 1990).

Spantide, spantide II, and a number of tachykinin antagonists derived from the sequence of SP share the common feature of D-Trp substitution for Phe^7 and Gly^9. A recent report has shown that these types of SP antagonists can be reduced in size to a tripeptide and still retain potent antagonist activity (FR113680, **12**) (Hagiwara et al., 1991; Jukic et al., 1991). **12** is one of the shortest SP antagonists known that still retains activity equivalent to longer peptides such as spantide and spantide II. This compound also shows similar selectivity for NK_1 vs the NK_2 and NK_3 receptors as the spantides; this observation would lead one to propose that FR113680 is binding in

D-Arg-Pro-Lys-Pro-Gln-Gln-D-Trp-Phe-D-Trp-Leu-Leu-NH_2

Spantide (10)

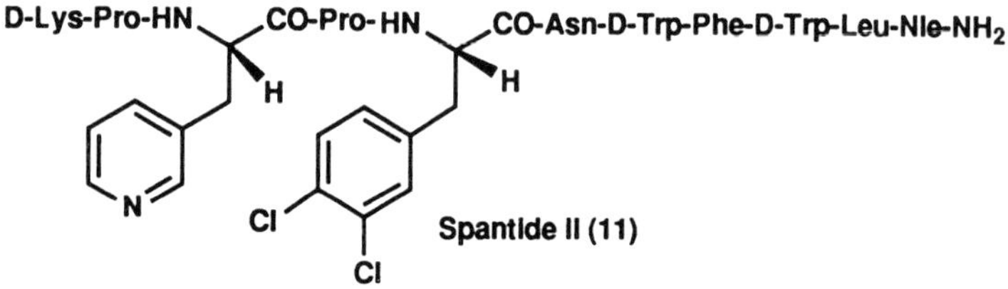

Spantide II (11)

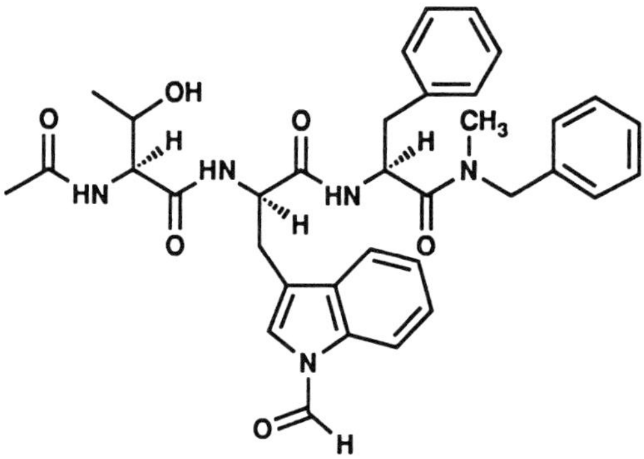

FR113680 (12)

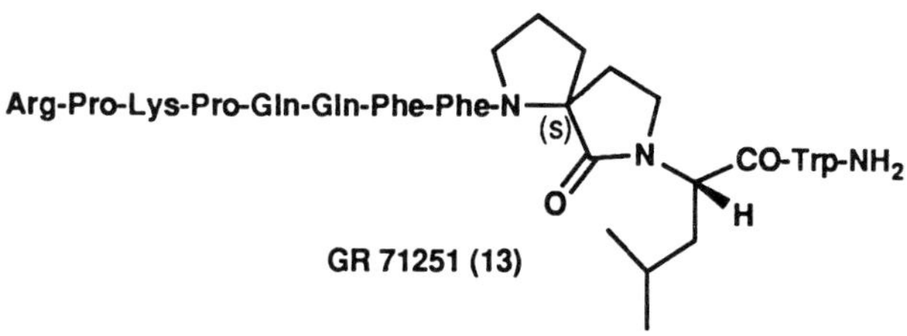

GR 71251 (13)

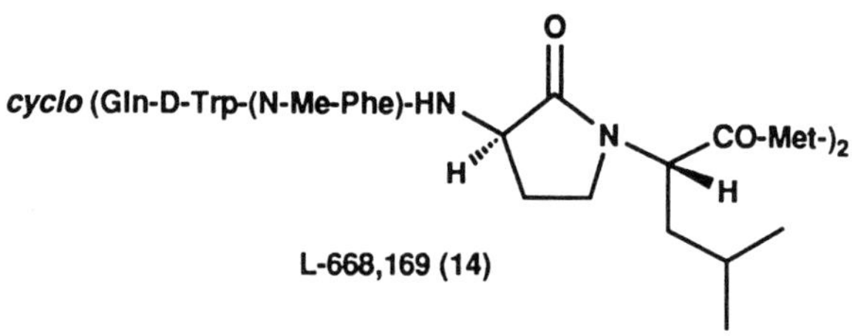

L-668,169 (14)

Fig. 2. SP antagonists.

a mode analogous to other D-Trp containing SP antagonists. The antagonists derived by D-Trp substitutions in the sequence of SP have been reviewed previously (Logan et al., 1991). Since SP, NKA, and NKB show activity at all of the tachykinin receptor classes, it is not surprising that D-Trp substitutions are not sufficient to yield receptor selective antagonists. More rigid conformational constraints coupled to other structural modifications are required to achieve potent and selective tachykinin antagonists.

1.2.1.2. CONFORMATIONALLY CONSTRAINED PEPTIDES. The combined use of more rigid conformational constraints and amino acid substitutions in the sequence of SP has yielded antagonists that do show marked selectivity for the NK_1 receptor. GR71251 (**13**) contains a spirolactam moiety that apparently diverts the peptide backbone into an antagonist conformation (Ward et al., 1990). These authors propose that the N-terminal residues bind in an analogous fashion to SP itself, but the spirolactam diverts the C-terminal residues into a nonactivating pocket of the receptor that favors bulky, hydrophobic residues. These conclusions are supported by the observation that compounds **13** and **13A** shown in Fig. 3 exhibit either agonist or antagonist activity dependent on the stereochemistry of the spirolactam moiety. The compound containing the R-spirolactam (**13A**) behaves as an agonist since its preferred extended conformation is proposed to position the C-terminus into a region of the receptor that causes signal transduction, whereas compound **13**, containing the S-spirolactam, is proposed to exist in a folded conformation that does not activate signal transduction. The differential binding of the C-terminus is also indicated by the fact that for GR71251, a C-terminal Trp is preferred, whereas this substitution for the C-terminal methionine is not tolerated in the agonist. These data further emphasize the proposed role of the N-terminus of SP for recognition and binding and the critical importance of the C-terminus for receptor activation. In this case, two classes of SAR data are important: the SAR of residues 1–8 in known agonists to enhance affinity and selectivity and the SAR of the C-terminus that will differ for agonists and antagonists since this region appears to be interacting at different sites on the receptor.

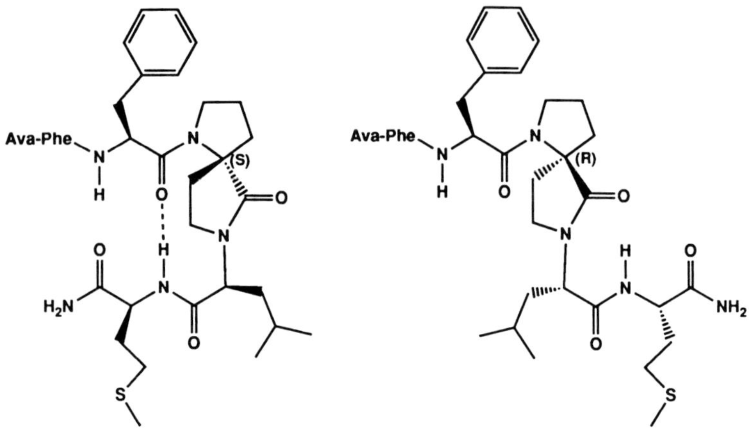

13: (S)-spirolactam (Antagonist) 13a: (R)-spirolactam (Agonist)

Fig. 3. Proposed antagonist and agonist conformations of spirolactam containing peptides (Ward et al., 1990).

As will be discussed in more detail in Section 2.2.2.1., a cyclic peptide NK_1 antagonist (compound **14**) has been prepared based on lactam analogs of SP (McKnight et al., 1988). It had been observed that the lactam constraint, as shown in compound **14** and compounds **28** and **29** (Fig. 4), decreased activity at the NK_1 receptor when inserted into SP peptides. This decrease in NK_1 activity is proposed to arise from the cyclic conformation that peptides containing this lactam constraint appear to adopt. As will be discussed in more detail, cyclic peptides do show decreased affinity for the NK_1 receptor with respect to the NK_2 receptor; however, L-668,169 (**14**) is a potent and selective NK_1 antagonist (pA_2 = 7.0 vs eledoisin in guinea pig ileum). This compound shows significantly weaker activity as either an NK_2 antagonist (pA_2 = 5.1 vs eledoisin in rat vas deferens) or an NK_3 antagonist (pA_2 = 5.6 vs eledoisin in rat portal vein).

1.2.1.3. Peptides Containing Amide Bond Replacements. This work then leads to the area of SP analogs in which amide bonds have been replaced with other functionalities (Fig. 5). Although these

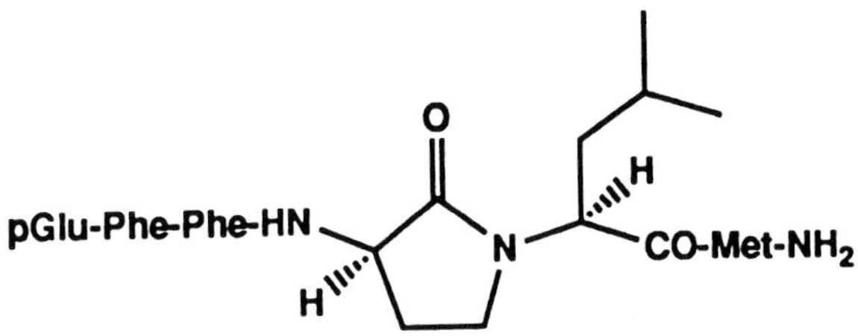

L-363,851 (28)

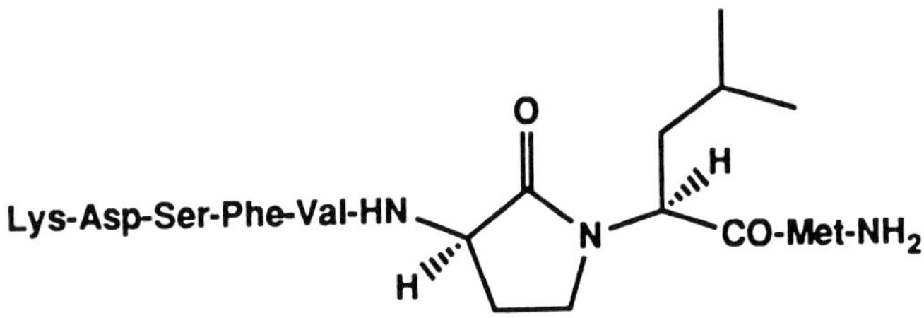

GR 64349 (29)

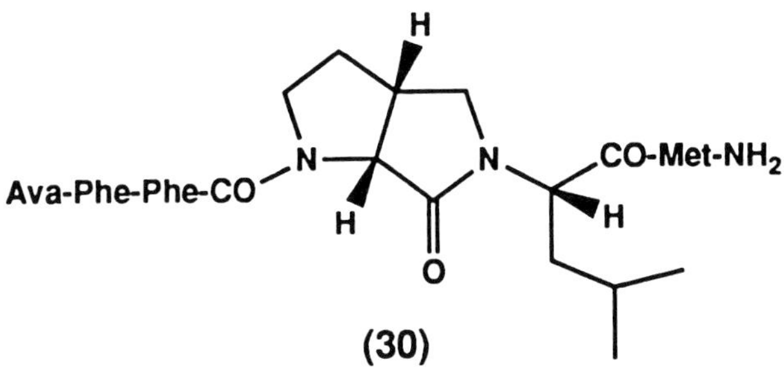

(30)

Fig. 4. NK_2 agonists containing lactam conformational constraints.

pGlu-Phe-Pheψ(trans-CH=CH)Gly-Leu-Leu-NH_2

(15)

pGlu-Pheψ(trans-CH=CH)Phe-Gly-Leu-Leu-NH_2

(16)

Arg-Pro-Lys-Pro-Gln-Gln-Phe-Pheψ(CF=CH)Gly-Leu-Met-NH_2

(17)

pGlu-Phe-Pheψ($COCH_2$)Gly-Leu-Met-NH_2

(18)

Arg-Pro-Lys-Gln-Gln-Phe-Phe-Gly-Leuψ(CH_2NH)Leu-NH_2

(19)

Ac-Arg-Phe-Phe-Sar-ψ(CH_2NH)-Leu-Met(O_2)-NH_2

(20)

Ac-Arg-Phe-ψ(CH_2NH)-Phe-Sar-Leu-Met(O_2)-NH_2

(21)

Fig. 5. SP analogs containing amide bond replacements.

compounds are presented in the section on antagonists, both agonists and antagonists will be considered. Indeed, as will be seen in several examples of SP analogs and also with NKA pseudopeptide analogs, agonist and antagonist activity can be exhibited by the same compound in different tissues. It will also be seen that when the amide bond replacement in an agonist results in an antagonist compound, this compound will show SAR that is different from the agonist. This would lead one to propose that these compounds may be interacting with additional sites on the receptor that are not accessed by agonists.

Early examples of amide bond replacements in SP are peptides **15** and **16** containing the *trans*-olefinic dipeptide isostere (Cox et al., 1980). In these two analogs of SP, the *trans* olefin replaces the Phe-Gly amide bond (compound **15**) and the Phe-Phe amide bond (compound **16**). In isolated guinea pig ileum, compounds **15** and **16** retained 24 and 100%, respectively, of the activity of the parent hexapeptide, pGlu-Phe-Phe-Gly-Leu-Leu-NH_2. More recently, the fluoroolefin dipeptide isostere (**17**) has also been shown to be equipotent to SP in a receptor binding assay in an unspecified tissue (Allmendinger et al., 1990). What can be safely gleaned from the scant biological data presented for these compounds is that the *trans* double bond is a good mimic for the amide bond at Phe^8-Gly^9 in SP, and that the amide bond itself probably plays a minor role in binding to the receptor via H-bonding and is not important for signal transduction. These results are also supported by the work of another group that prepared compound **18** (Ewenson et al., 1986). In this case, the Phe^8-Gly^9 amide bond has been replaced by the ketomethylene moiety that does not contain an amide NH that can participate in a hydrogen bond. Compound 18 is reported to be a full agonist in the isolated guinea pig ileum and to possess 70% biological activity relative to [$pGlu^6$]$SP_{6\text{-}11}$.

It had been initially reported that $\psi(CH_2NH)Leu^{11}$ analogs of SP are antagonists in isolated guinea pig pancreatic acini, although the affinity for the receptor is low (Qian et al., 1989). The $\psi(CH_2NH)$ moiety was introduced at positions 10, 8, and 7 in [Leu^{11}]-SP and the resulting pseudopeptides were reported to be antagonists, the most potent compound being compound **19** with an $IC_{50} = 7.1\ \mu M$ in SP stimulated amylase release from pancreatic acini. Subsequent studies, however, have demonstrated that in guinea pig ileum these compounds all behave as agonists with EC_{50} values in the μM range, more than 100-fold less active than SP (Zacharia et al., 1991). Other researchers have observed that compounds **20** and **21**, which are $\psi(CH_2NH)$ analogs of the SP agonist, [Ac-Arg^6,Sar^9,Met$(O_2)^{11}$]-SP_{6-11}, behave as agonists in dog carotid artery (Jukic et al., 1991). Although both compounds are agonists, compound **20** is 500-fold weaker and compound **21** is 20-fold weaker than SP. These results point out two important aspects of receptor SAR which will also be important for the NK_2 receptor: (1) substitution of an amide bond in SP with the reduced amide moiety is tolerated although affinity is greatly reduced; and (2) the decrease in affinity is probably due to the presence of a protonated amine functionality in the backbone. Since the lipophilic C-terminus is proposed to be very important in tachykinin binding and signal transduction, it is not too surprising that the $\psi(CH_2NH)$ moiety in this portion of SP has dramatic effects. The observation that NK_1 receptors in one tissue can respond to a ligand as an agonist, whereas the same ligand in another tissue can behave as an antagonist indicates that changes in the residues of the receptor important for receptor–ligand interactions may occur from tissue to tissue. Changes in a few critical residues of the NK_1 receptor could cause $\psi(CH_2NH)$-containing peptides to bind in either an agonist or antagonist conformation. The conformational flexibility of the $\psi(CH_2NH)$ moiety as compared to an amide bond confers mobility to peptides containing this amide bond replacement. Alternatively, specific interactions such as an ionic interaction with the protonated amine of the reduced dipeptide moiety could block signal transduction in a receptor even though the ligand may bind in the same manner as an agonist ligand. Further studies will be required to determine if the activity of these peptides is determined by the conformation of the bound ligand or the protonation state of the $\psi(CH_2NH)$ moiety. The $\psi(CH_2NH)$ analogs of tachykinin peptides will be discussed in more detail in the case of the NK_2 receptor.

2. NK_2 Receptor Ligands

2.1. Neurokinin A Agonists

2.1.1. SAR of the Endogenous Ligand

2.1.1.1. Peptide Length. The biological activities of NKA C-terminal partial sequences at monoreceptorial NK_2 preparations clearly show that the N-terminal tripeptide (His-Lys-Thr) can be removed without any significant loss of affinity (Osakada et al., 1986). Interestingly, the C-terminal heptapeptide, NKA_{4-10}, displays greater potency than NKA itself in some preparations, such as GPT and the rabbit pulmonary artery (Dion et al., 1987b). Also, the binding data show a dramatic drop in affinity when comparing the C-terminal octapeptide and the hexapeptide (Buck and Shatzer, 1988). As mentioned before (*see* Section 1.1.1.), the response of the NK_2 receptor to NKA and C-terminal fragments has been used in defining this tachykinin receptor type. As seen for SP, N-terminal fragments of NKA do not interact with the NK_2 receptor. In fact, NKA_{1-9} and NKA_{1-8} are devoid of biological activity (Rovero et al., 1989c), and shorter fragments, namely NKA_{1-7} and NKA_{1-5}, do not bind to the NK_2 receptor (Buck and Shatzer, 1988).

2.1.1.2. Systematic Amino Acid Substitutions. Five NKA analogs systematically substituted with Gly at positions 3 to 7 have been reported (Hashimoto et al., 1987a). The contractile activity of the analogs at the NK_2 receptor, measured in isolated rat vas deferens, indicates that Asp^4, Phe^6, and Val^7 are essential residues. The complete series of the Ala-substituted analogs has also been prepared and tested in a number of pharmacological assays (Rovero et al., 1989c). The results obtained at the NK_2 monoreceptorial preparation (rat pulmonary artery) confirm the crucial role of residues 4, 6, and 7 and of the C-terminal dipeptide, the relative affinities being 2.5, 0.01, 6, 0.3, and 0.8% for the analogs substituted with alanine at position 4, 6, 7, 9, and 10, respectively, as compared to NKA.

2.1.1.3. Amino Acid Substitutions That Enhance NK_2 Selectivity. In the search for amino acid substitutions which enhance the NK_2 selectivity of NKA, Regoli and coworkers (Drapeau et al.,

1987b) analyzed the role of Met^{10}, following the observation that oxidation of the Met in SP enhances the NK_1 selectivity of the peptide (*see* Section 1.1.3.). Contrary to the results obtained with SP, both the sulfoxide and the sulfone analogs of NKA display a marked reduction of affinity at the three tachykinin receptors (Drapeau et al., 1987b). Replacement of Met^{10} by the isosteric, but more hydrophobic Nle, brings about a moderate enhancement of selectivity. When the same modification is performed on the sequence of NKA_{4-10}, which is *per se* slightly more selective than NKA, the resulting peptide, $[Nle^{10}]NKA_{4-10}$, displays a good selectivity. It maintains almost 50% of affinity as compared to NKA in rat pulmonary artery (NK_2 receptor), but is practically inactive on the rat portal vein (NK_3 receptor) and about two orders of magnitude less active than NKA on the dog carotid artery (NK_1 receptor). This selective agonist has been subsequently used as a probe for the pharmacological characterization of several isolated tissues (Dion et al., 1987a).

Also working on the sequence of NKA_{4-10}, Rovero and coworkers (1989a) found that replacement of Gly in position 8 by β-Ala gives rise to a selective peptide that, as compared to $[Nle^{10}]NKA_{4-10}$, maintains better affinity for the NK_2 receptor. In fact, the affinity of $[\beta\text{-Ala}^8]NKA_{4-10}$ for the NK_2 receptor (rat vas deferens) is enhanced as compared to NKA_{4-10}, whereas affinity decreases about one order of magnitude at the NK_1 and NK_3 receptors (guinea pig ileum and rat portal vein, respectively). Moreover, the presence of a single unnatural residue is sufficient to confer some enzymatic stability to the peptide (Maggi et al., 1991a). In this case, instead of introducing conformational constraints, enhanced selectivity has been achieved by introducing a more flexible residue, namely β-Ala, in position 8. As previously seen (*see* Section 1.1.1.4.), the introduction of limited conformational constraints in the corresponding position of SP yielded NK_1 selective analogs. This point has been further investigated by performing a conformation–activity relationship study based on NMR analysis of NKA_{4-10} and some analogs replaced in position 8 (Saviano et al., 1991). The results of this study indicate that the preferred conformation of $[\beta\text{-Ala}^8]NKA_{4-10}$ is not compatible with models for NK_1 and NK_3 agonists proposed on the basis of rigid peptide agonists and may thus be considered as a likely bioactive conformation for NK_2 selective peptides.

2.1.2. *Conformationally Constrained Agonists*

2.1.2.1. PEPTIDES CONTAINING *N*-METHYL AMINO ACIDS. Although analogs of NKA containing *N*-methyl amino acid residues are discussed in this section as conformationally constrained peptides, the effect on activity observed on *N*-methylation of an amide bond can arise for several different reasons. In a conformational sense, the *N*-methyl amide bond can have an increased propensity to adopt a *cis* configuration. However, if the amide NH proton is involved in an intramolecular hydrogen bond that stabilizes a specific secondary structural feature such as a γ- or β-turn, then *N*-methylation destabilizes secondary structure and results in increased conformational mobility. The amide bond may also be involved in hydrogen bonding with the receptor; therefore, the effects of *N*-methylation may not be upon conformation, but instead upon receptor binding and/or signal transduction. With these caveats in mind, the SAR of peptides incorporating *N*-methyl amino acids must be viewed with caution.

The results obtained upon incorporation of *N*-methyl amino acids into NKA_{4-10} have shown a general decrease in activity (Rovero et al., 1987, 1989b). The data in Table 1 show that incorporation of *N*-methyl amino acid residues at positions 5,6, and 8 results in a fivefold decrease in activity, and substitution at position 7 causes a tenfold decrease. This study shows that substitution of *N*-methyl residues at positions 5-8 does not increase activity; therefore, a favorable conformational constraint has not been introduced. The effect of *N*-methylation at position 9 has also been studied (Chassaing et al., 1991). The fact that [*N*-Me-Leu^{10}]SP has very poor affinity for the NK_2 receptor led to the belief that *N*-methylation of the penultimate leucine of NKA would yield an inactive compound. Surprisingly, [*N*-Me-Leu^9]NKA (**27**) is equipotent in activity to NKA in rat pulmonary artery and shows equal affinity to NKA in binding to the NK_1 receptor in rat duodenum membranes (IC_{50} = 7.4 n*M* vs ^{3}H-NKA). As will be seen from the results obtained with peptides containing lactam constraints, the activity of **27** is not caused by stabilization of a *cis*

Table 1
N-Methyl Amino Acid Analogs of NKA

Compound	pD_2(RVD)[a]	RA[b]	EC_{50}(RPA)[c]
NKA_{4-10} (22)	5.98	100	—
[*N*-Me-Ser[5]]NKA_{4-10} (23)	5.28	19.9	—
[*N*-Me-Phe[6]]NKA_{4-10} (24)	5.27	19.5	—
[*N*-Me-Val[7]]NKA_{4-10} (25)	4.90	8.3	—
[*N*-Me-Gly[8]]NKA_{4-10} (26)	5.30	20.9	—
NKA	—	100	5 n*M*
[*N*-Me-Leu[9]]NKA_{4-10} (27)	—	100	5 n*M*

RVD, rat vas deferens; RA, relative affinity; RPA, rat pulmonary artery.
[a]Rovero et al., 1987.
[b,c]Chassaing et al., 1991.

amide bond, but probably is caused by the fact that this amide proton is not important for binding and occupies a region of the receptor that can tolerate the steric bulk of the methyl moiety.

Incorporation of an *N*-methyl residue at the C-terminus of NKA has not been reported. It is interesting that [*N*-Me-Leu[10]]SP and [*N*-Me-Met[11]]SP are both potent NK_1 agonists in guinea pig ileum (Lavielle et al., 1988a,b). Although these same SP analogs have very weak affinity for the NK_2 receptor, [*N*-Me-Leu[9]]NKA is equipotent to NKA, which might lead one to predict that [*N*-Me-Met[10]]NKA would also be active. Obviously, the SAR of the C-terminus of NKA could be studied further by the incorporation of *N*-methyl residues and lactam constraints.

2.1.2.2. Lactam Constraints. Peptides containing lactam constraints have been prepared (Fig. 4) and do provide more conformational information about NK_2 agonists than the *N*-methyl amino acid analogs of NKA, since the conformation of the amide bond is fixed by the lactam ring. The lactam constraint has been introduced into the SP agonist, [Glp[6]]SP_{6-11}, to yield L-363,851 (**28**). This substitution resulted in a sharp decrease in activity at the NK_1 receptor, but it was observed that L-363,851 is more active than [Glp[6]]SP_{6-11} at the NK_2 and NK_3 receptors (McKnight et al., 1991; Cascieri et al., 1986;

McKnight et al., 1987). The authors reasoned that this lactam constraint favors a folded peptide backbone that is not recognized as an agonist structure at the NK_1 receptor, but is preferred by the NK_2 and NK_3 receptors. This lack of activity at the NK_1 receptor is not due to steric interactions since both [Pro^9]SP (Drapeau et al., 1987a) and [Me-Leu^{10}]SP (Chaissang et al., 1991) retain full NK_1 activity. Introduction of this lactam constraint into the NKA sequence yielded GR64349 (**29**), a very potent and selective NK_2 agonist (Hagan et al., 1989). GR64349 is almost equipotent to NKA in contracting rat colon muscularis mucosae (EC_{50} = 3.7 n*M* vs 1.7 n*M* for NKA). However, GR64349 shows very weak activity in guinea pig ileum (NK_1) and rat portal vein (NK_3), indicating that this compound is selective for the NK_2 receptor. Compound **30** has appeared in the recent literature as another potent and selective NK_2 agonist containing a lactam constraint (Ward et al., 1990). In this compound, two SAR features that had been identified as important for NK_2 agonist activity have been combined: the R-γ-lactam of GR64349 and substitution of Gly^9 by D-Pro. This combination yielded **30** with potent activity in the rat colon muscularis mucosae (EC_{50} = 15 n*M* vs 1.7 n*M* for NKA) and weaker activity in both guinea pig ileum (NK_1; EC_{50} = 545 n*M* vs 5.6 n*M* for SP) and rat portal vein (NK_3; EC_{50} > 18,000 n*M* vs 6.3 n*M* for NKB). These results emphasize that NK_2 and NK_3 agonists may share a secondary structural motif such as a turn that is not recognized by the NK_1 receptor. It is also important to note that the NK_2 receptor is sterically tolerant, since the insertion of bulky lactam constraints does not prohibit binding.

2.2. Receptor Antagonists

2.2.1. Amino Acid Substitutions of the NKA Sequence

The development of NKA antagonists by the classical approach of amino acid substitutions was undertaken at Menarini Pharmaceuticals (Florence, Italy), starting from the observation that earlier SP antagonists also interact with the NK_2 receptor, albeit with low affinity (Regoli et al., 1985). Thus, the same key modifications, i.e., the introduction of two D-Trp residues, were applied to an NK_2 selective template, the C-terminal heptapeptide, NKA_{4-10}, yielding [D-Trp^{6,8}]NKA$_{4-10}$, a weak but selective NK$_2$ antagonist (Rovero et al., 1990b). A subsequent empirical SAR study revealed that each additional amino acid substitution dramatically affects affinity and selectivity. For instance, the introduction of a third D-Trp residue in position 9 yields an antagonist with a pA$_2$ close to 6 at the NK$_2$ receptor (rat vas deferens) (Rovero et al., 1990b). The activity of this peptide has been improved using a chemometric approach to quantitative structure–activity relationships (QSAR). This approach provides a strategy for selecting the most informative molecular structures in a series of analogs (Clementi et al., 1990, 1991). As a result of this study, two additional amino acid substitutions were suggested based on the template of NKA$_{4-10}$ bearing three D-Trp residues in positions 6, 8, and 9; Tyr in the place of Ser in position 5; and Arg in the place of Met in position 10. The resulting peptide, [Tyr^5, D-Trp^{6,8,9}, Arg^{10}]NKA$_{4-10}$ (MEN 10,207; compound **31** in Fig. 6), is a potent and selective NK$_2$ antagonist: pA$_2$ = 7.9 vs NKA in the rat pulmonary artery (NK$_2$); 5.2 vs SP in the dog carotid artery (NK$_1$), and 4.9 vs NKB in the rat portal vein (NK$_3$) (Rovero et al., 1990a). Interestingly, the lower affinity (pA$_2$ = 5.9) displayed by MEN 10,207 (**31**) in a pharmacological preparation that has been shown to contain mainly NK$_2$ receptors by the use of selective agonists, namely the hamster trachea, lead to the discovery of NK$_2$ receptor subtypes (Maggi et al., 1990; *see also* Chapter 14 in this book). In an attempt to define the structural requirements essential for subtype selectivity of MEN 10,207 (**31**), the role of the three D-Trp residues has been studied (Rovero et al., 1991). A series of analogs of compound **31** has been synthesized in which each D-Trp residue is replaced by either L-Trp or the naturally occurring residue in the corresponding position of NKA$_{4-10}$. The evaluation of biological activity of these analogs has shown the following:

1. The presence of three D-Trp residues is crucial both for affinity and selectivity;
2. The D-Trp residue in position 6 confers affinity, whereas the D-Trp residue in position 9 is important for selectivity; and
3. The third D-Trp (position 8) seems to have an intermediate function, being important both for affinity and selectivity.

Asp-Tyr-(D-Trp)-Val-(D-Trp)-(D-Trp)-Arg-NH_2

MEN 10,207 (31)

Asp-Tyr-(D-Trp)-Val-(D-Trp)-(D-Trp)-Lys-NH_2

MEN 10,376 (32)

Ac-Leu-Asp-Gln-Trp-Phe-Gly-NH_2

R 396 (33)

Boc-Arg-Ala-(D-Trp)-Phe-(D-Pro)-Pro-Nle-NH_2

GR 83704 (34)

Benzoyl-Ala-Ala-(D-Trp)-Phe-(D-Pro)-Pro-Nle-NH_2

GR 94800 (35)

GR100679 (36)

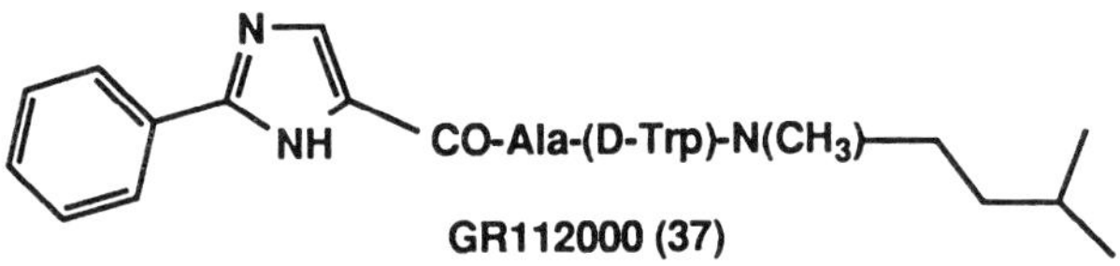

GR112000 (37)

Fig. 6. NK_2 selective antagonists.

A systematic study has also been reported on the role of the residue at position 10. This position had been indicated as important for selectivity by the chemometric approach (Rovero et al., 1992b). The analysis of the biological activity of a series of 12 analogs of

MEN 10,207, modified at position 10 by amino acid substitution, revealed an interesting correlation between lipophilicity of the C-terminal residue and subtype selectivity. In fact, in the presence of a C-terminal hydrophilic residue, the affinity is maximal for the NK_{2A} receptor subtype, whereas lipophilic, bulky, and aromatic residues slightly increase affinity for the NK_{2B} subtype. The importance of the C-terminal residue is further illustrated by the partial agonism of MEN 10,207 in vivo. However, substitution of Arg with Lys in position 10 as in MEN 10,376, [Tyr^5, D-Trp^{6,8,9}, Lys^{10}]NKA_{4-10} (**32**), reduces partial agonist activity 100-fold, but maintains the high affinity for the NK_2 receptor ($pA_2 = 8.1$ vs NKA in the rabbit pulmonary artery) (Maggi et al., 1991b).

R 396 (**33**) is an NK_2 antagonist that has been obtained by amino acid substitutions (Dion et al., 1990). This peptide was designed taking as a template the linear hexapeptide, Ac-Leu-Met-Gln-Trp-Phe-Gly-NH_2 (*see* Section 2.2.2.1.), which had been reported to be an NK_2 antagonist by the Merck Sharp and Dohme group (McKnight et al., 1988; Williams et al., 1988). The NK_2 selectivity of the original linear hexapeptide was improved by replacement of Met by Asp, a residue known to be important for affinity of NKA (Rovero et al., 1989). Interestingly, the pattern of NK_2 receptor subtype selectivity of R 396 (**33**) is opposite to that of MEN 10,207 (**31**) (Maggi et al., 1990), but the question of whether the differences of affinity of R 396 in different preparations reflect the existence of species-related differences or the existence of true receptor subtypes is still open (Maggi et al., 1993; *see also* Chapter 14 in this book). Recently, a SAR study of R 396, designed with the aim of elucidating the structural requirements responsible for subtype selectivity, has been reported (Rovero et al., 1992a). Even if no clearcut conclusions were drawn, the crucial role of the Asp^2 residue was established.

A new class of D-Trp containing NK_2 antagonists has been recently described (McElroy et al., 1992), taking as a template a poorly selective, SP-derived antagonist containing D-Trp residues in positions 7 and 9. The SAR of SP analogs suggested to these authors to replace D-Trp^9 by D-Pro, previously reported to enhance NK_2 selectivity in agonist sequences (Lee et al., 1986). Concomitant empirical replacement of Leu^{10} by Pro and introduction of a lipophilic N-terminal modification resulted in a potent and selective analog (GR83074; compound **34**) with $pK_B = 8.23$ in the rat colon muscularis mucosae vs NKA. Affinity for the NK_2 receptor was enhanced by over one log unit by further increasing lipophilicity in the N-terminal portion of the peptide. Thus, replacement of Boc-Arg^6 by PhCO-Ala yielded GR94800 (**35**) with $pK_B = 9.6$ vs NKA. Interestingly, these data show that, although the lipophilic N-terminal modification appears to enhance either NK_1 or NK_2 antagonist activity, depending on the substitution elsewhere in the molecule the replacement of D-Trp^9 by D-Pro is a key modification for introducing potent and selective NK_2 antagonist potency. Thus, in this case, similar SAR apply to both agonists and antagonists, as previously seen also in the case of the Asp residue of R 396 *(vide supra).* Subsequent work at Glaxo (Ware, UK) (Smith et al., 1993) has been directed toward the development of low molecular weight antagonists through the application of a deletion-optimization strategy to the heptapeptide, GR94800. Thus, deletion of the C-terminal tripeptide, D-Pro-Pro-Nle, and optimization of the resulting tetrapeptide yielded GR100679 (**36**), reported to have substantially the same pharmacological profile as the parent heptapeptide ($pK_B = 9.05$ in the rat colon vs NKA), but with a considerably reduced size. Further deletion of the Phe residue and introduction of a nonpeptidic, conformationally constrained group at the N-terminus gave a "dipeptoid" derivative (GR112000, **37**) that, in spite of reduced affinity for the NK_2 receptor ($pK_B = 7.64$), should be regarded as a promising lead toward the development of novel nonpeptidic antagonists.

2.2.2. *Conformationally Constrained Peptides*

2.2.2.1. Lactam Constraints. The potent NK_2 antagonist peptides that contain lactam conformational constraints are all structurally related, since these compounds evolved from a common structure. In the previous section on NK_2 agonists, L-363,851 (**28**, Fig. 4) was discussed as an NK_2 ligand containing a lactam constraint. This lactam was proposed to constrain the peptide in a cyclic conformation that could bind to the NK_2 and NK_3 receptors, but not the NK_1 receptor. This observation provided the impetus for the preparation of cyclic peptides as potentially potent and selective antagonists at the NK_2 receptor (McKnight, et al., 1987, 1988, 1991). L-659,837 (**39**, Fig. 7), which is a cyclic analog of L-363,851, shows significant NK_2 antagonist activity ($pA_2 = 6.7$ vs eledoisin in rat vas deferens). It was subsequently discovered that the lactam constraint is not required for activity. L-659,877 (**38**, Fig. 7), in which the lactam constraint is removed, is a more potent and selective antagonist ($pA_2 = 8.0$). A suprising result is that the cyclic peptide is not an absolute requirement for antagonist activity, since the linear analog of compound **38**, Ac-Leu-Met-Gln-Trp-Phe-Gly-NH_2, is an antagonist ($pA_2 = 6.8$ vs eledoisin in rat vas deferens), although weaker than the cyclic peptide **38**. Without further SAR and conformational studies, it is unclear if L-659,877 is more active because of a conformation that is different from L-659,837 or because the steric requirements of the lactam bridge in L-659,837 are a detriment to binding.

2.2.2.2. ψ(CH_2NR) Cyclic Peptide Antagonists. Although the next series of antagonists contains a reduced amide bond, they are more appropriately presented here since they are cyclic peptides modeled after the potent cyclic NKA antagonist, L-659,877 (**38**). Compound **40** (Table 2) is an analog of L-659,877 in which the Leu-Met dipeptide has been replaced with the Leuψ(CH_2NH)Leu reduced dipeptide moiety. **40** is a potent antagonist of NKA induced phosphatidyl inositol (PI) turnover in hamster urinary bladder ($pA_2 = 8.5$) (Harbeson et al., 1989). Some structure–activity data are available for this antagonist. Conservative amino acid substitutions for the Phe residue are tolerated as shown by the fully active Tyr analog, **41**. The Trp residue appears to be required for activity, since substitution by either Phe (**42**) or β-Nal (**43**) results in a significant loss in binding affinity. The secondary amine of the reduced amide bond can be substituted with alkyl groups. The *N*-methyl analog (**44**) retains equipotent antagonist activity, whereas the *N*-ethyl (**45**) and *N*-propyl (**46**) derivatives are approximately 3- and 20-fold less active than **40** as antagonists in hamster urinary bladder, respectively. These results indicate that this portion of the molecule must bind in a region

Compound	pIC_{50}[a]
cyclo (Leu-Met-Gln-Trp-Phe-Gly) L-659,877 (38)	8.0
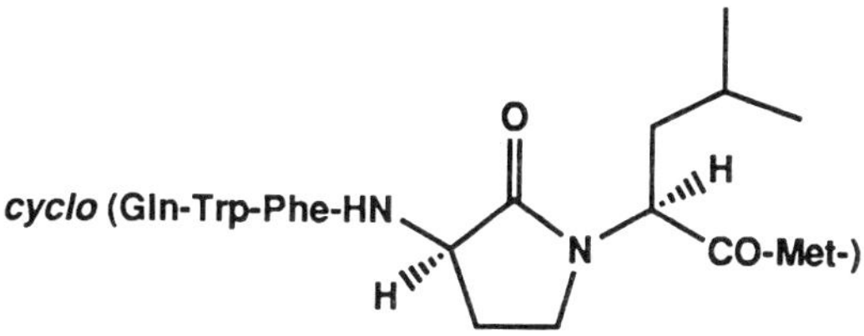 L-659,837 (39)	6.7

[a]versus [^{125}I-BH]-Eledoisin in HUB (McKnight et al., 1988).

Fig. 7. Cyclic peptide NK_2 antagonists

of the receptor that is tolerant of steric bulk. Steric bulk at this portion of the molecule does not appear to moderate activity, since a leucine side chain can be removed to yield **47**, which is still equipotent to **40**.

By analogy to the sequence of NKA, it would appear that the Gly-Leuψ(CH_2NH)Leu region of **40** might bind to the same region of the receptor as the C-terminus of NKA. However, the fact that this region of **40** is very tolerant of structural changes, whereas the C-terminus of NKA is not, leads to the proposal that these domains are binding to different regions of the receptor. Confirmation of this postulate will require much more SAR data on the cyclic peptide NK_2 antagonists and information on the structure of the receptor.

Table 2
Cyclic $\Psi(CH_2NR)$ Pseudopeptide NK_2 Antagonists

Peptide	NK_2 (HUB) IC_{50}(n*M*)[a]
cyclo [Leu-Met-Gln-Trp-Phe-Gly] (38)	2.1 ± 0.3
cyclo [Leu$\Psi(CH_2NH)$Leu-Gln-Trp-Phe-Gly] (40)	2.3 ± 0.2
cyclo [Leu$\Psi(CH_2NCH_3)$Leu-Gln-Trp-Phe-Gly] (44)	2.8 ± 0.3
cyclo [Leu$\Psi(CH_2NCH_2CH_3)$Leu-Gln-Trp-Phe-Gly] (45)	8 ± 0.6
cyclo [Leu$\Psi(CH_2N(CH_2)_2CH_3)$Leu-Gln-Trp-Phe-Gly] (46)	48 ± 5
cyclo [Leu$\Psi(CH_2NCH_3)$Leu-Gln-Trp-Tyr-Gly] (41)	1.9 ± 0.5
cyclo [Leu$\Psi(CH_2NCH_3)$Gly-Gln-Trp-Phe-Gly] (47)	2.1 ± 0.2
cyclo [Leu$\Psi(CH_2NCH_3)$Leu-Gln-Phe-Phe-Gly] (42)	1330 ± 235
cyclo [Leu$\Psi(CH_2NCH_3)$Leu-Gln-β-Nal-Phe-Gly] (43)	412 ± 77
cyclo [Leu$\Psi(CH_2NCH_3)$Leu-Gln-Trp-(D-Phe)-Gly] (48)	5.3 ± 1.4

HUB, hamster urinary bladder.
[a]Versus ^{125}I-NKA (Harbeson et al., 1991).

The conformation of **40** has been determined by the combined use of two-dimensional NMR and restrained molecular dynamics (Malikayil and Harbeson, 1992). The predominant conformation is characterized by two β-turns; one turn contains the sequence Trp-Phe-Gly-Leu, whereas the other turn contains the Leu$\psi(CH_2NH)$Leu-Gln-Trp sequence. Molecular dynamics predicted that substitution of D-Phe for Phe would be compatible with the proposed conformation of **40**, which was confirmed by the high affinity of **48** for the NK_2 receptor. Conformational studies on constrained cyclic peptides such as L-659,877 and **40** should prove useful in refining structural models of the NK_2 receptor.

2.2.3. Other Analogs of NKA

2.2.3.1. $\psi(CH_2NR)$ PSEUDOPEPTIDES. In contrast to the results obtained with $\psi(CH_2NH)$ reduced peptide analogs of SP, analogs of NKA that contain reduced amide bonds have been reported that bind with very high affinity to the NK_2 receptor. The only amide bond in the peptide backbone of NKA_{4-10} that will tolerate the $\psi(CH_2NH)$

replacement is the Leu^9-Met^{10} amide bond (Harbeson et al., 1989). Similar results have also been observed when the ψ(CH_2NH) moiety is inserted at positions 6, 7, and 8 of [Nle^{10}]NKA_{4-10} (Jukic et al., 1991). These three compounds are at least 100-fold less active as agonists than [Nle^{10}]NKA_{4-10} and are not reported to possess any antagonist activity (rat pulmonary artery). The most carefully studied compound in this class is **49**. In hamster urinary bladder, this compound retains significant affinity with respect to NKA_{4-10} (Table 3) and behaves as a competitive antagonist ($pA_2 = 6.7$) of NKA-stimulated PI turnover in this same tissue (Buck et al., 1990). Compound **49** has been the subject of several studies because of its proposed ability to discriminate between NK_2 receptor subtypes (*see* Chapter 4).

The pertinent SAR data for **49** are summarized in Table 2 (Harbeson et al., 1992). Although most of the structure modifications have been made at the ψ(CH_2NH) moiety, some amino acid substitutions have been made. When Leu^{10} of **49** is replaced with Met, the C-terminal residue of NKA, a twofold increase in binding affinity is observed. However, this increase in binding is accompanied by significant partial agonism in hamster urinary bladder (approximately 50% of the response to NKA_{4-10}). This result shows that at the C-terminus of these peptides structural modifications based on agonist SAR are incompatible with the enhancement of antagonist activity. Substitution of Gly^8 in NKA_{4-10} with β-Ala has been reported to enhance activity in rat vas deferens by 8.5-fold (Rovero et al., 1989a,b); however, the activity of [β-Ala^8]NKA_{4-10} in hamster urinary bladder is not known. This same substitution has been made in **49** and **61** to give **51** and **63**, respectively; however, no significant change in affinity has been observed. The effect of replacing the C-terminal residue of NKA_{4-10} with Phe has not been reported, but in a series of NK_2 antagonists based on the NKA_{4-10} sequence, substitution with Phe at the C-terminus does enhance binding affinity (Rovero et al., 1990b). These results agree with the data in Table 3 that show that the Phe^9 analog (**60**) is inactive, whereas the Phe^{10} analog (**61**) binds to the NK_2 receptor of hamster urinary bladder with approximately 13-fold greater affinity than **49**. In this case, it appears that the SAR results for the antagonists mentioned above may have

Table 3
Ψ(CH_2NR) Pseudopeptide Analogs of NKA_{4-10}

Structure	IC_{50}[a] (n*M* ± SEM)
Asp-Ser-Phe-Val-Gly-Leu-Met-NH_2 (NKA_{4-10})	6 ± 1
Asp-Ser-Phe-Val-Gly-Leu-Leu-NH_2	136 ± 20
Asp-Ser-Phe-Val-Gly-LeuΨ(CH_2NH)Leu-NH_2 (49)	115 ± 8
Asp-Ser-Phe-Val-GlyΨ(CH_2NH)Leu-Leu-NH_2 (50)	> 10^4
Asp-Ser-Phe-ValΨ(CH_2NH)Gly-Leu-Leu-NH_2 (51)	5662 ± 333
Asp-Ser-PheΨ(CH_2NH)Val-Gly-Leu-Leu-NH_2 (52)	> 10^4
Asp-SerΨ(CH_2NH)Phe-Val-Gly-Leu-Leu-NH_2 (53)	> 10^4
SucΨ(CH_2NH)Ser-Phe-Val-Gly-Leu-Leu-NH_2 (54)	1652 ± 175
Asp-Ser-Phe-Val-Gly-LeuΨ(CH_2NCH_3)Leu-NH_2 (55)	193 ± 14
Asp-Ser-Phe-Val-Gly-LeuΨ($CH_2NCH_2CH_3$)Leu-NH_2 (56)	277 ± 20
Asp-Ser-Phe-Val-Gly-LeuΨ($CH_2N(CH_2)_2CH_3$)Leu-NH_2 (57)	103 ± 13
Asp-Ser-Phe-Val-Gly-LeuΨ(CH_2N(Iva))Leu-NH_2 (58)	162 ± 21
Asp-Ser-Phe-Val-Gly-LeuΨ(CH_2NH)Met-NH_2 (59)	50 ± 4
Asp-Ser-Phe-Val-Gly-PheΨ(CH_2NH)Leu-NH_2 (60)	> 10^4
Asp-Ser-Phe-Val-Gly-LeuΨ(CH_2NH)Phe-NH_2 (61)	9 ± 2
Asp-Ser-Phe-Val-β-Ala-LeuΨ(CH_2NH)Leu-NH_2 (62)	108 ± 8
Asp-Ser-Phe-Val-β-Ala-LeuΨ(CH_2NH)Phe-NH_2 (63)	10 ± 1
Asp-Ser-Phe-Val-β-Ala-LeuΨ($CH_2N(CH_2)_2CH_3$)Phe-NH_2 (64)	1 ± 0.2

[a]Versus ^{125}I-NKA in hamster urinary bladder (Harbeson et al., 1992).

greater relevance to analogs of **49** than SAR based on agonist peptides. This is a very tenuous analogy to make, since the SAR reported for **49** has been obtained in hamster urinary bladder, whereas the other studies were performed in hamster trachea and rat vas deferens. However, all three tissues are proposed to represent the same NK_2 receptor subtype (NK_{2B}) (Guiliani et al., 1991). In addition to the previous cautionary statements concerning the application of agonist SAR to the design of antagonists, one must also be aware of the NK_2 receptor subtype, since the NK_{2A} and NK_{2B} receptor subtypes can show different SAR for the same series of antagonists (Rovero et al., 1992b).

The other structural modification of interest is alkylation of the secondary amine of the $\psi(CH_2NH)$ moiety to give *N*-alkyl $\psi(CH_2NR)$ analogs of NKA_{4-10}. Compound **49** has been converted to compounds **55**, **56**, **57**, and **58** by introduction of methyl, ethyl, propyl, and isovaleryl substituents, respectively. The methyl and ethyl substituents cause small decreases in receptor affinity, whereas the propyl substituent yields **57**, with affinity essentially equal to **49**. Further increasing of the size of the alkyl substituent to isovaleryl (**58**) only results in a small decrease in receptor affinity. The region of the NK_2 receptor that binds the $\psi(CH_2NR)$ moiety appears to be very large and tolerant of hydrophobic bulk. The effect of hydrophilic substituents has not been investigated. The most active compound is **64** in which the propyl substituent on the reduced amide bond has been combined with the β-Ala^8 and Phe^{10} substitutions to yield a compound with affinity equal to NKA. In two previous examples, compounds **62** and **63**, the β-Ala^8 modification did not enhance receptor affinity; therefore, it is probable that the very high affinity of **64** is owing to the combined effects of the propyl substituent and the Phe^{10} residue. Again, this indicates that for this series of antagonists the C-terminus is binding to a region on the receptor that favors bulky, hydrophobic residues. This region may also be distinct from the region of the receptor which binds the C-terminus of agonists. Only further SAR studies can shed light on this problem.

3. NK_3 Receptor Ligands

3.1. Receptor Agonists

3.1.1. NK_3 Selective Agonists

NK_3 receptor agonists have not been the object of as much attention as agonists for the NK_1 and NK_2 receptors; therefore, there is less literature on the SAR of NKB. As has been pointed out in Section 1.1.1.1., the complete sequence of NKB is required for full activity, and fragments are much less active. Because of the paucity of detailed SAR data on NKB, the best approach is to summarize the peptides that have been synthesized and that show selectivity as NK_3 agonists. A cautionary remark is warranted at this point. Rat portal vein has been the principal tissue used in identifying NK_3 ligands, and this tissue had been considered to be monoreceptorial. A recent publication (Chassaing et al., 1991), however, has suggested that rat portal vein also contains a small amount of NK_2 receptors that may have some relevance to the proposed selectivity of the compounds that have to date been assumed to be NK_3 selective.

The most active NK_3 selective agonists that have been reported are shown in Table 4. It is clear from these structures that at least one carboxylate-containing residue is required in the N-terminal region, and that two carboxylates (the pattern found in NKB) are also conducive to agonist activity. It is very apparent that a secondary amide at position 7 (with respect to NKB) is critical for NK_3 selectivity. This feature is contained in three of the NK_3 selective compounds shown in Table 4, one of which is senktide (**67**), an early NK_3 selective peptide (Wormser et al., 1986). Further examples of this structural feature are compounds **65** and **66**, which have been shown to be potent, selective NK_3 agonists (Regoli et al., 1988), and [*N*-Me-Val^7]-NKB and [Pro^7]NKB (Lavielle et al., 1988b), which are also selective agonists. Disulfide bridging of the N-terminal portion of the peptide is important for NK_3 selectivity as shown by **68** and also [Cys^{3,6}, Tyr^8]SP, which shows high affinity for the NK_3 receptor (Lavielle et al., 1988a). There is also some evidence that the NK_3 receptor does not appear to tolerate steric bulk at the Gly^8 position of NKB, since [Cys^{3,6},Tyr^8, Pro^9]SP is approximately 300-fold less active than [Cys^{3,6}, Tyr^8]SP.

3.2. Receptor Antagonists

3.2.1. NK_3 Selective Ligands

Although highly selective synthetic agonists for the NK_3 receptor have been described (*see* Section 3.1.1.), potent, selective antagonists have not been reported. A D-Trp containing analog of NKB, namely [D-Pro^2, D-Trp^{6,8}, Nle^{10}]NKB, has been reported to show some selective competitive antagonism vs NKB in the guinea pig ileum (Jacoby et al., 1986). On the other hand, the binding of the same peptide to the NK_3 receptor in a rat cerebral cortex membrane

Table 4
Agonists Selective for the NK_3 Receptor

	RPV	
	pD_2[a]	RA[b]
Arg-Pro-Lys-Pro-Gln-Gln-Phe-Phe-Gly-Leu-Met-NH_2 SP	5.82[c]	1.4
His-Lys-Thr-Asp-Ser-Phe-Val-Gly-Leu-Met-NH_2 NKA	6.45[c]	6
Asp-Met-His-Asp-Phe-Phe-Val-Gly-Leu-Met-NH_2 NKB	7.68[c]	100
β-Asp-Phe-Phe-(*N*-Me-Phe)-Gly-Leu-Met-NH_2 [β-Asp^4,*N*-Me-Phe^7]NKB_{4-10} (**65**)	7.56[d]	76
Asp-Met-His-Asp-Phe-Phe-(*N*-Me-Phe)-Gly-Leu-Met-NH_2 [*N*-Me-Phe^7]NKB (**66**)	8.30[d]	417
HOOC-$(CH_2)_2$-CO-Asp-Phe-(*N*-Me-Phe)-Gly-Leu-Met-NH_2 Senktide (**67**)	7.61[d]	85
Asp-Cys-His-Asp-Cys-Phe-Val-Gly-Leu-Met-NH_2 (Cys^2 S——S Cys^5) [Cys^{2,5}]NKB (**68**)	8.12[d]	272

RPV, rat portal vein.
[a]pD_2 = -log[EC_{50}].
[b]RA, relative affinity as percent of NKB.
[c]Regoli et al., 1989.
[d]Regoli et al., 1988.

preparation has been shown to be negligible (Buck and Shatzer, 1988). Two octapeptidic C-terminal partial sequences of NKB modified in position 6, namely [Gly^6]NKB_{3-10} and [Arg^3, D-Ala^6]NKB_{3-10}, have also been reported to weakly antagonize NKB in the guinea pig ileum preparation (Hashimoto et al., 1987b). No data concerning the selectivity of these peptides have been reported. Finally, an analog of the NK_2 selective agonist, [β-Ala^8]NKA_{4-10}, bearing a Trp residue in position 7, namely [Trp^7, β-Ala^8]NKA_{4-10}, has been reported to antagonize NKB in rat portal vein (Drapeau et al., 1990). The concomitant agonist activity at NK_1 and NK_2 receptors displayed at slightly higher concentrations limits the value of this peptide.

References

Allmendinger, T., Felder, E.,and Hungerbuhler, E. (1990) Fluoroolefin dipeptide isosteres-II. Enantioselective synthesis of both antipodes of the Phe-Gly dipeptide mimic. *Tet. Lett.* **31,** 7301–7304.

Buck, S. and Shatzer, S. A. (1988) Agonist and antagonist binding to tachykinin peptide NK-2 receptors. *Life Sci.* 42, 2701–2708.

Buck, S. H., Harbeson, S. L., Hassman III, C. F., Shatzer, S. A., Rouissi, N., Nantel, F., and van Giersbergen, P. L. M. (1990) [$Leu^9\psi(CH_2NH)Leu^{10}$] -Neurokinin A_{4-10} (MDL 28,564) distinguishes tissue tachykinin peptide NK-2 receptors. *Life Sci.* **47,** PL-37–PL-41.

Bury, R. W. and Mashford, M. L. (1976) Biological activity of C-terminal partial sequences of substance P. *J. Med. Chem.* **19,** 854–856.

Cascieri, M. A., Chicchi, G. G., Freidinger, R. M., Colton, C. D., Perlow, D. S., Williams, B., Curtis, N. R., McKnight, A. T., Maguire, J. J., Veber, D. F., and Liang, T. (1986) Conformationally constrained tachykinin analogues which are selective ligands for the eledoisin receptor. *Mol. Pharmacol.* **29,** 34–38.

Cascieri, M. A., Goldenberg, M. M., and Liang, T. (1981) Biological activity of substance P methyl ester. *Mol. Pharmacol.* **20,** 457–459.

Cascieri, M. A., Huang, R. R. C., Fong, T. M., Cheung, A. H., Sadowski, S., Ber, E., and Strader, C. (1992) Determination of the amino acid residues in substance P conferring selectivity and specificity for the rat neurokinin receptors. *Mol. Pharmacol.* **41,** 1096–1099.

Chassaing, G., Lavielle, S., Loeuillet, D., Robilliard, P., Carruette, A., Garret, C., Beaujouan, J. C., Saffroy, M., Petitet, F., Torrens, Y., and Glowinski, J. (1991) Selective agonists of NK_{-2} binding sites highly active on rat portal vein (NK_3 bioassay). *Neuropeptides* **19,** 91–95.

Clementi, S., Cruciani, G., Riganelli, D., Rovero, P., Pestellini, V., Maggi, C. A., and Baroni, M. (1990) Chemometric approach to a QSAR study of peptides behaving as NK_{-2} receptor antagonists. *Tetrahedron Computer Method* **3,** 379–387.

Clementi, S., Cruciani, G., Rovero, P., Pestellini, V., and Baroni, M. (1991) Chemometric investigation in peptide QSAR, in *QSAR: Rational Approach to the Design of Bioactive Compounds* (Silipo, C. and Vittoria, A., eds.), Elsevier, Amsterdam, pp. 353–356.

Convert, O., Duplaa, H., Lavielle, S., and Chassaing, G. (1991) Influence of the replacement of amino acid by its D-enatiomer in the sequence of substance P. 2. Conformational analysis by NMR and energy calculations. *Neuropeptides* **19,** 259–270.

Couture, R., Fournier, A., Magnan, J., St. Pierre, S., and Regoli, D. (1979) Structure-activity studies of substance P. *Can. J. Physiol. Pharmacol.* **57,** 1427–1436.

Cox, M. T., Gormley, J. J., Hayward, C. F., and Petter, N. N. (1980) *J. Chem. Soc. Chem. Commun.* 800.

Dion, S., D'Orléans-Juste, P., Drapeau, G., Rhaleb, N.-E., Rouissi, N., Tousignant, C., and Regoli, D. (1987a) Characterization of neurokinin receptors in various isolated organs by the use of selective agonists. *Life Sci.* **41,** 2269–2278.

Dion, S., D'Orléans-Juste, P., Rhaleb, N. E., Drapeau, G., Rovero, P., and Regoli, D. (1987b) Neurokinin fragments and analogues as selective activators of receptors, in *Substance P and Neurokinins* (Henry, J., Couture, R., Cuello, A. C., Pelletier, G., Quirion, R., and Regoli, D., eds.), Springer-Verlag, New York, pp. 129–131.

Dion, S., Rouissi, N., Nantel, F., Jukic, D., Rhaleb, N.-E., Tousignant, C., Telemaque, S., Drapeau, G., Regoli, D., Naline, E., Advenier, C., Rovero, P., and Maggi, C. A. (1990) Structure-activity study of neurokinins: antagonists for the neurokinin-2 receptor. *Pharmacology* **41,** 184–194.

Drapeau, G., D'Orléans-Juste, P., Dion, S., Rhaleb, N.-E., Rouissi, N.-E., and Regoli, D. (1987a) Selective agonists for substance P and neurokinin receptors. *Neuropeptides* **10,** 43–54.

Drapeau, G., Rovero, P., D'Orléans-Juste, P., Dion, S., Rhaleb, N. E., and Regoli, D. (1987b) Mammalian neurokinins: structure-activity study of the C-terminal methionine amide, in *Substance P and Neurokinins* (Henry, J., Couture, R., Cuello, A. C., Pelletier, G., Quirion, R. and Regoli, D. , eds.), Springer-Verlag, New York, pp. 132–134.

Drapeau, G., Rouissi, N., Nantel, F., Rhaleb, N.-E., Tousignant C., and Regoli, D. (1990) Antagonists for the neurokinin NK-3 receptor evaluated in selective receptor systems. *Regul. Pept.* **31,** 125–135.

Duplaa, H., Chassaing, G., Lavielle, S., Beaujouan, J. C., Torrens, Y., Saffroy, M., Glowinski, J., D'Orléans-Juste, P., Regoli, D., Carruette, A., and Garret, D. (1991) Influence of the replacement of amino acid by its D-enantiomer in the sequence of substance P. 1. Binding and pharmacological data. *Neuropeptides* **19,** 251–257.

Ewenson, A., Laufer, R., Chorev, M., Selinger, Z., and Gilon, C. J. (1986) Ketomethylene pseudopeptide analogues of substance P: synthesis and biological activity. *J. Med. Chem.* **29,** 1281.

Folkers, K., Hakanson, R., Horig, J., Jie-Cheng, X., and Leander, S. (1984) Biological evaluation of substance P antagonists. *Br. J. Pharmacol.* **83,** 449–456.

Folkers, K., Feng, D.-M., Asano, N., Hakanson, R., Weisenfeld-Hallin, Z., and Leander, S. (1990) Spantide II, an effective tachykinin antagonist having high potency and negligible neurotoxicity. *Proc. Natl. Acad. Sci. USA* **87,** 4833–4835.

Gilon, C., Halle, D., Chorev, M., Selinger, Z., and Byk, G. (1991a) Backbone cyclization: a new method for conferring conformational constraint on peptides. *Biopolymers* **31,** 745–750.

Gilon, C., Halle, D., Chorev, M., Selinger, Z., Goldshmith, R., and Byk, G. (1991b) Backbone-to-end cyclic tachykinins: A new approach to conformationally restricted peptides, in *Peptides 1990* (Giralt. E. and Andreu, D., eds.), ESCOM Science Publishers B.V., Leiden, pp. 404–406.

Guiliani, S., Baranti, G., Turini, D., Quartara, L., Rovero, P., Giachetti, A., and Maggi, C. (1991) NK_2 tachykinin receptors and contraction of circular muscle of the human colon: characterization of the NK_2 receptor subtype. *Eur. J. Pharmacol.* **203,** 365–370.

Hagan, R. M., Ireland, S. J., Jordan, C. C., Bailey, F., Stephens-Smith, M., Deal, M., and Ward, P. (1989) Novel, potent and selective agonists at NK-1 and NK-2 receptors. *Br. J. Pharmacol.* **98,** 717P.

Hagan, R. M., Ireland, S. J., Jordan, C. C., Beresford, I. J. M., Deal, M. J., and Ward, P. (1991) Receptor selective peptidase-resistant agonists at neruokinin NK-1 and NK-2 receptors: new tools for investigating neurokinin function. *Neuropeptides* **19,** 127.

Hagiwara, D., Miyake, H., Morimoto, H., Murai, M., Fujii, T., and Matsuo, M. J. (1991) The discovery of a tripeptide substance P antagonist and its structure-activity relationships. *Pharmacobio.-Dyn.* **14,** s-104.

Harbeson, S. L., Buck, S. H., Hassman III, C. F., and Shatzer, S. A. (1989) Synthesis and Biological Activity of [$\psi(CH_2NH)$] Analogs of Neurokinin A_{4-10}, in *Peptides: Chemistry, Structure and Biology: Proceedings of the Eleventh American Peptide Symposium* (Rivier, J. E. and Marshall, G. R., eds.), ESCOM Science Publishers B. V., Leiden, pp. 180,181.

Harbeson, S. L., Malikayil, J. A., and Buck, S. H. (1991) Cyclic $\psi(CH_2NR)$ Peptide Neurokinin A Antagonists: Structure-Activity and Conformational Studies, in *Peptides: Chemistry and Biology: Proceedings of the 12th American Peptide Symposium* (Smith, J. A. and Rivier, J. E., eds.), ESCOM Science Publishers B. V., Leiden, pp. 124,125.

Harbeson, S. L., Shatzer, S. A., Le, T.-B., and Buck, S. (1992) A new class of antagonists at the neurokinin A NK-2 receptor: $\psi(CH_2NR)$ reduced peptide bond analogues of neurokinin A_{4-10}. *J. Med. Chem.* **35,** 3949–3955.

Hashimoto, T., Kurosawa, K., and Uchida, Y. (1987a) Syntheses and biological activities of neurokinin A analogs substituted with glycine. *Bull. Chem. Soc. Jpn.* **60,** 3071–3073.

Hashimoto, T., Uchida, Y., Naminohira, S., and Sakai, T. (1987b) Tachykinin antagonist I: specific, competitive and tissue-selective neurokinin B antagonists on contractile activity in smooth muscles. *Japan J. Pharmacol.* **45,** 570–573.

Holzemann, G., Jonczyk, A., Eiermann, V., Pachler, K. G. R., Barnickel, G., and Regoli, D. (1991) Conformation-based design of two cyclic physalaemin analogues. *Biopolymers* **31,** 691–697.

Jacoby, H. I., Lopez, I., Wright, D., and Vaught, J. (1986) Differentiation of multiple neurokinin recpetors in the guinea pig ileum. *Life Sci.* **39,** 1995–2003.

Jukic, D., Mayer, M., Schmitt, P., Drapeau, G., Regoli, D., and Michelot, R. (1991) Synthesis and biological activities of neurokinin pseudopeptide analogues containing a reduced peptide bond. *Eur. J. Med. Chem.* **26,** 921–928.

Karagiannis, K., Manolopoulos, A., Stavropoulos, G., Polulos, C., Jordan, C. C., and Hagan, R. M. (1991) Synthesis of a potent agonist of substance P by modifying the methionyl and glutamyl residies of the C-terminal hexapeptide of substance P. *Int. J. Peptide Protein Res.* **38,** 350–356.

Laufer, R., Gilon, C., Chorev, M., and Selinger, Z. (1986) [pGlu^6, Pro^9]-substance P(6–11), a selective agonist for the substance P P-receptor subtype. *J. Med. Chem.* **29,** 1284–1288.

Lavielle, S., Chassaing, G., Besseyre, J., Julien, S., Loeuillet, D., Marquet, A., Beaujouan, J. C., Bergstrom, L., Torrens, Y., and Glowinski, J. (1988a) Tachykinin receptors, in *Peptides, Structure and Function. Proceedings of the 10th American Peptide Symposium* (Marshall, G. R., ed.), ESCOM Science Publishers B. V., Leiden, pp. 482,483.

Lavielle, S., Chassaing, G., Ploux, O., Loeuillet, D., Besseyre, J., Julien, S., Marquet, A., Convert, O., Beaujouan, J.-C., Torrens, Y., Bergstrom, L., Saffroy, M., and Glowinski, J. (1988b) Analysis of tachykinin binding site interactions using constrained analogues of tachykinins. *Biochem. Pharmacol.* **37,** 41–49.

Lee, C. M., Campbell, N. J., Williams B. J., and Iversen L. L. (1986) Multiple tachykinin binding sites in peripheral tissues and in brain. *Eur. J. Pharmacol.* **130,** 209–217.

Logan, M. E., Goswami, R., Tomczuk, B. E., and Venepalli, B. R. (1991) Recent advances in neurokinin receptor antagonists, in *Annu. Rep. in Med. Chem.* **26** (Bristol, J. A., ed.), Academic, New York, pp. 43–51.

Maggi, C. A., Patacchini, R., Giuliani, S., Rovero, P., Dion, S., Regoli, D., Giachetti, A., and Meli, A. (1990) Competitive antagonists discriminate between NK-2 tachykinin receptor subtypes. *Br. J. Pharmacol.* **100,** 588–592.

Maggi, C. A., Patacchini, R., Quartara, L., Rovero, P., and Santicioli, P. (1991a) Tachykinin receptors in the guinea-pig isolated bronchi. *Eur. J. Pharmacol.* **197,** 167–174.

Maggi, C. A., Giuliani, S., Ballati, L., Lecci, A., Manzini, S., Patacchini, R., Renzetti, A. R., Rovero, P., Quartara, L., and Giachetti, A. (1991b) In vivo evidence for tachykininergic transmission using a new NK-2 receptor-selective antagonist, MEN 10,376. *J. Pharmacol. Exp. Therap.* **257,** 1172–1178.

Maggi, C. A., Patacchini, R., Astolfi, M., Rovero, P., Giachetti, A., and Van Giersbergen, P. L. M. (1992) Affinity of R 396, an NK-2 tachykinin receptor antagonist, for NK-2 receptors in preparations from different species. *Neuropeptides* **22,** 93–98.

Maggi, C. A., Patacchini, R., Rovero, P., and Giachetti, A. (1993) Tachykinin receptors and tachykinin receptor antagonists. *J. Autonom. Pharmacol.* **13,** 23–93.

Malikayil, J. A., and Harbeson, S. L. (1992) Conformation of a neurokinin antagonist in solution as studied by NMR and restrained molecular dynamics. *Intl. J. Pept. Prot. Res.* **39,** 497–505.

McElroy, A. B., Clegg, S. P., Deal, M. J., Ewan, G. B., Hagan, R. M., Ireland, S. J., Jordan, C. C., Porter, B., Ross, B. C., Ward, P., and Whittington, A. R. (1992) Highly potent and selective heptapeptide antagonists for the neurokinin NK-2 receptor. *J. Med. Chem.* **35,** 2582–2591.

McKnight, A. T., Maguire, J. J., and Varney, M. A. (1987) Characterization of receptors for tachykinins in guinea-pig islolated trachea. *Br. J. Pharmacol.* **91,** 360P.

McKnight, A. T., Maguire, J. J., Williams B. J., Foster, A. C., Tridgett, R., and Iversen, L. L. (1988) Pharmacological specificity of synthetic peptides as antagonists at tachykinin receptors. *Regul. Pept.* **22,** 127.

McKnight, A. T., Maguire, J. J., Elliott, N. J., Fletcher, A. E., Foster, A. C., Tridgett, R., Williams, B. J., Longmore, J., and Iversen, L. L. (1991) Pharmacological specificity of novel, synthetic, cyclic peptides as antagonists at tachykinin receptors. *Br. J. Pharmacol.* **104,** 355–360.

Michelot, R., Mayer, M., Magneney, S., Pham Van Chuong, P., Schmitt, P., and Potier, P. (1988) Activity of the C-terminal part of substance P on guinea pig ileum and trachea preparations I. *N*-Acylated pentapeptides substance P(7–10). *Eur. J. Med. Chem.* **23,** 243–247.

Osakada, F., Kubo, K., Goto, K., Kanazawa, I., and Munekata, E. (1986) The contractile activities of neurokinin A, B and related peptides on smooth muscles. *Eur. J. Pharmacol.* **120,** 201–208.

Piercey, M. F., Dorbry-Schreur, P. H. K., Masiques, N., and Schroeder, L. A. (1985) Stereospecificity of substance P1 and substance P2 substance P receptors. *Life Sci.* **36,** 777–780.

Poulos, C., Brown, J. R., and Jordan, C. C. (1986) Synthesis and biological activity of substance P C-terminal hexapeptide analogues: structure-activity studies. *J. Med. Chem.* **29,** 1281.

Poulos, C., Stavropoulos, G., Brown, J. R., and Jordan C. C. (1987) Structure-activity studies on the C-terminal hexapeptide of substance P with modifications at the glutaminyl and methioninyl residues. *J. Med. Chem.* **30,** 1512–1515.

Qian, J.-M., Coy, D. H., Jiang, N.-Y., Gardner, J. D., and Jensen, R. T. (1989) Reduced peptide bond pseudopeptide analogues of substance P. A new class of substance P receptor antagonists with enhanced specificity. *J. Biol. Chem.* **264,** 16,667–16,671.

Regoli, D., Mizrahi, J., D'Orléans-Juste, P., Dion, S., Drapeau, G., and Escher, E. (1985) Substance P antagonists showing some selectivity for different receptor types. *Eur. J. Pharmacol.* **109,**121–125.

Regoli, D., Drapeau, G., Dion, S., and D'Orléans-Juste, P. (1987) Pharmacological receptors for substance P and neurokinins. *Life Sci.* **40,** 109–117.

Regoli, D., Drapeau, G., Dion, S., and Couture, R. (1988) New selective agonists for neurokinin receptors: pharmacological tools for receptor characterization. *Trends Pharmacol. Sci.* **9,** 290–295.

Regoli, D., Drapeau, G., Dion, S., and D'Orléans-Juste, P. (1989) Receptors for substance P and related neurokinins. *Pharmacology* **38,** 1–15.

Rovero, P., Pestellini, V., Patacchini, R., Santicioli, P., Maggi, C. A., and Meli, A. (1987) Synthesis and biological activity of *N*-methylated analogues of neurokinin A. *Neuropeptides* **10,** 355–359.

Rovero, P., Pestellini, V., Patacchini, R., Giuliani, S., Santicioli, P., Maggi, C. A., Meli, A., and Giachetti, A. (1989a) A potent and selective agonist for NK-2 tachykinin receptor. *Peptides* **10,** 593–595.

Rovero, P., Pestellini, V., Patacchini, R., Santicioli, Maggi, C.A., and Meli, A. (1989b) Conformationally constrained tachykinins: *N*-methylated analogues of neurokinin A. *Biopolymers* **28,** 65–67.

Rovero, P., Pestellini, V., Rhaleb, N.-E., Dion, S., Rouissi, N., Tousignant, C., Telemaque, S., Drapeau, G., and Regoli, D. (1989c) Structure-activity studies of neurokinin A. *Neuropeptides* **13,** 263–270.

Rovero, P., Pestellini, V., Maggi, C. A., Patacchini, R., Regoli, D., and Giachetti, A. (1990a) A highly selective NK-2 tachykinin receptor antagonist containing D-tryptophan. *Eur. J. Pharmacol.* **175,** 113–115.

Rovero, P., Pestellini, V., Patacchini, R., Giuliani, S., Maggi, C. A., Meli, A., and Giachetti, A. (1990b) Synthesis and biological activity of NK-2 selective tachykinin antagonists containing D-tryptophan. *Peptides* **11,** 619,620.

Rovero, P., Astolfi, M., Renzetti, A. R., Patacchini, R., Giachetti, A., and Maggi, C. A. (1991) Role of D-tryptophan for affinity of MEN 10207 tachykinin antagonist at NK-2 receptors. *Peptides* **12,** 1015–1018.

Rovero, P., Astolfi, M., Manzini, S., Jucik, D., Rouissi, N., Maggi, C. A., and Regoli, D. (1992a) Structure-activity study of R 396, an NK-2 tachykinin antagonist selective for the NK-2B receptor subtype. *Neuropeptides* **23,** 143–145.

Rovero, P., Quartara, L., Astolfi, M., Patacchini, R., Giachetti, A., and Maggi, C. A. (1992b) Structure-activity study of the C-terminal residue of MEN 10207 tachykinin antagonist. *Peptides* **13,** 207,208.

Saviano, G., Temussi, P. A., Motta, A., Maggi, C. A., and Rovero, P. (1991) Conformation-activity relationship of tachykinin neurokinin A (4–10) and of some [Xaa^8] analogues. *Biochemistry* **30,** 10,175–10,181.

Smith, P. W., McElroy, A. B., Pritchard, J. M., Deal, M. J., Ewan, G. B., Hagan, R. M., Ireland, S. J., Ball, D., Beresford, I., Sheldrik, R., Jordan, C. C., and Ward, P. (1993) Low molecular weight neurokinin NK-2 antagonists. *Biorg. Med. Chem. Lett.* **3,** 931–936.

Tschirhart, E., Schmitt, P., Bertrand, C., Mayer, M., Magneney, S., Landry, Y., and Michelot, R. (1989) Contractile activity of the *N*-acylated C-terminal part of substance P(7–11) in guinea pig trachea. *Naunyn Schmiedeberg's Arch. Pharmacol.* **340,** 107–110.

Ward, P., Ewan, G. B., Jordan, C. C., Ireland, S. J., Hagan, R. M., and Brown, J. R. (1990) Potent and highly selective neurokinin antagonists. *J. Med. Chem.* **33,** 1848–1851.

Watson, S. P., Sandberg, B. E. B., Hanley, M. R., and Iversen, L. L. (1983) Tissue selectivity of substance P alkyl esters: suggesting multiple receptors. *Eur. J. Pharmacol.* **87,** 77–84.

Williams, B. J., Curtis, N. R., McKnight, A. T., Maguire, J., Foster, A., and Tridgett, R. (1988) Development of NK_2 selective antagonists. *Regul. Pept.* **22,** 189.

Wormser, U., Laufer, R., Hart, Y., Chorev, M., Gilon, C., and Selinger, Z. (1986) Highly selective agonists for substance P receptor subtypes. *EMBO J.* **5,** 2805–2808.

Zacharia, S., Rossowski, W. J., Jiang, N.-Y., Hrbas, P., Ertan, A., and Coy, D. H. (1991) New reduced peptide bond substance P agonist and antagonists: effects on smooth muscle contraction. *Eur. J. Pharmacol.* **203,** 353–357.

CHAPTER 13

Pharmacological Characterization of Receptor Types

Domenico Regoli, Noureddine Rouissi, and Pedro D'Orléans-Juste

1. Introduction

The various biological actions of substance P and related tachykinins/neurokinins are mediated by three different receptor types, which were named NK_1, NK_2, and NK_3 during the Montreal Symposium in 1986 (Henry, 1987). The three receptors have been characterized pharmacologically (Regoli et al., 1987a, 1988, 1989), biochemically (Lavielle et al., 1988; Buck et al., 1991), and genetically (Nakanishi, 1991). In vivo activities have been attributed to each receptor by Maggi et al. (1987), Regoli et al. (1988), and Maggi et al. (1993). Cloning and sequencing of receptors and the study of receptor properties of mRNA expressed in special systems (Nakanishi, 1991) appear to validate the three physiological entities.

The Tachykinin Receptors Ed.: S. H. Buck © 1994 Humana Press Inc., Totowa, NJ

1.1. Characterization of Receptors Using Neurokinins and Tachykinins

It has been suggested (Regoli et al., 1988, 1989) that NK_1 is the functional site for SP, NK_2 for neurokinin A, and NK_3 for neurokinin B on the basis of the order of potency of naturally occurring peptides (mammalian neurokinins, tachykinins) shown below:

NK_1	SP > NKA ≥ NKB PHY > ELE ≥ KAS
NK_2	NKA > NKB > SP ELE > KAS > PHY
NK_3	NKB > NKA > SP KAS > ELE > PHY

Numeric values obtained with neurokinins and tachykinins in selected preparations that are currently used in our laboratory (Regoli et al., 1985, 1987b, 1991) for the characterization of the three functional sites are presented in Table 1 for the NK_1, Table 2 for the NK_2, and Table 3 for the NK_3 receptors.

From Table 1, it is evident that the order of potency of agonists given above is found in two isolated vessels, the dog carotid artery (D'Orléans-Juste et al., 1985) and the rabbit vena cava or jugular vein (Nantel et al., 1991). The same order of potency of neurokinins observed in the rabbit vena cava has been obtained by Yokota et al. (1989) in COS-7 cells transfected with NK_1 receptor mRNA (Fig. 1). Worthy of notice is the fact that apparent affinities of SP in the two preparations are very similar (Fig. 1). The untreated guinea pig ileum, used extensively in early studies (Lee et al., 1982; Pernow, 1983; Regoli et al., 1985), and even the preparation treated with a mixture of antagonists and inhibitors (Table 1) responds fairly well to NKA and NKB and shows very high sensitivity to tachykinins, ELE, and KAS. These high residual activities indicate the presence of NK_3 receptors in the untreated ileum and of functional NK_2 sites after elimination of the NK_3 sites by atropine. In accord with Maggi et al. (1990b), it is concluded that contractions of the longitudinal and circular muscle of the guinea pig ileum are mediated by all three

Table 1
Apparent Affinities (pD2) and Relative Activities (RA)
of Neurokinins and Tachykinins in NK_1 Receptor Preparations

	GPI, nt[a]		GPI, t		DCA[b]		RVC	
	pD_2	RA	pD_2	RA	pD_2	RA	pD_2	RA
Substance P	8.78[c]	100[d]	8.48	100	10.0	100	8.60	100
Neurokinin A	8.40	42	7.90	26	9.40	25	7.30	4.8
Neurokinin B	8.64	72	7.94	29	8.90	7.9	7.20	3.7
Physalaemin	9.04	182	8.80	209	10.0	100	8.90	162
Eledoisin	9.21	269	8.60	132	9.50	32	7.70	12
Kassinin	9.35	372	8.76	191	9.20	16	7.80	14

[a]GPI, guinea pig ileum, (nt) not treated, or (t) treated with atropine ($4.0 \times 10^{-6}M$), indomethacin ($3.6 \times 10^{-6}M$), and diphenhydramine ($3.9 \times 10^{-6}M$).

[b]DCA, dog carotid artery (relaxation); RVC, rabbit vena cava (contraction).

[c]pD_2, log of the concentration of peptide producing 50% of the maximum effect.

[d]RA, relative activity in percent of the effect of SP.

Table 2
Apparent Affinities (pD_2) and Relative Activities (RA)
of Neurokinins and Tachykinins in NK_2 Receptor Preparations

	RPA[a]		RD[b]		GPT[c]	
	pD_2	RA	pD_2	RA	pD_2	RA
Substance P	6.13[d]	0.8[e]	6.50	1.9	6.75	2.2
Neurokinin A	8.22	100	8.22	100	8.40	100
Neurokinin B	7.45	17	8.15	85	7.54	14
Physalaemin	6.02	0.6	6.52	2.0	7.37	9.3
Eledoisin	8.22	100	7.70	30	7.88	30
Kassinin	7.66	28	8.00	60	6.88	3.0

[a]RPA, rabbit pulmonary artery.

[b]RD, rat duodenum.

[c]GPT, guinea pig trachea.

[d]pD_2, as in Table 1.

[e]RA, relative activity expressed in percent of NKA.

Table 3
Apparent Affinities (pD2) and Relative Activities (RA)
of Neurokinins and Tachykinins in NK_3 Receptor Preparations

	HUB[a]		DUB[b]		RPV[c]	
	pD_2	RA	pD_2	RA	pD_2	RA
Substance P	5.57[d]	2[e]	6.10	4.3	5.82	1.4
Neurokinin A	7.40	159	7.20	54	6.45	5.9
Neurokinin B	7.20	100	7.47	100	7.68	100
Physalaemin	5.90	5.0	7.55	120	6.18	3.1
Eledoisin	7.06	72	6.90	27	7.11	27
Kassinin	7.90	501	6.25	6.0	7.20	33

[a]HUB, hamster urinary bladder.
[b]DUB, dog urinary bladder.
[c]RPV, rat portal vein.
[d]pD_2, as in Table 1.
[e]RA, relative activity in percent of NKB.

types of neurokinin receptors. The preparation has been recently studied by Petitet et al. (1992a,b) who have suggested the existence of a new tachykinin receptor in the guinea pig ileum on the basis of discrepancies between data of biological and binding assays. Such discrepancies are fairly common and should not be given too much attention. Use of septide and SP-O-CH_3 for receptor characterization should be discouraged, as pointed out by several authors (Regoli et al., 1988; Guard and Watson, 1991).

Apparent affinities and relative activities of neurokinins and tachykinins obtained in tissues that have been used to characterize the NK_2 functional site are summarized in Table 2. The order of potency for the NK_2 site has been established in the rabbit pulmonary artery (D'Orléans-Juste et al., 1986), a preparation that is currently used in several laboratories (Dion et al., 1987; Buck et al., 1991; Advenier et al., 1992; Emonds-Alt et al., 1992) for evaluating agonist and antagonist affinities on the NK_2 receptor type. This preparation is compared with two other isolated organs, the rat duodenum

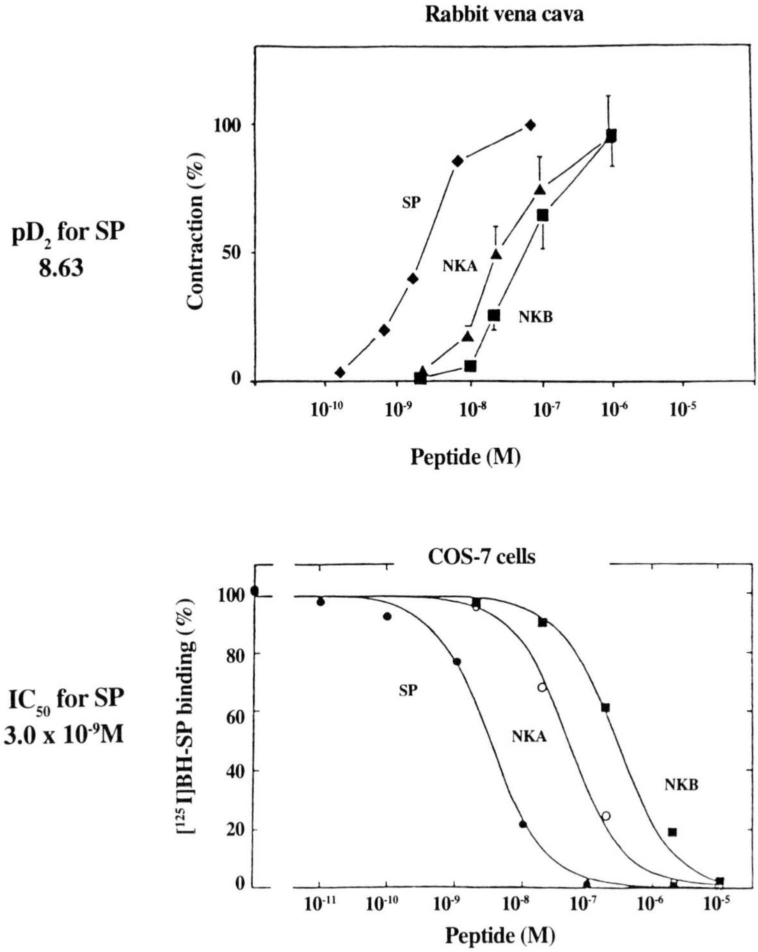

Fig. 1. Comparison of binding affinities of tachykinins to COS-7 cells (Yokota et al., 1989) and biological activities (contraction of rabbit vena cava) (Nantel et al., 1991).

and guinea pig trachea, which have been (Munekata et al., 1987) and still are used in other laboratories. Data in Table 2 indicate that the rat duodenum is still very sensitive to NKB and shows more sensitivity to KAS than to ELE. These data suggest that rat duodenum contains a large population of functional NK_3 sites and might therefore be inadequate for precise pharmacological studies on NK_2 receptors. The guinea pig trachea maintains some sensitivity to NK_1 agonists, particularly PHY, and is not a pure NK_2 receptor system. The presence of functional NK_1 receptor sites in this preparation can be better demonstrated by the selective agonists (*see* Section 2 and Table 5 later in this chapter) and has been suggested and discussed by several workers (Devillier et al., 1988; Maggi, 1990; Abelli et al., 1991; Ireland et al., 1991). The guinea pig trachea is therefore not to be considered as a preparation of choice for the characterization of NK_2 functional sites, either with agonists or with antagonists. Some years ago, a new receptor type, the NK_4, was proposed by McKnight et al. (1988, 1991) from data obtained in pharmacologic and biochemical assays on this tissue. This new receptor entity has not been confirmed; workers who have repeated the same experiments have reached the conclusion that the unusual behaviors of the guinea pig trachea could be explained by the presence of two functional sites, NK_1 and NK_2 (Ireland et al., 1988, 1991; Maggi et al., 1990a).

Data of pharmacological assays, directed toward the characterization of the NK_3 functional site, are presented in Table 3. Three isolated organs that have been used in pharmacological and biochemical studies on NK_3 sites have been compared, the hamster and dog urinary bladders and the rat portal vein. Orders of potency of agonists, indicated above as typical of the NK_3 functional site, have been determined on the rat portal vein, since the other two preparations appear to be complex multifunctional systems. The hamster and dog urinary bladders show unusual patterns and important discrepancies between neurokinins and tachykinins activities. The hamster urinary bladder is very sensitive to NKA and to KAS, whereas dog urinary bladder shows high sensitivity to NKB and PHY. These data suggest that hamster and dog urinary bladders are tissues whose responses are mediated by multiple receptors, NK_2 and NK_3 in the hamster urinary bladder and NK_1, NK_2, and NK_3 in the dog urinary bladder. Both tissues have been used to characterize receptors with binding assays, especially the hamster urinary bladder in which Buck and Shatzer (1988) described initially a new tachykinin binding site that showed high sensitivity to KAS, which for this reason was called SP-K. Discrepancies between neurokinins and tachykinins may in part be attributed to the presence of various peptidases in the isolated bladders, as described by Rouissi et al. (1990).

2. *Characterization of Receptors by Selective Agonists*

Primary structures of six selective agonists, two for each receptor site, are presented in Fig. 2. Results obtained with selective agonists, summarized in Tables 4, 5, and 6, confirm in large part the findings of the naturally occurring peptides. Six selective agonists, some of them analogs of the natural peptides, some others hexapeptide or heptapeptide C-terminal fragments, have been used to obtain a better characterization of the myotropic responses of the same preparations analyzed in Tables 1, 2, and 3.

Results summarized in Table 4 indicate that relative activities of NK_1 selective agonists differ between the two vascular preparations, $[Sar^9,Met(O_2)^{11}]$SP sulfone being almost three times and the NK_1 selective hexapeptide 1.5 times more active that SP on the dog carotid artery, while they are equivalent in the rabbit vena cava. Such differences are interesting, since they point to functional diversity between the NK_1 site that mediates release of NO and prostacyclin, smooth muscle relaxation, and peripheral vasodilatation on one site and the NK_1 site that contracts the venous smooth muscle on the other. Worthy of mention is the finding that $[Sar^9]$SP sulfone is almost eight times more potent that SP as hypotensive agent in the rat (Couture et al., 1989). Potencies of NK_1 selective agonists observed in the guinea pig ileum, nontreated or treated with antagonists, are similar to those of the dog carotid artery. Some important differences between the vascular and the intestinal preparations are, however, found with the NK_2 and NK_3 selective agonists that are almost inactive in the isolated vessels, while maintaining activites on the guinea pig ileum. As discussed above, both NK_3 and NK_2 receptors are present in this tissue together with the NK_1. NK_2 selective agonists, especially $[\beta Ala^8]$NKA(4–10), maintain some activity after elimination (by atropine) of the NK_3 functional site (*see also* Maggi et al., 1990b).

NK_1 Arg-Pro-Pro-Arg-Gln-Gln-Phe-Phe-**Sar**-Leu-$MetNH_2$

\+ \+ SO_2

Ac-Arg-Phe-Phe-**Sar**-Leu-$MetNH_2$

\+ SO_2

NK_2 Asp-Ser-Phe-Val-Gly-Leu-**Nle**-NH_2

--

Asp-Ser-Phe-Val-β-**Ala**-Leu-$MetNH_2$

--

NK_3 Asp-Met-His-Asp-Phe-Phe-**MePhe**-Gly-Leu-$MetNH_2$

-- --

Succ-Asp-Phe-**MePhe**-Gly-Leu-$MetNH_2$

-- --

Fig. 2. Primary structures and chemical features of tachykinin-receptor selective agonists.

NK_2 preparations are analyzed in Table 5, where the activities of the selective agonists on three NK_2 receptor systems have been put together for comparison. Thus, NK_1 selective agonists are inactive on the rabbit pulmonary artery and show very little activity on the rat duodenum, but are quite active (especially the hexapeptide) on the guinea pig trachea. NK_2 selective agonists are potent stimulants of the three preparations, [βAla^8]NKA(4–10) being more active than NKA, and [Nle^{10}]NKA(4–10) being approximately half as active as NKA. Some important differences are seen with [MePhe^7] NKB and (although much less evident) with senktide, which are inactive on the rat pulmonary artery and guinea pig trachea, but maintain strong

Table 4
Apparent Affinities (pD_2) and Relative Activities (RA) of Selective Agonists on NK_1 Receptor Preparations

	GPI, nt[a]		GPI, t		DCA		RVC	
	pD_2	RA	pD_2	RA	pD_2	RA	pD_2	RA
Substance P	8.78[b]	100[c]	8.48	100	10.00	100	8.63	100
[Sar^9,Met(O$_2$)^{11}]SP	8.91	135	8.51	107	10.45	282	8.60	93
Ac[Arg^6,Sar^6,Met (O$_2$)^{11}]SP(6–11)	9.52	550	8.83	224	10.20	158	8.66	107
[Nle^{10}]NKA(4–10)	7.99	16	7.43	8.9	7.00	0.1	5.37	0.06
[βAla^8]NKA(4–10)	7.00	1.6	7.12	23	6.71	0.5	4.93	0.02
[MePhe^7]NKB	8.76	96	6.59	1.2	7.15	0.1	5.63	0.1
Senktide	8.87	123	5.78	0.2	6.39	0.02	5.47	0.07

[a]abbreviations as in Table 1.
[b]pD_2, as in Table 1.
[c]RA, relative activity in percent of the effect of substance P.

Table 5
Apparent Affinities (pD_2) and Relative Activities (RA) of Selective Agonists on NK_2 Receptor Preparations

	RPA[a]		RD		GPT	
	PD_2	RA	PD_2	RA	pD_2	RA
Sar^9,Met(O$_2$)^{11}]SP	In.[b]		6.67	2.8	7.32	8.3
Ac[Arg^6,Sar^9,Met (O$_2$)^{11}]SP(6–11)	In.		6.63	2.6	7.79	25
Neurokinin A	8.22[c]	100[d]	8.22	100	8.40	100
[Nle^{10}]NKA(4–10)	7.90	48	8.02	63	7.99	39
[βAla^8]NKA(4–10)	8.60	240	8.27	112	8.85	282
[MePhe^7]NKB	5.24	0.10	8.66	275	6.02	0.42
Senktide	5.46	0.17	8.15	85	5.35	0.09

[a]Abbreviations as in Table 2.
[b]In., inactive at $10^{-5}M$
[c]pD_2, as in Table 1.
[d]Relative activity in percent of NKA.

Table 6
Apparent Affinities (pD2) and Relative Activities (RA) of Selective Agonists on NK_3 Receptor Preparations

	HUB[a]		DUB[b]		RPV[c]	
	pD_2[d]	RA[e]	pD_2	RA	pD_2	RA
[Sar[9],Met(O_2)[11]]SP	3.47	0.02	6.12	4.5	In.[f]	
Ac[Arg[6],Sar[9],Met(O_2)[11]]SP(6–11)	4.47	0.2	N.D.	N.D.	In.	
[Nle[10]]NKA(4–10)	7.09	78	6.73	18	5.68	1.0
[βAla[8]]NKA(4–10)	7.32	131	N.D.	N.D.	6.13	2.8
Neurokinin B	7.20	100	7.47	100	7.68	100
[MePhe[7]]NKB	6.14	8.7	5.46	1.0	8.3	417
Senktide	5.74	3.5	5.27	0.6	7.61	85

[a]HUB, hamster urinary bladder.
[b]DUB, dog urinary bladder.
[c]RPV, rat portal.
[d]pD_2; as in Table 1.
[e]RA, relative activity in percent of the effect of NKB.
[f]In. inactive at $10^{-5}M$; N.D., not determined.

activity on the rat duodenum. Results obtained with selective agonists validate therefore the conclusions drawn before, namely that the guinea pig trachea contains NK_1 and NK_2 and the rat duodenum NK_2 and NK_3 functional sites.

In the NK_3 systems of the hamster urinary bladder, dog urinary bladder, and rat portal vein, the selective agonists provide clear indication that the rat portal vein is a pure NK_3 system, while the hamster urinary bladder is predominantly a NK_2 (*see* residual activities of NK_2 selective agonists) and has some NK_3 sites, and the dog urinary bladder appears to contain two of the three neurokinin receptor sites (NK_1 and NK_2) and possibly some NK_3.

Selectivity of agonists was obtained by replacing Gly[9] or Phe[8] and modifying the C-terminal Met (Fig. 2). NK_1 receptor occupation is improved by stabilizing the C-terminal hexapeptide with Sar in position 9. Oxidation of Met[11] further improved affinity and selectivity (Dion et al., 1987; Drapeau et al., 1987). Selectivity for the NK_2 site is provided either with more freedom in position 9, obtained by replacing Gly with βAla (Rovero et al., 1989a,b), or by substituting Nle for the C-terminal Met (Drapeau et al., 1987; Dion et al., 1987). This second modification is, however, accompanied by loss of affinity. Occupation of the NK_3 site is favored by the use of a α-methyl Phe in position 8 both in the deca- and heptapeptide C-terminal sequences (Drapeau et al., 1987; Dion et al., 1987) (Fig. 2).

3. Characterization of Receptors by Antagonists

Recent developments in the field of neurokinin-receptor antagonists have been accomplished with the identification of new agents with short peptidic (tripeptide, dipeptide) sequences or nonpeptide structures. Some of these compounds have been found to antagonize efficiently and selectively the myotropic effects of neurokinins in vitro (Snider et al, 1991a,b; Garret et al., 1991; Jukic et al., 1991; Rouissi et al., 1993) and some effects in vivo (Rouissi et al., 1993). The primary structures of some compounds that have been widely studied are presented in Fig. 3, together with the structures of early peptidic compounds such as spantide and an octapeptide antagonist. Three antagonist categories have been considered on the basis of their chemical nature:

1. **Peptides** (the first generation), which are analogs of SP or of the C-terminal fragment of neurokinins;
2. **Peptidoids** (the second generation), identified by Fujisawa Pharmaceutical Co. (Japan) investigators (Fujii et al., 1992; Morimoto et al., 1992) and designed from the C-terminal sequences of octapeptide antagonists (Regoli et al., 1985); and
3. **Nonpeptides** (the third generation), which have been identified through extensive screening of numerous molecules by binding assays (Garret et al., 1991; Snider et al., 1991a; Emonds-Alt et al., 1992). The nonpeptidic compounds include two selective NK_1 and a selective NK_2 receptor antagonist.

SPANTIDE	D.Arg-Pro-Lys-Pro-Gln-Gln-D.Trp-Phe-D.Trp-Leu-Leu.NH_2
R-204	D.Pro-Gln-Gln-D.Trp-Phe-D.Trp-Leu-Nle.NH_2
MEN 10376	Asp-Tyr-D.Trp-Val-D.Trp-D.Trp-Lys.NH_2
R-396	Ac.Leu-Asp-Gln-Trp-Phe-Gly.NH_2
R-486	Asp-Ser-Phe-Trp-βAla-Leu-MetNH_2

R-544	Ac.Thr-D.Trp(CHO)-Phe.N.Me.Bz
R-545	Boc-D.Trp(CHO)-Phe.N.Me.Bz

CP 96545

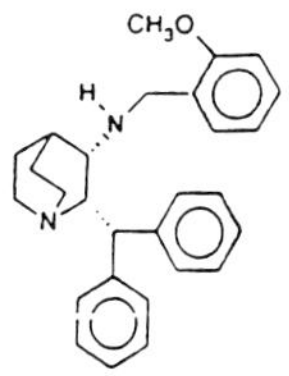

CP 96345

RP 67580

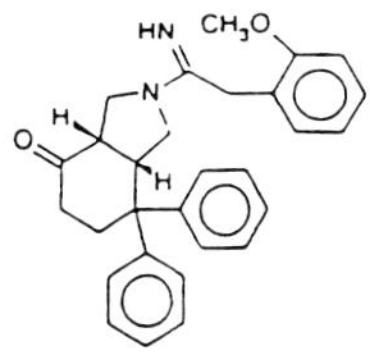

RP 67580

SR 48968

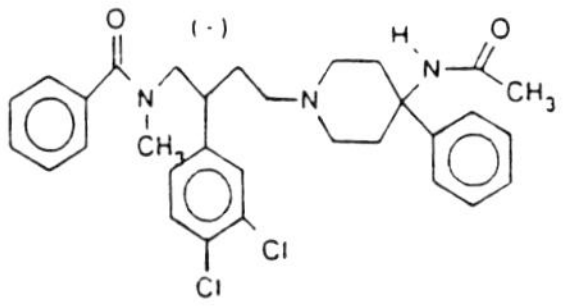

SR-48968

Fig. 3. Primary structures of peptide and peptidoid antagonists of tachykinin receptors. Structures and codes of nonpeptidic antagonists.

Results obtained with a series of antagonists (old and new) on the three neurokinin monoreceptor systems, the rabbit vena cava, the rat pulmonary artery, and the rat portal vein, are summarized in Table 7. As pointed out recently by Jukic et al. (1991), the peptidic com-

Table 7
Apparent Affinities (pA_2) of Tachykinin Antagonists in Three Monoreceptor Preparations

	Biological activities, pA_2		
Antagonist	RVC	RPA	RPV
Spantide	6.80[a]	5.80	6.20
[D-Pro[4],D-Trp[7,9] Nle[11]]SP(4–11)	7.00	5.10	5.40
R-544	7.30	5.40	5.80
R-545	6.77	4.77	5.52
CP 96345	9.52	<4.32	5.62
RP 67580	7.23	In.[b]	<6.0
SR 48968	6.08	9.78	In.
MEN 10376	In.	7.90	4.90
R-396	In.	5.60	In.
R-486	In.	In.	7.46

[a]pA_2, log of the concentration of antagonist that reduces the effect of a double dose of agonist to that of a single. pA_2 values were measured against SP on the RVC (rabbit vena cava), against NKA on the RPA (rabbit pulmonary artery), and against NKB on the RPV (rat portal vein).
[b]In., inactive at 10^{-5}*M*.

pounds (group 1.) are rather weak (pA_2 around 7.0) and show a certain selectivity for the NK_1 receptor. Peptidoids (group 2.), such as R-544 and R-545, show similar properties, although the dipeptidoid, FK 888 ([2-*N*-Me]indolil-CO-Hyp-NalNMeBz) (Fujii et al., 1992), has been found to exert a potent antagonism against SP on NK_1 receptors and to be quite selective for this functional site (Morimoto et al., 1992). Nonpeptide antagonists (group 3.) are extremely potent and show high selectivity. CP 96345 is an active agonist of SP in the rabbit vena cava at concentrations (pA_2 9.5) of less than 1 n*M* and is practically inactive on the other two systems, where concentrations of CP 96345 10,000 times higher may exert nonspecific effects (Constantine et al., 1991) (e.g., interference with Ca^{2+} channels [Guard and Watling, 1992]) rather that antagonism of NKA or NKB.

RP 67580 is less potent than CP 96345, but is equally selective for the NK_1 site (Garret et al., 1991). Worthy of mention is the fact that RP 67580 is as active as the peptidoids (e.g., R-544) on NK_1 functional sites and shows selectivity for the NK_1 receptor. This compound has been shown to be active in the rat, as well as in the guinea pig, and helped validate the suggestions that two NK_1 receptor subtypes may be present in different species (Gitter et al., 1991; Snider et al., 1991a; Beresford et al., 1991; Rouissi et al., 1993) (*see also* Chapter 14 by Maggi).

SR 48968 is a newly discovered compound (Emonds-Alt et al., 1992) extensively characterized pharmacologically (Advenier et al., 1992) and showing a very selective antagonistic activity against NKA in NK_2 systems, while having little active on NK_1 (weaker than on NK_2 by almost 3 log units) and inactive on NK_3. This finding is of particular importance because it allows for a clear and definite differentiation between the NK_2 and the NK_3 functional sites. SR 48968 appears to be a very useful new agent to determine the pathophysiological role of NK_2 receptors *(see below)*.

Two other compounds, MEN 10376 and R-396, are peptide antagonists showing fairly high affinities for the NK_2 receptor of the rat pulmonary artery (MEN 10376) and the hamster urinary bladder (R-396) (*see* Tables 7 and 8) and showing selectivity for NK_2 receptor subtypes (Maggi et al., 1990a). This topic is extensively discussed by Maggi in the next chapter.

R-486 (Asp-Ser-Phe-Trp-βAla-Leu-Met-NH_2) has been shown by Drapeau et al. (1990) to inhibit the effect of NKB on the NK_3 system of the rat portal vein. The antagonist is selective for NK_3 functional sites since it is inactive on the rat pulmonary artery (NK_2) and the rabbit vena cava (NK_1) (Table 7). In some NK_1 preparations, R-486 may even exert agonistic activity (Drapeau et al., 1990) (Table 8). R-486 is able to reduce the effects of NK_3 agonists on the rat duodenum (Table 8), which has been shown to respond to [$MePhe^7$]NKB. The presence of NK_3 functional sites in the rat duodenum is therefore confirmed by the use of the NK_3 antagonist. R-486 shows little activity in the hamster urinary bladder and is an agonist in the guinea pig trachea that contain little if any NK_3 sites.

Table 8
Apparent Affinities of Tachykinin Antagonists in Four Multireceptor Systems

	Biological activities, pA_2[a]				
Antagonist	GPI	RD	HUB	GPT*	GPT**
Spantide	6.70	5.20	<5.2	7.22	N.D.[b]
[D-Pro[4], D-Tlp$_{7,9}$ Nle[11]]SP(4–10)	7.10	5.10	<5.1	5.69	5.69
R-544	6.50	4.90	4.80	6.80	6.40
R-545	6.36	5.52	5.77	5.77	N.D.
CP96345	9.50	6.70	4.69	9.29	4.69
RP 67580	7.20	5.20	In.[c]	6.64	In.
SR 48968	5.50	6.60	7.45	5.76	7.88
MEN 10376	5.00	6.60	<5.6	5.64	7.03
R-396	4.90	6.90	7.50	In.	5.03
R-486	6.50	6.10	5.90	Ag.[d]	Ag.

[a]pA2, same as in Table 7. pA_2 values were measured against SP on the guinea pig ileum (GPI), against NKA in the rat duodenum (RD) and in the hamster urinary bladder (HUB), against Ac[Arg[6],Sar[9],Met(O_2)[11]]SP(6–11)(*) and against [βAla[8]]-NKA(4–10)(**) in the guinea pig trachea (GPT).

[b]N.D., not determined.

[c]In, inactive at 10^{-5}*M*.

[d]Ag, the compound shows some agonisitic activity.

Activities of antagonists, evaluated on the multireceptor systems of the guinea pig ileum, rat duodenum, hamster urinary bladder, and guinea pig trachea, are presented in Table 8. In the last preparation (the guinea pig trachea), the antagonists have been tested against two selective agonists, namely the NK_1 Ac[Arg[6],Sar[9],Met (O_2)[11]]SP(6–11) and the NK_2 [βAla[8]]NKA(4–10). The results obtained with the various compounds in the guinea pig ileum are similar to those observed in the NK_1 monoreceptor system of the rabbit vena cava (Table 7). Guinea pig ileum discriminates very well between the various compounds and shows high affinity for CP 96545; RP 67580 is almost 2.5 log units less active and shows similar affinity as the peptide and peptidoid antagonists. SR 48968 is very weak. On the contrary, the rat duodenum is clearly inadequate for antagonist evaluation, since it does not discriminate between the various compounds. Indeed, similar affinities are observed with CP 67580 (anti NK_1), SR 48968 (anti NK_2) and R486 (anti NK_3). Results obtained on the hamster urinary bladders suggest that this preparation is discriminative for the NK_{2B} functional site (Maggi et al., 1990a), and only some NK_2 receptor antagonists, such as SR 48968 and R-396, show pA_2 values of 7.5 or better. Results obtained on the guinea pig trachea against the NK_1 selective agonists are fairly close to those of the guinea pig ileum and rabbit vena cava. Indeed, CP 96345 is very active and some peptides (e.g., spantide) and peptidoids (e.g., R-544), as well as RP 67580, show similar affinities with pA_2 higher than 6.5; the other compounds are almost inactive.

When challenged with the NK_2-selective agonists, a completely different picture emerges in the guinea pig trachea, since CP 96345 and RP 67580 are very weak or inactive, whereas SR 48968 and MEN 10376 show pA_2 values higher than 7.0 (Table 8). These results, obtained on guinea pig trachea with antagonists, confirm the presence in this tissue of NK_1 and NK_2 receptors and demonstrate the usefulness of selective agonists for discriminating between the two receptor types in the same preparation.

4. Mediators and Second Messengers

Characterization of receptors by second messengers has been attempted for vasopressin by showing that V_1 vascular receptors act via Ca^{2+}, and V_2 renal receptors act via cAMP (Jard, 1988). Such an approach cannot be used for tachykinins, since the three tachykinin receptors activate the inositide-Ca^{2+}-calmodulin systems of a variety of cells with the final effect of inducing the release of mediators (NO, acetylcholine, noradrenaline, and so on) or provoking smooth muscle contraction. As shown in Table 9, the inhibitory effect of SP on vascular smooth muscle, which is responsible for vasodilation, is due to NK_1 receptors of the endothelium (dog carotid artery, rabbit pulmonary artery, and so on) that activate the IP_3-Ca^{2+}-calmodulin-mediated release of NO (Busse and Mulsch, 1990). NO enters the

Table 9
Tachykinin Mediators and Second Messengers

Preparation	Receptor	Target cell second messenger	Mediator	Effector cell second messenger	Effect	Ref.[b]
Dog carotid artery	NK_1	Endothelium, Ca^{2+}	NO[a]	cGMP	Relaxation	1,2
Rabbit vena cava	NK_1	Smooth muscle, Ca^{2+}		Ca^{2+}	Contraction	3
Rabbit pulmonary	NK_1	Endothelium, Ca^{2+}	NO	cGMP	Relaxation	4
artery	NK_2	Smooth muscle, Ca^{2+}		Ca^{2+}	Contraction	5
Rat portal vein	NK_3	Smooth mucle, Ca^{2+}		Ca^{2+}	Contraction	6
Guinea pig ileum	NK_1	Smooth muscle, Ca^{2+}		IP_3, Ca^{2+}	Contraction	7
	NK_3	Parasympathetic nerve, Ca^{2+}	Ach	Ca^{2+}	Contraction	8
Rat aorta	NK_2	Smooth muscle, Ca^{2+}		IP_3, Ca^{2+}	Mitogenesis	9
Rat ileum	NK_2	Smooth muscle, Ca^{2+}		IP_3, Ca^{2+}	Contraction	10

[a]EDRF, which is nitric oxide.

[b]1. Busse and Mulsch, 1990; 2. D'Orléans-Juste et al., 1985; 3. Nantel et al., 1991; 4. D'Orléans-Juste et al., 1986; 5. Nantel, 1991, M.Sc. Thesis, University of Sherbrooke; 6. Sutter, 1990; 7. Watson and Downes, 1983; 8. Guard and Watson, 1990; 9. Hultgardh-Nilsson et al., 1988; 10. Watson, 1984.

smooth muscle cell and activates the soluble guanylate cyclase that leads to increased production of cGMP and to smooth muscle relaxation (Ignarro, 1989). Contraction of smooth muscle and mitogenesis, on the other hand, depend on receptors that are probably localized on the surface of smooth muscle cells and act by activating the IP_3-Ca^{2+}-calmodulin system that leads to contraction of vascular (rabbit pulmonary artery, rat portal vein) and nonvascular (guinea pig ileum, rat ileum) smooth muscles (Table 9). This is true for NK_1 receptors of venous and intestinal tissues, NK_2 receptors of arterial and intestinal smooth muscles, as well as NK_3 receptors of venous smooth muscle on the autonomic parasympathetic nerve terminals (Table 9). In all cell types, the IP_3-Ca^{2+}-calmodulin system is activated, no matter what type of receptor may be present. It is therefore concluded that the three tachykinin receptors are pharmacologically distinct entities that utilize the same intracellular second messenger. This is consistent with the finding by Nakanishi (1991) who has demonstrated that the three tachykinin receptors belong to a superfamily (rhodopsin) of receptors that are coupled to G-proteins. The receptor is genetically differentiated and is determinant for selectivity and specificity of the biological actions, whereas the G-protein and the second messengers appear to be simple transducing systems (perhaps cell selective) that can be used by different receptor types.

5. Tachykinin Activities In Vivo

Tachykinins and their receptors have been implicated in a variety of biological actions in vivo (*see review by* Pernow, 1983; Maggi et al., 1987; Regoli et al., 1988). In vivo activities have been evaluated by applying exogenous peptides by the intravenous (iv), the intracerebral ventricular (icv), or the intrathecal (ith) routes in mice (Laneuville et al., 1988), rats, guinea pigs, other animals, and humans (Maggi et al., 1993). Such activities may not reflect the true physiological roles of neurotransmitters, such as tachykinins, and the discovery of specific and selective antagonists for the various receptor types now provides the essential tools required for evaluating the physiological and pathological implications of these peptides. Some of the in vivo activities observed in the rat and the guinea pig are analyzed in Table 10 by reporting results obtained with the natural peptides, as well as with selective agonists and with antagonists.

Thus, SP and the selective agonist, [Sar^{9}Met(O$_2$)^{11}]SP, have been shown to induce dose-dependent decrease of blood pressure in rats and guinea pigs when given intravenously. Opposite effects (hypertension and tachycardia) have been observed in the rat with the icv application of SP and NKA, and the effects have been blocked selectively by CP-96345 (SP) or by R-396 or SR 48968 (NKA) (Tschöppe et al., 1992). It appears that the peripheral vascular effects of SP and neurokinins are mediated by NK_1, whereas the central effects are mediated by NK_1 or NK_2 receptors (Laneuville et al., 1988; Unger et al., 1988; Couture et al., 1989; Tschöppe et al., 1992). NK_1 receptors are implicated in the potent stimulation of salivary secretion observed with SP in the rat and in other animals (Giuliani et al., 1988; Snider et al., 1991b). On the other hand, the airway resistance increase obtained in the guinea pig appears to be a NK_2 mediated phenomenon, while increase of plasma extravasation is a NK_1 receptor-mediated process in the airways and elsewhere (Ireland et al., 1988, 1991; Maggi, 1990; Maggi et al., 1990c; Abelli et al., 1989, 1991) in the rat and the guinea pig (Couture and Cuello, 1984; Nicolau et al., 1993). NK_3 receptors have been implicated in plasma extravasation in the rat skin (Couture and Kérouac, 1987) and the constriction of some vascular beds (e.g., the mesentery) (D'Orléans-Juste et al., 1991).

The most important central effect of SP is the facilitation of algesia that appears to be mediated by NK_1 and possibly NK_2 receptors (Laneuville et al., 1988). Recent data suggest that NK_1 receptors would favor the rapid hyperalgesia (Yamamoto and Yaksh, 1991) observed with intrathecal administration of SP, while NK_2 receptors may be implicated in the prolonged hyperalgesia observed with NKA and NK_2 agonists (Xu et al., 1991). It has also been suggested that NK_3 receptors may be present in the rat spinal cord, where they promote the release of opioids from the nerve terminals and mediate analgesia (Laneuville et al., 1988; Regoli et al., 1988; Helke et al., 1990).

Table 10
Characterization of Receptors for Tachykinins In Vivo

	Route	Natural agonist, MED	Selective agonist, MED[a]	Antagonist, MED	Receptor type	Ref.[d]
Hemodynamics						
Rat hypotension	iv	SP (1 nmol/kg)	[Sar^9,Met(O_2)^{11}]SP (1 nmol/kg)	RP 67580 (10 nmol/kg)	NK_1	1,2
Rat bradycardia	iv	SP (32.5 nmol/kg)	[Sar^9,Met(O_2)^{11}]SP (32.5 nmol/kg)		NK_1	1
Guinea pig hypotension	iv	SP (50 pmol/kg)	Ac[Arg^6,Sar^9,Met(O_2)^{11}]SP(6–11) (1.16 nmol/kg)	CP-96345 (10 nmol/kg)	NK_1	2,7
Rat hypertension and tachycardia	icv	SP (25 nmol/kg)		CP-96345 (5 nmol/kg)	NK_1	3
		NKA (25 nmol/kg)		R-396 (650 pmol/kg)	NK_2	
Salivary secretion						
Rat (increase)	iv	SP (10 nmol/kg)	[Sar^9,Met(O_2)^{11}]SP (1 nmol/kg)	CP-96345 (3–4 mg/kg)	NK_1	4,5
Airways						
Guinea pig (increase in resistance)	iv	NKA (1 nmol/kg)	[βAla^8]NKA(4–1O) (0.3 nmol/kg)	MEN 10207 (1 μmol/kg)	NK_2	6
		SP (10 nmol/kg)	[Sar^9,Met(O_2)^{11}]SP (1.0 nmol/kg)			
Plasma extravasation						
Rat (increase)	iv	NKB (50 nmol/kg)	[MePhe^7]NKB (50 nmol/kg)		NK_3[b]	8
		SP (0.1 nmol/kg)	[βAla^4,Sar^9,Met(O_2)^{11}]SP(4–11) (0.4 nmol/kg)		NK_1	9
Guinea pig (increase)	iv		[βAla^4,Sar^9,Met(O_2)^{11}]SP(4–11) (0.1 nmol/kg)		NK_1	9
Pain						
Rat hyperalgesia	ith	SP (6.5 nmol/kg)	[Sar^9,Met(O_2)^{11}]SP (6.5 nmol/kg)	CP-96345 (0.32 μmol/kg)[c]	NK_1	10,11
Rat analgesia	ith	NKB (8.0 nmol/kg)	[βAsp^4,MePhe]NKB(4–10) (5 nmol/kg)		NK_3	11

[a]MED, minimal effective dose or dose at which the effect was measured.
[b]Evans blue exudation in the ear.
[c]Administered iv or ith.
[d]1. Couture et al., 1989; 2. Rouissi et al., 1993; 3. Tschoppe et al., 1992; 4. Snider et al., 1992 ; 5. Maggi et al., 1991; 6. Maggi, 1990; 7. Rouissi et al., 1993a; 8. Jacques et al., 1989; 9. Abelli et al., 1991; 10. Picard et al., 1993; 11. Laneuville et al., 1988.

6. Conclusions

Classical criteria, applied to characterize and classify tachykinin receptors, have led to the identification of three functional sites that have been named NK_1, NK_2, and NK_3. Cloning and expression of receptors have validated this classification.

The first criterion, the order of potency of agonists, has been applied using the naturally occurring peptides that, however, have shown limitations in terms of selectivity. Use of selective agonists has contributed to make the first classification criterion more precise and useful. Utilization of selective agonists has also led to distinction of pharmacological preparations into tachykinin mono- or multireceptor systems.

The second criterion has been successfully applied thanks to the recent discovery of potent, specific, and extremely selective nonpeptide antagonists (CP-96345, RP-67580, SR-48968) that have confirmed the existence of mono- and multireceptor systems. Use of nonpeptide antagonists has contributed to validating the three functional entities and to identifying receptor subtypes related to species rather than function.

NK_1, NK_2, and NK_3 functional sites are membranous proteins of the rhodopsin family linked to G-proteins and to IP_3-Ca^{2+}-calmodulin effector. The three receptors subserve different biological effects in vivo, NK_1 being involved in arterial vasodilatation (via NO), plasma extravasation, exocrine (e.g., salivary) secretion, and spinal hyperalgesia, whereas NK_2 appears to mediate smooth muscle contraction in viscera and some vessels and to facilitate sympathetic tonus. NK_2 functional sites in the central nervous system appear to be involved in the regulation of cardiovascular functions and in prolonged hyperalgesia. NK_3 receptors facilitate the activity of the parasympathetic nerves and mediate venoconstriction in the splanchnic area, where they may induce plasma extravasation.

The various biological effects of tachykinins and neurokinins appear therefore to involve different receptor sites that can be selectively activated or blocked by antagonists. The future will tell us if some of these sites are important in pathophysiology and may become the targets of new drugs.

Abbreviations

SP: Substance P; NKA: neurokinin A; NKB: neurokinin B; PHY: physalaemin; ELE: eledoisin; KAS: kassinin; CP-96345: (2S,3S)-*cis*-2-(diphenylmethyl)-*N*-(2-methoxyphenyl)-methyl-1-azabicyclo[2,2,2]octan-3-amine]; RP-67580: [(3aR,7aR)-7,7-diphenyl-2-[imino-2-(2-methoxyphenyl)ethyl]perhydroisoindol-4-one]; SR-48968: (S)-*N*-methyl-*N*[4-acetylamino-4-phenylpiperidino)-2 (3,4-dichlorophenyl)butyl]benzamide.

Acknowledgments

The authors acknowledge the secretarial work of H. Morin and the technical assistance of M. Boussougou. Some of the data reported in the tables have been obtained by Q. T. Nguyen, and the peptides used have been prepared by D. Jukic. Experimental work has been performed with the financial help of the Medical Research Council of Canada (MRCC).

References

Abelli, L., Somma, V., Maggi, C. A., Regoli, D., Astolfi, M., Parlani, M., Rovero, P., Conte, B., and Meli, A. (1989) Effects of tachykinins and selective tachykinin receptor agonists on vascular permeability in the rat lower urinary tract: evidence for the involvement of NK-1 receptors. *J. Auton. Pharmac.* **9,** 253–263.

Abelli, L., Maggi, C. A., Rovero, P., Del Bianco, E., Regoli, D., Drapeau, G., and Giachetti, A. (1991) Effect of synthetic tachykinin analogs on airway microvascular leakage in rats and guinea pigs: evidence for the involvement of NK-1 receptors. *J. Auton. Pharmac.* **11,** 267–275.

Advenier, C., Rouissi, N., Nguyen, Q. T., Emonds-Alt, X., Breliere, J. C., Neliat, G., Naline, E., and Regoli, D. (1992) Neurokinin A (NK_2) receptor revisited with SR 48968, a potent non-peptide antagonist. *Biochem. Biophys. Res. Comm.* **184,** 1418–1424.

Beresford, I. J. M., Birch, P. J., Hagan, R. M., and Ireland, S. J. (1991) Investigation into species variants in tachykinin NK-1 receptors by use of the non-peptide antagonist, CP-96,345. *Br. J. Pharmacol.* **104,** 292–293.

Buck, S. H. and Shatzer, S. A. (1988) Agonist and antagonist binding to tachykinin peptide NK-2 receptors. *Life Sci.* **42,** 2701–2708.

Buck, S. H., Van Giersbergen, P. K. M., and Burcher, E. (1991) Tachykinins and their receptors: pharmacology biochemistry and molecular biology advance a neuropeptide story to the forefront of science. *Neurochem. Int.* **18,** 167–170.

Busse, R. and Mulsch, A. (1990) Calcium-dependent nitric oxyde synthesis in endothelial cytosol is mediated by calmodulin. *FEBS Lett.* **265,** 133–134.

Constantine, J. W., Lebel, W. S., and Woody, H. A. (1991) Inhibition of tachykinin-induced hypotension in dogs by CP-96,345, a selective blocker of NK-1 receptors. *Naunyn Schmiedeberg's Arch. Pharmacol.* **344,** 471–477.

Couture, R. and Cuello, C. A. (1984) Studies of the trigerminal antidronic vasodilatation and plasma extravasation in the rat. *J. Physiol.* **346,** 273–285.

Couture, R. and Kerouac, R. (1987) Plasma protein extravasation induced by mammalian tachykinins in rat skin: influence of anaesthetic agents and an acetylcholine antagonist. *Br. J. Pharmacol.* **91,** 265–273.

Couture, R., Laneuville, D., Guimond, C., Drapeau, G., and Regoli, D. (1989) Characterization of the peripheral action of neurokinins and neurokinin receptor selective agonists on the rat cardiovascular system. *Naunyn Schmiedeberg's Arch. Pharmacol.* **340,** 547–557.

Devillier, P., Advenier, C., Drapeau, G., Marsac, J., and Regoli, D. (1988) Comparison of the effect of the epithelium removal and of an eukephalinase inhibitor on the neurokinin-induced contractions of guinea pig isolated trachea. *Br. J. Pharmacol.* **94,** 675–684.

Dion, S., D'Orléans-Juste, P., Drapeau, G., Rhaleb, N.-E., Rouissi, N., Tousignant, C., and Regoli, D. (1987) Characterization of neurokinin receptors in various isolated organs by the use of selective agonists. *Life Sci.* **41,** 2269–2278.

D'Orléans-Juste, P., Dion, S., Mizrahi, J., and Regoli, D. (1985) Effects of peptides and nonpeptides on isolated arterial smooth muscles: role of endothelium. *Eur. J. Pharmacol.* **114,** 9–21.

D'Orléans-Juste, P., Dion, S, Drapeau, G., and Regoli, D. (1986) Different receptors are involved in the endothelium-mediated relaxation and the smooth muscle contraction of the rabbit pulmonary artery in response to substance P and related neurokinins. *Eur. J. Pharmacol.* 125, 37–44.

D'Orléans-Juste, P., Dion, S., Rhaleb, N.-E., Drapeau, G., and Regoli, D. (1987) Isolated arterial and venous vessels as selective pharmacological preparations for neurokinins, in *Substance P and Neurokinins* (Henry, J. L., Couture, R., Cuello, A. C., Pelletier, G., Quirion, R., and Regoli, D., eds.), Springer-Verlag, New York, pp. 125–128.

D'Orléans-Juste, P., Claing, A., Telemaque, S., Warner, T. D., and Regoli, D. (1991) Neurokinins produce selective venoconstriction via NK-3 receptors in the rat mesenteric vascular bed. *Eur. J. Pharmacol.* **204,** 329–334.

Drapeau, G., D'Orléans-Juste, P., Dion, S., Rhaleb, N.-E., Rouissi, N., and Regoli, D. (1987) Selective agonists for substance P and neurokinin receptors. *Neuropeptides* **10,** 43–54.

Drapeau, G., Rouissi, N., Nantel, F., Rhaleb, N.-E., Tousignant, C., and Regoli, D. (1990) Antagonists for the neurokinin NK-3 receptor evaluated in selective receptor systems. *Regul. Pept.* **31,** 125–135.

Emonds-Alt, X., Vilain, P., Goulaouic, P., Proietto, V., Van Broeck, D., Advenier, C., Naline, E., Neliat, G., Le Fur, G., and Brelière, J. C. (1992) A potent and selective non-peptide antagonist of the neurokinin A (NK-2) receptor. *Life Sci. Pharmacol. Lett.* **50,** PL101–PL106.

Fujii, T., Murai, M., Morimoto, H., Hagiwara, D., Miyake, H., and Matsuo, A. (1992) Effect of a novel substance P antagonist, FK 888 on airway constriction and airway edema in guinea pigs. *Neuropeptides* **22,** 24.

Garret, C., Carruette, A., Fardin, V., Moussaoui, S., Peyronel, J. F., Blanchard, J. C., and Laduron, P. J. (1991) Pharmacological properties of a potent and selective non-peptide substance P antagonist. *Proc. Natl. Acad. Sci. USA* **88,** 10,208–10,211.

Gitter, B. D., Waters, D. C., Bruns, R. F., Mason, N. R., Nixon, J. A., and Howbert, J. J. (1991). Species differences in affinities of neuropeptide antagonists for substance P receptors. *Eur. J. Pharmacol.* **197,** 237–238.

Guiliani, S., Maggi, C. A., and Meli, A. (1988) NK-1 receptors mediate the tachykinin stimulation of salivary secretion: selective agonists provide further evidence. *Eur. J. Pharmacol.* **150,** 377–379.

Guard, S. and Watson, S. P. (1991) Tachykinin receptor types: classification and membrane signalling mechanisms. *Neurochem. Intern.* **18,**149–165.

Guard, S. and Watling, K. J. (1992) Interaction of the non-peptide NK-1 tachykinin receptor antagonist (±) CP-96,345 with L-type Ca^{2+} channels in rat central cortex. *Br. J. Pharmacol.* **106,** 37P.

Helke, C. J., Krause, J. E., Mantyh, P. W., Couture, R., and Bannon, M. J. (1990) Diversity in mammalian tachykinin peptidergic neurons: multiple peptides, receptors and regulatory mechanisms. *FASEB J.* **4,** 1606–1614.

Henry, J. L. (1987) Discussion of nomenclature for tachykinins and tachykinin receptors, in *Substance P and Neurokinins* (Henry, J. L., Couture, R., Cuello, A.C., Pelletier, G., Quirion, R., and D. Regoli, eds), Springer-Verlag, New York, p. xvii.

Hultgardh-Nilsson, A., Nilsson, I., Jonzon, B., and Dalsgaard, C. J. (1988) Coupling between inositol phosphate formation and DNA synthesis in smooth muscle cells stimulated with neurokinin A. *J. Cell Physiol.* **137,** 141–145.

Ignarro, L. J. (1989) Endothelium-derived nitric oxide actions and properties. *FASEB J.* **3,** 31–36.

Ireland, S. J., Jordan, C. C., Stephens-Smith, M. L., and Ward, P. (1988) Receptors mediating the contractile response to neurokinin agonists in the guinea pig trachea. *Regul. Peptides* **22,** 93–98.

Ireland, S. J., Bailey, F., Cook, A., Hagan, R. M., Jordan, C. C. and Stephens-Smith, M. L. (1991) Receptors mediating tachykinin-induced contractile responses in guinea pig trachea. *Br. J. Pharmacol.* **103,** 1463–1469.

Jacques, L., Couture, R., Drapeau, G., and Regoli, D. (1989) Capillary permeability induced by intravenous neurokinins: receptor characterization and mechanism of action. *Naunyn Schmiedeberg's Arch. Pharmacol.* **340,** 170–179.

Jard, S. (1988) Mechanism of action of vasopressin and vasopressin antagonists. *Kidney Intern.* **34,** S38–S42.

Jukic, D., Rouissi, N., Laprise, R., Boussougou, M., and Regoli, D. (1991) Neurokinin receptor antagonists: old and new. *Life Sci.* **49,** 1463–1469.

Laneuville, O., Dorais, J., and Couture, R. (1988) Characterization of the effects produced by neurokinins and three agonists selective for neurokinin receptor subtypes in a spinal nociceptive reflex of the rat. *Life Sci.* **42,** 1295–1305.

Lavielle, S., Chassaing, G., Ploux, O., Loeuillet, D., Besseyre, J., Julien, S., Marquet, A., Convert, O., Beaujouan, J. C., Torrens, Y., Bergstrom, L., Saffroy, M., and Glowinski, J. (1988) Analysis of tachykinin binding site interactions using constrained analogs of tachykinins. *Biochem. Pharmacol.* **37,** 41–49.

Lee, C. M., Iversen, L. L., Hanley, M. R., and Sandberg, B. E. B. (1982) The possible existence of multiple receptors for substance P. *Naunyn Schmiedeberg's Arch. Pharmacol.* **318,** 281–287.

Maggi, C. A. (1990) Tachykinin receptors in the airways and lung: what should we block? *Pharmacol. Res.* **22,** 527–540.

Maggi, C. A., Giuliani, S., Santicioli, P., Regoli, D., and Meli, A. (1987) Mini review. Peripheral effects of neurokinins: functional evidence for the existence of multiple receptors. *J. Auton. Pharmacol.* **7,** 11–32.

Maggi, C. A., Patacchini, R., Giuliani, S., Rovero, P., Dion, S., Regoli, D., and Giachetti, A. (1990a) Competitive antagonists discriminate between NK-2 neurokinin receptor subtypes. *Br. J. Pharmacol.* **100,** 588–592.

Maggi, C. A., Patacchini, R., Giachetti, A., and Meli, A. (1990b) Tachykinin receptors in the circular muscle of the guinea pig ileum. *Br. J. Pharmacol.* **101,** 996–1000.

Maggi, C. A., Giuliani, S. Ballati, L., Rovero, P., Abelli, L., Manzini, S., Giachetti, A., and Meli, A. (1990c) *In vivo* pharmacology of [β-Ala^8]Neurokinin A(4–10), a selective NK-2 tachykinin receptor agonist. *Eur. J. Pharmacol.* **177,** 81–86.

Maggi, C. A., Giuliani, S., Ballati, L., Lecci, A., Manzini, S., Patacchini, R., Renzetti, A. R., Rovero, P., Quartara, L., and Giachetti, A. (1991) *In vivo* evidence for tachykininergic transmission using a new NK-2 receptor selective antagonist, MEN 10376. *J. Pharmacol. Exp. Therap.* **257,** 1172–1178.

Maggi, C. A., Patacchini, R., Rovero, P., and Giachetti, A. (1993) Tachykinin receptors and tachykinin receptor antagonists. *J. Autonom. Pharmacol.* **13,** 23–75.

Masu, Y., Hakayama, K., Tamaki, H., Harada, Y., Kuno, M., and Nakanishi, S. (1987) cDNA cloning of bovine substance K receptor through oocyte expression system. *Nature* **329,** 836–838.

McKnight, A. T., Maguire, J. J., Williams, B. J., Foster, A. C., Tridgett, R., and Iversen, L. L. (1988) Pharmacological specificity of synthetic peptides as antagonists at tachykinin receptors. *Regul. Pept.* **22,** 127.

McKnight, A. T., Maguire, J. J., Elliott, N. J., Fletcher, A. E., Foster, A. C., Tridgett, R., Willliams, B. J., Longmore, J., and Iversen, L. L. (1991) Pharmacological specificity of novel, synthetic, cyclic peptides as antagonists of tachykinin receptors. *Br. J. Pharmacol* **104,** 335–360.

Morimoto, H., Murai, M., Maeda, Y., Hagiwara, D., Miyake, H., Matsuo, M., and Fujii, T. (1992) FR 113680, a novel tripeptide substance P antagonist with NK-1 receptor selectivity. *Br. J. Pharmacol.* **106,**123–126.

Munekata, E., Kubo, K., Tanaka, H., and Osakada, F. (1987) Structure-activity studies of heptapeptide derivatives related to substance P, neurokinin A, B and other tachykinins on smooth muscles. *Peptides* **8,** 169–173.

Nakanishi, S. (1991) Mammalian tachykinin receptors. *Annu. Rev. Neurosci.* **14,** 123–136.

Nantel, F., Rouissi, N., Rhaleb, N.E., Jukic, D., and Regoli, D. (1991) Pharmacological evaluation of the angiotensin, kinin and neurokinin receptors on the rabbit vena cava. *J. Cardiovasc. Pharmacol.* **18,** 398–405.

Nicolau, M., Sirois, M. G., Bui, M., Plante, G. E., Sirois, P., and Regoli, D. (1993) Plasma extravasation induced by neurokinins in conscious rats, receptor characterization with agonists and antagonists. *Can. J. Physiol. Pharmacol.* **71,** 217–221.

Pernow, B. (1983) Substance P. *Pharmacol. Rev.* **35,** 85–147.

Petitet, F., Beaujouan, J. C., Saffroy, M., Torrens, Y. Lavielle, S., Chassaing, G., Loeuillet, D., and Glowinski, J. (1992a) Un nouveau récepteur des tachykinines révélé par des analogs de la substance P dans l'iléon de cobaye. *C.R. Acad. Sci. Paris* **314,** 299–303.

Petitet, F., Saffroy, M., Torrens, Y., Lavielle, S., Chassaing, G., Loeuillet, D., Glowinski, J., and Beaujouan, J. C. (1992b) Possible existence of a new tachykinin receptor subtype in the guinea pig ileum. *Peptides* **13,** 383–388.

Picard, P., Rouissi, N., Regoli, D., Gitter, B. D., Howbert, J. J., and Couture, R. (1993) Use of nonpeptide tachykinin receptor antagonists to substantiate the involvement of NK-1 and NK-2 receptors in a spinal nociceptive reflex in the rat. *Eur. J. Pharmacol.* **232,** 255–261.

Regoli, D., D'Orléans-Juste, P., Drapeau, G., Dion, S. and Escher, E. (1985) Pharmacological characterization of substance P antagonists, in *Tachykinin Antagonists* (Hakanson, R. and Sundler. F., eds.), Elsevier, Amsterdam, pp. 277–287.

Regoli, D., Drapeau, G., Dion, S., and D'Orléans-Juste, P. (1987a) Pharmacological receptors for substance P and neurokinins. *Life Sci.* **40,** 109–117.

Regoli, D., Drapeau, G., Dion, S., and D'Orléans-Juste, P. (1987b) Receptors for neurokinins in peripheral organs, in *Substance P and Neurokinins* (Henry, J. L., Couture, R., Cuello, A. C., Pelletier, G., Quirion, R., and Regoli, D., eds.) Springer-Verlag, New York, pp. 99–107.

Regoli, D., Drapeau, G., Dion, S., and Couture, R. (1988) New selective agonists for neurokinin receptors: pharmacological tools for receptor characterization. *Tr. Pharmacol. Sci.* **9,** 290–295.

Regoli, D., Drapeau, G., Dion, S., and D'Orléans-Juste, P. (1989) Receptors for substance P and related neurokinins. *Pharmacology* **38,** 1–15.

Regoli, D., Nantel, F., Tousignant, C., Jukic, D., Rouissi, N., Rhaleb, N.-E., Telemaque, S., Drapeau, G., and D'Orléans-Juste, P. (1991) Neurokinin agonists and antagonists. *Ann. NY Acad. Sci.* **632,** 170–183.

Rouissi, N., Claing, A., Nicolau, M., Jukic, D., D'Orléans-Juste, P., and Regoli, D. (1993a) Substance P (NK-1 receptor) antagonists: *in vivo* and *in vitro* activities in rats and guinea pigs. *Life Sci.* **52,** 1141–1147.

Rouissi, N., Jukic, D., D'Orléans-Juste, P., and Regoli, D. (1993b) A potent and long-acting NK-1 selective agonist. *Life Sci.* **52,** PL103–PL106.

Rouissi, N., Nantel, F., Drapeau, G., Rhaleb, N. E., Dion, S., and Regoli, D. (1990) Inhibitors of peptidases: how they influence the biological activities of substance P, neurokinins, bradykinin and angiotensin in guinea pig, hamster and rat urinary bladders. *Pharmacology* **40,** 196–204.

Rovero, P., Pestellini, V., Patacchini, R., Giuliani, S., Santicioli, P., Maggi, C. A., Meli, A., and Giachetti, A. (1989a) A potent and selective agonist for NK-2 tachykinin receptor. *Peptides* **10,** 593–595.

Rovero, P., Pestellini, V., Rhaleb, N-.E., Dion, S., Rouissi, N., Tousignant, C., Telemaque, S., Drapeau, G., and Regoli, D. (1989b) Structure-activity studies of neurokinin A. *Neuropeptides* **13,** 263–270.

Snider, R. J., Constantine, J. W., Lowe III, J. A., Longo, K. P., Lebel, W. S., Woody, H. A., Drozda, S. E., Desai, J. C., Vinick, F. J., Spencer, R. W., and Hess, H. J. (1991a) A potent non-peptide antagonist of the substance P (NK-1) receptor. *Science* **251,** 435–437.

Snider, R. M., Longo, K. P., Drozda, S. E., Lowe III, J. A., and Leeman, S. E. (1991b) Effect of CP96,345, a non-peptide substance P receptor antagonist, on salivation in rats. *Proc. Natl. Acad. Sci. USA* **88,** 10,042–10,044.

Sutter, M. C. (1990) *Pharmacol. Rev.* **42,** 287–325.

Tschöppe, C., Picard, P., Culman, J. Prat, A., Itoi, K., Regoli, D., Unger, T., and Couture, R. (1992) Use of selective antagonists to dissociate the central cardiovascular and behavioral effects of tachykinins on NK-1 and NK-2 receptors in the rat. *Br. J. Pharmacol.* **107,** 750–755.

Unger, T. L., Carolus, S., Demmert, G., Ganten, D., Lang, R. F., Maser-Gluth, C., Steinberg, H., and Veelken, R. (1988) Substance P induces a cardiovascular defense reaction in the rat: pharmacological characterization. *Circulation Res.* **63,** 812–820.

Watson, S. P. and Downes, C. P. (1983) Substance P-induced hydrolysis of inositol phospholipids in guinea pig ileum and rat hypothalamus. *Eur. J. Pharmacol.* **93,** 245–253.

Watson, S. P. (1984) The action of substance P on contraction, inositol phospholipids and adenylate cyclase in rat small intestine. *Biochem. Pharmcol.* **33,** 3733–3737.

Xu, X. J., Maggi, C. A., and Wiesenfeld-Hallin, Z. (1991) On the role of NK-2 tachykinin receptors in the mediation of spinal reflex excitability in the rat. *Neuroscience* **44,** 483–490.

Yamamoto, T. and Yaksh, T. L. (1991) Stereospecific effects of a non-peptidic NK-1 selective antagonist, CP-96,345: anti nociception in the absence of motor dysfunction. *Life Sci.* **49,** 1955–1963.

Yokota, Y., Sasai, Y., Tanaka, K., Fujiwara, T., Tsuchida, K., Shigemoto, R., Kakizuka, A., Ohkubo, H., and Nakanishi, S. (1989) Molecular characterization of a functional cDNA for rat substance P receptor. *J. Biol. Chem.* **264,** 17,649–17,652.

CHAPTER 14

Evidence for Receptor Subtypes/Species Variants of Receptors

Carlo Alberto Maggi

1. Introduction

Tachykinins are a family of peptides that share the common C-terminal sequence, Phe-Xaa-Gly-Leu-Met NH_2. Substance P (SP) was the first peptide of this family to be isolated by Von Euler and Gaddum in 1931, although its purification and determination of primary sequence did not occur until the early 1970s (Chang et al., 1971). In the intervening years, other members of the tachykinin family (e.g., physalaemin, eledoisin, and kassinin) were isolated from nonmammalian species (*see* Erspamer, 1981, for review). These discoveries prompted research about the possible existence of additional tachykinins in mammals. The existence of multiple tachykinin receptors was first suggested by the observation that some nonmammalian tachykinins (i.e., eledoisin and kassinin) possess higher potency than SP in certain mammalian tissues (Erspamer, 1981; Lee et al., 1982).

The Tachykinin Receptors Ed.: S. H. Buck © 1994 Humana Press Inc., Totowa, NJ

Between 1983 and 1984, several groups reported the isolation of two novel mammalian tachykinins, now termed neurokinin A (NKA, also known as substance K, neuromedin L, and neurokinin α) and neurokinin B (NKB, also known as neuromedin K or neurokinin β) (*see* Maggio, 1988, for review).

The discovery of NKA and NKB prompted an intense investigation about their pharmacological properties in comparison with SP, as well as to the analysis of the activity of their fragments (Buck et al., 1984; Nawa et al., 1984; Holzer-Petsche et al., 1985; Lee et al., 1986; Regoli et al., 1987; Maggi et al., 1985, 1987). These investigations, using bioassays, radioligand binding, and in vivo physiological experiments, have led to the recognition of the existence of three distinct types of tachykinin receptors that mediate the biological actions encoded by the common C-terminal sequence. After the meeting on Substance P and Neurokinins, held in Montreal in 1986 (Henry, 1987; *see* Burcher et al., 1991a, for further discussion), a common nomenclature was adopted to describe the three tachykinin receptors. The tachykinin receptor displaying higher affinity for SP, as compared to NKA or NKB, was termed NK_1, the receptor with higher affinity for NKA was termed NK_2, and that with higher affinity for NKB was termed NK_3 (Table 1).

The criteria of classification of the three receptors recognizing the C-terminal sequence of mammalian tachykinins have been reinforced by the synthesis of receptor-selective agonists modeled on the backbone of natural tachykinins. These ligands are highly selective for only one of the three known types of tachykinin receptors (e.g., Wormser et al., 1986; Dion et al., 1987; Drapeau et al., 1987; Regoli et al., 1988; Rovero et al., 1989; Lavielle et al., 1990). Although tachykinin receptor antagonists have been available since the early 1980s, the "first generation" of these ligands, the prototype of which is spantide I, has not been very useful for the task of discriminating between different tachykinin receptors because of their poor potency and selectivity. On the other hand, second and third generations of tachykinin receptor antagonists (Maggi et al., 1991a, 1993a, for reviews; Tables 2 and 3, Fig. 1) provide powerful tools to discriminate the three known types of tachykinin receptors.

Table 1
The Pharmacology of Tachykinin NK_1, NK_2, and NK_3 Tachykinin Receptors

	NK_1	NK_2	NK_3
Agonist order of potency	SP >NKA = NKB	NKA > NKB >> SP	NKB > NKA >> SP
Selective agonists	SP methylester [Sar^9] SP sulfone [Pro^9] SP sulfone [Pro^9] SP Septide GR 73,632	[Nle^{10}] NKA(4–10) [βAla^8] NKA (4–10) [Lys^5] NKA(4–10) [Lys^5, $MeLeu^9$, Nle^{10}] NKA(4–10) GR 64,349	Senktide [$MePhe^7$] NKB
Monoreceptorial bioassays for tachykinin receptors	Dog carotid artery Rabbit jugular vein Rabbit vena cava Mouse bronchus Guinea pig urethra	Rabbit pulmonary artery Guinea pig vas deferens Hamster trachea Human bronchus Human colon Human urinary bladder	Rat portal vein Rat vas deferens

Table 2
Amino Acid Sequence of Peptide Tachykinin Receptor Antagonists of Second Generation

NK_1 selective
L 668, 169: cyclo(Gln,D-Trp, (*N*Me)Phe(R)Gly[ANC-2]Leu,Met)$_2$
Spantide II: [D-NicLys^1,3-Pal,D-Cl_2Phe, Asn^6,D-Trp^{7,9}, Nle^{11}] SP
GR 71,251: [D-Pro^9[spiro-γ–lactam]Leu^{10}, Trp^{11}] SP
GR 82,334: [D-Pro^9[spiro-γ–lactam]Leu^{10}, Trp^{11}]physalaemin
FR113,680: Ac-Thr-D-Trp(CHO)-Phe-*N*MeBzl
FK 888: (2-(*N*-Me)indolil)-CO-Hyp-Nal-*N*MeBzl
NK_2 selective
L 659,877: cyclo (Gln-Trp-Phe-Gly-Leu-Met)
MEN 10,207: [Tyr^5, D-Trp^{6,8,9},Arg^{10}] NKA(4–10)
MEN 10,376: [Tyr^5, D-Trp^{6,8,9},Lys^{10}] NKA (4–10)
R 396: Ac-Leu-Asp-Gln-Trp-Phe-GlyNH_2
MDL 29,913: cyclo[Leu-ψ(CH_2NCH_3)-Leu-Gln-Trp-Phe-Gly]
GR 94,800: PhCO-Ala-Ala-D-Trp-Phe-D-Pro-Pro-NleNH_2
MEN 10,573: cyclo[Leu-ψ(CH_2NH)-Asp(OBzl)-Gln-Trp-Phe-βAla]
MEN 10,612: cyclo[Leu-ψ(CH_2NH)-Cha-Gln-Trp-Phe-βAla]
MEN 10,627: cyclo(Met-Asp-Trp-Phe-Dap-Leu)cyclo(2β–5β)

The isolation and cloning of the first tachykinin receptor (NK_2 from the bovine stomach, Masu et al., 1987) has been followed, in rapid sequence, by the isolation and cloning, from various species, of receptor proteins which fulfill the criteria used to define the three tachykinin receptors. All the tachykinin receptors isolated thus far belong to the "superfamily" of rhodopsin-like, G-protein-coupled receptors with seven hydrophobic transmembrane spanning segments that are thought to form the ligand (agonists and antagonists) recognition site. Much information is rapidly accumulating also on the mechanisms that regulate the expression of these receptors (Nakanishi, 1991, for review) and is extensively discussed in other sections of this book. The main second messenger system coupled to activation of the three tachykinin receptors is the stimulation of phospholipase C, leading to phosphoinositide breakdown and elevation

Table 3
Affinities of Receptor Antagonists of Second and Third Generation for Tachykinin NK_1, NK_2, and NK_3 Receptors

	NK_1 receptor,	NK_2 receptor		NK_3 receptor,
Antagonist	GPI[a]	RPA[b]	HT[c]	RPV[d]
L 668,169	6.44 (6.1 – 6.7)	Inactive[e]	6.16 (5.8 – 6.4)	Inactive[e]
Spantide II	7.08 (6.8 – 7.3)	5.43 (5.2 – 5.6)	6.00 (5.8 – 6.2)	Inactive[e]
GR 82,334	7.59 (7.2 – 8.0)	5.10 (4.9 – 5.3)	Inactive[e]	Inactive[e]
FR 113,680	6.61 (6.3 – 6.9)	5.37 (5.2 – 5.5)	5.21(4.9 – 5.5)	Inactive[e]
(±)CP96,345	8.11 (7.9 – 8.3)	Inactive[e]	Inactive[e]	Inactive[e]
RP 67,580	7.37 (6.9 – 7.9)	Inactive[f]	Inactive[f]	Inactive[f]
MEN 10,207	5.52 (5.3 – 5.8)	7.89 (7.7 – 8.1)	5.94 (5.7 – 6.2)	4.90 (4.7 – 5.1)
MEN 10,376	5.66 (5.4 – 5.9)	8.08 (7.8 – 8.3)	5.64 (5.5 – 5.8)	Inactive[e]
L 659,877	5.60 (5.2 – 5.9)	6.72 (6.5 – 7.0)	7.92 (7.8 – 8.0)	5.40 (5.1 – 5.7)
R 396	Inactive[e]	5.42 (5.2 – 5.6)	7.63 (7.3 – 7.9)	Inactive[e]
MDL 29,913	5.37 (5.2 – 5.6)	7.77 (7.6 – 7.9)	8.60 (8.4 – 8.8)	Inactive[e]
SR 48,968	Inactive[g]	9.60 (9.4 – 9.8)	8.50 (8.3 – 8.6)	Inactive[g]
MEN 10,573	6.37 (5.9 – 6.5)	7.31 (7.2 – 7.4)	8.66 (8.5 – 8.8)	Inactive[g]
MEN 10,612	6.09 (5.7 – 6.3)	7.41 (7.5 – 7.7)	9.06 (8.9 – 9.2)	Inactive[g]
MEN 10,627	6.10 (5.8 – 6.3)	8.1 (7.9 – 8.3)	10.1 (9.8 – 10.3)	Inactive[e]

[a]GPI = guinea pig ileum in the presence of atropine, SP methylester as agonist.
[b]RPA = endothelium–denuded rabbit pulmonary artery, neurokinin A as agonist.
[c]HT = hamster trachea, neurokinin A as agonist.
[d]RPV = rat portal vein, Arg–neurokinin B as agonist. Affinities are presented as pA_2 values. 95% c.l. in brackets.
[e]Inactive at 10 μ*M*.
[f]Inactive at 1 μ*M*.
[g]Inactive at 3 μ*M*.

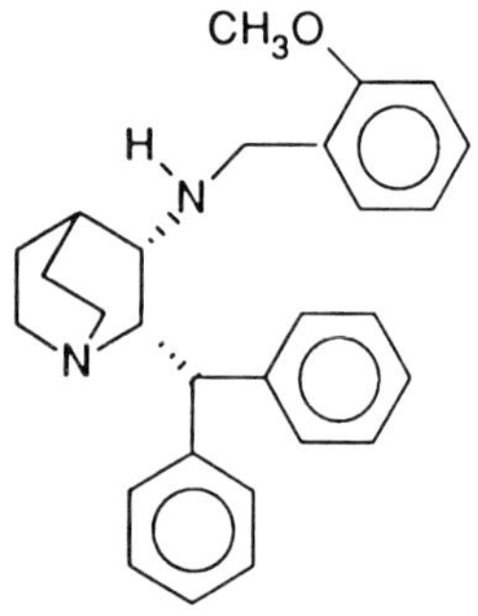

CP 96345

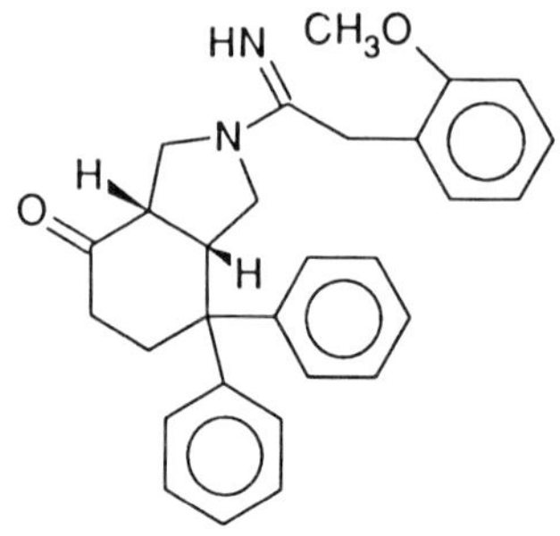

RP 67580

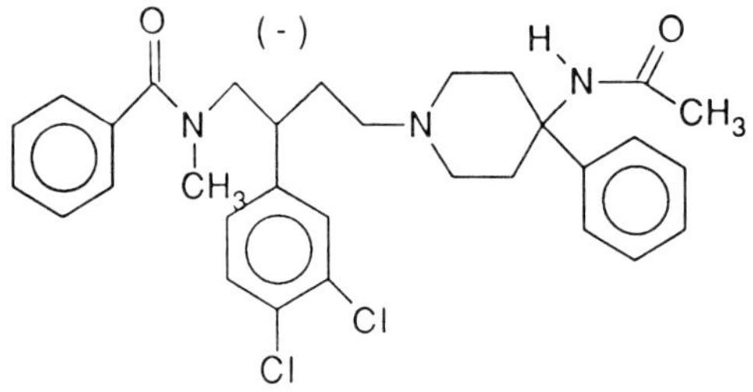

SR-48968

Fig. 1. Chemical structure of nonpeptide tachykinin receptor antagonists: CP 96,345 RP, 67,580 and SR 48,968.

of intracellular calcium (*see* Guard and Watson, 1991, for review). Evidence is also available that tachykinin receptors are coupled to adenylate cyclase stimulation and cAMP formation (Nakajima et al., 1992; Mitsuhashi et al., 1992). The three types of tachykinin receptor seem to use common second messenger systems, and no evidence has been presented thus far to suggest specialization at this level.

The aim of this chapter is to review current evidence indicating the existence of subtypes/species variants of tachykinin receptors. At present, only very limited evidence (Naminohira et al., 1989; Buell et al., 1992) is available to suggest the existence of such an heterogeneity for NK_3 receptors. The pharmacology of the NK_3 receptor is considerably underdeveloped compared to that of the other two receptor types (Table 1), and this drawback, particularly the lack of selective antagonists, has hampered the exploration of possible heterogeneity. On the other hand, there is mounting evidence indicating a striking pharmacological heterogeneity for both NK_1 and NK_2 receptors. As mentioned above, the criterion used to distinguish the three main *types* of tachykinin receptors was the order of potency of the natural tachykinins, SP, NKA, and NKB; this criterion has received support by the development of receptor selective agonists and antagonists and by the isolation and cloning of the three main types of tachykinin receptors. The use of the term *subtypes/species variants* to underline the pharmacological heterogeneity presented by, e.g., NK_2 receptors, is a direct consequence of the criterion used to define the NK_2 receptor *type*. The NK_2 receptor recognizes, with decreasing order of affinity, NKA, NKB, and SP, respectively. Possible subtypes/species variants of the NK_2 receptor are then defined because they meet this original criterion, whereas the introduction of other ligands (e.g., receptor selective competitive antagonists) reveals an additional pharmacological heterogeneity.

2. Problems Related to the Study of Tachykinin Receptor Heterogeneity and Classification

2.1. Degradation by Peptidases

Indications for heterogeneity within the established tachykinin receptor types originate from studies illustrating the differential ability of certain ligands (agonists or antagonists) to interact with tachykinin receptors that otherwise exhibit an identical pattern of recognition of natural ligands, the mammalian tachykinins. Until recently, the major part of ligands available for these studies has been represented by peptides (e.g., receptor antagonists bearing substitutions with nonnatural amino acids on the sequence of natural tachykinins). Tachykinin peptides are degraded by various peptidases such as "enkephalinase" or endopeptidase 24.11, angiotensin converting enzyme, aminopeptidases, and so on. This may occur to such an extent as to greatly influence the estimate of ligand affinities, resulting in inappropriate classification of receptors (e.g., Devillier et al., 1988; Ireland et al., 1988). Generally speaking, synthetic ligands acting as receptor selective agonists or antagonists are more resistant to degradation by peptidases than the natural peptides. Relatively minor modification on the backbone of natural tachykinins or their fragments (e.g., replacement of Gly in position 8 of the sequence of NKA(4–10) by βAla) may afford significant protection against degradation by peptidases as compared to the parent peptide (Patacchini et al., 1989). Similar considerations apply to peptide antagonists. It is evident, however, that a differential rate of peptide degradation in different preparations may lead to an incorrect estimate of ligand affinities. To minimize this problem, many of the recent studies on this topic have routinely adopted the use of peptidase inhibitors. Owing to the existence of multiple (and possibly some yet to be discovered) peptidases that degrade tachykinins, the possibility that differences in the affinities of peptide ligands in various preparations arise from and/or are influenced by different rates of degradation cannot be conclusively ruled out (*see* also Section 4.3.6. and 4.3.7.1.).

2.2. Natural Ligands for Tachykinin Receptors

It is important to note that *all natural tachykinins can act as full agonists at the three tachykinin receptor types,* although with different affinities. As an example, after transfection and functional expression of the cloned cDNA for NK_1, NK_2, and NK_3 receptors, it was shown that all natural tachykinins produce a full inhibition of binding of radiolabeled ligands from the three receptor sites (Ingi et al., 1991). All natural tachykinins, although with different potency, are able to activate fully second messenger systems (phosphatidylinositol hydrolysis) in cells transfected and stably expressing one of the three tachykinin receptors (Nakajima et al., 1992). A current interpretation of the "Montreal nomenclature" maintains that SP, NKA, and NKB *are* the endogenous ligands for NK_1, NK_2, and NK_3 receptors, respectively. Accordingly, several "receptor mismatches" could be identified (e.g., absence of NKB in the guinea pig ileum and presence of NK_3 receptors mediating neuronal stimulation). The assumption that SP is the natural ligand for NK_1 receptors, NKA for NK_2 receptors, and NKB for NK_3 receptors, respectively, reflects the frequent situation occurring when the natural tachykinin and its "preferred" receptor coexist in the same tissue. When more than one type of tachykinin receptor is expressed by target cells, examples are available of tachykininergic cotransmission involving a distinct role for, e.g., NK_1 and NK_2 receptors, putatively activated by physiological concentrations of SP and NKA, respectively (e.g., rat urinary bladder, Maggi et al., 1991b; guinea pig renal pelvis, Maggi et al., 1992c; human ileum, Maggi et al., 1992d). However, some of the purported examples of receptor mismatches arise from an unwarranted extension of the criteria used to define NK_1, NK_2, and NK_3 receptors, ignoring the extensive crosstalk between endogenous tachykinins and different tachykinin receptors.

Until now, only SP, NKA, and NKB have an established status of neurotransmitters, either in the central or peripheral nervous system. Other mammalian tachykinins, such as the N-terminally extended forms of NKA, neuropeptide K (NPK) and neuropeptide γ (NPγ), have been proposed to play a role as neurotransmitters (McDonald et al., 1989), but conclusive physiological evidence on this point is lacking. Pharmacological data are available to suggest that NPK or NPγ have higher affinity than NKA at certain NK_2 receptors, and the hypothesis has been advanced that they could be preferred ligands at different NK_2 receptor subtypes (*see* Burcher et al., 1991a, for further discussion on this point). As an example, NPγ is about 10 times more potent than NKA in contracting the guinea pig and human isolated bronchus, the rat gastric fundus, and dog urinary bladder (Black et al., 1990b; Burcher et al., 1991b; Burcher and Mussap, 1991; *see also* Takeda and Krause, 1989a,b). Takeda and Krause (1991) showed that NPγ and NPK are distinctly more potent than NKA in displacing the binding of [^{125}I]NPγ from rat duodenum membrane preparations, whereas no difference in potency of these ligands was detected in the rat vas deferens or urinary bladder. In other preparations (e.g., rabbit pulmonary artery, hamster trachea, and urinary bladder), the affinity of NPK and NPγ is comparable to that of NKA (Maggi et al., 1990a; Van Giersbergen et al., 1991,

1992). Since the rank order of potency of natural tachykinins is the original criterion for distinguishing between different receptor types, studies in this area are of great interest. It will be important, however, to totally exclude that the different potency of NPK and NPγ vs NKA, as detected in certain preparations, may reflect a different degree of inactivation by peptidases (*see* section 2.1.).

2.3. *Preparations*

Considerable efforts have been directed to the development of pharmacological preparations suitable to analyze the properties of only one of the three tachykinin receptors. The basic idea underlying this effort was that such "monoreceptorial" assays can provide simple systems to analyze in detail the pharmacology of the given tachykinin receptor and serve as experimental test objects for the development of new ligands with high potency and selectivity. Accordingly, various monoreceptorial bioassays have been developed in which the biological response to peptides of the tachykinin family is exclusively mediated by NK_1 (e.g., dog carotid artery), NK_2 (e.g., rabbit pulmonary artery), or NK_3 receptors (e.g., rat portal vein) (Table 1). These "monoreceptorial" assays have been extremely useful for the recognition of the pharmacological heterogeneity of NK_2 receptors (Maggi et al., 1990a; Buck et al., 1990), owing to the ability of novel ligands to discriminate between otherwise indistinguishable NK_2 receptor-mediated responses. The importance of such "monoreceptorial bioassays" should not be overestimated, however. If the tachykinin response in a given preparation strictly obeys the criteria used to define, e.g., NK_2 receptors, then this is considered a monoreceptorial assay for NK_2 receptor; if the response deviates from such a straightforward behavior, then the response is thought to be mediated by more than one tachykinin receptor simultaneously expressed in the preparation (in most instances NK_1 and NK_2). This kind of reasoning is somewhat vicious because it does not allow the incorporation of putative novel endogenous ligands for tachykinin receptors (Section 2.2.), nor does it take into account the possibility of the existence of totally new type(s) of tachykinin receptors, as suggested by certain data in, e.g., the guinea pig ileum (Petitet et al.,

1992a,b, *see* Section 3.3.3.2.). Overall, it is not possible to recommend a given preparation for studying the pharmacology of NK_1, NK_2, or NK_3 receptors. Owing to the striking pharmacological heterogeneity reviewed in this chapter, it appears wise that future studies in this field will compare the action of ligands in more than one preparation for the known types of tachykinin receptors and possibly in several preparations from different species.

2.4. Correlating Structure and Function

One of the major problems to be solved about the heterogeneity of tachykinin receptors is to establish to what extent this heterogeneity merely reflects the existence of species differences or might be related to the existence of intraspecies variants of the same receptor (subtypes). It is quite evident that the two possibilities are not mutually exclusive, i.e., *both* interspecies and intraspecies variants of tachykinin receptors could exist. The issue is also strictly related to the task of receptor classification and nomenclature to be used to label the discovered receptor heterogeneity. For quite a long time, the recognition of receptor heterogeneity has exclusively relied on functional criteria detecting differences in ligand affinities for the receptor, coupling to second messenger systems, and generation of the final response. The recent developments of molecular biology have had a tremendous impact on these topics, and the receptors of the tachykinin family are not an exception in this respect. Techniques available for the cloning and sequencing of neurotransmitter receptors offer the possibility of comparing receptors of a given neurotransmitter family for differences in their primary sequence and to correlate differences in structure to differences in function.

Such a revolution is leading to the concept that receptors should be primarily (if not solely) classified on the basis of their amino acid sequences and structure. Criteria have been proposed, even quantitatively, that put forward the importance of sequence homology for the subclassification of neurotransmitter receptors isolated and cloned from different or from the same species (e.g., Hartig et al., 1992). The term receptor *subtypes* should be reserved to label the existence of intraspecies variants of the same receptor, i.e., two or more different genes encoding receptor proteins for the same transmitter. On the other hand, species differences involve the presence of a single receptor gene within a given species, and the corresponding receptors are termed species homologues of the same receptor type. A classification based on a purely structural approach does not require the existence of a pharmacological heterogeneity in terms of ligand recognition. If differences exist in primary sequence, then the receptors *are* different. It is expected that differences in structure will be paralleled by some difference in function, including the ability to recognize agonists/antagonists. The problem arising when attempting to correlate structure to function on the basis of sequence homologies is that it is almost impossible to decide *a priori* whether even a single amino acid substitution in the primary sequence may or may not have an impact on the functional properties (ligand recognition, coupling to second messenger systems, and so on) of the receptor protein. Therefore, the structural approach needs to be integrated with knowledge on the function of the cloned receptor expressed in appropriate systems (e.g., oocytes) and evaluation of its ability to recognize ligands and to generate second messenger and functional responses. Examples are known of intraspecies variants of the same receptor (subtypes) that show relatively important differences in their primary sequence, but little or no difference in their pharmacological properties: the human 5-$HT_{1D\beta}$ and the human 5-$HT_{1D\alpha}$ receptors possess a 23% difference in their transmembrane region sequences (thought to be pivotal for ligand recognition properties) and yet they are essentially indistinguishable in terms of pharmacological binding properties for 19 different ligands (Hartig et al., 1992).

With regard to tachykinin receptors, the correlation between changes in amino acid sequences and pharmacological differences is becoming elucidated, especially for the NK_1 receptor. As will be discussed in Sections 3.1. and 4.1., it appears likely that certain species-related variations in the amino acid sequences of the seven transmembrane hydrophobic segments of the NK_1 and NK_2 receptors provide some structural basis for the corresponding species-related pharmacological heterogeneity.

The primary sequence of the various forms of tachykinin receptors isolated from the various mammalian species all differ, to various extents, in their length/primary sequence. Tachykinin receptor proteins have been isolated and cloned from various species; in spite of differences in their primary sequence, receptor proteins isolated from different species meet the order of potency of natural tachykinin agonist activity used to define NK_1, NK_2, or NK_3 receptors and represent *de facto* species variants of the corresponding tachykinin receptor. As will be discussed in the next sections, it is evident that at least part of the pharmacological heterogeneity thus far described for NK_1 and NK_2 receptors could be accounted for by species differences.

In each mammalian species examined thus far, only one copy of the NK_1 or NK_2 receptor gene has been identified. According to the most conservative point of view, this would mean that only species homologues exist for tachykinin receptors, but not receptor subtypes. The latter require the demonstration that structurally different forms of the given receptor are expressed in the same species and exhibit different functional properties in terms of ligand recognition or coupling to second messenger.

The fact that a single copy of the various tachykinin receptors exists in the genoma of various species does not exclude *a priori* the existence of the intraspecies heterogeneity required for the recognition of receptor subtypes, however. The following possibilities should be considered in future investigations in this field:

1. A first possibility is the existence of allelic forms of the genes encoding the tachykinin receptors. For such a possibility to be functionally relevant, it would require that the allelic forms of the given receptor present changes in key positions affecting receptor function. This possibility is already known for other genes, such as that responsible for cystic fibrosis. In this case, a single mutation of the gene encoding a chloride channel leads to remarkable changes in function (Riordan et al., 1989). The existence of allelic forms of the NK_2 receptor isolated from the human trachea has been reported (Graham et al., 1991), although the possible functional consequences of these changes on receptor function are unknown.

2. The three main types of tachykinin receptors all belong to the superfamily of rhodopsin-like G-protein-coupled receptors. For receptors of this family, the most common form of gene organization is intronless, i.e., the sequence encoding the receptor protein is uninterrupted. On the other hand, introns alternate to exons in the genes encoding the mammalian tachykinin receptors. Such a gene organization offers the possibility of generating multiple forms of the receptor protein from a single gene through an alternative splicing process. A prerequisite for such a mechanism to lead to intraspecies receptor heterogeneity is that the multiple forms of the receptor protein generated in this way are functionally capable of receptor function. Some evidence for the production of "short" forms of the NK_1 and NK_2 receptors has been obtained (Tsuchida et al., 1990; Fong et al., 1992a) and will be further discussed in Sections 3.1 and 4.1. The importance of splicing mechanisms for the generation of multiple forms of functionally distinct receptor proteins has been described for various dopamine receptors that, like tachykinin receptors, alternate introns and exons in their gene organization (e.g., Civelli et al., 1991; Sibley and Monsma, 1992; Van Tol et al., 1992).
3. Further possible mechanisms leading to intraspecies heterogeneity are linked to posttranslational modifications of the receptors and/or in a different ability of the same receptor to recognize ligands because of an interaction with the microenvironment of the cell membrane in which it is inserted (*see* Appell et al., 1992, for further discussion on this point). To date, there is no evidence to support the idea that such mechanisms may lead to intraspecies heterogeneity of tachykinin receptors. Some reports indicate that the extracellular sequence and glycosylation are important for binding to tachykinin receptors. Fong et al. (1992b,c) showed that mutants of the rat NK_1 receptor in which the first 13 or 27 amino acid residues at the N-terminal end had been replaced by other amino acids show a reduced affinity for SP binding. Other extracellular regions of the NK_1 receptor appear to be involved in binding of antagonists (Fong et al., 1992c; Gether et al., 1992). Takamatsu et al. (1991) showed that concanavalin A pretreatment reduces the affinity of SP binding to rat cerebral cortex NK_1 receptor. Since glycosylation sites are present on the extracellular domain of all tachykinin receptors (Nakanishi, 1991), the possibility exists that

posttranslational modification at this level generates pharmacological heterogeneity of tachykinin receptors.

3. Heterogeneity of the Tachykinin NK_1 Receptor

3.1. Molecular Biology

NK_1 receptor is defined as the mediator of those biological activities encoded by the C-terminal sequence of tachykinins for which SP is a better ligand relative to NKA or NKB. A 407-amino acid-residue receptor protein fulfilling these characteristics has been isolated from rat brain and salivary glands (Yokota et al., 1989; Hershey and Krause, 1990), from human IM-9 lymphoblast cell line (Takeda et al., 1991; Gerard et al., 1991), human lung (Hopkins et al., 1991; Gerard et al., 1991), guinea pig uterus (Gorbulev et al., 1992), and mouse genoma (Sundelin et al., 1992). The two receptors cloned from human tissues are identical. A feature common to many NK_1 receptor-mediated events is desensitization that is more intense than for NK_2 or NK_3 receptor-mediated responses. This may be related to the presence of a large number of potential targets for phosphorylation in the C-terminal, cytoplasmic region of the receptor (Hershey and Krause, 1990).

As will be more extensively discussed in Section 3.3.1., the introduction of nonpeptide antagonists has enabled the recognition of a remarkable species heterogeneity of the NK_1 receptor. Thus, the affinity of certain receptor antagonists, such as CP 96,345 (Snider et al., 1991; Fig. 1), is about 100-fold higher for the human or guinea pig than for the rat or mouse NK_1 receptor (e.g., Gitter et al., 1991), whereas RP 67,580 (Garret et al., 1991, 1992; Fig. 1) is more potent at the rat or mouse than human or guinea pig NK_1 receptor. The cloned human and rat NK_1 receptors have a high degree (about 95%) of homology; only 21 out of 407 amino acid residues differ between the species, the most frequent changes being present in the N- and C-terminal sequences. As pointed out by Gerard et al. (1991), one of the most interesting changes occurs at position 97 (second extracellular loop) where Val/Glu replacement occurs. The guinea pig NK_1 receptor bears a Glu residue in position 97 like the human NK_1 receptor (Gorbulev et al., 1992), whereas the mouse NK_1 receptor bears a Val residue in 97 like the rat NK_1 receptor (Sundelin et al., 1992). Fong et al. (1992b,c) showed that a mutant rat NK_1 receptor in which Glu was present instead of Val in position 97 exhibits a changed recognition affinity for NKB, but not SP or NKA. These findings, which indicate that the extracellular domain of NK_1 receptor participates in the ligand binding site for peptides of the tachykinin family (Fong et al., 1992b), also suggest that the amino acid residue in position 97 (Glu in human and guinea pig, Val in rat and mouse) may be of importance for the species-related variations in the pharmacology of the NK_1 receptor.

The comparison of amino acid sequences of human and guinea pig vs rat and mouse NK_1 receptor in the context of the species-related differences in pharmacology becomes even more interesting when focusing on the amino acids putatively assigned to the seven hydrophobic transmembrane segments (abbreviated as M1–M7) that, by analogy with other G-protein coupled receptors, should form the ligand recognition site. According to the amino acid assignment published in Gerard et al. (1991),158 out of 407 amino acid residues compose the seven transmembrane segments of rat and human NK_1 receptor. The homology at this level is 96.2% because only 6 amino acid residues are different between the two species (Table 4). With one exception (position 266 in M6, which is Val in rat and Ile in mouse), the mouse NK_1 receptor (Sundelin et al., 1992) presents the same amino acid residues of the rat NK_1 receptor at these variant positions in transmembrane segments (Table 4). In sharp contrast, the guinea pig NK_1 receptor (Gorbulev et al., 1992), which is 96.5% identical to the human NK_1 receptor in its overall sequence, is 100% identical to the human receptor in the amino acid sequences assigned to M1–M7 (Table 4). Therefore, the clustering of various species (guinea pig and human vs rat and mouse) with regard to heterogeneity in amino acid sequence at M1–M7 level corresponds to the clustering of species-related differences in affinities of competitive receptor antagonists. These considerations strongly suggest that some of the variant positions in M1–M7 and/or position 97 of the

Table 4
Comparison of Amino Acid Residues
at Species-Related Variant Positions
of the Seven Hydrophobic Transmembrane Segments (Ml-M7)
of the Tachykinin NK_1 Receptor[a]

Transmembrane segment	Amino acid position	Rat	Mouse	Human	Guinea pig
Ml	54	Ile	Ile	Met	Met
M2	80	Cys	Cys	Ser	Ser
M3	116	Leu	Leu	Val	Val
M4	152	Phe	Phe	Cys	Cys
M6	266	Val	Ile	Ile	Ile
M7	290	Ser	Ser	Ile	Ile

[a]Based on NK_1 receptor sequences published by Gerard et al. (1991, rat and human), Sundelin et al. (1992, mouse), Gorbulev et al. (1992, guinea pig), and amino acid assignment to the seven hydrophobic transmembrane segments published by Gerard et al. (1991).

second extracellular loop are responsible for the species-related pharmacological heterogeneity of the tachykinin NK_1 receptor. Recent data presented by Fong et al. (1992c) at the meeting on Substance P and Related Peptides held in Shizuoka, Japan, November 3–6, 1992, indicated that species-related variations in postitions 116 (Leu in rat and mouse, Val in human and guinea pig) and 290 (Ser in rat and mouse, Ile in human and guinea pig) are determinant for the different profile of species-related variations in affinity of CP 96,345 and RP 67,580 binding to the NK_1 receptor. As discussed in Section 3.3.1., these species-related differences in NK_1 receptor pharmacology are restricted to antagonists because no differences of this type have been described with regard to agonist affinities.

Two isoforms of the human NK_1 receptor, possibly generated through pre-mRNA alternative splicing, have been isolated (Fong et al., 1992a). One form has the same length (407 amino acids) of the NK_1 receptor previously cloned from human tissues; the second isoform (311 amino acids) contains a very short C-terminal sequence, extending only 7 amino acid residues after the end of the seventh transmembrane segment. The short form of the human NK_1 receptor binds SP with lower affinity (10 times) than the long one and is much less effective than the long form in stimulating an electrophysiological response in transfected cells (*Xenopus* oocytes). Owing to the importance of the C-terminal sequence for coupling with G-proteins, the short form of the NK_1 receptor may have an inappropriate coupling with effector systems in *Xenopus* oocytes, also influencing the binding affinity of the agonist (Fong et al., 1992a). The authors concluded that, until an appropriate effector can be identified for the short form of human NK_1 receptor, the possibility of physiological irrelevance cannot be ruled out (Fong et al., 1992a).

3.2. Pharmacology

The pharmacology of the NK_1 receptor has advanced considerably in recent years, leading to the development of several receptor-selective agonists and antagonists. From the pharmacological point of view, several bioassay preparations that can be considered as monoreceptorial NK_1 tissues are available. Several compounds have been described and characterized as selective NK_1 receptor agonists, including SP methylester (Watson et al., 1983), septide (Wormser et al., 1986), [Sar^9]SP sulfone, [Pro^9]SP sulfone (Dion et al., 1987), [Pro^9]SP (Petitet et al., 1991), and GR 73,632 (Hagan et al., 1989) (Table 1). The availability of various agonist ligands is important in this field because some of the reported cases indicating heterogeneity of the NK_1 receptor are based on differential agonist potencies and/or differential blockade of various agonists by different antagonists *(see below)*.

With regard to competitive antagonists, several compounds are also available (Tables 2–3). These include linear undecapeptides like GR 82,334 (Hagan et al., 1991) and spantide II (Folkers et al., 1990; Maggi et al., 1991c), the tripeptide FR 113,680 (Hagiwara et al., 1991, 1992a,b; Morimoto et al., 1992), the dipeptide FK 888 (Fujii et al., 1992), cyclic peptides like L668,169 (McKnight et al., 1988), and nonpeptide ligands like CP 96,345 (Snider et al., 1991; Lowe et al., 1992) and RP 67,580 (Garret et al., 1991). For the latter nonpeptide ligands, recognition of the NK_1 receptor is stereospecific, and the active isomer is about 1000 times more potent than the inactive one. CP 96,345, the first potent nonpeptide antagonist to be described (Snider et al., 1991), also possesses calcium channel blocking properties at relatively low concentrations that may complicate the interpretation of experimental results particularly in those species (rat, mouse) in which its affinity for NK_1 receptor is not very high (Schmidt et al., 1992). The various NK_1 receptor antagonists available possess quite a large spectrum of relative potencies at NK_1 receptors in, e.g., the guinea pig ileum (Table 3) where they possess selectivity for NK_1 over NK_2 or NK_3 receptors, with the exception of L668,169, which also blocks with similar NK_2 receptor mediated responses in the hamster isolated trachea (Table 3).

3.3. Evidence for Heterogeneity

3.3.1. Species-Related Heterogeneity

The most striking and clear examples of pharmacological heterogeneity of the NK_1 receptors have become apparent after the discovery of the nonpeptide NK_1 receptor antagonists, CP 96,345 and RP 67,580. CP 96,345 possesses much greater affinity (about 100 times) for the NK_1 receptor expressed in bovine, guinea pig, rabbit, and human brain as compared to that expressed in rat, mouse, or chicken brain (Beresford et al., 1991; Gitter et al., 1991; Watling et al., 1991; Snider et al., 1991a). Gitter et al. (1991) showed that CP 96,345 is about 100 times more potent in inhibiting ^{125}I-SP binding in human astrocytoma U373 (IC_{50} = 0.40 n*M*) and B lymphoblastoid (IM9) cell lines (IC_{50} = 0.35 n*M*), guinea pig brain (IC_{50} = 0.32 n*M*) and lung (IC_{50} = 0.34 n*M*), or rabbit brain (IC_{50} = 0.54 n*M*) than in mouse (IC_{50} = 32 n*M*), rat (IC_{50} = 35 n*M*), or chicken (IC_{50} = 156 n*M*) brain. Watling et al. (1991) similarly reported a 100-fold higher affinity of CP 96,345 in inhibiting ^{125}I-SP binding from guinea pig striatum (pIC_{50} = 8.2) and human caudate nucleus (pIC_{50} = 8.72) than from rat striatum (pIC_{50} = 6.91). Likewise, Beresford et al. (1991) reported a 100-fold higher affinity of CP 96,345 in inhibiting SP from rabbit (pIC_{50} = 8.62), guinea pig (pIC_{50} = 8.50), human (pIC_{50} = 8.46), and bovine (pIC_{50} = 8.86) brain than from rat (pIC_{50} = 6.77) or mouse (pIC_{50} 6.92) brain.

Table 6
Affinities (pK_B, Mean Values)
of Various NK_1 Receptor Selective Antagonists at NK_1 Receptors
in Peripheral Preparations from Various Mammalian Species

Antagonist	GPI[b]	GPVD[c]	RJV[d]	RUB[e]
(±)CP 96,345	8.11	7.79	8.33	6.16
	(7.9–8.3)	(7.4–8.2)	(8.2–8.5)	(5.9–6.4)
Spantide II	7.08	6.96	6.81	Inactive[f]
	(6.8–7.3)	(6.4–7.5)	(6.6–7.0)	
FR 113,680	6.61	6.36	7.12	Inactive[f]
	(6.3–6.9)	(5.8–6.9)	(6.9–7.3)	
L668,169	6.44	6.03	6.27	Inactive[f]
	(6.1–6.7)	(5.7–6.3)	(6.1–6.5)	

[a]Data from Patacchini et al., 1992.
[b]GPI = guinea pig ileum.
[c]GPVD = guinea pig vas deferens.
[d]RJV = rabbit jugular vein.
[e]RUB = rat urinary bladder. 95% c.l. in brackets.
[f]Inactive as agonist or antagonist up to 10 μM.

Remarkable species-related differences in the affinities of NK_1 receptor antagonists were also demonstrated by Appell et al. (1992) using three chemically unrelated classes of compounds. The authors compared the affinities of 5 heterosteroids, 3 cyanines, and 3 peptides in inhibiting [^{125}I]Bolton-Hunter SP binding in membranes of human astrocytoma cell line and urinary bladder, guinea pig forebrain and ileum, and rat forebrain and duodenum. In the 6 assays, the affinity of SP itself or of peptide antagonists of the first generation (such as spantide) was quite similar, failing to reveal any kind of heterogeneity. By contrast, the heterosteroids showed three orders of magnitude higher affinity for rat as compared to the human or guinea pig NK_1 receptor. A similar trend was observed for the cyanine compounds, whereas the peptide compounds were about 10 times more potent at human and guinea pig than at rat NK_1 receptors. Using these ligands, no difference was apparent between central and peripheral NK_1 receptors of each species.

Therefore, the introduction of CP 96,345 and RP 67,580 has differentiated the NK_1 receptor into two broad categories that can be roughly defined as follows: human, bovine, guinea pig, and rabbit NK_1 receptor for which CP 96,345 is markedly more potent than RP 67,580 and rat or mouse NK_1 receptor for which RP 67,580 is more potent than CP 96,345.

3.3.2. *Evidence for Intraspecies Heterogeneity*

3.3.2.1. CENTRAL and PERIPHERAL NK_1 RECEPTOR. A few studies have suggested a possible pharmacological heterogeneity of the central vs peripheral NK_1 receptor. Lew et al. (1990) reported differences in the affinities of some ligands in displacing ^{125}I-Bolton-Hunter [Sar^9]SP sulfone from rat brain and salivary glands. SP or unlabeled [Sar^9]SP sulfone show similar apparent affinities for brain or salivary NK_1 receptors, but other ligands (SP methylester, kassinin, NKB) are distinctly more potent (at least 10 times) in the salivary glands than in the brain. Similarly, Fardin and Garret (1991) observed a greater affinity (about 50 times) of SP methylester for NK_1 receptors in guinea pig brain than for those in the guinea pig ileum. In contrast, Tousignant et al. (1991) concluded that the central (rat brain) and peripheral (guinea pig ileum) NK_1 receptors are identical. Metabolic breakdown by peptidases may account for observed differences in the affinities of peptide ligands between central and peripheral NK_1 receptors (Lew et al., 1990; Fardin and Garret, 1991). To address the latter point, Clerc et al. (1992) have compared the stability of SP methylester and septide in rat brain and submaxillary gland and in guinea pig brain and ileum vs the ability of these ligands to displace binding of [^{3}H]SP in membrane preparations of the same tissues. These investigations revealed that SP methylester is extensively degraded to the free acid derivative in the four preparations tested. Consequently, the different K_i values measured for SP methylester in various preparations do not necessarily reflect different affinities for the NK_1 receptor. In contrast, septide was fairly stable in the same assay conditions (70–90% of the peptide still present after 70 min incubation) in both central and peripheral tissues. Clerc et al. (1992) found septide about 10 times more potent in displacing the binding of [^{3}H]SP in the rat submaxillary gland (K_i = 161 n*M*) than in rat brain (K_i = 1157 n*M*) and about 3 times more potent in the guinea pig ileum (K_i = 41 n*M*) than guinea pig brain (K_i = 142 n*M*). The same group also reported similar affinities for the novel nonpeptide NK_1 receptor antagonist, RP 67,580, in rat brain (K_i = 3.3 n*M*) and submaxillary gland (K_i = 2.9 n*M*) (Fardin et al., 1992b).

A separate line of evidence suggesting differences between central and peripheral NK_1 receptor was presented by Itoi et al. (1992). These authors investigated the central cardiovascular (tachycardia, hypertension) and behavioral responses to icv administration of SP or NKA in conscious rats. L 668,169, which has been characterized as a NK_1 receptor selective antagonist in peripheral preparations (McKnight et al., 1988), inhibited the response to icv administered NKA, but not SP, whereas L659,877, an NK_2 receptor selective antagonist at the peripheral level (McKnight et al., 1988), inhibited SP-, but not NKA-mediated responses. It should be noted, however, that L668,169 significantly interacts with NK_{2B} receptors in hamster trachea (*see* Section 4.3.4.) with potency slightly lower than that measured at peripheral NK_1 receptors in the guinea pig ileum (Patacchini et al., 1992). In a subsequent study, Tschope et al. (1992) reported that CP 96,345 attenuated the central responses to SP, but not to NKA, whereas R 396, an NK_{2B} receptor antagonist, inhibited selectively NKA-induced cardiovascular changes.

Overall, these results could be indicative of a heterogeneity of central vs peripheral NK_1 receptor, especially in the rat, and/or of a differential distribution of NK_1 receptor subtypes in central vs peripheral preparations, but this evidence is not yet conclusive.

Since identical clones of the NK_1 receptor have been found in the rat brain and salivary glands (Hershey and Krause, 1990), the above-mentioned discrepancies may be explained by differences in receptor glycosylation (Burcher et al., 1991a). Analogous examples (muscarinic M_2 receptor) of receptors in which sialic acid residues are directly involved in the formation of super high-affinity agonist-receptor complexes are known (Gies and Landry, 1988). Liu and Quirion (1991) reported that various carbohydrate moieties, including β-galactose and *N*-acetylglucosamine, are present on NK_1 receptors from the porcine brain. The affinity of SP is reportedly identical for the glycosylated and nonglycosylated forms of the NK_1 receptor, however (Kage et al., 1991). The possible relationship of the findings discussed in this section to the novel, septide-sensitive tachykinin receptor discussed in the next section remains to be determined.

3.3.2.2. A Novel, Septide-Sensitive Tachykinin Receptor

3.3.2.2.1. Septide-sensitive receptor in the guinea pig ileum. The ability of L- or D-Pro^9-substituted derivatives of SP(6–11) to discriminate between tachykinin receptors in different preparations was originally reported by Piercey et al. (1985). Owing to the tendency of N-terminal Gln to spontaneously undergo cyclization to Glp, the data reported by Piercey et al. (1985) on the activity of [L-Pro^9]SP(6–11), which bears an N-terminal Gln, can be considered as the first report on the activity of septide *(see below)*. In subsequent years, Lee and coworkers (1986) stressed the importance of measuring the [Glp^6, Pro^9]SP(6–11)/[Glp^6,D-Pro^9]SP(6–11) potency ratio for discriminating between NK_1 and NK_2 receptors (e.g., Lee et al., 1987; Suman-Chauhan et al., 1990). Wormser et al. (1986) introduced the term *septide* to label [Glp^6,Pro^9]SP(6–11) as a ligand selective for NK_1 receptors. They showed that septide is as potent as SP in producing atropine-resistant contraction of the guinea pig ileum, a typical NK_1 receptor assay, whereas it was about 1000 times less potent than NKB in producing acetylcholine release in the same preparation, a typical NK_3 receptor-mediated response. The same group also reported a negligible, if any, interaction on NK_2 receptors in rat isolated vas deferens (Laufer et al., 1988). The high potency of septide at NK_1 receptors was confirmed in other typical NK_1 receptor bioassays such as the endothelium-dependent relaxation of dog carotid artery (Dion et al., 1987) and enhancement of electrically-evoked contractions of guinea pig vas deferens (Hall and Morton, 1992), in which septide was as potent or even more potent than SP itself.

Petitet et al. (1992a,b) have reinvestigated the actions of septide in the guinea pig ileum and compared its ability to produce NK_1 receptor mediated contraction vs its ability to displace SP from NK_1 receptor binding sites. They found that various NK_1 receptor agonists, among which are SP methylester, septide, and [Pro^9]SP sulfone, are almost equipotent in contracting the ileum (EC_{50} = 2–3.5 n*M*); however, whereas SP and [Sar^9]SP sulfone were potent inhibitors of the binding of [^{3}H-Pro^9]SP sulfone in membranes of the guinea pig ileum (IC_{50} = 1.1–1.2 n*M*), septide or SP methylester were much weaker in this respect (IC_{50} = 250–260 n*M*). Similar results were obtained when using [^{3}H]SP as a ligand. Even more interestingly, the NK_1 receptor selective antagonist, GR 71,251 was distinctly more potent in antagonizing contraction of the ileum produced by septide (pK_B = 7.9) than those produced by [Pro^9]SP sulfone (pK_B = 6.5) (Petitet et al., 1992a,b). On this basis, Petitet et al. (1992a,b) proposed the existence of a novel, septide-sensitive tachykinin receptor in the guinea pig ileum. More recently, the same group showed that the antagonist, [D-Pro^9, Pro^{10}, Trp^{11}]SP, is more potent in antagonizing the response to septide (pA_2 = 6.5) than that to [Pro^9]SP (pA_2 < 5) in the ileum, while a related antagonist, [D-Pro^9, MeLeu^{10}, Trp^{11}]SP is equipotent (pA_2 = 6.6) against the two agonists (Chassaing et al., 1992).

In retrospect, a discrepancy between the high potency of septide in acting as a NK_1 receptor agonist in bioassay conditions (Wormser et al., 1986; Dion et al., 1987; Laufer et al., 1988; Hall and Morton, 1992) and in inhibiting SP binding at the NK_1 receptor was also evident in previous literature data. Thus, Lee et al. (1986) reported the activity of [Glp^6,Pro^9]SP(6–11) and of [Glp^6,D-Pro^9]SP(6–11) in various binding assays for tachykinin receptors. On the basis of their findings, the L-Pro analog (septide) is selective for NK_1 receptors, but its affinity is markedly lower than that of SP. The IC_{50} of SP in displacing ^{125}I-SP binding in rat salivary glands and guinea pig urinary bladder was 0.64 and 0.5 n*M*, respectively, as opposed to values of 188 and 43.5 n*M* for septide, respectively (Lee et al., 1986).

Data indicative that the pharmacology of the NK_1 receptor in the guinea pig ileum is more complicated than previously believed have also been obtained by Ireland et al. (1992). These authors showed that the affinity of GR 82,334, a peptide NK_1 antagonist derived from GR 71,251, is about 2 log units higher when measuring its ability to block SP methylester-induced contraction (pK_B = 7.64) than when measuring its ability to inhibit binding of [^{3}H]SP (pK_i = 5.58). Such a difference was not evident for the NK_1 receptor expressed by a human astrocytoma cell line when measuring binding of [^{3}H]SP vs NK_1 receptor-stimulated rise of intracellular calcium (pK_i = 7.70 vs pK_B = 7.73). The possible relevance of the findings of Ireland et al. (1992) to the septide-sensitive receptor described by Petitet et al. (1992a,b) remains to be determined. Recently, we have investigated the ability of (±) CP 96,345 to antagonize septide- and [Sar^9]SP sulfone-induced contractions in the circular muscle of the guinea pig ileum (Maggi et al., 1993b). We found that (±) CP 96,345 antagonizes with high affinity and in a competitive manner the response produced by both agonists, but the affinity measured against septide (pA_2 = 9.49, 95% c.l. 8.95–10.04; slope of Schild plot –0.87) is significantly higher than that measured against [Sar^9]SP sulfone (pA_2 = 8.14, 95% c.l. 7.61–8.67; slope of Schild plot –1.03). In a subsequent study, a similar pattern of higher affinity toward septide than toward [Sar^9]SP sulfone was detected for various NK_1 receptor antagonists, such as GR 82,334 and FK 888, in the circular muscle of the guinea pig ileum and colon (Maggi et al., 1994). Notably, no case of an antagonist showing higher affinity for classical than for septide-like agonist has been reported thus far.

3.3.2.2.2. Septide-sensitive receptor in the guinea pig bronchus. "Abnormal" (i.e., deviating from the expected behavior for an NK_1 receptor selective agonist) data with septide have been obtained in the guinea pig isolated bronchus (Maggi et al., 1991f). We found that the affinity of the NK_2 receptor selective antagonists, MEN 10,207 and MEN 10,376, in blocking septide-induced contraction of the bronchus is moderately, but distinctly, higher than their affinity toward SP methylester. For example, the pA_2 value of MEN 10,207 was 5.05 ± 0.10 against SP methylester and 5.97 ± 0.13 against septide; in the same preparation, the affinity of MEN 10,207 for NK_2 receptors using [βAla^8]NKA(4–10) as selective agonist was higher (pA_2 = 7.36 ± 0.11) than that measured against SP methylester or septide (Maggi et al., 1991f). A recent study (Fardin et al., 1992c) showed that septide is at least 100 times less potent than SP (IC_{50} values of 0.5 and 59 n*M* for SP and septide, respectively) in displacing binding of [^{3}H]SP in guinea pig lung membranes. In contrast, septide is about 20 times more potent in producing contraction of the guinea pig isolated main bronchus in the presence of peptidase inhibitors (EC_{50} = 10 and 0.5 n*M* for SP and septide, respectively; Maggi et al., 1991f).

The affinity of MEN 10,207 and MEN 10,376 toward septide and [Pro⁹]SP in the guinea pig ileum has been recently assessed by Lavielle and coworkers (personal communication). The pA_2 value of MEN 10,207 was about one log unit higher against septide (pA_2 = 6.7) than against [Pro^9]SP (pA_2 = 5.5), whereas the difference was less evident for MEN 10,376 (pA_2 = 5.9 and 5.5 against septide and [Pro^9]SP, respectively).

In the guinea pig bronchus, the responses to SP methylester and septide also differ in relation to the effect of the receptor alkylating agent, phenoxybenzamine (PBZ, Fig. 2). PBZ-pretreatment decreased the maximal response to SP methylester with little effect on affinity, while producing a rightward displacement with no reduction of maximal response to [βAla^8]NKA(4–10). The effect of PBZ on the response to septide was somehow mediated by both a rightward displacement and some depression of the maximum (Fig. 3; Maggi et al., 1991f).

3.3.2.2.3. Septide-sensitive receptor in the rat urinary bladder. We have addressed the possibility that septide stimulates a subtype of NK_1 receptor by studying the effect of three competitive NK_1 receptor antagonists, the peptide, GR 82,334, and the nonpeptides (±)CP 96,345 and RP 67,580, toward septide- and [Sar^9]SP-sulfone-induced contraction of the rat isolated urinary bladder (Meini et al., 1993, 1994). With both agonists, the rank order of antagonist potency was RP 67,580 > GR 82,334 > (±)CP 96,345, which is consistent with the profile of species-related variations of antagonist potency at the rat NK_1 receptor. However, the pA_2 values of the three antagonists indicate a significantly higher affinity for the septide- than [Sar^9]SP sulfone-induced response. As an example, the pA_2 of GR 82,334 toward septide and [Sar^9]SP sulfone (95% c.l. in brackets) was 7.01(6.89–7.13) and 5.93 (5.82–6.04), respectively, and the corresponding values for RP 67,580 were 7.57 (7.38–7.76) and 6.99 (6.77–7.20).

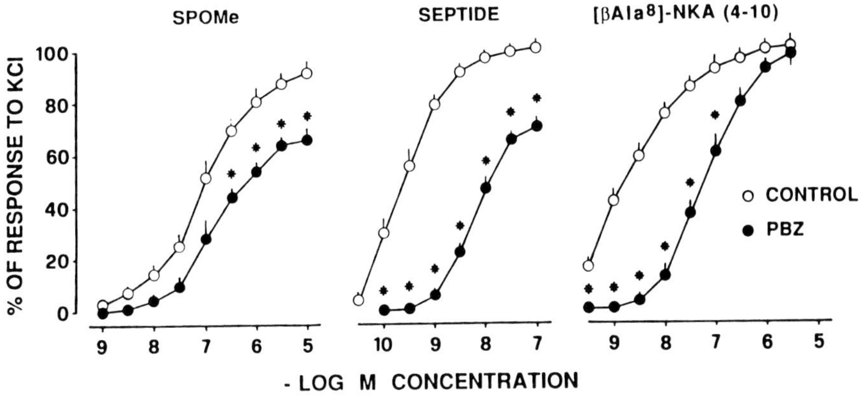

Fig. 2. Effect of phenoxybenzamine on the contractile response of the guinea pig isolated bronchi to substance P methylester (SPOMe), septide, and [βAla[8]]NKA(4–10). For each agonist, empty circles represent the response in control preparations and filled circles the response obtained after phenoxybenzamine (PBZ, 0.1 m*M* for 30 min) treatment. Reprinted with permission from Maggi et al. (1991f).

The response to both agonists was unchanged by L659,877 and SR 48,968, which potently block the NK_2 receptor in the rat urinary bladder (Meini et al., 1993, 1994). These data, although confirming and extending the findings of Petitet et al. (1992a,b), also indicate that the septide-sensitive receptor that is present in both the guinea pig ileum and bronchus and rat urinary bladder could indeed be a subtype of the NK_1 receptor because antagonists like RP 67,580 and (±)CP 96,345 are more potent toward septide than against the highly selective NK_1 receptor agonist, [Sar[9]]SP sulfone.

3.3.2.2.4. Septide-sensitive receptor: general considerations. Overall, the findings of Petitet et al. (1992a,b) have been confirmed and extended to other preparations to indicate the existence of unusual characteristics in the pharmacology of septide, which cannot be easily accommodated in the pharmacology of the "classical" NK_1 receptor (i.e., labeled with high potency by SP). On these bases, a septide-sensitive tachykinin receptor appears to be present not only in the guinea pig ileum, but also in guinea pig airways and rat urinary

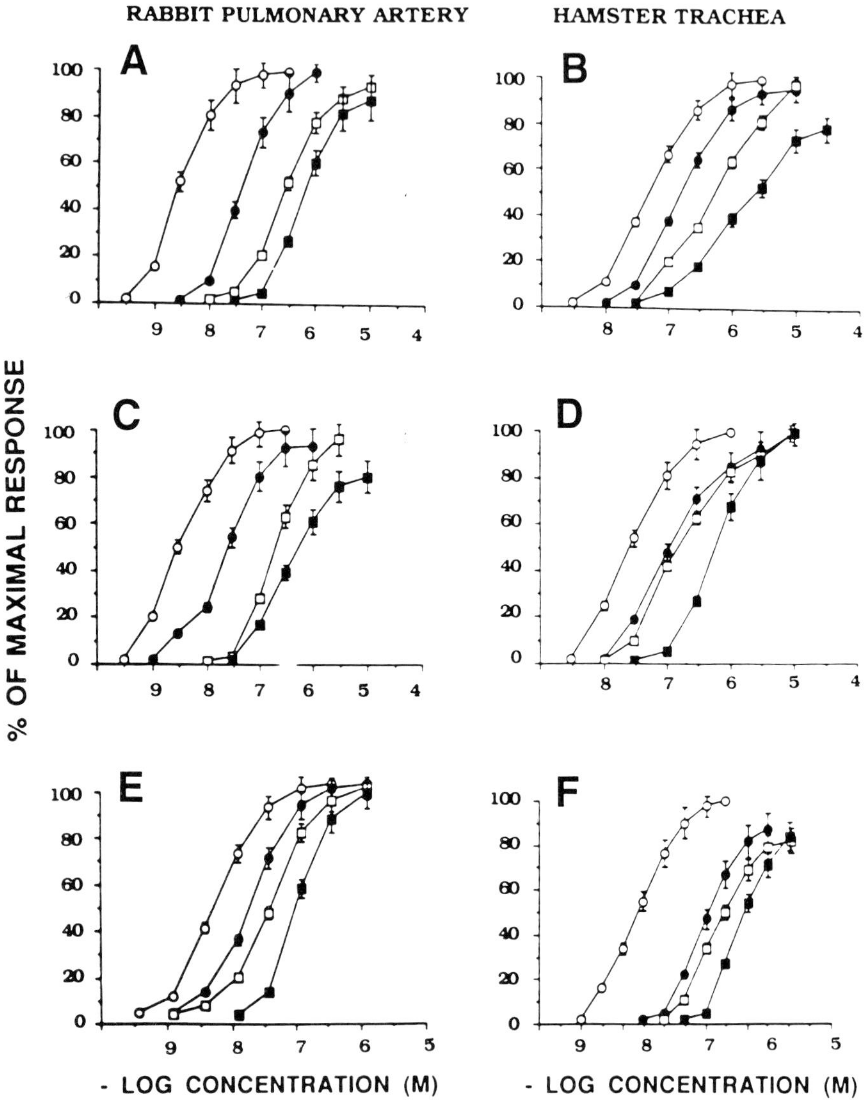

Fig. 3. Effect of peptide NK_2 receptor antagonists MEN 10,207 (panels A and B), MEN 10,208 (panels C and D), and R 396 (panels E and F) on the contractile response to NKA in the endothelium-deprived rabbit pulmonary artery (panels A, C, and E) or hamster trachea (panels B, D, and F). Reprinted with permission from Maggi et al. (1990a). In the original publication, MEN

bladder. Notably, high affinity competitive antagonists for the NK_1 receptor, like CP 96,345, GR 82,334, and RP 67,580, have higher affinity for the septide-sensitive receptor than for the classical NK_1 receptor labeled by agonists which, like [Sar^9]SP sulfone, bind to SP-labeled NK_1 classical receptors with high affinity. It appears from these results that many of the studies performed in various preparations about the physiology and pharmacology of NK_1 receptors should be reconsidered in the light of these novel findings. Furthermore, three main questions are open to investigations as follows:

1. Which is the rank order of potency of natural tachykinins at the septide-sensitive receptor? Since septide has a relatively poor affinity for NK_1 receptors labeled by SP (Lee et al., 1986; Petitet et al., 1992a), it may be argued that SP is not necessarily a high affinity ligand for the septide-sensitive receptor. Knowledge of the rank order of potency of natural tachykinins at the septide-sensitive site will also be important in terms of receptor nomenclature. Since competitive antagonists with high affinity for the NK_1 receptors have high potency toward septide-mediated biological effects, it might be argued that the septide-sensitive receptor is an NK_1 receptor subtype. However, NK_1 receptors have been originally defined on the basis of the rank order of potency of natural tachykinins, and a reply to this question will be crucial for any classification or nomenclature proposal. The results of a recent study have suggested that NKB could be better ligand than SP at the septide-sensitive receptor in the rat urinary bladder since, after occlusion of NK_2 receptors with the potent nonpeptide antagonist, SR 48,968, the affinity of RP 67,580 toward NKB was distinctly higher than that measured toward SP (Meini et al., 1994).
2. Is the septide-sensitive receptor, which is present in guinea pig and rat preparations, also present in human tissues? A reply to this

---

10,207, MEN 10,208, and R 396 are indicated as Peptide I, Peptide II, and Peptide III, respectively. In each panel, empty circles indicate the response to NKA in the absence of antagonists. Concentrations of antagonists tested were 0.1, 0.3, and 1 μ*M* in A; 3, 10, and 30 μ*M* in B; 0.3, 1, and 3 μ*M* in C; 3, 10, and 30 μ*M* in D; 10, 30, and 50 μ*M* in E; and 1, 3, and 10 μ*M* in F. Reprinted with permission from Maggi et al. (1990a).

question is important not only in terms of basic research, but also for applicative purposes. Very little information is available on this topic at present. Lee et al. (1992) showed that septide is about 300 times less potent than SP in displacing binding of ^{125}I-SP from NK_1 receptors on a human astrocytoma (U 373) cell line, but only 42 times less potent than SP in producing phosphoinositol accumulation.

3. Which are the molecular bases underlying the peculiar pharmacological properties of peptides of the septide family (Petitet et al., 1992a,b)? Does a separate gene exist to encode a distinct receptor protein different from NK_1, NK_2, or NK_3 receptors? Does the septide-sensitive receptor originate from alternative splicing/post-translational modification of the NK_1 receptor? A reply to these questions is crucial not only for classification/nomenclature purposes, but also for the final demonstration that the unusual pharmacology of septide and other peptides of this family is really linked to the existence of a "novel" type of tachykinin receptor, as originally proposed by Petitet et al. (1992a,b).
4. Since the recognition epitopes for certain NK_1 receptor antagonists by the NK_1 receptor protein appear to be distinct from the agonist binding site(s) (Gether et al., 1993), the possibility of an allosteric modulation between agonist and antagonist binding sites should be taken into serious consideration for the NK_1 receptor. It may be that the recognition epitopes of the NK_1 receptor protein for SP and classical NK_1 receptor agonists do not overlap with the epitopes recognizing the septide-like agonists, whereby septide and other agonists with a similar profile are more easily displaced by antagonists. This interpretation may explain why several NK_1 antagonists were in general more potent in blocking septide than classical NK_1 receptor agonists, while no example of the converse pattern has been reported yet. The available data from mutation experiments indicate that septide and SP recognize similar epitopes of the NK_1 receptor protein (Fong et al., 1992d). However, a systematic study of the effect of mutations of different regions of the NK_1 receptor protein on the binding of septide vs SP is not available.

3.3.2.3. Mouse and Rat Spinal Cord. Functional evidence suggesting the existence of NK_1 receptor subtypes in the mouse spinal cord derives from an investigation on the differential ability of spantide I and other peptide antagonists containing D-His to antagonize the licking, biting, and scratching behavioral response to intrathecal administration of SP, physalaemin, and septide (Sakurada et al., 1991). Spantide I was equieffective in antagonizing the response to SP, physalaemin, and septide, while [D-Phe^7,D-His^9]-SP(6–11) inhibits the response to SP, but not that to physalaemin and septide. On these bases, subtypes of the NK_1 receptor, termed NK_{1A} and NK_{1B} have been postulated to exist in the mouse spinal cord (Sakurada et al., 1991). Even in this case, the possible relationship of these findings to the septide-sensitive NK_1 receptor described in previous sections remains to be established.

Indirect evidence suggesting heterogeneity of NK_1 receptors in rat spinal cord has been obtained by our group while investigating the ability of various NK_1 receptor selective antagonists (GR 82,334, RP 67,580, and CP 96,345) to block certain micturition-related reflexes following the intrathecal administration of the antagonists at the level of the rat lumbosacral spinal cord. Basically, two models have been employed in these studies (Lecci et al., 1992; Maggi et al., 1992a): (1) the chemonociceptive vesico-vesical micturition reflex elicited by topical application of capsaicin onto the rat urinary bladder; and (2) the mechanoreceptive vesico-vesical micturition reflex elicited by bladder distension. In the former model, GR 82,334 at a dose of 1 nmol/rat completely blocked the response to capsaicin, whereas a tenfold larger dose was required to produce a similar effect with RP 67,580. GR 82,334 did not affect the mechanoreceptive micturition reflex at 1 nmol/rat, but was effective at 10 nmol/rat. For RP 67,580, the same dose effective toward capsaicin also affected the distension-evoked micturition reflex. In both models, CP 96,345 was effective only at a high dose (100 nmol/rat), in agreement with its poor affinity for rat NK_1 receptor. The effect of RP 67,580 was not shared by the NK_1 receptor blocking inactive enantiomer, RP 68,651. None of the treatments affected the somato-vesical excitatory reflex elicited by noxious perineal pinching, indicating the absence of neurotoxic or unspecific effects at the dose tested. Overall, the results of these experiments suggest that different subtypes of the NK_1 receptors may be involved in the processing of chemonociceptive and mechanoreceptive afferent input from the urinary bladder to the spinal cord.

3.3.2.4. Rabbit Smooth Muscle. A case for intraspecies heterogeneity of the rabbit NK_1 receptor has been raised by Beresford et al. (1992). They reported that the affinity of CP 96,345 in blocking the endothelium-dependent relaxation produced by SP methylester in rabbit thoracic aorta ($pK_B = 8.81$) is more than one log unit greater than the affinity measured toward contractions produced by the same agonist in the rabbit isolated iris sphincter muscle ($pK_B = 7.39$, Table 5). Similar quantitative differences have been found for the peptide NK_1 receptor antagonist, GR 82,334 (pK_B = 7.34 and 6.11 in the rabbit aorta and iris sphincter, respectively) (Beresford et al., 1992). These data are suggestive that intraspecies differences in the pharmacology of the NK_1 receptor may exist in rabbit (i.e., NK_1 receptor subtypes). The NK_1 receptor pharmacology of the rabbit iris sphincter seems quite complicated, however. In fact, the contractile response to SP is barely affected by NK_1 receptor antagonists like GR 71,251 or CP 96,345, and the half-time for relaxation of SP-induced contraction is very prolonged after washout compared to other NK_1 agonists (Hall et al., 1992). Interestingly, after exposure to PBZ, the response to SP washed out with a dramatically reduced half-time, and the postalkylation response then became sensitive to NK_1 receptor antagonists (Hall et al., 1992). A full characterization of the responses with other NK_1 receptor agonists and antagonists seems important, including agonists and antagonists used in the studies of Petitet et al. (1992a,b), to assess whether the results in the rabbit iris sphincter could involve the septide-sensitive receptor described in the guinea pig ileum.

4. Heterogeneity of the Tachykinin NK_2 Receptor

4.1. Molecular Biology

The NK_2 receptor is defined as the mediator of those biological activities encoded by the C-terminal sequence of tachykinins for which NKA is a more potent ligand than NKB and SP, in that order.

The first NK_2 receptor has been isolated from bovine stomach as a 384 amino acid residues protein (Masu et al., 1987). Other forms of the NK_2 receptor have been isolated from the rat stomach (Sasai and Nakanishi, 1989; 390 amino acid residues), human trachea (Gerard et al., 1990; 398 amino acid residues, *see also* Graham et al., 1991), human jejunum (Kris et al., 1991; 398 amino acid residues), and from mouse genoma (Sundelin et al., 1992; 384 amino acid residues).

All of these receptor forms meet the original criterion used to define the NK_2 receptor, i.e., when expressed in appropriate systems (oocytes, and so on) they recognize natural tachykinins with the rank order of potency NKA > NKB > SP. All of the forms of NK_2 receptor isolated from different species differ to some extent in their length and primary sequence. It appears that rat and mouse NK_2 receptors are more similar between them than to the human or bovine receptor. Thus, the bovine and rat NK_2 receptor share 86% homology in their primary sequence (Sasai and Nakanishi, 1989), whereas human and bovine NK_2 receptor are nearly 90% identical (Kris et al., 1991). Mouse and rat NK_2 receptor share 94% homology, but only 85% identity to the sequence of the bovine or human receptor (Sundelin et al., 1991). The reported sequences of human trachea and human jejunum NK_2 receptor (Gerard et al., 1990; Kris et al., 1991) differ by only one amino acid in the intracellular C-terminal tail.

As will be extensively discussed in Section 4.3., there is much pharmacological evidence indicating heterogeneity of the tachykinin NK_2 receptor that is largely species-related, but may also involve intraspecies heterogeneity. The putative NK_{2A} receptor, as defined below, seems to be preferentially expressed in smooth muscles of certain species (rabbit, guinea pig, bovine, human), whereas the putative NK_{2B} receptor is preferentially expressed in hamster and rat smooth muscles (Section 4.3.). No information is available to assign the mouse NK_2 receptor to one of these two categories. By analogy with the species-related differences in amino acid sequence of the hydrophobic transmembrane segments for the NK_1 receptor, the variant positions of M1–M7 of the NK_2 receptor (as assigned by Gerard et al., 1990) of the human, rat, bovine, and mouse are presented in Table 7.

Table 7
Comparison of Amino Acid Residues at Species-Related Variant Positions of the Seven Hydrophobic Transmembrane Segments (M1–M7) of the Tachykinin NK_2 Receptor[a]

Transmembrane segment	Amino acid position	Human	Rat	Bovine	Mouse
M1	37	Ala	Ala	Thr	Ala
	38	Thr	Thr	Ala	Thr
	49	Thr	Thr	Met	Thr
	53	Ile	Thr	Thr	Thr
M2	73	Val	Ile	Val	Ile
	88	Ala	Thr	Ala	Thr
M4	151	Val	Ile	Val	Val
	164	Ser	Ser	Phe	Ser
M5	200	Val	Val	Ile	Val
	202	Ile	Phe	Ile	Phe
	203	Ala	Val	Ala	Val
	211	Ala	Leu	Val	Val
	215	Val	Gly	Val	Ala
M6	251	Thr	Ala	Thr	Ala
	257	Leu	Leu	Val	Val

[a]Based on NK_2 receptor sequences published by Gerard et al. (1990, rat, bovine, and human) and Sundelin et al. (1992, mouse) and amino acid assignment to the seven hydrophobic transmembrane segments published by Gerard et al. (1990).

Unfortunately, the amino acid sequences of rabbit and hamster NK_2 receptor (the two species for which pharmacological differences are more striking) are not available for comparison. According to the assignment done by Gerard et al. (1990), 162 amino acid residues compose the seven hydrophobic transmembrane segments of the NK_2 receptor. Of these, species-related variations occur at 15 out of 162 positions (Table 7), indicating an overall 90.7% homology in M1–M7 regions of the NK_2 receptors in the four species.

The most striking difference occurs between the rat and bovine NK_2 receptor, which differ at 14 out of 15 variant sites in M1–M7 regions (Table 7). The least divergence is between rat and mouse NK_2 receptor that differ at only 4 out of 15 positions. The human NK_2 receptor differs from the rat and bovine NK_2 receptor at 9 and 8 out of 15 positions, respectively. As mentioned above, the pharmacological data indicate that the human NK_2 receptor is more similar to the bovine (NK_{2A}) than the rat (NK_{2B} receptor). If such differences were entirely species-related, it would be logical to speculate that species-related variations in M1–M7 sequence are important for determining this pharmacological heterogeneity. It is therefore of interest to see which of the 15 variant positions are common in the human and bovine receptor and different in the rat receptor. These 7 positions (underscored in Table 7) could thus provide the structural bases for species-related heterogeneity of the NK_2 receptor. It is important to note, however, that the rat and human NK_2 receptor share 6 amino acids at M1–M7 that differ from the bovine NK_2 receptor. Since it is not possible to decide *a priori* the relative importance of such variations of the ligand recognizing properties of the whole receptor protein, the significance of these species-related variations awaits single point mutation studies.

Until now, only one copy of the NK_2 receptor gene has been detected in the genoma of the various species examined. This would exclude the intraspecies heterogeneity that is required for the recognition of receptor subtypes. It is therefore possible that the heterogeneity revealed by the pharmacological approach is entirely related to the existence of species variants of the NK_2 receptor rather than to the existence of true receptor subtypes (intraspecies variants).

The organization of the NK_2 receptor gene indicates alternative splicing and/or posttranslational modifications as potential mechanisms leading to intraspecies receptor heterogeneity, as discussed in Section 2.4. Such a possibility has been described recently for the human NK_1 receptor, which has a similar gene organization to that of the NK_2 receptor (Fong et al., 1992a, *see* Section 3.1.). Tsuchida et al. (1990) have described the existence, in rat tissues, of a short form of the NK_2 receptor, which lacks the first two transmembrane domains, and proposed that it may originate through the use of heterogeneous transcription-initiation sites, rather than through alternative splicing. It is not known, however, whether these relatively abundant truncated forms of the NK_2 receptor are translated into a functional receptor (Tsuchida et al., 1990).

4.2. *Pharmacology*

The pharmacology of the NK_2 receptor includes a number of selective agonists and antagonists and sensitive "monoreceptorial" bioassays. Several synthetic peptides have been described as NK_2 receptor selective agonists, such as [Nle^{10}]NKA(4–10) (Dion et al., 1987), [βAla^{8}]NKA(4–10) (Rovero et al., 1989; Maggi et al., 1990b), GR 63,349 (Hagan et al., 1989), [Lys^{5}]NKA(4–10) (Lavielle et al., 1988), [Lys^{5}, MeLeu^{9}, Nle^{10}]NKA(4–10) (Chassaing et al., 1991), and, in certain preparations, the pseudopeptide, MDL 28,564 (Buck et al., 1990; Patacchini et al., 1991) (Table 1).

Various antagonists have been developed in the past few years which act as selective antagonists of the tachykinin NK_2 receptor (Table 2). Data on some of these compounds are shown in Table 3. Several of the available ligands suitable for achievement of a selective antagonism of the NK_2 receptor are peptides: the linear peptides such as MEN 10,207 and MEN 10,376 (Rovero et al., 1990, 1991a,b; Maggi et al., 1991d), R 396 (Dion et al., 1990a), and GR 94,800 (Hagan, 1992; McElroy et al., 1992); the cyclic peptides or pseudopeptides, L659,877 (McKnight et al., 1988, 1991; Williams et al., 1988), MDL 29,913 (Van Giersbergen et al., 1991), MEN 10,573, and MEN 10,612 (Quartara et al., 1992); the polycyclic peptide MEN 10,627 (Patacchini et al., 1993) (Tables 2 and 3); and the nonpeptide agent, SR 48,968 (Emonds-Alt et al., 1992; Fig. 1, Table 3). As will be evident below, the bulk of evidence for heterogeneity of NK_2 receptors relies on the different affinities of the various antagonists in different preparations, the only exception being the pseudopeptide, MDL 28,564, which acts as an agonist at some NK_2 receptors and an antagonist at others (Section 4.3.3.).

4.3. Evidence for Heterogeneity

4.3.1. Earlier Findings Indicating NK_2 Receptor Heterogeneity

The first indication that NK_2 receptors may be heterogeneous was obtained by McKnight et al. (1988) using L659,877, a cyclic hexapeptide (McKnight et al., 1991). L659,877 potently antagonizes NK_2 receptor agonists in the rat isolated vas deferens (pA_2 = 8), whereas it is weakly active or inactive in the guinea pig isolated trachea. As the action of tachykinins in these two preparations was believed to be mediated only by NK_2 receptors, McKnight et al. (1988) proposed the existence of a novel type of tachykinin receptor, putatively termed NK_4, in the guinea pig trachea. However:

1. The demonstration that both NK_1 and NK_2 receptors mediate tachykinin-induced contraction of the guinea pig airways, whereas only NK_2 receptors are present in the rat vas deferens;
2. The use of an agonist (eledoisin) which can act at both NK_1 and NK_2 receptors in the experiments of McKnight et al. (1988); and
3. The recognition that the activity of SP in guinea pig airways is strongly influenced by peptidases (Devillier et al., 1988; Ireland et al., 1988, 1991),

have lead to the rejection of the NK_4 receptor hypothesis.

Seen retrospectively, the results of McKnight et al. (1988) may be interpreted as the first indication of NK_2 receptor heterogeneity, reflecting the presence of different NK_2 receptor forms in different preparations.

4.3.2. Evidence with Antagonists

Data from various groups have established that NK_2 receptor antagonists discriminate, in terms of relative potencies, between heterogeneous forms of the NK_2 receptor. Our group, working on a series of NKA(4–10) analogs in which the insertion of multiple D-Trp conferred NK_2 receptor antagonist properties (Rovero et al., 1990a, 1991a,b), observed that some analogs (e.g., MEN 10,207 and MEN 10,376) possess much higher affinity for NK_2 receptors present in certain preparations (e.g., the endothelium-deprived rabbit pulmonary artery) than for those expressed in others (e.g., the hamster isolated trachea) (Maggi et al., 1989a, 1990a). On the other hand, R 396, a linear hexapeptide developed by the group of Regoli (Dion et al., 1990a), displays higher affinity for NK_2 receptors in the hamster trachea than for those of the rabbit pulmonary artery (Figs. 3 and 4). Having excluded a number of possible explanations (e.g., local metabolism by peptidases, differences in penetration rate in the two tissues), we interpreted the data as indicative of the existence of NK_2 receptor subtypes (Maggi et al., 1990a). It should be noted that the rank order of potency of agonists would lead to the conclusion that NK_2 receptors are homogeneous in the two preparations, i.e., no evidence for receptor heterogeneity was detected using natural or synthetic tachykinin receptor agonists, including NPK and NPγ.

The ability of NK_2 receptor antagonists to recognize with very different affinities two heterogeneous forms of the NK_2 receptor has also been demonstrated using radioligand binding techniques (Van Giersbergen et al., 1991a). NK_2 receptors expressed in the hamster urinary bladder, bovine stomach, and a fibroblast cell line transfected with mRNA encoding the NK_2 receptor isolated from bovine stomach bind various tachykinin receptor agonists with similar affinity. Clear distinctions became apparent when employing antagonists as inhibiting agents. Whereas MEN 10,207 is more potent (about 100 times) in the cell line transfected with bovine NK_2 receptor than in the hamster urinary bladder, other NK_2 selective antagonists, such as L659,877, are significantly more potent in the hamster urinary bladder (*see also* data from Aharony and Conner, 1992; Aharony et al., 1992) than on the bovine stomach NK_2 receptor. Furthermore, differences in the coupling to G-protein between the two receptors have been reported by Van Giersbergen et al. (1991a). In a subsequent study (Henderson et al., 1992), MEN 10,207 was found to antagonize competitively with high affinity ($pA_2 = 8$) the PI accumulation produced by stimulation of the bovine NK_2 receptor expressed in transfected fibroblasts.

The notion of the existence of pharmacological heterogeneity within the NK_2 receptor has been extended by Ireland et al. (1991) who found that the peptide antagonists, L659,877 and R 396, distin-

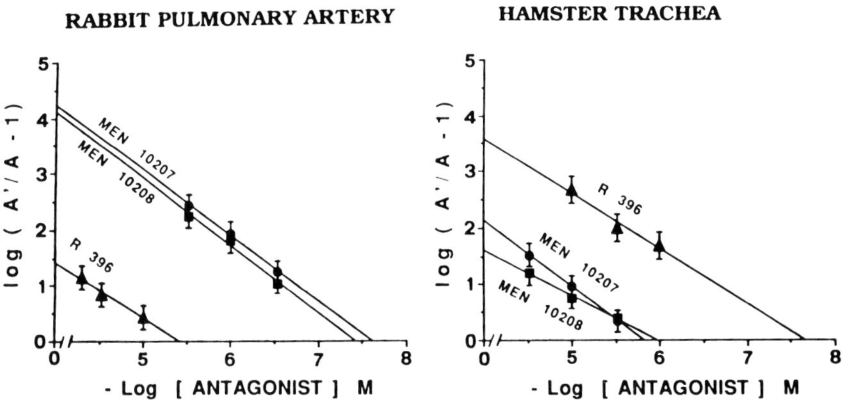

Fig. 4. Schild plots relative to the antagonist properties of MEN 10,207, MEN 10,208, and R 396 toward NKA in the endothelium-deprived rabbit pulmonary artery or hamster trachea (*see* Fig. 3). Reprinted with permission from Maggi et al. (1990a). In the original publication, MEN 10,207, MEN 10,208, and R 396 are indicated as Peptide I, Peptide II and Peptide III, respectively.

guish between high-affinity NK_2 receptors in the rat colon muscularis mucosa and low-affinity NK_2 receptors in the guinea pig and rabbit trachea and aorta. Ireland et al. (1991) concluded that NK_2 receptors in guinea pig and rabbit trachea and aorta resemble those described in the rabbit pulmonary artery, but are pharmacologically distinct from those present in the hamster trachea and rat colon.

Further evidence for the existence of multiple forms of NK_2 receptor originates from the work of Burcher and coworkers (Geraghty et al., 1991; Burcher et al., 1991a; Xiao et al., 1992). They found that ^{125}I-iodohistidyl NKA fails to label NK_2 receptors in the guinea pig lung and rabbit pulmonary artery. By contrast, this ligand binds with high affinity to classical NK_2 receptors in the hamster, rat, and dog urinary bladder and rat vas deferens. The classical NK_2 receptor in these preparations is recognized with high-affinity by antagonists like L659,877 and MDL 29,913. The hypothesis has been advanced (Geraghty et al., 1991) that steric hindrance produced by the large iodine atom on the N-terminal histidine prevents binding to the nonclassical NK_2 receptor present in the rabbit pulmonary artery and guinea pig bronchus.

4.3.3. Evidence with MDL 28,564

The second line of evidence leading to the recognition of a remarkable pharmacological heterogeneity within the NK_2 receptor came from the work done with a pseudopeptide analog of NKA(4–10), MDL 28,564. MDL 28,564, although a weaker agonist than NKA at NK_2 receptors, is probably the most selective NK_2 receptor ligand available, showing negligible interaction with NK_1 or NK_3 receptors (Harbeson et al., 1990). Buck et al. (1990) reported that MDL 28,564 behaves as a full agonist at certain NK_2 receptors (e.g., in the guinea pig trachea and rabbit pulmonary artery), whereas it behaves as a competitive antagonist and is devoid of agonist activity at other NK_2 receptors (e.g., those present in rat vas deferens) and proposed the existence of NK_2 receptor subtypes. The diverging agonist/antagonist character of the action of MDL 28,564 was also demonstrated in the rabbit pulmonary artery (full agonist action) and hamster trachea (no detectable agonism, but competitive antagonism toward NKA), the two preparations in which putative NK_2 receptor subtypes had been proposed on the basis of different antagonist affinities (Patacchini et al., 1991; Fig. 5). It is worth noting that the affinity of MDL 28,564 at NK_2 receptors in the rabbit pulmonary artery and hamster trachea is very close (Patacchini et al., 1991). Therefore, the chemical modifications introduced in this pseudopeptide ligand critically affect its ability to stimulate different receptor forms.

Using PBZ-alkylation technique, it has been possible to establish that the diverging agonist/antagonist character of MDL 28,564 at the proposed NK_2 receptor subtypes is not related to tissue factors, but involves a variable intrinsic efficacy of this ligand at different NK_2 receptors. Thus, the intrinsic efficacy of MDL 28,564 in stimulating the NK_2 receptor in the endothelium-deprived rabbit pulmonary artery is only 2.2 times lower than that of NKA, whereas the agonist activity is virtually undetectable at the NK_2 receptor of the hamster trachea, in which this pseudopeptide is unable to elicit agonism at a concentration sufficient to occupy 98% of total NK_2 receptors (Patacchini et al., 1991).

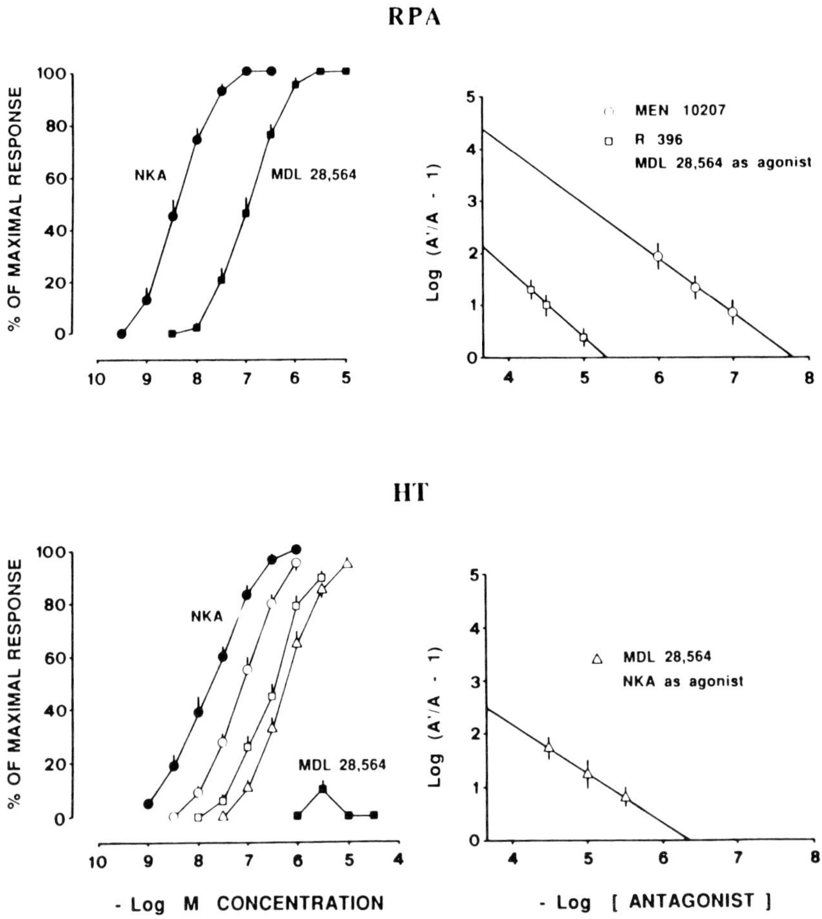

Fig. 5. Diverging agonist/antagonist character of the action of the pseudopeptide, MDL 28,564, in the rabbit pulmonary artery (RPA) and hamster trachea (HT). In the RPA, MDL 28,564 acts as a full agonist as compared to NKA response, and MDL 28,564-induced contraction is competitively antagonized by MEN 10,207 and R 396 with affinity (Schild plots in upper right panel) not significantly different from those measured against NKA. In the HT, MDL 28,564 is barely effective as an agonist (filled squares, lower left panel), but competitively antagonizes the response to NKA (lower left panel). Schild plot for MDL 28,564 antagonism toward NKA is shown in lower right panel. Reprinted with permission from Patacchini et al. (1991).

4.3.4. Two Broad Categories of NK_2 Receptor: NK_{2A} and NK_{2B}

The next logical step was to determine whether the two pharmacological criteria identified for distinguishing heterogeneity of NK_2 receptors, as defined in Maggi et al. (1990a) and Buck et al. (1990), could be rationalized into a simple pattern. The outcome of these studies was that, for those tissues in which MEN 10,207 or its analogs possess high affinity, MDL 28,564 behaves as a full agonist (NK_{2A} receptor), whereas for those tissues in which MEN 10,207 and its analogs possess low affinity, MDL 28,564 is almost devoid of agonist activity, but acts as a competitive antagonist (NK_{2B} receptor) (Patacchini et al., 1991; Maggi et al., 1991e,f, 1992b). Extending the investigation to a larger series of NK_2 receptor selective antagonists, the following rank orders of potencies were defined: a) MEN 10,207 (or analogs such as MEN 10,376) > L 659,877 > R 396 at NK_{2A} receptors, and b) L 659,877 > R 396 > MEN 10,207 (or analogs) at NK_{2B} receptors (Tables 8 and 9). These two criteria, i.e., rank order of antagonist potency and the diverging agonist/antagonist character of MDL 28,564, hold true in a number of preparations bearing the NK_2 receptor (Maggi et al., 1991e,f, 1992b,c,d; Table 9).

On the basis of data presented in Table 9, it appears that the NK_2 receptor mediating smooth muscle contraction in various rabbit, guinea pig, and human preparations is recognized by competitive antagonists with a rank order of potency that fits in the NK_{2A} receptor category, while various rat and hamster smooth muscles fit into the NK_{2B} receptor category. Likewise, MDL 28,564 acts as a full or near-to-full agonist at various rabbit, guinea pig, and human preparations listed in Table 9, whereas it is an antagonist in the rat and hamster preparations. Using such bioassay conditions, no examples have been reported thus far of two preparations from the same species that show the two distinct patterns of antagonist affinities. As will be discussed in Section 4.3.7., there are reasons to believe that the pharmacology of the rat and hamster NK_2 receptors also differ to some extent and that the same may occur for the rabbit and human NK_2 receptors. Overall, the rabbit and hamster smooth muscle NK_2 receptors show the largest pharmacological differences, especially in terms

Table 8
Pharmacological Criteria for Distinguishing NK_{2A} and NK_{2B} Receptors

NK_{2A} receptor, criteria for classification	NK_{2B} receptor, criteria for classification
1) MEN 10,376 > L659,877 > R 396	1) L659,877 > R 396 > MEN 10,376
2) MDL 28,564 agonist	2) MDL 28,564 antagonist
Rabbit pulmonary artery	Hamster trachea
Rabbit vas deferens	Hamster stomach
Rabbit bronchus	Hamster colon
Guinea pig bronchi	Hamster urinary bladder
Guinea pig renal pelvis	Rat colon
Guinea pig gallbladder	Rat urinary bladder
Human ileum (circular muscle)	Rat vas deferens
Human colon (circular muscle)	Rat ileum and colon
Bovine urinary bladder	Rat stomach
Bovine stomach	

of antagonist affinities. This is evident when considering, e.g., the pA_2 values of various NK_2 receptor competitive antagonists in the endothelium-deprived rabbit pulmonary artery and hamster trachea (Table 3). In absolute values, the most potent antagonist for the two assays are the nonpeptide ligand, SR 48,968, for the rabbit pulmonary artery ($pA_2 = 9.60$) and the cyclic pseudopeptide, MEN 10,612, ($pA_2 = 9.06$) for the hamster trachea. With regard to discriminative ability between these NK_2 receptors, other ligands are superior; thus, MEN 10,376 is the most selective ligand for the rabbit pulmonary artery (pA_2 difference between the two preparation is 2.43 log units) and R 396 is the most selective antagonist for the hamster trachea receptor (2.21 log units difference in pA_2 values).

The discovery of the first potent nonpeptide antagonist for NK_2 receptors, SR 48,968 (Emonds-Alt et al., 1992), also has a bearing on the topics dealt with in this section. In particular, previous evidence on the pharmacological heterogeneity of the NK_2 receptors relied on

Table 9
Affinities of Various Antagonists Selective for the NK_2 Tachykinin Receptor in Various Preparations from Different Species[a]

Preparation	MEN 10,376	MEN 10,207	L659,877	R 396	Source
Rabbit pulmonary artery	8.08	7.89	6.72	5.42	Maggi et al., 1990a, 1991d
Rabbit vas deferens	7.87	NT	6.86	5.48	Maggi et al., 1992b
Rabbit bronchus	8.19	NT	6.97	5.56	Maggi et al., 1992b
Guinea pig bronchus	7.27	7.36	6.57	5.35	Maggi et al., 1991f
Guinea pig gallbladder	6.77	6.26	6.29	5.26	Patacchini and Maggi, 1992
Guinea pig renal pelvis	7.41	NT	7.15	6.43	Maggi et al., 1992c
Human ileum	6.93	7.25	6.64	6.10	Maggi et al., 1992d and unpublished
Human colon	7.34	7.45	7.00	5.41	Giuliani et al., 1991
Hamster trachea	5.64	5.95	7.92	7.63	Maggi et al., 1990a, 1991d
Hamster urinary bladder	NT	NT	8.01	7.49	Maggi et al., unpublished
Hamster stomach	5.84	NT	8.24	7.46	Maggi et al., 1992b

Hamster colon	6.07	NT	7.59	7.47	Maggi et al., 1992b
Rat vas deferens	6.67	6.80	7.90	6.20	Maggi et al., 1991e and unpublished
Rat urinary bladder	6.20	5.82	7.75	6.45	Maggi et al., unpublished
Rat stomach	6.59	NT	7.75	6.53	Maggi et al., unpublished

[a]All numbers are pA_2 values calculated using the constrained plot method (slope constrained at -1) after having verified the competitive nature of antagonism (slope of Schild plot not signiffcantly different from unity) in the various preparations. In those preparations in which only NK_2 receptors mediate contraction to tachykinins (rabbit pulmonary artery, guinea pig gallbladder, human colon, hamster trachea, and urinary bladder, rat vas deferens), neurokinin A was used as agonist. In other preparations containing more than one tachykinin receptor (rabbit and guinea pig bronchus, guinea pig renal pelvis, human ileum circular muscle, and rat urinary bladder), the selective NK_2 receptor agonist, [βAla[8]]NKA(4–10), was used. The same was done in other preparations (rabbit vas deferens, hamster stomach and colon, and rat stomach) in which the possible presence of tachykinin receptors other than NK_2 was not investigated.

In all preparations from rabbit, guinea pig, and human, the pseudopeptide, MDL 28,564, displayed a clearcut agonist character that was otherwise absent or negligible in hamster and rat tissues.

Nt, not tested.

the use of peptide ligands. Although the above-mentioned experiments had been performed in the presence of some peptidase inhibitors, the possibility of a differential degradation of the antagonists by peptidases resistant to the inhibitors used may not be totally excluded. SR 48,968 (Fig. 1), which possesses a very strong affinity for NK_2 receptors expressed by various mammalian species, displays the highest affinity for the NK_{2A} receptor category as opposed to the NK_{2B} category. Data reported recently by Advenier et al. (1992) show a remarkably high affinity of SR 48,968 for rabbit and guinea pig NK_2 receptor (pA_2 values between 9.8–10.5), somewhat lower affinity for rat NK_2 receptor (pA_2 values 9.41 and 9.58 in the rat vas deferens and rat colon), and a further drop of affinity in hamster preparations (pA_2 values of 8.66 and 8.11 in the hamster trachea and urinary bladder). We have substantially confirmed this NK_{2A} preferring pattern of activity for SR 48,968 (e.g., pA_2 values of 9.60 and 9.52 in the rabbit pulmonary artery and bronchus, respectively, vs pA_2 values of 8.50 and 8.31 in the hamster trachea and rat vas deferens, respectively, Table 3; Patacchini et al., 1992).

Summarizing this section, evidence obtained in different laboratories and using various ligands of both peptide and nonpeptide nature indicate that the tachykinin NK_2 receptor, as originally defined on the basis of the rank order of potency of natural agonists NKA > NKB >> SP, is pharmacologically heterogeneous. Such a heterogeneity is very evident when using NK_2 receptor selective antagonists and seems largely species-dependent, as discussed in Section 3.1. for the NK_1 receptor. The notation, $NK_{2A/2B}$, has been introduced to distinguish the pharmacological characteristics of the NK_2 receptor found in bovine, rabbit, guinea pig, and some human tissues (NK_{2A}) vs the pharmacological characteristics found in hamster and rat tissues (NK_{2B}). Seen retrospectively, the NK_{2B} category represents the classical NK_2 receptor defined in the rat vas deferens and hamster urinary bladder (Buck et al., 1984; Lee et al., 1982, 1986; Van Giersbergen et al., 1992), whereas the NK_{2A} category expands and overlaps to the previously proposed (McKnight et al., 1988) NK_4 receptor. The next sections will address some other issues related to the heterogeneity of the tachykinin NK_2 receptor.

4.3.5. *Evidence for Intraspecies Heterogeneity*

A major point open to investigation is whether the disclosed heterogeneity of NK_2 receptors is ascribable to the existence of true NK_2 receptor *subtypes* (i.e., different forms of the same receptor alternatively expressed on different target cells of the same species) or simply involves species-related isoforms of the NK_2 receptor. The more straightforward demonstration of the first case would be the isolation of mRNA for different forms of the NK_2 receptor from the same species and characterization of their pharmacological profiles. Such evidence is not available. A possibility to be considered is that, in the same species, multiple forms of the NK_2 receptor originate from posttranslational modifications and/or alternative splicing of the products of one gene. The progress in molecular biology of tachykinin receptors is expected to provide further insight on this matter, but efforts to unravel the possible coexistence of NK_2 receptor subtypes in the same species have been made with more classical pharmacological approaches, as follows.

4.3.5.1. Rat Urinary Bladder. In membranes from the rat urinary bladder, using ^{125}I-iodohistidyl NKA as ligand, Nimmo et al. (1992) detected two populations of binding sites as indicated by a biphasic displacement of the ligand by MEN 10,207. They described the presence of both high- and low-affinity sites (IC_{50} = 0.89 n*M* and 0.28 μ*M*, respectively), which account for 16 and 84% of total NK_2 binding sites, respectively. At autoradiography, NK_2 receptors in the rat bladder are densely expressed on smooth muscle, whereas a minor population is present at the epithelial level. The latter disappear in the bladder of capsaicin-pretreated rats, probably reflecting a population of tachykinin receptors expressed on primary afferent nerves (Morrison et al., 1990). These may account for the facilitation of the micturition reflex observed after application of low concentrations of exogenous tachykinins (*see* Maggi et al., 1986, 1987). The "epithelial" NK_2 receptor labeling was abolished by extremely low concentrations (10 n*M*) of MEN 10,207, which affects only slightly the smooth muscle receptor population. Even at 1 μ*M*, MEN 10,207 did not completely inhibit the smooth muscle labeling (Nimmo et al., 1992). The NK_2 receptor for which MEN 10,207 possesses low-affinity is most likely that responsible for the direct smooth muscle contraction produced by tachykinin agonists. Thus, Flowers et al. (1991) reported a pA_2 of 5.8 for MEN 10,207 as a competitive antagonist of contraction produced by the NK_2 receptor agonist, [βAla[8]]NKA (4–10), in the rat isolated bladder. Together, these findings indicate that the rat urinary bladder contains NK_2 receptors having high (n*M*) and low (μ*M*) affinity for MEN 10,207, which may reflect the contemporary expression of the NK_2 receptor subtypes on different cell types in the same organ of the same species.

4.3.5.2. Rat Spinal Cord. In various rat peripheral tissues, the NK_2 receptor mediating smooth muscle contraction (urinary bladder, ileum, and so on) has been characterized as belonging to the NK_{2B} category. Likewise, a displacement profile typical of the NK_{2B} receptor (L659,877 > R 396 >> MEN 10,207 or MEN 10,376) has been obtained in binding experiments on membranes from rat small intestine (Brown et al., 1992). Thus, MEN 10,207 and MEN 10,376 act as competitive NK_2 receptor antagonists in various rat smooth muscles, but with low affinity (μ*M* concentration) (*see* Section 4.3.5.1. and Table 9). Xu et al. (1991) studied the ability of intrathecally (it) administered MEN 10,207 to antagonize the facilitatory action on a nociceptive flexor reflex produced by it administration of SP and NKA in rats. They found that very low doses of MEN 10,207 (70–700 pmol) specifically block the facilitation produced by NKA without affecting the response to SP. These findings, besides providing pharmacological evidence for the presence of NK_2 receptors in the rat spinal cord, indicate that MEN 10,207 possesses high affinity for these receptors, in sharp contrast with the low-affinity displayed by this antagonist in rat smooth muscle preparations.

4.3.5.3. Guinea Pig Alveolar Macrophages. Experiments on guinea pig isolated smooth muscles (bronchi, renal pelvis, gallbladder, circular muscle of the proximal colon) have indicated that the NK_{2A} receptor category prevails in this species (Maggi et al., 1991f; Patacchini et al., 1991; Patacchini and Maggi, 1992; Giuliani et al., 1991; Maggi et al., 1992c; Table 9). Guinea pig alveolar macrophages are also stimulated by tachykinins, chiefly through the NK_2

receptor (Brunelleschi et al., 1990). However, in these cells, MDL 28,564 fails to elicit agonism and behaves as an antagonist of NKA. When compared for their ability to antagonize NKA-stimulated superoxide anion production, R 396 (pK_B = 10.13) was more potent than L659,877 (pK_B = 9.27) or MEN 10,376 (pK_B = 8.8) (Brunelleschi et al., 1992). Therefore, the rank order of potency of the antagonists and the activity of MDL 28,564 are in keeping with the presence of NK_{2B} receptors in guinea pig alveolar macrophages. It should be noted, however, that the affinities of antagonists in this assay were much higher than expected on the basis of their affinities determined in isolated organs or binding assays.

The pieces of evidence described in this section are suggestive of an intraspecies heterogeneity of NK_2 receptor and the existence of pharmacologically distinct patterns of activity that may be related to the above-mentioned NK_{2A} and NK_{2B} receptor categories. Each one of the above-mentioned pieces of evidence needs to be substantiated, however, by additional data. It would be important, for example, to unravel a functional correlate for the MEN 10,207 high-affinity binding site described by Nimmo et al. (1992) in the rat urinary bladder. A complete pharmacological analysis with the various tools employed to define the NK_{2A} and NK_{2B} receptor categories is needed to substantiate the hypothesis that a NK_{2A} receptor mediates nociceptor reflex facilitation in the rat spinal cord, which is suggested from the results of Xu et al. (1991). Finally, comparative binding studies between the pharmacological characteristics of the NK_2 receptor expressed on guinea pig alveolar macrophages vs other guinea pig preparations are needed for a firm conclusion to be drawn about the possible existence of NK_{2B} receptors on guinea pig alveolar macrophages. It is also important to consider that the outcome of these studies might reveal the existence of an intraspecies heterogeneity of the NK_2 receptor that does not fit into the $NK_{2A/2B}$ scheme. If this were the case, then the criteria used to define these receptor categories could simply reflect the existence of species-related differences in the pharmacology of the NK_2 receptor, while intraspecies variants (subtypes) could require a new set of criteria for their definition.

4.3.6. NK_2 Receptor in Human Smooth Muscles

In the human gut (circular muscle of the ileum and colon), the NK_2 receptor mediating smooth muscle contraction belongs to the NK_{2A} category, as indicated by the rank order of antagonist potency (MEN 10,207 = MEN 10,376 > L659,877 > R 396) and by the full agonist character of MDL 28,564 (Giuliani et al., 1991; Maggi et al., 1992d; Table 9; Fig. 6).

In the human isolated bronchus. in which tachykinin contractility is sustained by the NK_2 receptor (Naline et al., 1989), the classification of the receptor subtype involved is not straightforward. Some data obtained with MEN 10,207 (pA_2 5.8, Dion et al., 1990a; pA_2 = 6.7, Sheldrick et al., 1992), MEN 10,376 (pA_2 = 6.3, Ellis and Undem, 1992), and R 396 (pA_2 = 6.8, Dion et al., 1990b) are not compatible with the profile observed for the NK_{2A} receptor. Contrasting data have been reported on the affinity of L659,877 for NK_2 receptor in the human bronchus. Rhoden and Barnes (1990) reported a pA_2 of 6.98, whereas Dion et al. (1990b) reported a pA_2 of 5.9. The affinity value reported by Rhoden and Barnes (1990) would be consistent with an NK_{2B} receptor hypothesis, but that found by Dion et al. (1990b) does not fit in this schematization. Also, Burcher et al. (1991b) reported that MDL 29,913, which is more potent that L659,877 at NK_{2B} receptor in hamster trachea (*see* Table 9), is barely effective in the human isolated bronchus. A systematic comparative study of the affinities of various NK_2 receptor antagonists in the human isolated bronchus has not been performed in a single laboratory, and discrepant results have been obtained in different laboratories with the same antagonist *(see below)*.

A complex situation also occurs in the human isolated urinary bladder in which the response to tachykinins is entirely mediated through NK_2 receptors (Maggi et al., 1988). Giuliani et al. (1992) reported a pattern of antagonist affinities for NK_2 receptor in the human urinary bladder with the following pA_2 values (in parentheses): L659,877 (7.10) = MEN 10,376 (7.08) = MEN 10,207 (6.81) > R 396 (6.30). MDL 28,564 showed a partial agonist activity, not exceeding 50% of maximal NKA response. These experiments involved antagonism of NKA-induced contraction of the human iso-

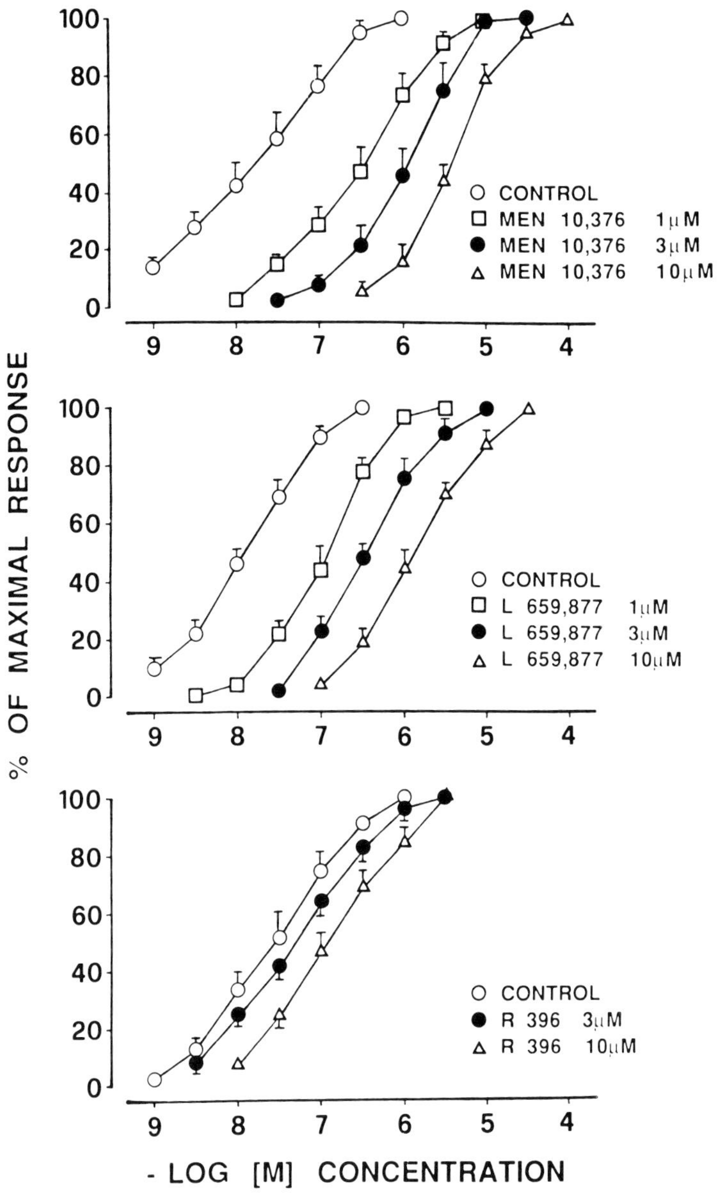

Fig. 6. Concentration-dependent displacement of the curve to the selective NK_2 receptor agonist, [βAla[8]]NKA(4–10), by the NK_2 receptor antagonists, MEN 10,376, L659,877, and R 396 in the circular muscle from human colon. Rank order of potency of antagonists indicate the presence of a tachykinin receptor similar to that responsible for contraction of the rabbit pulmonary artery. Reprinted with permission from Giuliani et al. (1991).

lated detrusor muscle in the presence of peptidase inhibitors (captopril, bestatin, and thiorphan). Subsequent experiments revealed the existence in this preparation of an additional peptidase activity (presumably aminopeptidase M), which is inhibited by amastatin. In the presence of amastatin, the following pA_2 values (in parentheses) have been determined against NKA: MEN 10,376 (7.90) > L659,877 (7.08) > R 396 (5.80) (Giuliani et al., 1993). Therefore, in the presence of amastatin, the affinity of MEN 10,376 was increased by about one log unit, and the rank order of antagonist potency becomes fully consistent with an NK_{2A} profile.

From the above, it appears that with the peptide ligands used to define NK_{2A} and NK_{2B} receptors, the human ileum and colon fit into the NK_{2A} receptor category. Although many of the above reported experiments using peptide ligands have been performed in the presence of peptidase inhibitors (e.g., captopril, bestatin, and thiorphan, 1 μ*M* each, in the experiments performed in our laboratories), degradation by peptidases may not be entirely ruled out. In this respect, recent findings obtained with SR 48,968 (Advenier et al., 1992) are of great interest. This ligand possesses a distinctly higher affinity for NK_2 receptors expressed in rabbit than hamster tissues (*see,* e.g., Table 3), thus ruling out any residual hypothesis that this pharmacological heterogeneity may be an artifact linked to peptide degradation. Advenier et al. (1992) reported pA_2 values of 9.40, 9.59, and 9.36 for SR 48,968 in the human bronchus, colon, and small intestine, respectively, and concluded that the same receptor (NK_{2A}) mediates contraction to tachykinins in these human preparations.

This conclusion is also supported by data obtained with the novel cyclic pseudopeptides, MEN 10,573 and MEN 10,612 (Table 2; Quartara et al., 1992), which are potent NK_2 receptor antagonists with some preference for the NK_{2B} receptor category. Data shown in Table 10, including also values obtained with the cyclic peptide L659,877 and the cyclic pseudopeptide MDL 29,913, indicate that the affinites of these antagonists in the human isolated urinary bladder, colon, and ileum are closer to the affinity measured in the rabbit pulmonary artery than that measured in the hamster isolated trachea (Giuliani, Maggi, et al., unpublished data).

Table 10
Affinities (pA_2 Values) of Selective NK_2 Receptor Antagonists (Cyclic Peptides and Pseudopeptides) and the Nonpeptide Compound SR 48,968 for NK_2 Receptor in Rabbit Pulmonary Artery (RPA), Hamster Trachea (HT), Human Urinary Bladder (HUB), and Circular Muscle of Human Colon (HUC) and Ileum (HUI)[a]

Antagonist	RPA	HT	HUB	HUC	HUI
L 659,877	6.72 ± 0.15	7.92 ± 0.03	7.10 ± 0.17	7.00 ± 0.10	6.83 ± 0.12
MDL 29,913	7.77 ± 0.08	8.60 ± 0.11	7.62 ± 0.11	7.83 ± 0.12	7.67 ± 0.18
MEN 10,573	7.31 ± 0.04	8.66 ± 0.12	7.76 ± 0.04	8.01 ± 0.11	7.72 ± 0.03
MEN 10,612	7.41 ± 0.04	9.06 ± 0.04	7.81 ± 0.06	8.06 ± 0.05	8.14 ± 0.10

[a]Each value is mean ± SEM of at least 6 determinations. NKA was used as agonist in the RPA, HT, HUB, and HUC; [βAla^8]NKA(4–10) was used in the HUI.

This conclusion has been fully supported by a recent study showing that, when investigated in the presence of amastatin, the rank order of potency of various NK_2 receptor antagonists in the human bronchus becomes fully consistent with the presence of NK_{2A} receptor (Astolfi et al., 1994).

Summarizing this section, it appears now established that the NK_2 receptor expressed in smooth muscle of human colon (Giuliani et al., 1991), ileum (Maggi et al., 1992d), urinary bladder (Giuliani et al., 1993), and bronchus (Astolfi et al., 1994) fulfills the criteria used to define the NK_{2A} receptor.

4.3.7. Further Complexities

With the increasing number of studies on NK_2 receptor subtypes/species variants, certain discrepancies have emerged, suggesting a more complex pharmacological heterogeneity than that recognized by the above discussed criteria.

4.3.7.1. DIFFERENCES BETWEEN RABBIT AND BOVINE NK_2 RECEPTOR. A first case for further heterogeneity within the NK_{2A} receptor category has been raised by Henderson et al. (1992). It had been previously shown (Maggi et al., 1990a) that the hepta- and octapeptide derivatives of NKA(4–10), MEN 10,207 and MEN 10,208 (referred to as Peptide I and II in Maggi et al., 1990a), possess the same affinity for NK_2 receptor mediating contraction of the rabbit isolated pulmonary artery. MEN 10,207 also possesses high affinity for the NK_2 receptor from bovine stomach (Van Giersbergen et al., 1991) on which basis the bovine and rabbit NK_2 receptor have been assigned to the same (NK_{2A}) receptor category. Extending the comparison to other NK_{2A} receptor bearing preparations (rabbit bronchus, guinea pig bronchus, human colon, and ileum), it was found again that the affinities of these two linear peptides are comparable (Maggi et al., 1992f).

However, Henderson et al. (1992) showed that MEN 10,208 is about 30 times less potent at NK_2 receptors in the bovine stomach when measuring either binding or the ability of the two antagonists to prevent phosphoinositol hydrolysis in a fibroblast cell line transfected with the bovine NK_2 receptor. On this basis, the proposal was advanced of a further heterogeneity between bovine stomach and rabbit pulmonary artery NK_2 receptor (Henderson et al., 1992). However, recent studies (Patacchini and Maggi, 1993) have shown that, when measured in the presence of the aminopeptidase inhibitor, amastatin, the affinity of MEN 10,208 for NK_2 receptor in the rabbit pulmonary artery drops by about one order of magnitude (pA_2 values of 7.47 and 5.94 in the absence and presence of amastatin, respectively), whereas the affinity of MEN 10,207 (pA_2 values of 7.80 and 7.68 in the absence and presence of amastatin, respectively) or MEN 10,376 (pA_2 values of 8.10 and 8.14 in the absence and presence of amastatin, respectively) is unchanged. On the other hand, amastatin did not affect the potency of the three antagonists in the hamster isolated trachea. On this basis, it appears conceivable that the true affinity of MEN 10,208 in the rabbit pulmonary artery had been overestimated (Maggi et al., 1990a) because of conversion, through an amastatin-sensitive peptidase, of MEN 10,208 to MEN 10,207 through the removal of the C-terminal Asp residue of the octapeptide.

4.3.7.2. DIFFERENCES BETWEEN HAMSTER AND RAT NK_2 RECEPTOR. On the basis of the rank order of antagonist potency and lack of significant agonist action of MDL 28,564, the NK_2 receptor in rat and hamster species has been attributed to the NK_{2B} receptor category. Some pharmacological differences between the two species have been identified, however. The linear hexapeptide, R 396, shows a tenfold higher affinity for the NK_2 receptor expressed in hamster smooth muscles (trachea, stomach, and colon) than for that expressed in rat smooth muscles (vas deferens, urinary bladder, stomach, and colon) as determined by bioassay and radioligand binding techniques (Maggi et al., 1992e). In contrast, the cyclic hexapeptide, L659,877, shows an even pattern of affinities (pA_2 values between 7.5–8) in the above-mentioned preparations. Since neither metabolism by peptidases nor differences in tissue penetration are likely to account for the behavior of R 396, the observed discrepancy might be indicative of further heterogeneity within the NK_{2B} receptor category.

Data in Table 11 illustrate further differences between the NK_2 receptor expressed in hamster trachea vs rat vas deferens as follows:

1. The linear heptapeptides, MEN 10,207 and MEN 10,376, are more potent at the rat vas deferens than hamster trachea NK_2 receptor (about 5 and 10 times, respectively) (Maggi et al., 1992f; Table 11). Interestingly, MEN 10,282, which differs from MEN 10,207 because it bears an L-Trp instead of D-Trp in position 9, and MEN 10,432, which bears Orn in position 10 instead of a basic residue (Arg in MEN 10,207, Lys in MEN 10,376), do not significantly discriminate between the hamster trachea and rat vas deferens NK_2 receptor.
2. The linear pseudopeptide, MDL 28,564, which acts as a competitive antagonist in both preparations, is about 10 times more potent at the hamster trachea (pA_2 6.21 ± 0.09) than at the rat vas deferens NK_2 receptor (pA_2 4.83 ± 0.10) (Maggi et al., 1992f).
3. As mentioned above, the cyclic hexapeptide, L659,877, does not distinguish between rat and hamster NK_2 receptor, and the same applies to its pseudopeptide derivative, MDL 29,913 (Table 11; Maggi et al., 1992f). The cyclic pseudopeptides, MEN 10,573 and MEN 10,612, however, show a slight but distinctly higher affinity for the hamster as compared to rat NK_2 receptor.

Overall, these findings indicate the existence of several differences between rat and hamster NK_2 receptors, which are recognized by some but not all ligands including linear peptides (R 396, MEN 10,207 and MEN 10,376), pseudopeptides (MDL 28,564), and cyclic pseudopeptides (MEN 10,573 and MEN 10,612).

4.3.7.3. DIFFERENCES BETWEEN RABBIT, GUINEA PIG, AND HUMAN NK_2 RECEPTOR. As described in section 4.3.4., the rabbit, guinea pig, and human NK_2 receptors fulfill the criteria used to define the NK_{2A} receptor category.

Some differences emerge between these species when considering the data shown in Table 12, as follows (Maggi et al., 1992f; Quartara et al., 1993):

1. MEN 10,282, a close analog of MEN 10,207 that bears an L-Trp in position 9 instead of D-Trp, is about 10 times less potent than MEN 10,207 in the rabbit pulmonary artery and main bronchus, but it is

Table 11
Comparison of Antagonist Potencies at NK_2 Receptors in Hamster Isolated Trachea and Rat Isolated Vas Deferens

Antagonist	Hamster trachea (pA_2)	Rat vas deferens (pA_2)	Difference (log unit)
Linear peptides and pseudopeptides			
R 396	7.63 ± 0.10	6.29 ± 0.15	+ 1.34[a]
MDL 28,564	6.21 ± 0.09	4.83 ± 0.10	+ 1.38[a]
MEN 10,207	5.95 ± 0.05	6.70 ± 0.15	– 0.75
MEN 10,282	6.12 ± 0.13	6.44 ± 0.06	– 0.32
MEN 10,376	5.64 ± 0.09	6.66 ± 0.18	– 1.02[a]
MEN 10,432	5.64 ± 0.05	5.90 ± 0.08	– 0.26
Cyclic peptides and pseudopeptides			
L 659,877	7.92 ± 0.03	7.93 ± 0.08	– 0.01
MDL 29,913	8.60 ± 0.11	8.45 ± 0.09	+ 0.15
MEN 10,573	8.66 ± 0.12	8.14 ± 0.10	+ 0.52
MEN10,612	9.06 ± 0.04	8.42 ± 0.10	+ 0.64[a]
Nonpeptide			
SR 48,968	8.50 ± 0.06	8.31 ± 0.11	+ 0.19

Neurokinin A was used as agonist in both preparations. Each value is mean ± SEM of at least 6 determinations.

[a]Significant difference between pA_2 values, $p < 0.05$.

equipotent or slightly more potent than MEN 10,207 in the guinea pig bronchus, human colon, and human ileum (Patacchini et al., 1991a; Maggi et al., 1992f).

2. MEN 10,392, a derivative of MEN 10,376 bearing Leu in position 10 instead of Lys, is distinctly less potent than MEN 10,376 in rabbit preparations and in the human ileum, although being equipotent in the guinea pig bronchus and human colon (Maggi et al., 1992f).
3. The N-terminally truncated derivatives of MEN 10,207/10,376, the hexapeptide MEN 10,412, and the pentapeptide, MEN 10,414, are at least 10 times less potent than, e.g., MEN 10,376, in rabbit

Table 12
Affinities (Expressed as pA_2 Values) of Linear Hepta-, Hexa-, and Pentapeptide Derivatives of NKA(4–10) NK_2 Receptor Antagonists at NK_2 Receptors in Various Isolated Smooth Muscle Preparations from different Species

Antagonist	RPA[a]	RB[b]	GPB[c]	HUC[d]	HUI[e]
MEN 10,207	7.89 ± 0.12	7.39 ± 0.09	7.28 ± 0.08	7.45 ± 0.13	7.29 ± 0.13
MEN 10,282	6.47 ± 0.13	6.46 ± 0.13	7.64 ± 0.06	7.74 ± 0.10	6.79 ± 0.02
MEN 10,376	8.07 ± 0.11	7.85 ± 0.09	7.43 ± 0.09	7.34 ± 0.03	6.93 ± 0.05
MEN 10,392	7.10 ± 0.19	7.22 ± 0.05	7.25 ± 0.12	7.20 ± 0.10	5.98 ± 0.19
MEN 10,412	6.75 ± 0.04	6.73 ± 0.11	6.48 ± 0.21	7.23 ± 0.16	6.51 ± 0.19
MEN 10,414	6.29 ± 0.10	6.33 ± 0.18	6.51 ± 0.09	7.24 ± 0.02	6.99 ± 0.22

MEN 10,207 = $[Tyr^5$, D-$Trp^{6,8,9}$, $Arg^{10}]$NKA(4–10).
MEN 10,282 = $[Tyr^5$, D-$Trp^{6,8}$, L-Trp^9, $Arg^{10}]$NKA(4–10).
MEN 10,376 = $[Tyr^5$, D-$Trp^{6,8,9}$, $Lys^{10}]$NKA(4–10).
MEN 10,392 = $[Tyr^5$, D-$Trp^{6,8,9}$, $Leu^{10}]$NKA(4–10).
MEN 10,412 = $[Tyr^5$, D-$Trp^{6,8,9}]$NKA(4–9).
MEN 10,414 = $[Tyr^5$, D-$Trp^{6,8}]$NKA(4–8).
Each value is mean ± SEM of at least 5 determinations.
[a]RPA = rabbit pulmonary artery (endothelium-deprived).
[b]RB = rabbit bronchus.
[c]GPB = guinea pig bronchus.
[c]HUC = human colon (circular muscle).
[d]HUI = human ileum (circular muscle). Activity of the antagonists was investigated after 15 min contact time in the presence of peptidase inhibitors (captopril, bestatin, and thiorphan; 1 μ*M*)15 min before using NKA (RPA and HUC) or $[\beta Ala^8]$NKA(4–10) (GPB, RB, and HUI) as agonists.

and guinea pig preparations, but are equipotent to MEN 10,376 in the human colon and ileum (Quartara et al., 1993).

The quantitative pharmacological differences described in this section have been obtained in the presence of a mixture of peptidase inhibitors (captopril, bestatin, and thiorphan; 1 μ*M*). However, the possibility that these differences originate from a differential degradation of linear peptides tested in different preparations cannot be completely ruled out at present.

5. Conclusions

5.1. NK_1 Receptor

The available information on the heterogeneity of the NK_1 receptor can be outlined as follows:

1. Structurally different forms of the NK_1 receptor protein exist in the four species thus far examined (human, rat, mouse, guinea pig). These are *de facto* species variants of the NK_1 receptor. Competitive antagonists, especially CP 96,345 and RP 67,580, discriminate with very different potencies the NK_1 receptor expressed in different species.
2. The organization of the NK_1 receptor gene lends itself to the possibility that multiple forms of the receptor protein are edited from the same gene, thus leading to intraspecies receptor heterogeneity, but this possibility has not yet been substantiated.
3. Pharmacological evidence suggesting the existence of heterogeneous forms of the NK_1 receptor has been presented. This involves possible differences between rat central and peripheral (salivary glands) NK_1 receptor, as recognized by septide and other agonists (Lew et al., 1990; Clerc et al., 1992); possible heterogeneity of NK_1 receptor in mouse and rat spinal cord (Sakurada et al., 1991; Maggi et al., 1992a); and possible heterogeneity of NK_1 receptors in different rabbit smooth muscle preparations (Beresford et al., 1992). At present, the evidence gathered in these studies cannot be considered as conclusive. The molecular bases of the heterogeneity disclosed with the pharmacological approach need to be elucidated.

4. The significance of the findings of Petitet et al. (1992) about the presence of a "novel," septide-sensitive tachykinin receptor needs careful evaluation. The evidence for the existence of such a receptor originates from discrepancies in the pharmacology of ligands, such as septide, which have been previously characterized as NK_1 receptor selective. In various preparations, high-affinity NK_1 receptor antagonists are more potent towards septide than classical NK_1 receptor agonists. Thus, the septide-sensitive tachykinin receptor could represent a variant of the NK_1 receptor, but further studies are needed to elucidate this point.

5.2. NK_2 Receptor

The information gathered up to now on the heterogeneity of the NK_2 receptor can be outlined as follows:

1. Structurally different forms of the NK_2 receptor protein exist in the species examined thus far (bovine, human, rat, mouse). These are *de facto* species variants of the NK_2 receptor.
2. The organization of the NK_2 receptor gene lends itself to the possibility that multiple forms of the receptor protein are edited from the same gene, thus leading to intraspecies receptor heterogeneity, but this possibility has not yet been substantiated.
3. Competitive antagonists of both peptide and nonpeptide nature discriminate with very different potencies the NK_2 receptor expressed in different species.
4. The pseudopeptide, MDL 28,564, shows a diverging agonist/antagonist character at NK_2 receptors expressed in different species.
5. Combining the two criteria, two broad categories of the NK_2 receptor have been distinguished on a pharmacological basis, which have been termed NK_{2A} and NK_{2B}. On the other hand, further minor differences exist in the pharmacology of rat vs hamster and human vs rabbit NK_2 receptor (Section 4.3.7.). As outlined in Section 4.3.5., there are suggestions for a further intraspecies heterogeneity of the NK_2 receptor. If conclusively proven, these may or may not be accommodated into the $NK_{2A/2B}$ receptor categories, and the $NK_{2A/2B}$ subclassification may have to be abandoned because it is insufficient to account for all species-related differences in NK_2

receptor pharmacology. At present, it cannot be decided whether the $NK_{2A/2B}$ schematization merely reflects the existence of species-related variations or may also incorporate the intra-species heterogeneity required for recognition of receptor subtypes. For this reason, the more neutral expression "receptor category" has been used here in reviewing the status of research in this field.

Acknowledgments

I wish to thank S. H. Buck for the gift of MDL 28,564 and MDL 29,913, K. Folkers for the gift of spantide II, X. Emonds-Alt and G. Le Fur for the gift of SR 48,968, C. Garret for the gift of RP 67,580, D. Regoli for the gift of R 396, and G. Viti for the synthesis of (±)CP 96,345. A special thanks also to various researchers of the Pharmacology and Chemistry Departments of A. Menarini Pharmaceuticals and related companies (M. Astolfi, S. Giuliani, A. Lecci, S. Meini, R. Patacchini, L. Quartara, A. R. Renzetti, P. Rovero, P. Santicioli) and to A. Giachetti and A. Meli for encouragement and support.

References

Advenier, C., Rouissi, N., Nguyen, Q. T., Emonds-Alt, X., Breliere, J. C., Neliat G., Naline, E., and Regoli, D. (1992) Neurokinin A (NK-2) receptor revisited with SR 48,968 a potent nonpeptide antagonist. *Biochem. Biophys. Res. Commun.* **184,** 1418–1424.

Aharony, D. and Conner, G. E. (1992) Pharmacologic characterization of binding of the novel ligand [4,5-^{3}H-Leu^9]NKA to NK-2 receptors on hamster urinary bladder. *FASEB J.* **6,** A1277.

Aharony, D., Conner, G. E., and Woodhouse, D. P. (1992) Pharmacologic characterization of the novel ligand [4,5-^{3}H-Leu]neurokinin A binding to NK-2 receptors on hamster urinary bladder. *Neuropeptides* **23,** 121–130.

Appell, K. C., Fragale, B. J., Loscig, J., Singh, S., and Tomczuk, B. E. (1992) Antagonists that demonstrate species differences in neurokinin 1 receptors. *Mol. Pharmacol.* **41,** 772–778.

Astolfi, M., Treggiari, S., Giachetti, A., Meini, S., Maggi, C. A., and Manzani, S. (1994) Characterization of the tachykinin NK_2 receptor in human bronchus: influence of amastatin-sensitive metabolic pathways. *Br. J. Pharmacol.,* in press.

Beresford, I. J. M., Birch, P. J., Hagan, R. M., and Ireland, S. J. (1991) Investigation into species variants in tachykinin NK-1 receptors by use of the nonpeptide antagonist, CP 96345. *Br. J. Pharmacol.* **104,** 292,293.

Beresford, I. J. M., Ireland S. J., Stables, J. M., Stubbs, C. M., Ball, D., Hagan R. M., and Birch, P. J. (1992) Investigation into differences in tachykinin NK-1 receptors between and within species using a peptide and a nonpeptide NK-1 receptor antagonist. *Br. J. Pharmacol.* **106,** 92P.

Black, J. L., Johnson, P. R. A., Alouan, L. A., and Burcher, E. (1990b) Comparative effects of neurokinin A (NKA) and neuropeptide γ (NPγ in human isolated bronchus. *Clin Exp. Pharmacol. Physiol.* **Suppl. 17,** 5.

Brown, M. C. S., Biagi, M., Renzetti, A. R., Rovero, P., Criscuoli, M., and Mizrahi, J. (1992) [^{125}I] histidine neurokinin A binds selectively to NK-2 receptors of the B type in rat small intestine smooth muscle membranes. *Eur. J. Pharmacol.* **227,** 163–171.

Brunelleschi, S., Ceni, E., Fantozzi, R., and Maggi, C. A. (1992) Evidence for tachykinin NK-2B-like receptors in guinea-pig alveolar macrophages. *Life Sci. Pharmacol. Lett.* **51,** PL177–181.

Brunelleschi, S., Vanni, L., Ledda, F., Giotti, A., Maggi, C. A., and Fantozzi, R. (1990) Tachykinins activate guinea-pig alveolar macrophages: involvement of NK-2 and NK-1 receptors. *Br. J. Pharmacol.* **100,** 417–420.

Buck, S. H., Burcher, E., Shults, C. W., Lovenberg, W., and O'Donohue, T. L. (1984) Novel pharmacology of substance K binding sites: a third type of tachykinin receptor. *Science* **226,** 987–989.

Buck, S. H., Harbeson, S. L., Hassmann, III, C F., Shatzer, S. A., Rouissi, N., Nantel, F., and Van Gierbergen, P. L. M. (1990) [Leu$^9\Psi CH_2NH_2$)Leu^{10}]-neurokinin A(4–10) (MDL 28,564) distinguishes tissue tachykinin peptide NK-2 receptors. *Life Sci. Pharmacol. Lett.* **47,** PL37–PL41.

Buell, G., Schulz, M. F., Arkinstall, S. J., Maury, K., Missotten, M., Adami, N., Talabot, F., and Kawashima, E. (1992) Molecular characterisation, expression and localisation of human neurokinin 3 receptor. *FEBS Lett.* **299,** 90–95.

Burcher, E. and Mussap, C. J. (1991) Neuropeptide γ-preferring subtypes of NK-2 receptors in rat fundus and dog urinary bladder. *Soc. Neurosci. Abstr.* **17,** 805.

Burcher, E., Mussap, C. J., Geraghty, D. P., McClure Sharp, J. M., and Watkins, D. J. (1991a) Concepts in characterization of tachykinin receptors. *Ann. NY Acad. Sci. USA* **632,** 123–136.

Burcher, E., Alouan, L. A., Johnson, P. R. A., and Black. J. L. (1991b) Neuropeptide γ, the most potent contractile tachykinin in human isolated bronchus, acts via a 'non-classical' NK-2 receptor. *Neuropeptides* **20,** 79–82.

Chang, M. M., Leeman, S. E., and Niall, H. D. (1971) Amino acid sequence of substance P. *Nature* **232,** 86,87.

Chassaing, G., Lavielle, S., Brunissen, A., Carruette A., Garret, C., Petitet, F., Saffroy, M., Beaujouan, J. C., Torrens, Y., and Glowinski, J. (1992) [Pro^9]SP and [pGlu^6, Pro^9]SP(6–11) interact with two different receptors in the guinea-pig ileum as demonstrated with new SP antagonists. *Neuropeptides* **23,** 73–79.

Chassaing, G., Lavielle, S., Loeuillet, D., Robilliard, P., Carruette, A., Garret, C., Beaujouan, J. C., Saffroy, M., Petitet, F., Torrens, Y., and Glowinski, J. (1991) Selective agonists of NK-2 binding sites highly active on rat portal vein (NK-3 bioassay). *Neuropeptides* **19,** 91–95.

Civelli, O., Bunzow, J. R., Grandy, D. K., Zhou, Q. Y., and Van Tol, H. H. M. (1991) Molecular biology of the dopamine receptors. *Eur. J. Pharmacol. Mol. Pharmacol.* **207,** 277–286.

Clerc, F. F., Foucault, F., Ridoux-Silly, L., and Fardin, V. (1992) Affinity and stability of some substance P agonists in rat and guinea-pig: a comparative study in central and peripheral tissues. *Neuropeptides* **22,** 14.

Devillier, P., Advenier, C., Drapeau, G., Marsac, J., and Regoli, D. (1988) Comparison of the effects of epithelium removal and of an enkephalinase inhibitor on the neurokinin-induced contractions of guinea-pig isolated trachea. *Br. J. Pharmacol.* **94,** 675–684.

Dion, S., D'Orléans-Juste, P., Drapeau, G., Rhaleb, N. E., Rouissi, N., Tousignant, C., and Regoli, D. (1987) Characterization of neurokinin receptors in various isolated organs by the use of selective agonists. *Life Sci.* **41,** 2269–2278.

Dion, S., Rouissi, N., Nantel, F., Drapeau, G., Regoli, D., Naline, E., and Advenier, C. (1990b) Receptors for neurokinins in human bronchus and urinary bladder are of the NK-2 type. *Eur. J. Pharmacol.* **178,** 215–219.

Dion, S., Rouissi, N., Nantel, F., Jukic, D., Rhaleb, N. E., Tousignant, C., Telemaque, S., Drapeau, G., Regoli, D., Naline, E., Advenier, C., Rovero, P., and Maggi C. A. (1990a) Structure-activity study of neurokinins. Antagonists for the NK-2 receptor. *Pharmacology* **41,** 184–194.

Drapeau, G., D'Orléans-Juste, P., Dion, S., Rhaleb, N. E., and Regoli, D. (1987) Specific agonists for neurokinin B receptors. *Eur. J. Pharmacol.* **136,** 401–403.

Ellis, J. L. and Undem, B. J. (1992) Tachykinin receptor subtypes in human bronchi. *Am. Rev. Resp. Dis.* **145,** A260.

Emonds-Alt, X., Vilain, P., Goulaouic, P., Proietto, V., Van Broeck, D., Advenia, C., Naline, E., Neliat, G., Le Fur, G., and Breliere, J. C. (1992) A potent and selective nonpeptide antagonist of the neurokinin A (NK-2) receptor. *Life Sci. Pharmacol. Lett.* **50,** PL101–106.

Erspamer V. (1981) The tachykinin peptide family. *Trends Neurosci.* **4,** 297–299.

Fardin, V., Flamand, O., Foucault, F., Jolly, A., Bock, M. D., and Garret, C. (1992a) RP 67,580, a nonpeptide substance P antagonist: comparison with other antagonists and affinties in different animal species. *Neuropeptides* **22,** 22.

Fardin, V., Foucault, F., Bock, M. D., and Garret, C. (1992b) Binding profile of RP 67,580, a new nonpeptide SP antagonist: comparison with CP 96,345. *Br. J. Pharmacol.,* in press.

Fardin, V. and Garret, C. (1991) Species differences between [^{3}H] substance P binding in rat and guinea-pig shown by the use of peptide agonists and antagonists. *Eur. J. Pharmacol.* **20,** 1231–1234.

Fardin, V., Vaissade, F., Jolly, A., Flamand, O., and Garret, C. (1992c) Distribution and characterization of tachykinin binding in bronchopulmonary tract of rat and guinea-pig. *Regul. Pept.* **Suppl. 1,** S54.

Flowers, J. M., Hall, J. M., and Morton, I. K. M. (1991) NK_1 and NK_2 neurokinin receptors characterized in the rat isolated urinary bladder preparation. *Br. J. Pharmacol.* **104,** 149P.

Folkers, K., Hakanson, R., Horig, H., Xu, J. C., and Leander, S. (1984) Biological evaluation of SP antagonists. *Br. J. Pharmacol.* **83,** 449–456.

Folkers, K., Feng, D. M., Asano, N., Hakanson, R., Wiesenfeld-Hallin, Z., and Leander, S. (1990) Spantide II, an effective tachykinin antagonist having high potency and negligible neurotoxicity. *Proc. Natl. Acad. Sci. USA* **87,** 4833–4835.

Fong, T. M., Anderson, S., Yu, H., Huang, R. R. C., and Strader, C. D. (1992a) Differential activation of intracellular effector by two isoforms of human neurokinin 1 receptor. *Mol. Pharmacol.* **41,** 24–30.

Fong, T. M., Yu, H., Huang, and Strader, C. D. (1992b) The extracellular domain of substance P(NK-1) receptor comprises part of the ligand binding site. *Biophys. J.* **62,** 59,60.

Fong, T. M., Yu, H., Huang, R. R. C., and Strader, C. D. (1992c) The extracellular domain of the NK-1 receptor is required for high affinity binding of peptides. *Regul. Pept.* **Suppl. 1,** S57.

Fong, T. M., Yu, H., Huang, R. R. C., and Strader, C. D. (1992d) The extracellular domain of the neurokinin-1 receptor is required for high-affinity binding of peptides. *Biochemistry* **31,** 11,806–11,811.

Fujii, T., Murai, M., Morimoto, H., Hagiwara, D., Miyake, H., and Matsuo, A. (1992) Effect of novel SP antagonist, FK 888 on airway constriction and airway edema in guinea-pigs. *Neuropeptides* **22,** 24.

Garret, C., Carrueue, A., Fardin, V., Moussaoui, S., Peyronel, J. F., Blanchard, J. C., and Laduron, P. M. (1991) Pharmacological properties of a potent and selective nonpeptide substance P antagonist. *Proc. Natl. Acad. Sci. USA* **88,** 10,208–10,211.

Garret, C., Carruette, A., Fardin, V., Moussaoui, S., Peyronel, J. F., Blanchard, J. C., and Laduron, P. M. (1992) RP 67,580 a potent and selective nonpeptide SP antagonis. *C. R. Acad. Sci. Paris* **314,** 199–204.

Geraghty, D. P., Mussap, C. J., and Burcher, E. (1991) Radioiodinated substance P, neurokinin A and eledoisin bind predominantly to NK-1 receptors in guinea-pig lung. *Mol. Pharmacol.* **41,** 147–153.

Gerard, N. P., Eddy R. L., Shows, T. B., and Gerard, C. (1990) The human neurokinin A (substance K) receptor. *J. Biol. Chem.* **265,** 20,455–20,462.

Gerard, N. P., Garraway, L. A., Eddy, R. L., Shows, T. B., Iijima, H., Paquet, J. L., and Gerard, C. (1991) Human substance P receptor (NK-1): organization of the gene, chromosome localization and functional epression of cDNA clones. *Biochemistry* **30,** 10,640–10,646.

Gether, U., Marray, T., Schwartz, T. W., and Johansen, T. E. (1992) Stable expression of high affinity NK-1 (substance P) and NK-2 (neurokinin A) but low affinity NK-3 (neurokinin B) receptors in transfected CHO cells. *FEBS Lett.* **296,** 241–244.

Gether, U., Yokota, Y., Emonds-Alt, X., Breliere, J. C., Lowe, J. A., Snider, R. M., Nakanishi, S., and Schwartz, T. W. (1993) Two nonpeptide tachykinin antagonists act through epitopes on corresponding segments of the NK1 and NK2 receptors. *Proc Natl. Acad. Sci. USA.* **90,** 6194–6199.

Gies, J. P. and Landry, Y. (1988) Sialic acid is selectively involved in the interaction of agonists with M2 muscarinic acetylcholine receptors. *Biochem. Biophys. Res. Comm.* **150,** 673–680

Gitter, B. D., Waters, D. C., Bruns, R. F., Mason, N. R., Niton, J A., and Howbert, J. J. (1991) Species differences in affinities of nonpeptide antagonists for substance P receptors. *Eur. J. Pharmacol.* **197,** 237,238.

Giuliani, S., Barbanti, G., Turini, D., Quartara, L., Rovero, P., Giachetti, A., and Maggi, C. A. (1991) NK-2 tachykinin receptors and contractions of circular muscle of the human colon: characterization of the NK_2 receptor subtype. *Eur. J. Pharmacol.* **203,** 365–370.

Giuliani, S., Patacchini, R., Barbanti, G., Turini, D., Giachetti, A., and Maggi, C. A. S. (1992a) Characterization of the tachykinin NK-2 receptor in the human urinary bladder by means of selective antagonists. *Neuropeptides* **22,** 26.

Giuliani, S., Lecci, A., and Maggi, C. A. (1992b) Tachykinin receptor antagonists and non-cholinergic activation of motility in the guinea-pig proximal colon. *Neuropeptides* **22,** 26.

Giuliani, S., Patacchini, R., Barbanti, G., Turini, D., Rovero, P., Quartara, L., Giachetti, A., and Maggi, C. A. (1993) Characterization of the tachykinin NK_2 receptor in the human urinary bladder by means of selective receptor antagonists and peptidase inhibitors. *J. Pharmacol. Exp. Ther.* **267,** 590–595.

Gorbulev, V., Akhundova, A., Luzius, H., and Fahrenholz, F. (1992) Molecular cloning of substance P cDNA from guinea-pig uterus. *Biochim. Biophys. Acta* **113,** 199–202.

Graham, A., Hopkins, B., Powell, S. J., Danks, P., and Briggs, I. (1991) Isolation and characterization of the human lung NK-2 receptor gene using rapid amplification of cDNA ends. *Biochem. Biophys. Res. Commun.* **177,** 8–16.

Guard, S. and Watson, S. P. (1991) Tachykinin receptor types: classification and membrane signaling mechanisms. *Neurochem. Int.* **18,** 149–165.

Hagan, R. M. (1992) Tachykinin receptor antagonists. *Neuropeptides* **22,** 27.

Hagan, R. M., Ireland, S. J., Bailey, F., McBride, C., Jordan, C. C., and Ward, P. (1991) A spirolactam conformationally constrained analogue of physalaemin which is a peptidase resistant, selective neurokinin NK-1 receptor antagonist. *Br. J. Pharmacol.* **102,** 168P.

Hagan, R. M., Ireland, S. J., Jordan, C. C., Bailey, F., Stephens-Smith, M., and Ward, P. (1989) Novel, potent and selective agonists at NK-1 and NK-2 receptors. *Br. J. Pharmacol.* **98,** 71P.

Hagan, R. M., Ireland, S. J., Bailey, F., McBride, C., Jordan, C. C., and Ward, P. (1991) A spirolactam conformationally-constrained analogue of physalaemin which is a peptidase resistant, selective NK-1 receptor agonist. *Br. J. Pharmacol.* **102,** 168P.

Hagan, R. M., Ireland, S. J., Jordan, C. C., Beresford, I. J. M., Stephens-Smith, M., Ewan, G., and Ward, P. (1990) GR71,251 a novel, potent and highly selective antagonist at neurokinin NK-1 receptor. *Br. J. Pharmacol.* **99,** 62P.

Hagiwara, D., Miyake, H., Morimoto, H., Murai, M., Fujii, T., and Matsuo, M. (1991) The discovery of a tripeptide substance P antagonist and its structure-activity relationship. *J. Pharmacobiol. Dyn.* **14,** s104.

Hagiwara, D., Miyake, H., Morimoto, H., Murai, M., Fujii, T., and Matsuo, M. (1992a) Studies on neurokinin antagonists. 1. The design of novel tripeptides possessing the glutaminyl-D-tryptophylphenylalanine sequence as substance P antagonists. *J. Med. Chem.* **35,** 2015–2025.

Hagiwara, D., Miyake, H., Morimoto, H., Murai, M., Fujii, T., and Matsuo, M. (1992b) Studies on neurokinin antagonists. 2. Design and structure-activity relationships of novel tripeptide substance P antagonists, N^a-[N^a-(N^a-aceyl-L-threonyl)-N^1-formyl-D-tryptophyl]-*N*-methyl-*N*-(phenyl-methyl)-L-phenylalaninamide and its related compounds. *J. Med. Chem.* **35,** 3184–3191.

Hall, J. M., Mitchell, D., and Morton, I. K. M. (1992) Does substance P mediate NANC nerve-mediated responses to field stimulation and ligands in the rabbit isolated sphincter pupillae? *Regul. Pept.* **Suppl. 1,** S67.

Hall, J. M. and Morton, I. K. M. (1992) Novel selective agonists and antagonists confirm neurokinin NK-l receptors in guinea-pig vas deferens. *Br. J. Pharmacol.* **102,** 511–517.

Harbeson, S. L., Buck, S. H., Hassmann, III, C. F., and Shatzer, S. A. (1990) Synthesis and biological activity of [$\psi(CH_2NH)$] analogs of neurokinin A(4–10) in *Peptides, Chemistry, Structure and Biology* (Rivier, J. and Marshall, G. R., eds), ESCOM, Leiden, pp. 180,181.

Hartig, P. R., Branchek, T. A., and Weinshank, R. L. (1992) A subfamily of $5\text{-}HT_{1D}$ receptor genes. *Trends Pharmacol. Sci.* **13,** 152–159.

Henderson, A. K., Yamamura, H. I., Maggi, C. A., Buck, S. H., Van Giersbergen, P. L. M., and Roeske, W. R. (1992) Demonstration of a neurokinin A receptor subtype in transfected fibroblasts, *Eur. J. Pharmacol. Mol. Pharmacol.* **225,** 175–178.

Henry, J. L. (1987) Discussion of nomenclature for tachykinins and tachykinin receptors, in *Substance P and Neurokinins* (Henry, J. L., Couture, R., Cuello, A. C., Pelletier, G., Quirion, R., and Regoli, D., eds.), Springer-Verlag, New York, xvii.

Hershey, A. D. and Krause, J. E. (1990) Molecular characterization of a functional cDNA encoding the rat substance P receptor. *Science* **247,** 958–962.

Holzer-Petsche, U., Schimek, E., Amann, R., and Lembeck, F. (1985) In vivo and in vitro actions of mammalian tachykinins. *Naunyn Schmiedeberg's Arch. Pharmacol.* **330,** 130–135.

Hoplcins, B., Powell, S. J., Danks, P., Briggs, I., and Graham, A. (1991) Isolation and characterization of the human lung NK-1 receptor cDNA. *Biochem. Biophys. Res. Commun.* **180,** 1110–1117.

Ingi, T., Kitajima, Y., Minamitake, Y., and Nakanishi, S. (1991) Characterization of ligand binding properties and selectivities of three rat tachykinin receptors by transfection and functional expression of their cloned cDNAs in mammalian cells. *J. Pharmacol. Exp. Ther.* **259,** 968–975.

Ireland, S. J., Bailey, F., Cook, A., Hagan, R. M., Jordan, C. C., and Stephens-Smith, M. (1991) Receptors mediating tachykinin-induced contractile responses in guinea-pig trachea. *Br. J. Pharmacol.* **103,** 1463–1469.

Ireland, S. J., Beresford, I. J. M., and Stables, J. (1992) Comparison of the apparent affinity of GR 82,334 at NK-1 receptors in various tissue preparations. *Br. J. Pharmacol.* **106,** 93P.

Ireland, S. J., Jordan, C. C., Stephens-Smith, M. L., and Ward, P. (1988) Receptors mediating the contractile response to neurokinin agonists in the guinea-pig trachea. *Regul. Pept.* **22,** 93.

Itoi, K., Tschope, C., Jost, N., Culman, J., Lebrun, C., Stauss, B., and Unger, T. (1992) Idenification of central tachykinin receptor subclass involved in substance P - induced cardiovascular and behavioral responses in conscious rats. *Eur. J. Pharmacol.* **219,** 435–444.

Jukic, D., Rouissi, N., Laprise, R., Boussougou, M., and Regoli, D. (1991) Neurokinin receptors antagonists: old and new. *Life Sci.* **49,** 1463–1469.

Kage, R., Hershey, A. D., Krause, J. E., Boyd, N. D., and Leeman, S. E. (1991) Characterization of the substance P receptor in tunicamycin-treated transfected cells using a photoaffinity analog. *Soc. Neurosci. Abst.* **17,** 805.

Kris, R. M., South, V., Saltzman, A., Felder, S., Ricca, G. A., Jaye, M., Huebner, K., Kagan, J., Croce, C. M., and Schlessinger, J. (1991) Cloning and expression of the human substance K receptor and analysis of its role in mitogenesis. *Cell Growth Differentiation* **2,** 15–22.

Laufer, R., Gilon, C., Chorev, M., and Selinger Z. (1988) Desensitization with a selective agonist discriminates between multiple tachykinin receptors. *J. Pharmacol. Exp. Ther.* **245,** 639–643.

Lavielle, S., Chassaing, G., Loeuillet, D., Convert, O., Torrens, Y., Beaujouan, J. C., Saffroy, M., Petitet, F., Bergstrom, L., and Glowinski, J. (1990) Selective agonists of tachykinin binding sites. *Fundam. Clin. Pharmacol.* **4,** 257–268.

Lavielle, S., Chassaing, G., Ploux, O., Loeuillet, D., Besseyre, J., Julien, S., Marquet, A., Convert, O., Beaujouan, J. C., Torrens, Y., Bergstrom, L., Saffroy, M., and Glowinski, J. (1988) Analysis of tachykinin binding site interactions using constrained analogues of tachykinins. *Biochem. Pharmacol.* **37,** 41–49.

Lecci, A., Giuliani, S., and Maggi, C. A. (1992) Tachykinin receptor antagonists and modulation of micturition reflex in rats. *Neuropeptides* **22,** 40.

Lecci, A., Giuliani, S., Patacchini, R., Viti, G., and Maggi, C. A. (1991) Role of NK-l tachykinin receptors in thermonociception: effect of (±)- CP 96,345, a nonpeptide substance P antagonist on the hot plate test in mice. *Neurosci. Lett.* **129,** 299–302.

Lee, C. M., Iversen, L. L., Hanley, M. R., and Sandberg, B. E. B. (1982) The possible existence of multiple receptors for substance P. *Naunyn Schmiedeberg's Arch. Pharmacol.* **318,** 281–288.

Lee, C. M., Campbell, N. J., Williams, B. J., and Iversen, L. L. (1986) Multiple tachykinin binding sites in peripheral tissues and in brain. *Eur. J. Pharmacol.* **130,** 209–217.

Lee, C. M., Campbell, N. J., Williams, B. J., and Iversen, L. L. (1987) Multiple tachykinin binding sites in rat brain and peripheral tissues, in *Substance P and Neurokinins* (Henry, J. L., ed), Springer-Verlag, New York, pp. 44–46

Lee, C M., Tung, W. L., and Young, J. D. (1992) Tachykinin-stimulated inositol phospholipid hydrolysis and taurine release from human astrocytoma cells. *J. Neurochem.* **59,** 406–414.

Lembeck, F., Donnerer, J., Tsuchiya, M., and Nagahisa, A. (1992) The nonpeptide tachykinin antagonist, CP96,345 is a potent inhibitor of neurogenic inflammation. *Br. J. Pharmacol.* **105,** 527–530.

Lew, R., Geraghty, D. P., Drapeau, G., Regoli, D., and Burcher, E. (1990) Binding characteristics of [^{125}I]Bolton-Hunter [Sar^9,Met(O$_2$)^{11}] substance P, a new selective radioligand for the NK-l receptor. *Eur. J. Pharmacol.* **184,** 97–108.

Liu, Y. F. and Quirion, R. (1991) Presence of various carbohydrate moieties including β-galactose and N-acetylglucosamine residues on solubilized porcine brain neurokinin-A/substance P receptors. *J. Neurochem.* **57,** 1944–1950.

Longmore, J., Neal, R., Razzaque, Z., and Hill, R. G. (1992) Characterization of neurokinin receptors in guinea-pig urinary bladder smooth muscle: use of selective neurokinin antagonists. *Neuropeptides* **22,** 41,42.

Lowe, III, J. A., Drozda, S. E., Snider, R. M., Longo, K. P., Zorn, S. H., Morrone, J., Jackson, E. R., Mc Lean, S., Bryce, D. K., Bordner, J., Nagahisa, A., Kanai, Y., Suga, O., and Tsuchiya, M. (1992) The discovery of (2S,3S)-cis-(diphenylmethyl)-*N*-[(2-methoyphenyl)methyl]-1-azabicyclo[2.2.2]-octan-3-amine as a novel nonpeptide substance P antagonist. *J. Med. Chem.* **35,** 2591–2600.

Maggi, C. A. and Patacchini, R. (1992) Tachykinin NK-l receptor in the guinea-pig isolated proximal urethra. Characterization by receptor selective agonists and antagonists. *Br. J. Pharmacol.* **106,** 888–892.

Maggi, C. A., Giuliani, S., Santicioli, P., Regoli, D., and Meli, A. (1985) Comparison of the effects of substance P and substance K on blood pressure, salivation and urinary bladder motility in urethane anesthetized rats. *Eur. J. Pharmacol.* **113,** 291–294.

Maggi, C. A., Santicioli, P., Giuliani, S., Regoli, D., and Meli, A. (1986) Activation of micturition reflex by substance P and substance K: indirect evidence for the existence of multiple tachykinin receptors in the rat urinary bladder. *J. Pharmacol. Exp. Ther.* **238,** 259–266.

Maggi, C. A., Giuliani, S., Santicioli, P., Regoli, D., and Meli, A. (1987) Peripheral effects of neurokinins: functional evidence for the existence of multiple receptors. *J. Autonom. Pharmacol.* **7,** 243–255.

Maggi, C. A., Santicioli, P., Patacchini, R., Cellerini, M., Turini, D., Barbanti, G., Beneforti, P., Rovero, P., and Meli, A. (1988) Contractile response of the human isolated urinary bladder to neurokinins: involvement of NK-2 receptors. *Eur. J. Pharmacol.* **145,** 335–340.

Maggi, C. A., Patacchini, R., Rovero, P., and Meli, A. (1989) The hamster isolated trachea: a new preparation for studying NK-2 receptors. *Eur. J. Pharmacol.* **166,** 435–440.

Maggi, C. A., Patacchini, R., Giuliani, S., Rovero, P., Dion, S., Regoli, D., Giachetti, A., and Meli A. (1990a) Competitive antagonists discriminate between NK-2 tachykinin receptor subtypes. *Br. J. Pharmacol.* **100,** 588–592.

Maggi, C. A., Giuliani, S., Ballati, L., Rovero, P., Abelli, L., Manzini, S., Giachetti, A., and Meli, A. (1990b) In vivo pharmacology of [βAla[8]] neurokinin A (4–10), a selective NK-2 tachykinin receptor agonist. *Eur. J. Pharmacol.* **177,** 81–86.

Maggi, C. A., Patacchini, R., Rovero, P., Giuliani, S., and Giachetti, A. (1991a) Tachykinin receptor antagonists and potential clinical applications at peripheral level. *Biochem. Soc. Trans.* **19,** 909–912.

Maggi, C. A., Patacchini, R., Santicioli, P., and Giuliani, S. (1991b) Tachykinin antagonists and capsaicin-induced contraction of the rat isolated urinary bladder: evidence for tachykinin-mediated cotransmission. *Br. J. Pharmacol.* **103,** 1535–1541.

Maggi, C. A., Patacchini, R., Feng, D. M., and Folkers, K. (1991c) Activity of spantide I and II at various tachykinin receptors and NK-2 tachykinin receptor subtypes. *Eur. J. Pharmacol.* **199,** 127–129.

Maggi, C. A., Giuliani, S., Ballati, L., Lecci, A., Manzini, S., Patacchini, R., Renzetti, A. R., Rovero, P., Quartara, L., and Giachetti, A. (1991d) In vivo evidence for tachykininergic trasmission using a new NK-2 receptor selective antagonist. MEN 10376. *J. Pharmacol. Exper. Ther.* **257,** 1172–1178.

Maggi, C. A., Patacchini, R., Astolfi, M., Rovero, P., Giuliani, S., and Giachetti, A. (1991e) NK_2 receptor agonists and antagonists. *Ann. NY Acad. Sci.* **632,** 184–191.

Maggi, C. A., Patacchini, R., Quartara, L., Rovero, P., and Santicioli, P. (1991f) Tachykinin receptors in the guinea-pig isolated bronchi. *Eur. J. Pharmacol.* **197,** 167–174.

Maggi, C. A., Lecci, A., Garret, C., and Giuliani, S. (1992a) Spinal effects of selective NK-1 and NK-2 receptor antagonists on bladder motility in anesthetized rats. *Regul. Pept.* **Suppl. 1,** S103.

Maggi, C. A., Eglezos, A., Quartara, L., Patacchini, R., and Giachetti, A. (1992b) Heterogeneity of NK-2 tachykinin receptors in hamster and rabbit smooth muscles. *Regul. Pept.* **37,** 85–93.

Maggi, C. A., Patacchini, R., Eglezos, A., Quartara, L., Giuliani, S., and Giachetti A. (1992c) Tachykinin receptors in guinea-pig renal pelvis: activation by exogenous and endogenous tachykinins. *Br. J. Pharmacol.* **107,** 27–33.

Maggi, C. A., Giuliani, S., Patacchini, R., Santicioli, P., Theodorsson, E., Barbanti, G., Turini, D., and Meli, A. (1992d) Tachykinin antagonists inhibit nerve-mediated contractions in the circular muscle of the human ileum: involvement of NK-2 receptors. *Gastroenterology* **102,** 88–96.

Maggi, C. A., Patacchini, R., Astolfi, M., Rovero, P., Giachetti, A., and Van Giersbergen, P. L. M. (1992e) Affinity of R 396, an NK-2 tachykinin receptor antagonist, for NK_2 receptors in preparations from different species. *Neuropeptides* **22,** 93–98.

Maggi, C. A., Giuliani, S., Patacchini, R., Quartara, L., Rovero, P., Renzetti, A. R., Mizrahi, J., and Giachetti, A. (1992f) Heterogeneity of tachykinin NK-2 receptors in rabbit, guinea-pig and human smooth muscles. *Neuropeptides* **23,** 181–186.

Maggi, C. A., Patacchini, R., Rovero, P., and Giachetti, A. (1993a) Tachykinin receptors and receptor antagonists. *J. Autonom. Pharmacol.* **13,** 23–93.

Maggi, C. A., Patacchini, R., Meini, S., and Guiliani, S. (1993b) Evidence for the presence of a septide-sensitive tachykinin receptor in the circular muscle of the guinea-pig ileum. *Eur. J. Pharmacol.* **235,** 309–311.

Maggi, C. A., Patacchini, R., Meini, S., Quartara, L., Sisco, A., Potier, E., Giuliani, S., and Giachetti, A. (1994) Comparison of tachykinin NK_1 and NK_2 receptors in the circular muscle of the guinea-pig ileum and proximal colon. *Br. J. Pharmacol.,* in press.

Maggio, J. E. (1988) Tachykinins. *Annu. Rev. Neurosci.* **11,** 13–21.

Masu, Y., Nakayama, K., Tamaki, H., Harada, Y., Kuno, M., and Nakanishi, S. (1987) cDNA cloning of bovine substance K receptor through oocyte expression system. *Nature* **329,** 836–838.

McDonald, M. R., Takeda, J., Rice, C. M., and Krause, J. E. (1989) Multiple tachykinins are produced and secreted upon posttranslational processing of the three substance P precursor proteins, α-, β-, and γ-preprotachykinin. *J. Biol. Chem.* **264,** 15,578–15,592.

McElroy, A. B., Clegg, S. P., Deal, M. J., Ewan, G. B., Hagan, R. M., Ireland S. J., Jordan, C. C., Porter, B., Ross, B. C., Ward, P., and Whittington, A. R. (1992) Highly potent and selective heptapeptide antagonists of the neurokinin NK-2 receptor. *J. Med. Chem.* **35,** 2582–2591.

McKnight, A. T., Maguire, J. J., Elliott, N. J., Fletcher, A. E., Foster, A. C., Tridgett, R., Williams, B. J., Longmore, J., and Iversen, L. L. (1991) Pharmacological specificity of novel, synthetic, cyclic peptides as antagonists at tachykinin receptors. *Br. J. Pharmacol.* **104,** 335–360.

McKnight, A. T., Maguire, J. J., Williams, B. J., Foster, A. C., Tridgett, R., and Iversen, L. L. (1988) Pharmacological specificity of synthetic peptides as antagonists at tachykinin receptors. *Regul. Pept.* **22,** 127.

Meini, S., Patacchini, R., and Maggi, C. A. (1993) Tachykinin NK_1 receptor subtypes in the rat urinary bladder. *Br. J. Pharmacol.* **108,** 177P.

Meini, S., Patacchini, R., and Maggi, C. A. (1994) Tachykinin NK_1 receptor subtypes in the rat urinary bladder. *Br. J. Pharmacol.,* in press.

Mitsuhashi, M., Ohashi, Y., Shichijo, S., Christian, C., Sudduth-Klinger, J., Harrowe, G., and Payan, D. G. (1992) Multiple intracellular signaling pathways of the substance P receptor. *FASEB J.* **6,** A1562.

Morimoto, H., Murai, M., Maeda, Y., Hagiwara, D., Miyake, H., Matsuo, M., and Fujii, T. (1992) FR 113,680: a novel tripeptide substance P antagonist with NK-1 receptor selectivity. *Br. J. Pharmacol.* **106,** 123–126

Morrison, J. F. B., Nimmo, A. J., and Whitaker, E. M. (1990) The effect of neonatal capsaicin treatment on the distribution of neurokinin binding sites in rat bladder. *J. Physiol. (Lond.)* **423,** 79P.

Nakajima, Y., Tsuchida, K., Negishi, M., Ito, S., and Nakanishi, S. (1992) Direct linkage of three tachykinin receptors to stimulation of both phosphatidylinositol hydrolysis and cyclic AMP cascades in transfected chinese hamster ovary cells. *J. Biol. Chem.* **267,** 2437–2442.

Nakanishi, S. (1991) Mammalian tachykinin receptors. *Annu. Rev. Neurosci.* **14,** 123–136.

Naline, E., Devillier, P., Drapeau, G., Toty, L., Bakdach, H., Regoli, D., and Advenier, C. (1989) Characterization of neurokinin effects and receptor selectivity in human isolated bronchi. *Amer. Rev. Resp. Dis.* **140,** 679–686.

Naminohira, S., Uchida, Y., Okimura, K., Kurosawa, K., Sakai, T., Sakura, N., and Hashirnoto, T. (1989) Tissue selectivity of novel specific and competitive antagonists of neurokinin B on isolated smooth muscles. *Adv. Exp. Biol. Med.* **247,** 401–406.

Nawa, H., Doteuchi, M., Igano, K., Inouye, K., and Nakanishi, S. (1984) Substance K a novel mammalian tachykinin that differs from substance P in its pharmacological profile. *Life Sci.* **34,** 1153–1160.

Nimmo, A., Carstairs, J. R., Maggi, C. A., and Morrison, J. F. B. (1992) Evidence for the co-existence of multiple NK-2 tachykinin receptor subtypes in rat bladder. *Neuropeptides* **22,** 48.

Patacchini, R., Astolfi, M., Quartara, L., Rovero, P., Giachetti, A., and Maggi, C. A. (1991) Further evidence for the existence of NK-2 tachykinin receptor subtypes. *Br. J. Pharmacol.* **104,** 91–96.

Patacchini, R. and Maggi, C. A. (1992) Effect of newly developed tachykinin agonists and antagonists on the guinea-pig isolated gallbladder. *J. Pharmacol. Exp. Ther.* **261,** 191–194.

Patacchini, R. and Maggi, C. A. (1993) Amastatin interferes with the antagonist properties of MEN 10,208 in the rabbit pulmonary artery but not in the hamster trachea. *Eur. J. Pharmacol.* **236,** 31–37.

Patacchini, R., Maggi, C. A., and Giachetti, A. (1992) Activity of SR 48,968 at smooth muscle NK-2 tachykinin receptors of hamster, rat and rabbit species. *Br. J. Pharmacol.* **107,** 152P.

Patacchini, R., Maggi, C. A., Rovero, P., Regoli, D., Drapeau, G., and Meli, A. (1989) Effect of thiorphan on tachykinin-induced potentiation of nerve-mediated contractions of the rat isolated vas deferens. *J. Pharmacol. Exp. Ther.* **250,** 678–681.

Patacchini, R., Quartara, L., Astolfi, M., Pavone, V., Pedono, C., Lombardi, A., Giachetti, A., and Maggi, C. A. (1993) MEN 10,627, a potent polycyclic peptide antagonist at tachykinin NK_2 receptors. *Br. J. Pharmacol.,* in press.

Patacchini, R., Santicioli, P., Astolfi, M., Rovero, P., Viti G., and Maggi, C. A. (1992) Activity of peptide and nonpeptide antagonists at peripheral NK-1 tachykinin receptors. *Eur. J. Pharmacol.* **215,** 93–98.

Petitet, F., Saffroy, M., Torrens, Y., Lavielle, S., Chassaing, G., Loeuillet, D., Glowinski, J., and Beaujouan, J. C. (1992a) Possible existence of a new tachykinin receptor subtype in the guinea-pig ileum. *Peptides* **13,** 383–388.

Petitet, F., Beaujouan, J. C., Saffroy, M., Torrens, Y., Lavielle, S., Chassaing, G., Loeuillet, D., and Glowinski, J. (1992b) A new tachykinin receptor revealed by substance P analogues in the guinea-pig ileum. *Acad. Sci. Paris* **314,** Series III 299–303.

Piercey, M. F., Dobry-Scheurer, P. J. K., Masiques, N., and Scroeder, L. A. (1985) Stereospecificity of SP1 and SP2 substance P receptors. *Life Sci.* **36,** 777–780.

Quartara, L., Patacchini, R., Giachetti, A., and Maggi, C.A. (1992) Novel cyclic pseudopeptides with high affinity for tachykinin NK-2 receptor. *Br. J. Pharmacol.*, in press.

Quartara, L., Patacchini, R., Giuliani, S., Renzetti, A. R., Rovero, P., and Maggi, C. A. (1993) N-terminal truncated analogs of MEN 10,376 as tachykinin NK-2 receptor antagonists. *Life Sci.* **51,** 1923–1936.

Regoli, D., Drapeau, G., Dion, S., and Couture, R. (1988) New selective agonists for neurokinin receptors: pharmacological tools for receptor characterization. *Trends Pharmacol. Sci.* **9,** 290–295.

Regoli, D., Drapeau, G., Dion, S., and D'Orléans-Juste, P. (1987) Pharmacological receptors for substance P and neurokinins. *Life Sci.* **40,** 109–117.

Rhoden, K. J. and Barnes, P. J. (1990) Classification of tachykinin receptors in guinea-pig and human airway smooth muscle. *Amer. Rev. Resp. Dis.* **141,** A726.

Riordan, J. R., Rommens, J. M., Kerem, B. S., Alon, N., Rozmahel, R., Grzeczak, Z., Zielenski, J., Lok, S., Plavsci, N., Chou, J. L., Drumm, M. L., Iannuzzi, M. C., Collins, F. S., and Tsui, L. C. (1989) Identification of the cystic fibrosis gene: cloning and characterization of complementary DNA. *Science* **245,** 1066–1073

Rovero, P., Pestellini, V., Rhaleb, N. E., Dion, S., Rouissi, N., Tousignant, C., Telemaque, S., Drapeau, G., and Regoli, D. (1989) Structure-activity studies of neurokinin A. *Neuropeptides* **13,** 263–270.

Rovero, P., Pestellini, V., Maggi, C. A., Patacchini, R., Regoli, D., and Giachetti, A. (1990) A highly selective NK-2 tachykinin receptor antagonist containing D-tryptophan. *Eur. J. Pharmacol.* **175,** 113–115.

Rovero, P., Quartara, L., Astolfi, M., Patacchini, R., Giachetti, A., and Maggi, C. A. (1991a) Structure-activity study of the C-terminals residue of MEN 10207 tachykinin antagonist. *Peptides* **13,** 207,208.

Rovero, P., Astolfi, M., Renzetti, A. R., Patacchini, R., Giachetti, A., and Maggi, C. A. (1991b) Role of D-tryptophan for affinity of MEN 10207 tachykinin antagonists at NK-2 receptors. *Peptides* **12,** 1015–1018.

Rubino, A., Thomann, H., Henlin, J. M., Schilling, W., and Criscione, L. (1992) Endothelium-dependent relaxant effect of neurokinins on rabbit aorta is mediated by NK-1 receptor. *Eur. J. Pharmacol.* **212,** 237–240.

Sakurada, T., Yamada, T., Tan-No, K., Manome, Y., Sakurada, S., Kisara, K., and Ohba, M. (1991) Differential effects of substance P analogs on NK-1 receptor agonists in the mouse spinal cord. *J. Pharmacol. Exp. Ther.* **259,** 205–210.

Sasai, Y., and Nakanishi, S. (1989) Molecular characterization of rat substance K receptor and its mRNA. *Biochem. Biophys. Res. Commun.* **165,** 605–702.

Schmidt, A. W., McLean, S., and Heym, J. (1992) The substance P receptor antagonist CP 96,345 interacts with calcium channels. *Eur. J. Pharmacol.* **215,** 351,352.

Sibley, D. R. and Monsma, F. J. (1992) Molecular biology of dopamine receptors. *Trends Pharmacol. Sci.* **13,** 61–69.

Snider, R. M., Constantine, J. W., Lowe, III, J. A., Longo, K. P., Lebel, W. S., Woody, H. A., Drozda, S. E., Desai, M. C., Vinick, F. J., Spencer, R. W., and Hess, H. J. (1991) A potent nonpeptide antagonist of the substance P (NK-1) receptor. *Science* **251,** 435–437.

Suman-Chauhan, N., Guard, S., Williams, B. J., and Watling, K. J. (1990) Pharmacological characterization of tachykinin-stimulated inositol phospholipid hydrolysis in peripheral tissues. *Br. J. Pharmacol.* **101,** 1001–1005.

Sundelin, J. B., Provvedini, D. M., Wahlestedt, C., Laurell, H., Pohl, J. S., and Peterson, P. A. (1992) Molecular cloning of the murine substance K and substance P receptor genes. *Eur. J. Biochem.* **203,** 625–631.

Takahashi, K., Tanaka, A., Hara, M., and Nakanishi, S. (1992) The primary structure and gene organization of human substance P and neuromedin K receptors. *Eur. J. Biochem.* **204,** 1025–1033.

Takamatsu, H., Tani, Y., Akiyarna, M., Nakata, Y., and Segawa, T. (1991) Characterization of the carbohydrate chain on the substance P receptor in the rat brain cortex: effects of lectins on [^{3}H]substance P binding. *J. Neurochem.* **56,** 1452–1454.

Takeda, Y., Chou, K. B., Takeda, J., Sachais, B. S., and Krause, J. E. (1991) Molecular cloning, structural characterization and functional expression of the human substance P receptor. *Biochem. Biophys. Res. Commun.* **179,** 1232–1240.

Takeda, Y. and Krause, J. E. (1989a) Characterization of a high affinity neuropeptide γ binding site in rat duodenum, urinary bladder and vas deferens. *Soc. Neurosci. Abstr.* **15,** 346.

Takeda, Y. and Krause, J. E. (1989b) Neuropeptide K potently stimulates salivary gland secretion and potentiates substance P-induced salivation. *Proc. Natl. Acad. Sci. USA* **86,** 392–396.

Takeda, Y. and Krause, J. E. (1989c) γ-Preprotachykinin(72-92)-peptide amide potentiates substance P induced salivation. *Eur. J. Pharmacol.* **161,** 267–271.

Takeda, Y. and Krause, J. E. (1991) Pharmacological and molecular biological studies on the diversity of rat tachykinin NK-2 receptor subtypes in rat CNS, duodenum, vas deferens and urinary bladder. *Ann. NY Acad. Sci.* **632,** 479–482.

Tousignant, C., Guillemette, G., and Regoli, D. (1991) Binding sites for [^{3}H][Sar^9]SP in rat brain and guinea-pig ileum. *Brain Res.* **560,** 1–11.

Tschope, C., Culman, J., Picard, P., Pratl, A., Regoli, D., Couture, R., and Unger, T. (1992) Use of selective antagonists to dissociate the central cardiovascular and behavioral effects of tachykinins on NK-l and NK-2 receptors in the rat. *Regul. Pept.* **Suppl. 1,** S165.

Tsuchida, K., Shigemoto, R., Yokota, Y., and Nalcanishi, S. (1990) Tissue distribution and quantitation of the mRNA for three rat tachykinin receptors. *Eur. J. Biochem.* **193,** 751–757.

Van Giersbergen, P. L. M., Shatzer, S. A., Burcher, E., Buck, S. H. (1992) Comparison of the effects of neuropeptide K and neuropeptide γ with neurokinin A at NK-2 receptors in the hamster urinary bladder. *Naunyn Schmiedeberg's Arch. Pharmacol.* **345,** 51–56.

Van Giersbergen, P. L. M., Shatzer, S. A., Henderson, A. K., Lai, J., Nakanishi, S., Yamamura, H. I., and Buck, S. H. (1991) Characterization of a tachykinin peptide NK-2 receptor transfected into murine fibroblast B82 cells. *Proc. Natl. Acad. Sci. USA* **88,** 1661–1665.

Van Tol, H. H. M., Wu, C. M., Guan, H. C., Ohara, K., Bunzow, J. R., Civelli, O., Kennedy, J., Seeman, P., Niznik, H. B., and Jovanovic V. (1992) Multiple dopamine D4 receptor variants in the human population. *Nature* **358,** 149–152.

Von Euler, U. S. and Gaddum, J. H. (1931) An unidentified depressor substance in certain tissue extracts. *J. Physiol. (Lond.)* **72,** 74–86.

Watling, K. J., Guard, S., Howson, W., and Walton, L. (1991) Inhibition of binding to tachykinin NK-l receptor sites in rat, guinea-pig and human brain by the nonpeptide substance P antagonist CP 96345. *Br. J. Pharmacol.* **104,** 27P.

Watson, S. P. (1983) Pharmacological characterization of a substance P antagonist, [D-Arg^{1}, D-Pro^{2}, D-Trp^{7,9}, Leu^{11}] substance P. *Br. J. Pharmacol.* **80,** 205–209.

Williams, B. J., Curds, N. R., McKnight, A. T., Maguire, J., Foster, A., and Tridgett, R. (1988) Development of NK-2 selective antagonists. *Regul. Pept.* **22,** 189.

Wormser, U., Laufer, R., Hart, Y., Chorev, M., Gilon, C., and Selinger, Z. (1986) Highly selective agonists for substance P receptor subtypes. *EMBO J.* **5,** 2805–2808.

Xiao, X. H., Mussap, C. J., and Burcher, E. (1992) Characterization of the tachykinin NK-2 receptor subtype in the rabbit pulmonary artery. *Peptides* **13,** 281–285.

Xu, X. J., Maggi, C. A., and Wiesenfeld-Hallin, Z. (1991) On the role of NK-2 tachykinin receptors in the mediation of spinal reflex excitability in the rat. *Neuroscience* **44,** 483–490.

Yokota, Y., Sasai, Y., Tanaka, K., Fujiwara, T., Tsuchida, K., Shigemoto, R., Kakizuka, A., Ohkubo, H., and Nakanishi, S. (1989) Molecular characterization of a functional cDNA for rat substance P receptor. *J. Biol. Chem.* **264,** 17,469–17,652.

CHAPTER 15

Role of Receptors in Nociception

Raymond G. Hill

1. Introduction

Nociception depends on the synaptic transmission of action potentials generated in high threshold sensory afferent fibers within the peripheral tissues, into the dorsal horn of the spinal cord and, thereafter, up the neuroaxis to the level of consciousness where, in humans, the sensation of pain is experienced. Those sensory fibers involved in nociception are primarily the smaller, unmyelinated ones, although it is clear that the smallest unmyelinated (Aδ) fibers also contribute and even Aβ fibers may be involved in nociceptive responses under particular circumstances (Hill, 1986; Cervero and Laird, 1991; Handwerker and Kobal, 1993). The unmyelinated fibers are of greatest interest in the present context as these are the ones that contain substance P (SP) and other peptides as part of their transmitter complement (Salt and Hill, 1983; Duggan and Weihe, 1991). Sensory nerve fibers are pseudounipolar, and peptides synthesized in dorsal root ganglion neurons are transported both centrally and peripherally and can be released from terminals within the dorsal horn of the spinal cord and/or within peripheral tissues (Fig. 1; Hill,

The Tachykinin Receptors Ed.: S. H. Buck © 1994 Humana Press Inc., Totowa, NJ

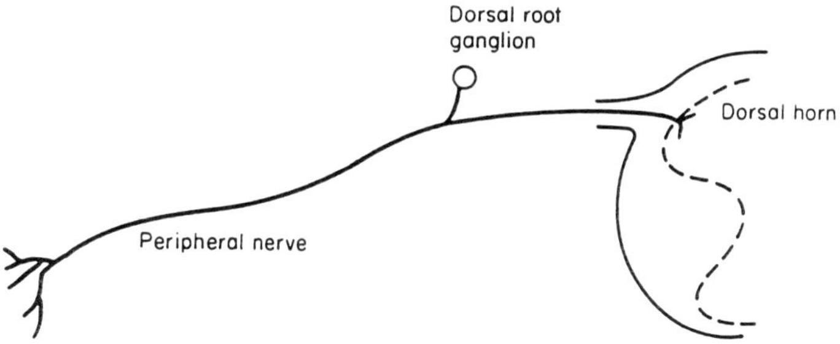

Fig. 1. Cartoon illustrating the pseudounipolar nature of peptide containing sensory neurons. Activation of the fiber by noxious stimulation within the region innervated by its terminals in the periphery evokes peptide release peripherally (the axon reflex) and also release from terminals within the dorsal horn of the spinal cord.

1986; Salt and Hill, 1983). The putative peripheral role of tachykinin receptors in nociception is dealt with briefly later in this chapter, but a detailed account of the peripheral function of tachykinin receptor activation can be found in the chapter by Hall elsewhere in this volume. It is important to note that, although emphasis in this chapter falls on tachykinins and their receptors, many other peptides are found in unmyelinated primary sensory fibers, sometimes coexisting, and sensory neurotransmission is now thought of as a multitransmitter phenomenon (Salt and Hill, 1983; Duggan and Weihe, 1991), (*see* Table 1). It is as yet unresolved whether fibers containing multiple sensory transmitters can function in such a way that only a single substance is released, but this seems unlikely. It is therefore necessary to consider whether blockade of the action of a single peptide, such as SP, would significantly effect the efficiency of a synaptic relay between unmyelinated sensory fibers and secondary sensory neurons in the dorsal horn of the spinal cord. The availability of peptide receptor antagonist compounds now allows this to be experimentally tested.

Recent studies on the plasticity of sensory neurotransmission have emphasized that the role of nociceptive fibers is much more than a simple carriage of a message from A to B, and maintained activa-

Table 1
Peptides Found in Mammalian Primary Afferents[a]

Atrial natriuretic factor	Neuropeptide Y
Angiotensin II	Oxytocin
Bombesin	Dynorphins
Cholecystokinin	Enkephalins
Calcitonin gene related peptide	β-Endorphin
Endothelin	Somatostatin
Galanin	Tachykinins
Growth hormone releasing factor	Vasoactive intestinal peptide
Vasopressin	

[a]This table illustrates the range of peptides found in the primary afferent fibers projecting to the dorsal horn of the spinal cord. The information is taken from a more detailed table of spinal transmitter candidates in a recent review (Duggan and Weihe, 1991).

tion of these fibers can lead to reprogramming of central circuits and changes in responsiveness to subsequent noxious stimuli (Fig. 2; and *see* Cervero and Laird, 1991). Evidence presented later suggests that activation of tachykinin receptors may be more important in the situation that Cervero and Laird have termed phase 2 pain (i.e., following prolonged noxious stimulation) than in phase 1 pain (following a brief noxious stimulus). It is now apparent that the stimulus needed to produce an increase in sensitivity and magnitude of pain perception in man, if sufficiently well maintained, does not itself need to be of noxious intensity (Cervero et al., 1993).

The suggestion that SP (Otsuka and Yoshioka, 1993) has a specific role in nociception can be traced back to the observation of Lembeck (1953) that bioassay revealed more tachykinin activity in extracts from spinal dorsal roots than from ventral roots, but the first clear association between the presence of SP and an animal's ability to respond to noxious stimulation came from experiments with capsaicin, where depletion of the primary afferent content of the peptide correlated with a reduction in the behavioral response to noxia (Gasparovic et al., 1964). It is now realized that capsaicin is not selective for SP, or even for tachykinins as a class, but will deplete

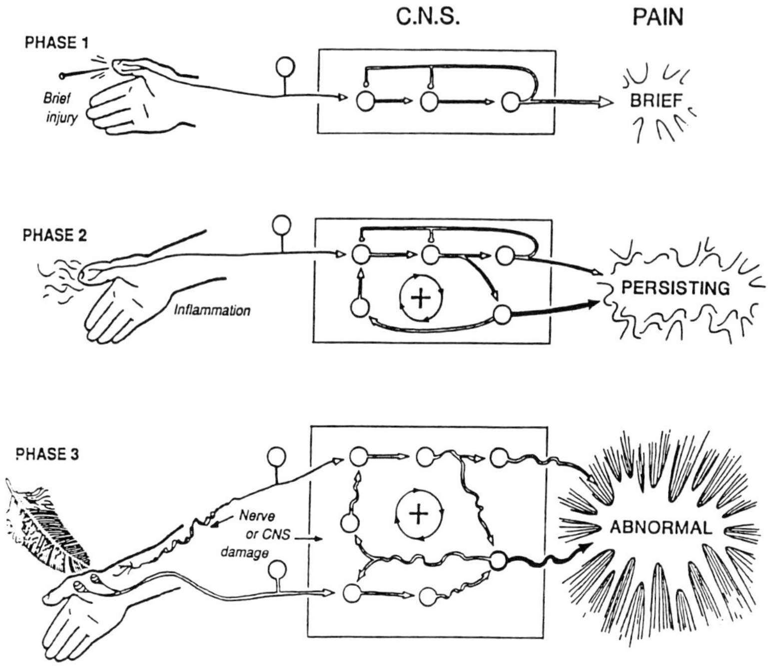

Fig. 2. This diagram illustrates the consequences of different ways of activating nociceptive sensory nerve fibers on the pain response experienced by the individual. Phase 1 shows the response to a brief stimulus, such as may be experienced as a normal protective signal. The stimulus is removed as soon as pain is registered. Evidence presented in this chapter suggests that activation of tachykinin receptors is unlikely to be an important part of this phase of pain perception. Phase 2 shows the situation after a damaging stimulus that produces prolonged activation of nociceptor fibers and also turns on a population of nociceptor fibers that are normally silent. This raises the central nervous system (CNS) to a higher state of excitability that may last for long periods in the absence of any further stimulation. Evidence is accumulating that tachykinin receptor activation may be involved in this process *(see later)*. Phase 3 suggests the state of affairs that might occur in an abnormal pain syndrome. In these chronic conditions, spontaneous pain can arise and there are often greatly reduced pain thresholds. It is not yet known whether tachykinin receptor mechanisms are likely to be involved in such pain states. Diagram from Cervero and Laird (1991), which should be consulted for details. Reprinted with permission.

small primary afferents of most, if not all, of their peptide content (Weihe, 1990) and, additionally, that there is a poor correlation between the extent of SP depletion and the modification of nociception (*see* Salt and Hill, 1983, for discussion of this point).

An additional complication is that neurokinin A (NKA) has as good a claim to be a primary afferent transmitter as has SP (Weihe, 1990; Duggan and Weihe, 1991; Otsuka and Yoshioka, 1993). Briefly, only preprotachykinin A is found in primary afferents and, as SP and NKA are the major posttranslational processing products of this precursor, either of these peptides (but not the preprotachykinin B product, NKB) is likely to have a role in transmitting nociceptive messages into the CNS. Immunohistochemical studies suggest that some afferent fibers contain SP alone, whereas others contain both SP and NKA, and both may coexist with other peptides (Weihe, 1990). It is important to note that evidence is accumulating for the coexistence of excitatory amino acids and tachykinins in unmyelinated fibers (Duggan and Weihe, 1991), and the pharmacological consequences of this will be discussed later. Release experiments (*see* Table 2 in Duggan and Weihe, 1991) suggest that cutaneous noxious stimuli will release NKA into the dorsal horn of the spinal cord under most experimental circumstances, but that the release of SP requires stimuli of a higher intensity, probably maintained for a longer period. Ligand binding studies (*see* other chapters of this volume) have shown that NK_1 and NK_2 receptors predominate within the dorsal horn of the spinal cord, consistent with the suggestion that SP and NKA are the important messengers here (as these peptides are likely to be the endogenous ligands for these sites). It has been shown that there are regional differences in the binding of different tachykinins within the spinal cord in that SP (presumably NK_1) binding is predominantly found within laminae I and II and around the central canal (lamina X) (Shults et al., 1984). Recent studies with a rabbit polyclonal antibody to rat NK_1 receptors showed that rat NK_1 sites in the superficial laminae of the dorsal horn are largely on dendrites, whereas in deep dorsal horn and in ventral horn they are mostly somatic (Moussaoui et al., 1992). Recently (Schafer et al., 1993), it has been shown that unilateral inflammation of a hind paw in the rat causes a selective increase in NK_1 receptor mRNA in a distinct population of ipsilateral superficial and deep dorsal horn neurons. This indicates that receptor distribution in the absence of a painful input may give an erroneous estimate of the target neuron population. Eledoisin binding (which may be to NK_3 receptors, although this ligand is somewhat unselective) was restricted to lamina I and the outer part of lamina II, although it should be noted that the distribution of eledoisin binding changes with development, being widespread in the neonate rat but circumscribed in the adult (Beresford et al., 1992). Yashpal et al. (1990) have described binding of labeled NKA to spinal cord in the adult rat, suggesting the presence of NK_2 receptors, but the precise location of these sites is incompletely described as yet.

The only sure way to establish the role of particular tachykinin receptors in the nociceptive process is by appropriate antagonist pharmacology. This has been fraught with difficulty in the past as the only available antagonists were themselves peptides and therefore difficult to deliver in an intact and active form to the site at which they were expected to act. Additionally, some of the most commonly used peptide antagonists have been shown to have agonist properties (e.g., Salt et al., 1982) or to be frankly neurotoxic (Hokfelt et al., 1981; Rodriguez et al., 1983). Recently, nonpeptide antagonists for neurokinin receptors have been introduced (*see* other chapters of this volume) that are highly potent and selective and that can be given systemically to animals. In particular, CP-96,345 and CP-99,994, which are selective for human-type NK_1 receptors (*see* Chapter 4; Desai et al., 1992; Sachais et al., 1993), and RP-67,580, which is selective for rodent type NK_1 receptors (Beaujouan et al., 1993), are the most used nonpeptide NK_1 antagonists, and their species selectivity is now well documented (Fong et al., 1992; Barr and Watson, 1993). A single nonpeptide NK_2 antagonist is currently available—SR-48,968 (Emonds-Alt et al., 1993). No nonpeptide, selective NK_3 receptor antagonists are available at the present time, although SR-48,968 may be a chemical lead to such agents as it is capable of blocking both NK_2 and NK_3 receptors in the guinea pig (Petitet et al., 1993).

These new agents have stimulated interest in a more objective appraisal of the role of NK_1 and NK_2 receptors in nociception.

2. Peripheral Mechanisms of Nociception

There is no direct evidence for potentiation of nociception or for a direct excitation of nociceptor fibers in the periphery by SP or any other tachykinins, but there is abundant evidence that the release of these peptides from unmyelinated afferents in the peripheral tissues is an important part of the overall inflammatory reaction (Saria, 1984; Hanesch et al., 1992). Injection of exogenous peptides suggests that the activation of NK_1 receptor leads to neurogenic extravasation, but that activation of NK_2 or NK_3 receptors does not (Hirayama et al., 1993; O'Shaughnessy and Connor, 1993). Experiments with selective NK_1 antagonists show that extravasation produced by electrical stimulation of nerve trunks is completely blocked, suggesting an exclusive role for the endogenous ligand for NK_1 receptors (presumably SP) in this phenomenon (*see,* for example, Hirayama et al., 1993; Moussaoui et al., 1993a,b; Shepheard et al., 1993). Depletion of SP from its fibers by neonatal capsaicin treatment abolishes nerve stimulation-induced extravasation yet produces hypersensitivity to exogenous SP (Moussaoui et al., 1993a). The anatomical location of SP-containing fibers in the periphery is consistent with a permissive role in nociception by release of directly algogenic substances or by a general vasodilatation that allows access of algogens to a larger volume of tissue. This vasodilatation also appears to be primarily via NK_1 receptors (Beattie et al., 1993), and it is interesting to note that two of the cardinal features of migraine headache, namely neurogenic inflammation and vasodilatation, can be produced by activation of NK_1 receptors on dural vessels following electrical stimulation of the rat trigeminal ganglion (Shepheard et al., 1993). Recently, it has been shown that intravenous injection of capsaicin to the anesthetized rat will also produce neurogenic inflammation in the meninges and that this can be blocked with RP-67,580 (Moussaoui et al., 1993b).

3. Electrophysiological Study of Tachykinins in the Spinal Cord In Vitro

Otsuka and colleagues first showed that SP would depolarize the ventral root of an isolated spinal cord preparation from an immature rat (Konishi and Otsuka, 1974). This initial observation led to studies in which the slow depolarization of the ventral root produced by sustained electrical stimulation of the dorsal root in the same segment was found to resemble that produced by application of exogenous SP to the solution bathing the preparation. It was therefore suggested that the activation of C-fibers by such intense stimulation might reflect the situation after a painful event in the peripheral receptive field of the neurons being studied and that SP might function as a transmitter in pain perception.

This contention was strengthened by the demonstration that a peptide SP receptor antagonist, (D-Arg^1 D-Pro^2 D-$Trp^{7,9}$ SP), would block both the depolarization produced by exogenous SP and by the sustained dorsal root stimulation (Yanagisawa et al., 1982), although other laboratories (e.g., Salt et al., 1982) have found it difficult to reproduce this finding. More recently, single neuron intracellular recordings have been made from isolated spinal cord preparations or from slices of this tissue. Again, it has been shown that the late slow excitatory postsynaptic potential (epsp) generated in dorsal horn neurons by high intensity electrical stimulation of a dorsal rootlet resembles the action of exogenous SP, and that peptide SP receptor antagonists can sometimes reverse the effects of both of the above (e.g., Urban et al., 1985). It has also been shown that poly- and monoclonal antibodies to SP will block the slow synaptic depolarization of dorsal horn neurons in rat cord slices following electrical stimulation of a dorsal rootlet (Randic et al., 1986).

SP, NKA, and a number of other tachykinins have been shown to excite and/or depolarize neurons in isolated spinal cord preparations, and it is likely that most of these responses are expressed via

NK_1 and NK_2 receptors that are well represented in this tissue, with a smaller contribution from NK_3 sites. The most reproducible activation of dorsal horn neurons by bath applied SP has been seen in studies on the larger laminae IV and V neurons, and the selective NK_1 agonist, [Sar^9 Met $(O_2)^{11}$]SP, was the most potent ligand tested (Morris et al., 1992a). It has proved difficult to demonstrate that the smaller neurons of the substantia gelatinosa are excited by tachykinins (Bleazard and Morris, 1993) even though this region of the dorsal horn has an abundant innervation with fibers that can be shown immunocytochemically to contain SP and other tachykinins.

It has been suggested that more than one mechanism underlies the SP response that can be recorded from dorsal horn neurons, and, in particular, SP has been shown to stimulate the entry of Ca^{2+} through large cation channels (Murase et al., 1986) and to inhibit an outward K^+ conductance (Nowak and MacDonald, 1981). It is possible that these responses may be the result of activation of two separate receptor types, and it has been shown in experiments on acutely isolated neurons from rat dorsal horn that SP can trigger calcium entry through voltage dependent channels, but only in a subpopulation of neurons (Rusin et al., 1992).

Limited studies have been made in spinal cord slices with the nonpeptide NK_1 antagonist, CP-96,345, but it appears that the success rate for antagonizing excitation produced by bath applied SP on rat dorsal horn neurons is not markedly higher than with the previously employed peptide agents (Morris et al., 1992b). This may, of course, be attributable to the low-affinity of this particular antagonist for rat NK_1 receptors (e.g., *see* Beaujouan et al., 1993). It is interesting to note, however, that when SP-induced calcium entry was studied on acutely isolated rat dorsal horn neurons, CP-96,345 blocked this effect with an ED_{50} of approx 100 n*M* (consistent with the affinity of this agent for the rat NK_1 receptor), suggesting that, when depolarization is a consequence of Ca^{2+} channel opening, then this may be operated via NK_1 receptors (Rusin et al., 1993).

Interesting interactions have been recorded between the effects of excitatory amino acids and of SP on acutely isolated rat dorsal horn neurons studied under voltage clamp that may reflect the probable coexistence and release of these substances by nociceptive unmyelinated afferent fibers (Rusin et al., 1992). The simultaneous application of *N*-methyl-D-aspartate (NMDA) and SP reversibly enhanced the initial component of the NMDA current, and this effect persisted for up to 20 min following the SP application. This action of SP was blocked by the peptide NK_1 receptor antagonist, spantide II (Folkers et al., 1990), although the intracellular mechanisms involved in the potentiation may be complex. This effect appears to be selective for the NMDA receptor (Randic et al., 1990), and activation of this receptor has been suggested to be the main determinant of the long duration of the dorsal root evoked ventral root potential (VRP; Brugger et al., 1990), although it appears that a component of the late depolarizing response of deep dorsal horn neurons persists in the presence of blockade of both NMDA and non-NMDA excitatory amino acid receptors (Gerber et al., 1991), suggesting an independent peptide-operated component. Prolonged ventral root depolarizations evoked by dorsal root stimulation in rat isolated spinal cord have been found to be resistant to application of the NK_1 antagonist, CP-96,345, although reduction in this potential was seen after application of the peptide NK_2 antagonist, MEN-10,376, or the NMDA antagonist, APV (Nagy et al., 1993; Thompson et al., 1993a). When this experiment was repeated in spinal cords taken from animals in which thermal and mechanical hyperalgesia had been produced by UV irradiation, the amplitude of the VRP was reduced by CP-96,345 (Thompson et al., 1993b). This demonstration that baseline nociceptive reflexes may only require the activation of NK_2 and NMDA receptors, but that a component operated through NK_1 receptors becomes relevant in the presence of peripheral inflammation, can be supported by observations made in vivo.

The majority of the work described above was conducted in tissues obtained from the spinal cords of young animals (typically 6–20 d old). It should be noted that studies on the pharmacology of tachykinin receptors in immature spinal cord or slices may give results that are poorly predictive of the situation in the adult as receptor distribution and density changes with age (*see,* e.g., Ireland et al., 1992).

4. Electrophysiological Studies on Spinal Cord In Vivo

4.1. Single Neuron Recordings

The first in vivo electrophysiological evidence suggesting that SP might have a specific role in nociception came from the iontophoretic studies of Henry (1976) in the cat spinal cord. He showed that SP would excite those dorsal horn neurons that also responded to noxious stimuli applied to their cutaneous receptive fields, but that this peptide did not excite neurons that were only excited by low-intensity cutaneous stimuli. These difficult and time consuming experiments have never been exactly duplicated by others, but it is noteworthy that not all neurons found by Henry to respond to noxious stimuli were excited by SP, and that some multireceptive neurons were excited by SP (Henry, 1976; *see also* Randic and Miletic, 1977; Wright and Roberts, 1980). Under different experimental conditions, it has been shown that cat low- and high-threshold dorsal horn neurons, and also motoneurons, were excited by SP (Zieglgansberger and Tulloch, 1979). In subsequent studies on the cat trigeminal complex, a more exact correlation was seen with only those nucleus caudalis neurons that were excited by cutaneous noxia or by electrical stimulation of tooth pulp being excited by SP (Andersen et al., 1978; Henry et al., 1980) although this selectivity may be species dependent as it is not mirrored in the rat, where nociceptive and nonnociceptive neurons are both readily excited by SP applied iontophoretically (Hill et al., 1980; Crozier et al., 1981; Salt et al., 1983). Recently, weak excitation of nociceptive dorsal horn neurons by iontophoretic SP has been observed in the rhesus monkey (Dougherty and Willis, 1991).

Evidence for SP as a sensory transmitter in the dorsal horn of the spinal cord has been gained from studies in which an enhanced sensitivity to microiontophoretically applied SP resulted following either surgical section of dorsal roots (Wright and Roberts, 1980) or a chemical denervation (by giving capsaicin to a neonatal rat that was subsequently studied as an adult) (Hill et al., 1980). Sensitivity to glutamate was unaffected in the latter experiments. Neonatal capsaicin treatment does not affect the levels of mRNA coding for NK_1 receptor synthesis (Sivam and Krause, 1992), although the binding of SP is increased in rat dorsal horn ipsilateral to chronic constriction injury of sciatic nerve (Aanonsen et al., 1992).

Using iontophoretic application of SP and NKA to cat dorsal horn neurons, Salter and Henry (1991) demonstrated that NK_1 receptor agonists preferentially excited nociceptive neurons, whereas NK_2 receptor agonists excited neurons responding to both noxious and nonnoxious stimuli. Fleetwood-Walker et al. (1987) reported that the NK_1 selective agonist, [Met-OMe^{11}]-SP, excited neurons in laminae IV and V of cat spinal cord but when ejected in the superficial laminae of the cord it only exhibited inhibitory effects, and these were restricted to nonnociceptive neurons. In contrast, the NK_2 agonist, NKA, produced a selective facilitation of the responses to noxious heat without affecting the responses to nonnoxious stimuli. In subsequent studies, this group has been able to show blockade by selective peptide antagonists of the actions of NK_2 receptor agonist excitation of dorsal horn neurons and related antinociceptive effects of these agents against brief nociceptive responses evoked by thermal, mechanical, and chemical noxious stimuli applied to the peripheral receptive fields of neurons in both rat (Fleetwood-Walker et al., 1991) and cat (Fleetwood-Walker et al., 1990) dorsal horn. Recent experiments have shown that the peptide NK_2 antagonist, L-659,874, will attenuate the sustained nociceptive response of multireceptive dorsal horn neurons in the rat evoked by peripheral application of mustard oil (Munro et al., 1993). When the peptide NK_1 antagonist, [D-Pro^2-D-Trp^{7,9}]-SP, was tested against the A- and C-fiber mediated firing of dorsal horn neurons in the rat, it showed some selective reduction of C-fiber evoked responses, but with a bell-shaped dose response curve (Kellstein et al., 1990).

Newer nonpeptide NK_1 agents, such as CP-96,345, will apparently readily block both the excitation produced by iontophoretic SP and that produced by nociceptive inputs to dorsal horn neurons in both the cat (Radhakrishnan and Henry, 1991) and the rhesus monkey (Dougherty et al., 1993), and it has been questioned whether the peptide and nonpeptide SP receptor antagonists actually bind to different subtypes of the NK_1 receptor (Tousignant et al., 1992 and other chapters of this volume), although recent site-directed mutagenesis studies suggest that alternative binding sites are available within the clonal NK_1 receptor (Fong et al, 1992). Radhakrishnan and Henry (1991) studied the effects of CP-96,345 on the responses of cat dorsal horn neurons to iontophoretic SP and to peripheral thermal or mechanical noxious stimuli. On 6 of 8 nociceptive neurons examined, SP also increased neuronal firing rate and, following the slow iv injection of the antagonist, the response to SP and to noxious heat was blocked with the predominant effect being seen on the after-discharge, whereas effects on the response to mechanical noxia were less clear cut. In a parallel study, also in the cat, De Koninck and Henry (1991), recording from dorsal horn neurons, observed that noxious stimuli or repetitive C-fiber intensity stimuli to a peripheral nerve would evoke a characteristic late, slow epsp associated with a conductance decrease. Doses of CP-96,345 of 0.5 to 2.0 mg/kg iv blocked this late, slow epsp, but had no effect on the depolarizing response to single intense stimuli even when these were recorded at C-fiber latencies. These observations are therefore consistent with the data obtained from in vitro studies, suggesting that activation of NK_1 receptors may have a role in the nociceptive responses of dorsal horn neurons to intense and maintained stimuli and allow for the possibility that NK_2 receptor activation is more likely to have a role in the response to a brief noxious event.

In the rhesus monkey spinal cord, wide dynamic range and nociceptive specific spinothalamic tract (STT) cells were only weakly excited by iontophoretic SP, but showed an enhanced response when coapplied with NMDA. This enhancement was blocked by iontophoretic [D-Pro^2, D-Trp^{7,9}]-SP (Dougherty and Willis, 1991). More recently, it has been shown that application of CP-96,345 from a microdialysis fiber positioned in the spinal cord dorsal horn would attenuate the nociceptive responses of monkey STT cells and prevent the response facilitation produced by intradermal capsaicin injection (Dougherty et al., 1993). In these experiments, the action of

CP-96,345 closely resembled that of the NMDA receptor antagonist, AP7, in its action in depressing the nociceptive responses when applied to the same neurons.

4.2. Nociceptive Spinal Reflexes

Characteristic flexion reflexes have long been known to be evoked by noxious stimuli, and these reflexes are facilitated by the presence of tissue inflammation (Woolf, 1983; Ferrell et al., 1988) or by high intensity (i.e., sufficient to excite C-nociceptor fibers) electrical conditioning stimuli applied to peripheral nerves. Intrathecal administration of SP to a decerebrate, spinalized rat was found to facilitate the hamstring flexion reflex for 5 min when given in a dose of 10 ng (Wiesenfeld-Hallin and Duranti, 1987). Initial studies with peptide tachykinin antagonists such as spantide on this reflex were inconclusive as spantide itself was found to facilitate the reflex and summate with the effects of SP rather than producing blockade (*see also* Salt et al., 1982). The nontoxic peptide NK_1 antagonist, spantide II (Folkers et al., 1990), has been found to block the SP-induced facilitation of the flexion reflex when this drug was given intrathecally to the decerebrate rat (Wiesenfeld-Hallin et al., 1990a,b), although this agent did show weak facilitatory effects on the reflex when given alone. The antagonism seen was selective as reflex facilitation produced by the primary afferent peptides somatostatin, vasoactive intestinal peptide (VIP), calcitonin gene related peptide (CGRP), and galanin was not blocked by spantide II. Facilitation of the flexor reflex evoked by electrical stimulation of either the sural nerve (Wiesenfeld-Hallin et al., 1990b) or the gastrocnemius muscle nerve (Wiesenfeld-Hallin et al., 1991) was shown to be blocked by spantide II. More recently, similar studies with CP-96,345, again given intrathecally, have demonstrated that this agent will block the flexor reflex facilitation produced either by C-fiber conditioning stimulation or by intrathecal SP (Xu et al., 1992b). It is noteworthy that high doses of CP-96,345 were given in these experiments, and that this agent has a relatively poor separation between its affinity for rat NK_1 receptors and for L-type calcium channels (Schmidt et al., 1992). Recently, it has also been shown to block Na^+ currents in rat cortical neurons when applied at micromolar concentrations (Caeser et al., 1993). Similar experiments with RP-67,580, a rodent NK_1 selective antagonist, and its less active enantiomer, RP-68,651, have confirmed that blockade of NK_1 receptors in the spinal cord of an anesthetized rat will prevent the facilitation of flexor reflexes but, interestingly, will not reduce the size of the baseline reflex or windup (Laird et al., 1993; Fig. 3). In this latter study, it was striking that low doses of drug (3–30 μg/kg) adequate to maintain plasma concentrations in the region of 3–20 n*M* would block facilitation, and this is consistent with its affinity of 4 n*M* at the rodent NK_1 receptor. The inactive enantiomer, RP-68,651, was without effect on flexor reflex facilitation. Recently, a similar flexor reflex preparation has been developed in the anesthetized rabbit, which has NK_1 receptors with similar pharmacology to those in humans, and CP-99,994 (Desai et al., 1992) has been shown to attenuate the facilitation produced by electrical stimulation of the skin (ID_{50} = 2.7 μg/kg iv) with no effects on the baseline reflex (Boyce et al., 1993). NKA has also been shown to facilitate flexor reflexes when given intrathecally (Xu and Wiesenfeld-Hallin, 1992), and this facilitation, but not that produced by SP, is blocked by the peptide NK_2 antagonist, MEN-10,207, also given intrathecally (Xu et al., 1991). MEN-10,207 given by itself facilitated the reflex, possibly indicating partial agonism. High doses appeared to completely block the baseline reflex (700 pmol–7 nmol), whereas lower doses (70–700 pmol) blocked the facilitation produced by gastrocnemius nerve stimulation. Visceral reflexes have been less well studied but, in the rat, distention of the bladder evokes reflexes that the NK_1 antagonists, GR-82,334, RP-67,580, and CP-96,345, will block, but that are insensitive to the NK_2 antagonists, MEN-10,376 and SR-48,968, all antagonists being given intrathecally (Lecci et al., 1993).

It is striking that the most intense blockade of the nociceptive flexor reflex is produced by NMDA-receptor antagonists such as MK-801, and that this reduces the baseline reflex, windup, and facilitation (Xu et al., 1992a). MK-801, when given together with CP-96,345, both in subthreshold doses, strongly blocked windup and facilitation *(ibid)*. There is thus a separation of the effects of NK_1

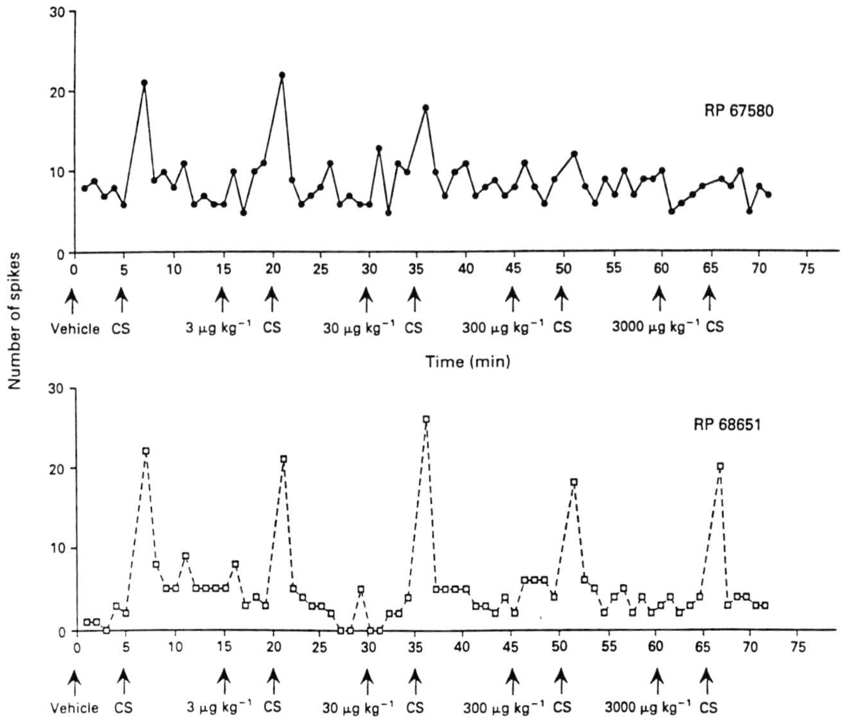

Fig. 3. Results from an experiment on nociceptive reflex facilitation in an anesthetized, spinalized rat. Single unit recordings of motor unit activity were made from biceps femoris/semitendinous muscle with a bipolar, concentric needle electrode. Numbers of spikes generated in response to high-intensity electrical stimulation of the receptive field of the motor unit (within the territory of the sural nerve) are plotted on the vertical axis of the two graphs shown. Single shocks (1 ms, 2 × threshold [T]) once/min were applied continuously throughout the experiments and generated the baseline response level. A conditioning stimulus (CS; 1 ms, 2 × T, 1 Hz for 20 shocks) was applied every 15 min, and the response to the baseline stimulus 30 s after termination of the CS was taken as the conditioned response (the peaks in the traces). Sequential rising doses of either the NK_1 antagonist, RP-67,580 (upper graph), or its inactive enantiomer, RP-68,651 (lower graph), were given at 15 min intervals. Five min before the CS, RP-67,580 produced a dose-related reduction in the facilitated reflex, but did not affect the baseline response level and, in contrast, RP-68,651 was inactive over the dose range tested.

This experiment suggests that facilitation of a spinal flexor reflex by a noxious conditioning stimulus is operated by activation of NK_1 receptors, whereas the baseline flexor reflex is not. Substance P, acting at NK_1 receptors, is thus involved in the increase in excitability after intense and/or prolonged noxious stimulation, but not in the processing of brief, noxious stimuli (*see* text and Laird et al., 1993, for details).

and NK_2 receptor activation on the flexor reflex, consistent with the suggestion that NK_1 receptors (and presumably synaptic release of SP) are involved in the facilitated reflex, but have little effect on the baseline response. NK_2 receptor involvement is more difficult to characterize given the limited published work to date.

Interaction between the action of glutamate and SP at NMDA and NK_1 receptors, respectively, is likely to be important in nociception. Most recently, it has been claimed that inhibition of NO synthase with L-NAME will block the responses of nociceptive neurons in cat dorsal horn to iontophoretically applied NMDA, SP, and a noxious cutaneous stimulus, inferring that NO is the final common pathway for nociception in these cells (Radhakrishnan and Henry, 1993). This intriguing observation awaits independent confirmation.

5. Electrophysiological Studies on Tachykinin Receptors at Supraspinal Sites

Comparatively little work has been done on the role of tachykinin receptors in nociceptive relays at supraspinal sites. Areas such as the locus ceruleus, which are known to contain neurons excited by peripheral noxious stimuli or iontophoretic SP and possess tachykinin binding sites, have been used to investigate the action of NK_1 receptor agonists and antagonists. Studies with peptide NK_1 antagonists have been equivocal, and a report that [D-Pro^2, D-Trp^{7,9}]-SP would block SP excitation of locus ceruleus neurons (Engberg et al., 1981) did not extrapolate to other brain areas in the rat (Salt et al., 1982). More recently, it has been shown that CP-96,345 will readily antagonize the excitation of locus ceruleus neurons by SP in a slice preparation (McLean et al., 1991). No studies have been made on the actions of NK_1 antagonists against nociceptive responses of these neurons.

There is at present only a single report of the action of a nonpeptide NK_2 antagonist on the nociceptive responses of single neurons. In studies on neurons in the posterior thalamus of the anesthetized rat, Santucci et al. (1993) showed that iv administration of SR-48,968 (0.125–0.5 mg/kg) would inhibit the neuronal excitation produced by thermal, but not by mechanical, noxious cutaneous stimuli. This action was enantioselective, as SR-48,965, which has 2000-fold less activity as an NK_2 receptor antagonist, was ineffective against either stimulus.

NK_3 receptor agonists are known to excite supraspinal neurons in a number of areas of the brain, but it has proved difficult to obtain reproducible responses to NK_2 receptor agonists (*see,* for example, Norris et al., 1993).

6. Behavioral Studies on Nociception

A number of recent reviews have dealt in detail with the enormous literature on behavioral studies with SP and its congeners, and it is not proposed that an exhaustive description of this topic will be given here. Many experiments have been performed in which motor behaviors have been evoked by intrathecal or intracisternal/ventricular administration of SP and other tachykinin receptor agonists in rodents (*see* Vaught, 1988), and, in particular, the production of caudally directed biting and scratching behavior has been assumed to be indicative of a nociceptive role for tachykinins. This behavior is better interpreted as a general stimulant action on motor outflow and by itself cannot be used to associate activation of any of the tachykinin receptors with nociception. Agonists at both NK_1 and NK_2 receptors will cause this behavior and facilitate or reduce reaction time to noxious stimuli. The action of these agonists can be blocked by peptide NK_1 and NK_2 receptor antagonists injected by the same route (*see* Picard et al., 1993) but, interestingly, agonists at NK_3 receptors do not reduce nociceptive response latency (Laneuville et al., 1988). The most convincing evidence for a role in nociception comes from experiments in which nociceptive thresholds to mechanical and thermal stimuli are reduced by central administration of tachykinins with both NK_1 and NK_2 agonists having the ability to produce this effect, but rank order of potency suggesting that NK_1 receptor activation may be the most effective potentiator (Cridland and Henry, 1986; Picard et al., 1993). It should be noted that such potentiation is not exclusive to tachykinins, and that excitatory amino acid receptor agonists can produce a similar effect (Mjellem-Joly et al., 1992). The receptor selectivity of the tachykinin effects has been adequately confirmed by the use of antagonists, but the literature is confused by numerous studies in the rat with these peptide antagonists that have subsequently been shown to be neurotoxic in this species (*see* Rodriguez et al., 1983; Vaught, 1988). More recently, experiments with the nonpeptide antagonists, CP-96,345 and SR-48,968, convincingly show that potentiation of nociception can occur following activation of either NK_1 or NK_2 receptors (Picard et al., 1993). The well controlled study by Yashpal et al. (1993) demonstrates that the facilitation of tail flick response in the rat by intrathecal SP or by immersion of the tail in hot water, but, importantly, not the baseline reflex, is attenuated by CP-96,345 at a dose of 5 mg/kg sc, but not by its inactive enantiomer, CP-96,344. In addition, the second phase of the formalin paw response was reduced with doses of 10 or 20 mg/kg of CP-96,345.

Demonstration of antinociceptive actions of tachykinin receptor antagonists in the absence of sensitization or potentiation with agonists is more problematic, and no convincing studies have been published using peptide antagonists largely because of the motor impairment produced as a consequence of the neurotoxicity mentioned above (*see* Rodriguez et al., 1983; Vaught, 1988). The newer nonpeptide antagonists have been used in a number of studies, and CP-96,345 has been found active in the mouse and rat formalin paw test (Yamamoto and Yaksh, 1991; Birch et al., 1992; Sakurada et al., 1993) and in mouse acetic acid writhing test (Nagahisa et al., 1992a) and mouse hot plate tests (Lecci et al., 1991), but in cases where enantioselectivity has been checked the putative inactive enantiomer appears as active as CP-96,345 (Nagahisa et al., 1992b). This paradoxical observation is explained by significant Ca^{2+} L-channel and Na^+ current blockade with concentrations of CP-96,345 sufficient to block rodent NK_1 receptors, and the demonstration that similar doses of systemic calcium channel blockers or lidocaine will produce an equivalent degree of antinociception (Nagahisa et al., 1992b). The rodent NK_1 receptor selective agent, RP-67,580, was found to be antinociceptive in the phenylbenzoquinone (PBQ) writhing and formalin paw tests in mice (Garret et al., 1991), but was inactive in tests using thermal or mechanical noxious stimuli. Subsequent studies have shown that this substance is as likely to be antinociceptive in the gerbil (which has human type NK_1 receptors) as it is in rats or mice (Rupniak et al., 1993), and enantiomeric selectivity cannot be demonstrated. High concentrations (10 μ*M*) of RP-67,580 have now been shown to have ion channel blocking actions similar to those seen with CP-96,345 at tenfold lower concentrations (Rupniak et al., 1993), and action at Ca^{2+} channels is also suggested by its ability to displace verapamil binding to rat heart with IC_{50} of 210 n*M* (S. McLean, personal communication).

It is noteworthy that the doses used in most of these behavioral studies are very much higher than those needed to block facilitated flexor reflexes in electrophysiological studies or to reduce neurogenic extravasation.

7. Prognosis for Analgesic Actions in Humans with Tachykinin Receptor Antagonists

The unequivocal behavioral evidence for an analgesic action of NK_1 antagonists is somewhat sparse, and even those studies where enantioselectivity has been demonstrated appear to require extremely high doses of the antagonists relative to the amounts theoretically necessary to saturate the receptor. It must be remembered, however, that the acute antinociceptive tests used in animals bear little, if any, resemblance to the situation that obtains in a clinical pain state, and, in particular, the stimuli used are unlikely to be of a sufficient intensity to cause maximal release of SP within the spinal dorsal horn or to cause upregulation of SP release or NK_1 receptor expression. A definitive study on nociception in the presence of a maintained inflammatory state remains to be done.

Electrophysiological studies show convincingly that a facilitated nociceptive response (but not either the baseline or partially wound up response) is extremely sensitive to NK_1 antagonists, and attenuation of the response can be demonstrated with tissue concentrations of drug barely adequate to saturate NK_1 receptors and almost a thousandfold lower than those needed to reduce the behavioral responses described above.

Peripheral neurogenic extravasation is also extremely sensitive to blockade with NK_1 antagonists, needing doses of drug close to those found effective in electrophysiological studies, although the significance of this phenomenon to nociception remains to be demonstrated.

It is appropriate to conclude that NK_1 antagonists may well be antinociceptive in humans and by a uniquely subtle mechanism that would have little or no effect on baseline protective reflexes. The test of this hypothesis awaits appropriate clinical trial. Actions of NK_2 receptors in nociception remain to be substantiated, and the role of NK_3 receptors, if any, is essentially unstudied at this time.

References

Aanonsen, L. M., Kajander, K. C., Bennett, G. J., and Seybold, V. S. (1992) Autoradiographic analysis of ^{125}I-substance P binding in rat spinal cord following chronic constriction injury of the sciatic nerve. *Brain Res.* **596,** 259–268.

Andersen, R. K., Lund, J. P., and Puil, E. (1978) Enkephalin and substance P effects related to trigeminal pain. *Can. J. Phys. Pharm.* **56,** 216–222.

Barr, A. J. and Watson, S. P. (1993) Non-peptide antagonists, CP-96,345 and RP67580, distinguish species variants in tachykinin NK_1 receptors. *Br. J. Pharmacol.* **108,** 223–227.

Beattie, D. T., Stubbs, C. M., Connor, H. E., and Feniuk, W. (1993) Neurokinin-induced changes in pial artery diameter in the anaesthetized guinea-pig. *Br. J. Pharmacol.* **108,** 146–149.

Beaujouan, J. C., Heuillet, E., Petitet, F., Saffroy, M., Torrens, Y., and Glowinski, J. (1993) Higher potency of RP67580 in the mouse and the rat compared with other non-peptide and peptide tachykinin NK_1 antagonists. *Br. J. Pharmacol.* **108,** 793–800.

Beresford, I. J. M., Ireland, S. J., Stables, J., and Hagan, R. M. (1992) Ontogeny and characterisation of ^{125}I-Bolton Hunter-eledoisin binding sites in rat spinal cord by quantitative autoradiography. *Neuroscience* **46,1,** 225–232

Birch, P. J., Harrison, S. M., Hayes, A. G., Rogers, H., and Tyers, M. B. (1992) The non-peptide NK_1 receptor antagonist, (±)-CP-96,345, produces antinociceptive and anti-oedema effects in the rat. *Br. J. Pharmacol.* **105,** 508–510.

Bleazard, L. and Morris, R. (1993) Paradoxical lack of action of neurokinin agonists on subsbntia gelatinosa neurons of the rat spinal cord: An *'in vitro'* study. *Proc. 32nd Int. Phys. Congr.* **209,** 1P.

Boyce, S., Laird, J. M. A., Tattersall, F. D., Rupniak, N. M. J., Hargreaves, R. J., and Hill, R. G. (1993) Antinociceptive effects of NK_1 receptor angonists: Comparison of behavioral and electrophysiological tests. *7th World Congress on Pain* (abstract, 641).

Brugger, F., Evans, R. H., and Hawkins, N. S. (1990) Effects of N-methyl-D-aspartate antagonists and spantide on spinal reflexes and responses to substance P and capaicin in isolated spinal cord preparations from mouse and rat. *Neuroscience* **36, 3,** 611–622.

Caeser, M., Seabrook, G. R., and Kemp, J. A. (1993) Block of voltage-dependent sodium currents by the substance P receptor antagonist (±)-CP-96,345 in neurons cultured from rat cortex. *Br. J. Pharmacol.* **109,** 918–924.

Cervero, F. and Laird, J. M. A. (1991) One pain or many pains? A new look at pain mechanisms. *NIPS* **6,** 268–273.

Cervero, F., Gilbert, R., Hammond, R. G. E., and Tanner, J. (1993) Development of secondary hypelgesia following non-painful thermal stimulation of the skin: A psychophysical study in man. *Pain* **54,** 181–189.

Cridland, R. A. and Henry, J. L. (1986) Comparison of the effects of substance P, neurokinin A, physalaemin and eledoisin in facilitating a nociceptive reflex in the rat. *Brain Res.* **381,** 93–99

Crozier, C. S., Hill, R. G., and Salt, T. E. (1981) The effects of capsaicin pre-treatment on substance P levels and sensory responses of neurons in the rat trigeminal nucleus caudalis. *J. Physiol.* **324,** 78P.

De Koninck, Y. and Henry, J. L. (1991) Substance P-mediated slow excitatory postsynaptic potential elicited in dorsal horn neurons *in vivo* by noxious stimulation. *Proc. Natl. Acad. Sci. USA* **88,** 11,344–11,348.

Desai, M. C., Lefkowitz, S. L., Thadeio, P. F., Longo, K. P., and Snider, R. M. (1992) Discovery of a potent substance P antagonist: Recognition of the key molecular determinant. *J. Med. Chem.* **35,** 4911–4913.

Dougherty, P. M., Palecek, J., Paleckova, V., Sorkin, L. S., and Willis, W. D. (1993) The role of NMDA, non-NMDA, and NK_1 receptors in the excitation of spinothalamic tract neurons in anaesthetized monkeys. *J. Physiol.* **459,** 209P.

Dougherty, P. M. and Willis, W. D. (1991) Enhancement of spinothalamic neuron responses to chemical and mechanical siimuli following combined micro-iontophoretic application of N-methyl-D-aspartic acid and substance P. *Pain* **47,** 85–93.

Duggan, A. W. and Weihe, E. (1991) Central transmission of impulses in nociceptors: Events in the superficial dorsal horn, in *Towards a New Pharmacotherapy of Pain: Beyond Morphine* (Basbaum, A. and Besson, J. M., eds.), Wiley, New York, pp. 35–67.

Emonds-Alt, X., Golliot, F., Pointeau, P., Le Fur, G., and Breliere, J. C. (1993) Characterisation of the binding sites of [^{3}H]SR-48,968, a potent non-peptide radioligand antagonist of the neurokinin-2 receptor. *Biochem. Biophys. Res. Comm.* **191,** 1172–1177

Engberg, G., Svensson, T. H., Rosell, S., and Folkers, K. (1981) A synthetic peptide as an antagonist of substance P. *Nature* **293,** 222,223.

Ferrell, W. R., Wood, L., and Baxendale, R. H. (1988) The effect of acute joint inflammation on flexion reflex excitability in the decerebrate, low spinal cat. *Q. J. Exp. Physiol.* **73,** 95–102.

Fleetwood-Walker, S. M., Mitchell, R., Hope, P. J., El-Yassir, N., and Molony, V. (1987) The roles of tachykinin and opioid receptor types in nociceptive and non-nociceptive processing in superficial dorsal horn, in *Fine Afferent Nerve Fibers and Pain* (Schmidt, R. F., Schaible, H. G., and Vahle-Hinz, C., eds.), VCH, Weinheim, pp. 239–247.

Fleetwood-Walker, S. M., Mitchell, R., Hope, P. J., El-Yassir, N., Molony, V., and Bladon, C. M. (1990) The involvement of neurokinin receptor subtypes in somatosensory processing in the superficial dorsal horn of the cat. *Brain Res.* **519,** 169–182.

Fleetwood-Walker, S. M., Hope, P. J., Parker, R. M. C., and Mitchell, R. (1991) Effects of antagonists selective for neurokinin-1 and neurokinin-2 receptors on somatosensory responses of rat dorsal horn neurons. *Br. J. Pharmacol.* **104,** 472P.

Folkers, K., Feng, D. M., Asano, N., Hakanson, R., Wiesenfeld-Hallin, Z., and Leander, S. (1990) Spantide II, an effective tachykinin antagonist having high potency and negligible neurotoxicity. *Proc. Natl. Acad. Sci. USA* **87,** 4833–4835.

Fong, T. M., Yu, H., and Strader, C. (1992) Molecular basis for the species selectivity of the neurokinin-1 receptor antagonists CP-96,345 and RP-67,580. *J. Biol. Chem.* **267,** 25,668–25,671.

Garret, C., Carruette, A., Fardin, V., Moussaoui, S., Peyronel, J. F., Blanchard, J. C., and Laduron, P. (1991) Pharmacological properties of a potent and selective non-peptide substance P antagonist. *Proc. Natl. Acad. Sci. USA* **88,** 10,208–10,212.

Gasparovic, K., Hadzovic, S., and Stern, P. (1964) Contribution to the theory that substance P has a transmitter role in sensory pathways. *Med. Exp.* **10,** 303–306.

Gerber, G., Ceme, R., and Randic, M. (1991) Participation of excitatory amino acid receptors in the slow excitatory synaptic transmission in rat spinal dorsal horn. *Brain Res.* **561,** 236–251.

Handwerker, H. O. and Kobal, G. (1993) Psychophysiology of experimentally induced pain. *Physiol. Rev.* **73, 3,** 639–671.

Hanesch, U., Heppelmann, B., and Schmidt, R. F. (1992) Neurokinin A-like immunoreactivity in articular afferents of the cat. *Brain Res.* **586,** 332–335.

Henry, J. L. (1976) Effects of substance P on functionally identified units in cat spinal cord. *Brain Res.* **114,** 439–451.

Henry, J. L., Sessle, B. J., Lucier, G. E., and Hu, J. W. (1980) Effects of substance P on nociceptive and non-nociceptive trigeminal brain stem neurons. *Pain* **8,** 33–45.

Hill, R. G. (1986) Current perspectives on pain. *Sci. Prog. Ox.* **70,** 95–107.

Hill, R.G., Hoddinott, M. L., and Keen, P. M. (1980) Action of substance P on trigeminal nucleus caudalis neurons in capsaicin-treated rats, in *Neuropeptides and Neural Transmission* (Ajmone-Marsan, C., and Traczyck, W. Z., eds.), Raven, New York, pp. 31–41.

Hirayama, Y., Lei, Y. H., Barnes, P. J., and Rogers, D. F. (1993) Effects of two novel tachykinin antagonists, FK224 and FK888, on neurogenic airway plasma exudation, bronchoconstriction and systemic hypotension in guinea-pigs *in vivo*. *Br. J. Pharmacol.* **108,** 844–851.

Hokfelt, T., Vincent, S., Hellsten, L., Rosell, S., Folkers, K., Markey, K., Goldstein, M., and Cuello, C. (1981) Immunohistochemical evidence for a 'neurotoxic' action of (D-Pro^2,D-Trp^{7,9})-substance P, an analogue with substance P antagonistic activity. *Acta Physiol. Scand.* **113,** 571–573.

Ireland, S. J., Wright, I. K., and Jordan, C. C. (1992) Characterization of tachykinin-induced ventral root depolarization in the neonatal rat isolated spinal cord. *Neuroscience* **46,** 217–223.

Kellstein, D. E., Price, D. D., Hayes, R. L., and Mayer, D. J. (1990) Evidence that substance P selectively modulates C-fiber-evoked discharges of dorsal horn nociceptive neurons. *Brain Res.* **526,** 291–298.

Konishi, S. and Otsuka, M. (1974) Excitatory action of hypothalamic substance P on spinal motoneurons of newbom rats. *Nature* **252,** 734,735.

Laird, J. M. A., Hargreaves, R. J., and Hill, R. G. (1993) Effect of RP67580, a non-peptide neurokinin-1 receptor antagonist, on facilitation of a nociceptive spinal flexion reflex in the rat. *Br. J. Pharmacol.* **109,** 713–718.

Laneuville, O., Dorais, J., and Couture, R. (1988) Characterization of the effects produced by neurokinins and three agonists selective for neurokinin receptor subtypes in a spinal nociceptive reflex of the rat. *Life Sci.* **42,** 1295–1305.

Lecci, A., Giuliani, S., Garret, C., and Maggi, C. A. (1993) Evidence for a role of tachykinins as sensory transmitters in the activation of micturition reflex. *Neuroscience* **54,** 827–837.

Lecci, A., Giuliani, S., Patacchini, R., Viff, G., and Maggi, C. A. (1991) Role of NK_1 tachykinin receptors in thermonociception: effect of (±)CP-96,345, a non-peptide substance P antagonist, on the hot plate test in mice. *Neurosci. Lett.* **129,** 299–302.

Lembeck, F. (1953) Zur frage der zentralen ubertragung afferenter inpulse III. Mitteilung. Das vorkommen und die bedeutung der substanz P in den dorsalen wurzeln des ruckenmarks. *Arch. Exp. Pth. Pharmak.* **219,** 197–213.

McLean, S., Ganong, A. H., Seeger, T. S., Bryce, K. G., Reynolds, L. S., Siok, C. J., Lowe, J. A., and Heym, J. (1991) Activity and distribution of binding sites in brain of a non-peptide substance P (NK_1) receptor antagonist. *Science* **251,** 437–439.

Mjellem-Joly, N., Lund, A., Berge, O. G., and Hole, K. (1992) Intrathecal co-administration of substance P and NMDA augments nociceptive responses in the formalin test. *Pain* **51,** 195–198.

Morris, R., Bleazard, L., and Hill, R. G. (1992a) The responses of neurons, in the deep dorsal horn of rat spinal cord slices *in vitro*, to the application of neurokinin agonists are correlated with their responses to peripheral nerve stimulation. *J. Physiol.* **452,** 252P.

Morris, R., Bleazard, L., and Hill, R. G. (1992b) Effects of the non-peptide NK_1 antagonist, racemic CP-96,345, on NK_1 agonist evoked responses of rat spinal dorsal horn neurons *in vitro*. *Neurosci. Lett.* **42,** S1–S64.

Moussaoui, S. M., Hermans, E., Mathieu, A. M., Bonicl, B., Clerc, F., Guinet, F., Garret, C., and Laduron, P. M. (1992) Polyclonal antibodies against the rat NK_1 receptor: Characterisation and localisation in the spinal cord. *NeuroReport* **3,** 1073–1076.

Moussaoui, S. M., Montier, F., Carruette, A., Blanchard, J. C., Laduron, P. M., and Garret, C. (1993a) A non-peptide NK_1-receptor antagonist, RP67580, inhibits neurogenic inflammation postsynaptically. *Br. J. Pharmacol.* **109,** 259–264.

Moussaoui, S. M., Philippe, L., Le Prado, J., and Garret, C. (1993b) Inhibition of neurogenic inflammation in the meninges by a non-peptide NK_1 receptor antagonist, RP 67580. *Eur. J. Pharmacol.* **238,** 421–424.

Munro, F. E., Fleetwood-Walker, S. M., Parker, R. M. C., and Mitchell, R. (1993) Mustard oil-evoked activation of rat dorsal horn neurons is inhibited by NK_2 but not NK_1 receptor antagonists. *Br. J. Pharmacol.* **108,** 243P.

Murase, K., Ryu, P. D., and Randic, M. (1986) Substance P augments a persistent slow inward calcium-sensitive current in voltage-clamped spinal dorsal horn neurons of the rat. *Brain Res.* **365,** 369–376.

Nagahisa, A., Yoshihito, K., Suga, O., Tanaguchi, K., Tsuchiya, M., Lowe, J. A., and Hess, H. J. (1992a) Antiinflammatory and analgesic activity of a non-peptide substance P receptor antagonist. *Eur. J. Pharmacol.* **217,** 191–195.

Nagahisa, A., Asai, R., Kanai, Y., Murase, A., Tsuchiya-Nakagaki, M., Nakagaki, T., Shieh, T. C., and Taniguchi, K. (1992b) Non-specific activity of (±)CP-96,345 in models of pain and inflammation. *Br. J. Pharmacol.* **107,** 273–275.

Nagy, I., Maggi, C. A., Dray, A., Woolf, C. J., and Urban, L. (1993) The role of neurokinin and N methyl-D-aspartate receptors in synaptic transmission from capsaicin-sensitive primary afferents in the rat spinal cord *in vitro. Neuroscience* **52,** 1029–1037.

Norris, S. K., Boden, P. R., and Woodruff, G. N. (1993) Agonists selective for tachykinin NK_1 and NK_3 receptors excite subpopulations of neurons in the rat medial habenula nucleus *in vitro. Eur. J. Pharmacol.* **234,** 223–228.

Nowak, L. M. and MacDonald, R. L. (1981) Substance P decreases a potassium conductance of spinal cord neurons in cell culture. *Brain Res.* **214,** 416–423.

O'Shaughnessy, C. T. and Connor, H. E. (1993) Neurokinin NK_1 receptors mediate plasma protein extravasation in guinea pig dura. *Eur. J. Pharmacol.* **236,** 319–321.

Otsuka, M., and Yoshioka, K. (1993) Neurotransmitter functions of mammalian tachykinins. *Physiol. Rev.* **73, 2,** 229–308.

Petitet, F., Beaujouan, J. C., Saffroy, M., Torrens, Y., and Glowinski, J. (1993) The non-peptide NK_2 antagonist SR-48,968 is also a NK_3 antagonist in the guinea pig but not in the rat. *Biochem. Biophys. Res. Commun.* **191,** 180–187.

Picard, P., Boucher, S., Regoli, D., Gitter, B. D., Howbert, J. J., and Couture, R. (1993) Use of non-peptide tachykinin receptor antagonists to substantiate the involvement of NK_1 and NK_2 receptors in a spinal nociceptive reflex in the rat. *Eur. J. Pharmacol.* **232,** 255–261.

Radhakrishnan, V. and Henry, J. L. (1991) Novel substance P antagonist, CP-96,345, blocks responses of cat spinal dorsal horn neurons to noxious cutaneous stimulation and to substance P. *Neurosci. Lett.* **132,** 39–43.

Radhakrishnan, V. and Henry, J. L. (1993) L-NAME blocks responses to NMDA, substance P and noxious cutaneous stimuli in cat dorsal horn. *NeuroReport* **4,** 323–326.

Randic, M., Hecimovic, H., and Ryu, P. D. (1990) Substance P modulates glutamate-induced currents in acutely isolated rat spinal dorsal horn neurons. *Neurosci. Lett.* **117,** 74–80.

Randic, M. and Miletic, V. (1977) Effect of substance P in cat dorsal horn neurons activated by noxious stimuli. *Brain Res.* **128,** 164–169.

Randic, M., Ryu, P. D., and Urban, L. (1986) Effects of polyclonal and monoclonal antibodies to substance P on slow excitatory transmission in rat spinal dorsal horn. *Brain Res.* **383,** 15–27.

Rodriguez, R. E., Salt, T. E., Cahusac, P. M. B., and Hill, R. G. (1983) The behavioural effects of intrathecally administered [D-Pro^2,D-Trp^{7,9}]-substance P, an analogue with presumed antagonist actions, in the rat. *Neuropharmacology* **22,** 173–176.

Rupniak, N. M. J., Boyce, S., Williams, A. R., Cook, G., Longmore, J., Seabrook, G. R., Caeser, M., Iversen, S. D., and Hill, R. G. (1993) Antinociceptive actions of NK_1 receptor antagonists: Non-specific effects of racemic RP-67,580. *Br. J. Pharmacol.* **110,** 1607–1613.

Rusin, K. I., Bleakman, D., Chard, P. S., Randic, M., and Miller, R. J. (1993) Tachykinins potentiate *N*-methyl-D-aspartate responses in acutely isolated neurons from the dorsal horn. *J. Neurochem.* **60, 3,** 952–960.

Rusin, K. I., Ryu, P. D., and Randic, M. (1992) Modulation of excitatory amino acid responses in rat dorsal horn neurons by tachykinins. *J. Neurophys.* **68,** 265–286.

Sachais, B. S., Snider, R. M., Lowe, J. A., and Krause, J. E. (1993) Molecular basis for the species selectivity of the substance P antagonist CP-96,345. *J. Biol. Chem.* **268,** 2319–2323.

Sakurada, T., Katsumata, K., Yogo, H., Tan-No, K., Sakurada, S., and Kisara, K. (1993) Antinociception induced by CP-96,345, a non-peptide NK_1 receptor antagonist, in the mouse fomalin and capsaicin tests. *Neurosci. Lett.* **151,** 142–145.

Salt, T. E., De Vries, G. J., Rodriguez, R. E., Cahusac, P. M. B., Morris, R., and Hill, R. G. (1982) Evaluation of [D-Pro^2, D-Trp^{7,9}]-substance P as an antagonist of substance P responses in the rat central nervous system. *Neurosci. Lett.* **30,** 291–295.

Salt, T. E. and Hill, R. G. (1983) Neurotransmitter candidates of somatosensory primary afferent fibers. *Neuroscience* **10,** 1083–1103.

Salt, T. E., Morris, R., and Hill, R. G. (1983) Distribution of substance P-responsive and nociceptive neurons in relation to substance P-immunoreactivity within the caudal trigeminal nucleus of the rat. *Brain Res.* **273,** 217–228.

Salter, M. W. and Henry, J. L. (1991) Responses of functionally identified neurons in the dorsal horn of the cat spinal cord to substance P, neurokinin A and physalaemin. *Neuroscience* **43,** 601–610.

Santucci, V., Gueudet, C., Emonds-Alt, X., Breliere, J. C., Soubrie, P., and Le Fur, G. (1993) The NK_2 receptor antagonist SR-48,968 inhibits thalamic responses evoked by thermal but not mechanical nociception. *Eur. J. Pharmacol.* **237,** 143–146.

Saria, A. (1984) Substance P in sensory nerve fibers contributes to the development of oedema in the rat hind paw after thermal injury. *Br. J. Pharmacol.* **82,** 217–222.
Schäfer, M. K.-H., Nohr, D., Krause, J. E., and Weihe, E. (1993) Upregulation of neurokinin-1 receptor gene expression in dorsal horn neurons evoked by unilateral peripheral inflammation in the rat. *Br. J. Pharmacol.* **109,** 22P.
Schmidt, A. W., McLean, S., and Heym, J. (1992) The substance P receptor antagonist CP-96,345 interacts with Ca^{2+} channels. *Eur. J. Pharmacol.* **219,** 491,492.
Shepheard, S. L., Williamson, D. J., Hill, R. G., and Hargreaves, R. J. (1993) The non-peptide neurokinin-1 receptor antagonist, RP67580, blocks neurogenic plasma extravasation in the dura mater of rats. *Br. J. Pharmacol.* **108,** 11,12.
Shults, C. W., Quirion, R., Chronwall, B., Chase, T. N., and O'Donohue, T. L. (1984) A comparison of the anatomical distribution of substance P and substance P receptors in the rat central nervous system. *Peptides* **5,** 1097–1128.
Sivam, S. P. and Krause, J. E. (1992) Tachykinin systems in the spinal cord and basal ganglia: Influence of neonatal capaicin treatment or dopaminergic intervention on levels of peptides, substance P-encoding mRNAs, and substance P receptor mRNA. *J. Neurochem.* **59,** 2278–2284.
Thompson, S. W. N., Dray, A., and Urban, L. (1993a) The contribution of tachykinin receptor activation to C fiber evoked responses in the neonatal rat spinal cord *in vitro*. *J. Physiol.* **459,** 464P.
Thompson, S. W. N., Urban, L., and Dray, A. (1993b) NMDA and tachykinin receptor-mediated contributions to the C-fiber-evoked response in the neonatal rat spinal cord *in vitro* are enhanced following peripheral inflammation. *Br. J. Pharmacol.* **108,** 22P.
Tousignant, C., Chretien, L., Guillemette, G., and Regoli, D. (1992) Characterization of NK receptors in guinea pig and rat brain membranes with NK_1 peptides and a non-peptide antagonist. *Brain Res.* **596,** 243–250.
Urban, L., Willetts, J., Randic, M., and Papka, R. E. (1985) The acute and chronic effects of capsaicin on slow excitatory transmission in rat dorsal horn. *Brain Res.* **330,** 390–396.
Vaught, J. L. (1988) Substance P antagonists and analgesia: A review of the hypothesis. *Life Sci.* **43,** 1419–1431.
Weihe, E. (1990) Neuropeptides in primary afferent neurons, in *The Primary Afferent Neuron* (Zenker, W. and Neuhuber, W., eds.), Plenum, New York, pp. 127–159.
Wiesenfeld-Hallin, Z. and Duranti, R. (1987) D-Arg^1, D-$Trp^{7,9}$, Leu^{11}-substance P (spantide) does not antagonize substance P-induced hyperrecitability of the nociceptive flexion withdrawal reflex in the rat. *Acta Physiol. Scand.* **129,** 55–59.
Wiesenfeld-Hallin, Z., Xu, X. J., Kristensson, K., Hakanson, R., Feng, D. M., and Folkers, K. (1990a) Antinociceptive and substance P antagonistic effects of intrathecally injected spantide II in rat: no signs of motor impairment or neurotoxicity. *Regul. Pept.* **29,** 1–11.
Wiesenfeld-Hallin, Z., Xu, X. J., Hakanson, R., Feng, D. M., and Folkers, K. (1990b) The specific antagonistic effect of intrathecal spantide II on substance P and C-fiber conditioning stimulation-induced facilitation of the nociceptive flexor reflex in rat. *Brain Res.* **526,** 284–290.

Wiesenfeld-Hallin, Z., Xu, X. J., Hahnson, R., Feng, D. M., and Folkers, K. (1991) Low dose intrathecal morphine facilitates the spinal flexor reflex by releasing neuropeptides in rats with intact and sectioned peripheral nerves. *Brain Res.* **551,** 157–162.

Woolf, C. J. (1983) Evidence for a central component of postinjury pain hypersensitivity. *Nature* **306,** 686–688.

Wright, D. M. and Roberts, M. H. T. (1980) Responses of spinal neurons to a substance P analogue, noxious pinch and bradykinin. *Eur. J. Pharmacol.* **64,** 165–167.

Xu, X. J. and Wiesenfeld-Hallin, Z. (1992) Intrathecal neurokinin A facilitates the spinal nociceptive flexor reflex evoked by thermal and mechanic stimuli and synergistically interacts with substance P. *Acta Physiol. Scand.* **144,** 163–168.

Xu, X. J., Maggi, C. A., and Wiesenfeld-Hallin, Z. (1991) On the role of NK_2 tachykinin receptors in the mediation of spinal reflex excitability in the rat. *Neuroscience* **44, 2,** 483–490.

Xu, X. J., Dalsgaard, C. J., and Wiesenfeld-Hallin, Z. (1992a) Spinal substance P and *N*-methyl-D-aspartate receptors are coactivated in the induction of central sensitization of the nociceptive flexor reflex. *Neuroscience* **51,** 641–648.

Xu, X. J., Dalsgaard, C. J., and Wiesenfeld-Hallin, Z. (1992b) Intrathecal CP-96,345 blocks reflex facilitation induced in rats by substance P and C-fiber-conditioning stimulation. *Eur. J. Pharmacol.* **216,** 337–344.

Yamamoto, T. and Yaksh, T. L. (1991) Stereospecific effects of a nonpeptidic NK_1 selective antagonist, CP-96,345: antinociception in the absence of motor dysfunction. *Life Sci.* **49,** 1955–1963.

Yanagisawa, M., Otsuka, M., Konishi, S., Akagi, H., Folkers, K., and Rosell, S. (1982) A substance P antagonist inhibits a slow reflex response in the spinal cord of the newborn rat. *Acta Physiol. Scand.* **116,** 109–112.

Yashpal, K., Dam, T. V., and Quirion, R. (1990) Quantitative autoradiographic distribution of multiple neurokinin binding sites in rat spinal cord. *Brain Res.* **506,** 259–266.

Yashpal, K., Radhakrishnan, V., Coderres, T. J., and Henry, J. L. (1993) CP-96,345, but not its stereoisomer, CP-96,344, blocks the nociceptive responses to intrathecally administered substance P and to noxious thermal and chemical stimuli in the rat. *Neuroscience* **52,** 1039–1047.

Zieglgansberger, W. and Tulloch, I. F. (1979) Effects of substance P on neurons in the dorsal horn of the spinal cord of the cat. *Brain Res.* **166,** 273–282.

CHAPTER 16

Pharmacological Characterization of Receptors in the Spinal Cord of the Newborn Rat

Masanori Otsuka, Hidenori Suzuki, Mitsuhiko Yanagisawa, Rumiko Hosoki, Jian-Zhong Guo, and Koichi Yoshioka

1. Introduction

The mammalian central nervous system (CNS) contains a considerable amount of neurokinin A (NKA) (Kanazawa et al., 1984; Arai and Emson, 1986; Takano et al., 1986; Tateishi et al., 1989). In contrast, blot-hybridization and RNase-protection analyses detected substantial amounts of mRNAs for NK_1 and NK_3 receptors, whereas mRNA for NK_2 receptors was undetectable in the rat CNS (Tsuchida et al., 1990). This constitutes a remarkable transmitter-receptor mismatch, since NKA is generally assumed to be an endogenous ligand for the NK_2 receptor. Does NKA act on NK_1 or NK_3 receptors? In several places in the CNS, a definite excitatory action of NKA was observed (Matsuto et al., 1984; Innis et al., 1985; Kalivas et al., 1985; Xu et al., 1991). Furthermore, autoradiographic studies detected

The Tachykinin Receptors Ed.: S. H. Buck © 1994 Humana Press Inc., Totowa, NJ

NKA binding sites in some regions in the CNS of the rat, e.g., cerebral cortex, solitary nucleus, and substantia gelatinosa of the spinal cord (Mantyh et al., 1984b, 1989; Quirion and Dam, 1985). These facts suggest the existence of tachykinin receptors responding to NKA.

Tachykinin receptor subtypes have been characterized in peripheral tissues based on potencies of various agonists and antagonists, as well as on analyses of dose-response curves to agonists. In the CNS preparations in vitro and in vivo, however, with electrophoretic or intracerebroventricular application of drugs, it is difficult to obtain concentration-response curves to various agonists in the absence or presence of antagonists, to analyze the antagonism on neurons with controlled concentrations of drugs, and to characterize the receptor subtypes. In this respect, the isolated spinal cord of the newborn rat provides an excellent opportunity for pharmacological characterization of tachykinin receptor types in the CNS because it is possible to record dose-response curves with controlled concentrations of agonists and antagonists on motoneurons after suppressing synaptic transmission with tetrodotoxin (TTX). We have previously shown in the isolated spinal cord of the newborn rat that NKA and a selective NK_1 agonist, acetyl-Arg^6-septide *(see below)*, induce depolarizing responses of motoneurons in the presence of TTX, and that the actions of both NKA and acetyl-Arg^6-septide are antagonized by a presumptive NK_1 antagonist, spantide, with almost the same pA_2 values. In contrast, the depolarizing action of substance P (SP) in the presence of TTX was much less sensitive to antagonistic action of spantide (Yanagisawa and Otsuka, 1990). These results suggest the existence on neonatal rat spinal motoneurons of an atypical type of tachykinin receptor, which is different from the classical NK_1, NK_2, or NK_3 receptor types. In this chapter, we present our results on the effects of tachykinins, related agonists, and antagonists on neurons in the isolated spinal cord of the newborn rat, and based on these results in comparison with the results obtained in peripheral tissues, we discuss tachykinin receptor types in the neonatal rat spinal cord and more generally in the CNS.

2. Methods

2.1. Electrophysiological Experiments

We used the isolated spinal cord preparation of newborn rats. The methods are the same as described previously (Akagi et al., 1985; Otsuka and Yanagisawa, 1988). The spinal cord below the thoracic region was isolated from 1- to 5-d-old Wistar rats, hemisected, placed in a bath of 0.2 mL volume, and perfused with artificial cerebrospinal fluid (CSF) (for composition, *see* Yanagisawa and Otsuka, 1990). The temperature was kept at 27°C. The medium was saturated with a gas mixture of 95% O_2 and 5% CO_2 before perfusion. Potentials were recorded extracellularly from L3-5 ventral roots with a tightly fitting suction electrode on a dc pen recorder. Agonists, antagonists, and other drugs were dissolved in artificial CSF and applied by perfusion.

2.2. Experiments on Tachykinin-Evoked Acetylcholine Release

Two hemisected spinal cords were fixed in a perfusion chamber of 1-mL volume and were continuously perfused with artificial CSF containing eserine at 5 μM (for composition, *see* Kobayashi et al., 1991) at a rate of 0.7 mL/min. The medium was thoroughly oxygenated by bubbling with a mixture of 95% O_2 and 5% CO_2 in the reservoir and the perfusion chamber. The temperature of the perfusion chamber was kept at 27°C. Three-minute fractions were collected through a glass wool filter into test tubes. The collected samples were frozen, lyophilized, dissolved in phosphate buffer, and centrifuged. Twenty microliters of each supernatant was submitted to acetylcholine (ACh) analysis using HPLC equipped with a chromatographic column, an immobilized enzyme column, and an electrochemical detection system (Fujimori and Yamamoto, 1987).

2.3. Experiments with Peripheral Tissues

Guinea pig ileum, rat vas deferens, and rat portal vein were used as model tissues for NK_1, NK_2, and NK_3 receptors, respectively. They were suspended in a glass organ-bath containing Tyrode solution (guinea pig ileum and rat vas deferens) or Krebs-Henseleit solution (rat portal vein) at 37°C, and their mechanical responses were recorded isotonically on a kimograph.

2.4. Drugs

Acetyl-[Arg^6,Pro^9]SP_{6-11}, which is a water soluble derivative of septide and hereafter referred to as acetyl-Arg^6-septide (Papir-Kricheli et al., 1987), was a gift of Z. Selinger, Department of Biological Chemistry, the Hebrew University of Jerusalem, Israel; GR71251 was a gift of R. M. Hagan, Department of Neuropharmacology, Glaxo Group Research Ltd., UK; spantide ([D-Arg^1,D-$Trp^{7,9}$,Leu^{11}]SP) was kindly synthesized and supplied by M. Fujino, Takeda Chemical Industries Ltd., Japan; SP and NKA were purchased from Peptide Institute Inc., Osaka, Japan.

3. Results

In the isolated spinal cord preparation of the newborn rat, NKA induced depolarizing responses of the ventral root in a dose-dependent manner in normal artificial CSF. The value of pD_2 (negative logarithm of EC_{50}) was 7.24 ± 0.12. After addition of TTX (0.3 μ*M*), the concentration-response curve was shifted to the right, with pD_2 value of 6.33 ± 0.14 (Table 1). Figure 1A shows the results of analysis of antagonism between NKA and spantide on neonatal rat spinal motoneurons in the presence of TTX. The concentration-response curve was shifted to the right by 1 and 2 μ*M* spantide. From the Schild plot shown in Fig. 1Ab, pA_2 of spantide against NKA was estimated at 6.49 ± 0.11 (Table 2). The estimated slope of Schild plot was 1.20 ± 0.10, suggesting competitive antagonism by spantide.

Figure 1B shows the results of analysis of antagonism between acetyl-Arg^6-septide and spantide on motoneurons in the presence of TTX. The results suggest competitive antagonism between acetyl-Arg^6-septide and spantide, with an estimated pA_2 value of 6.51 ± 0.04 and the slope of Schild plot of 1.23 ± 0.14. That the pA_2 value for the antagonism between NKA and spantide was almost the same as that between acetyl-Arg^6-septide and spantide (Table 2), and that the

Table 1
pD_2 Values of SP, NKA, and Acetyl-Arg^6-Septide in Various Tissues[a]

Tissue	SP	NKA	Acetyl-Arg^6-septide
Neonatal rat spinal cord			
in the absence of TTX	7.57 ± 0.08 (4)	7.24 ± 0.12 (3)	7.48 ± 0.23 (2)
in the presence of TTX (0.3 μ*M*)	6.22 ± 0.08 (4)	6.33 ± 0.14 (3)	6.72 ± 0.17 (4)
Guinea pig ileum	8.69 ± 0.07 (13)	7.72 ± 0.12 (4)	9.09 ± 0.45 (5)
Rat vas deferens	4.92 ± 0.02 (3)	6.96 ± 0.06 (9)	5.32 (1)
Rat portal vein	5.66 ± 0.12 (5)	6.95 ± 0.23 (3)	5.77 (1)

[a]Each value is mean ± SEM (number of preparations).

slope of Schild plot was close to unity in both cases, suggest that NKA, acetyl-Arg^6-septide, and spantide bind to the same and single population of tachykinin receptors.

Similar results to those with spantide were obtained with another tachykinin antagonist, GR71251, which is known as an NK_1 antagonist (Ward et al., 1990). The pA_2 value and the slope of Schild plot for the antagonism between NKA and GR71251 in the presence of TTX were 6.71 ± 0.10 and 1.01 ± 0.01, respectively, and those for the antagonism between acetyl-Arg^6-septide and GR71251 were 6.66 ± 0.05 and 1.01 ± 0.01 (Table 2). These results again support that NKA, acetyl-Arg^6-septide, and GR71251 bind to the same subpopulation of tachykinin receptors.

GR71251 also exhibited an antagonistic action against NKA and acetyl-Arg^6-septide in the absence of TTX. Under such conditions, the depolarizing responses to tachykinin agonists are much larger than those in the presence of TTX and represent summation of their direct action on motoneurons and transynaptic action via interneurons (Yanagisawa and Otsuka, 1990). Therefore, interpretation of the concentration-response curves is complicated. However, the apparent pA_2 values for the antagonism between GR71251 and NKA or acetyl-Arg^6-septide obtained in the absence of TTX were very similar to those obtained in the presence of TTX (Table 2; Guo et al., 1993). These results suggest that the subpopulation of tachykinin

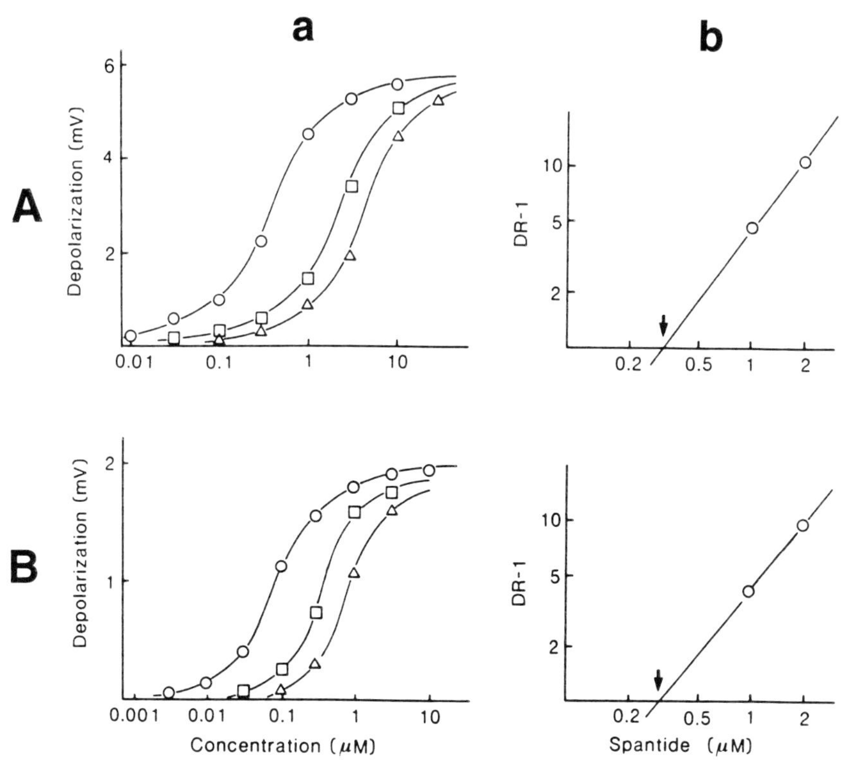

Fig. 1. Effects of spantide on the responses of motoneurons to NKA **(A)** and acetyl-Arg[6]-septide **(B)**. Potentials were recorded extracellularly from L5 ventral roots of isolated hemisected spinal cords of 1-d-old rats. NKA or acetyl-Arg[6]-septide was applied by perfusion for 30 s at 10-min intervals. Spantide was applied by continuous perfusion during the periods of application of agonists and intervals. **(a)** Concentration-response curves to NKA and acetyl-Arg[6]-septide. Ordinate scale, peak amplitude of depolarization induced by the agonists. Abscissa scale, logarithmic concentration of the agonists. (s) In artificial CSF containing TTX (0.3 μ*M*); (h) and (n) after addition of spantide at 1 and 2 μ*M*, respectively, in the presence of TTX (0.3 μ*M*). **(b)** Schild plots obtained from the results in (a). Ordinate, dose ratio–1 in logarithmic scale. Abscissa, concentration of spantide in logarithmic scale. Arrows indicate K_B values. (Modified from Figs. 1 and 2 in Yanagisawa and Otsuka, 1990).

Table 2
pA_2 Values of Spantide and GR71251 Estimated Against SP, NKA, and Acetyl-Arg^6-Septide in Various Tissues[a]

Antagonist	Tissue	Agonist		
		SP	NKA	Acetyl-Arg^6-septide
Spantide	Neonatal rat spinal cord			
	in the absence of TTX	5.40 (1)		
	in the presence of TTX (0.3 μ*M*)		6.49 ± 0.11 (3)	6.51 ± 0.04 (3)
	Guinea pig ileum	6.28 ± 0.05 (3)	6.72 (1)	7.16 (1)
	Rat vas deferens		5.55 ± 0.03 (4)	
	Rat portal vein	5.33 ± 0.75 (2)		
GR71251	Neonatal rat spinal cord			
	in the absence of TTX	6.05 ± 0.11 (5)	6.73 ± 0.01 (4)	6.73 ± 0.07 (4)
	in the presence of TTX (0.3 μ*M*)		6.71 ± 0.10 (4)	6.66 ± 0.05 (4)
	Guinea pig ileum	7.50 ± 0.15 (3)	7.83 ± 0.01 (2)	7.33 ± 0.10 (2)
	Rat vas deferens		5.43 (1)	
	Rat portal vein	5.05 (1)	5.08 (1)	

[a]Each value is mean ± SEM (number of preparations).

receptors to which NKA, acetyl-Arg^6-septide, spantide and GR71251 bind with high affinities is distributed not only on motoneurons, but also on spinal interneurons, presumably in the dorsal horn. This subtype of tachykinin receptors will be provisionally referred to as "NKA-preferring NK_1 receptors" *(see below)* because spantide (Buck and Shatzer, 1988; Barthó et al., 1989), GR71251 (Ward et al., 1990) and acetyl-Arg^6-septide (Papir-Kricheli et al., 1987) are thought to be the ligands with preference to NK_1 receptors.

That NKA binds to a subtype of NK_1 receptors and that its action is antagonized by NK_1 antagonists might at first appear odd, but may be understandable in view of the fact that NKA and acetyl-Arg^6-septide have potent contracting actions on the guinea pig ileum, an NK_1 preparation, and their actions are antagonized by spantide and GR71251 (Tables 1 and 2). However, the results with SP suggest that the NKA-preferring NK_1 receptors in the neonatal rat spinal cord are considerably different from classical NK_1 receptors in peripheral tissues or those expressed in oocytes injected with mRNAs for NK_1 receptors (Ingi et al., 1991). Namely, as shown in Fig. 2, the depolarizing action of SP on neonatal rat motoneurons in the presence of TTX was depressed by spantide at a low concentration range of SP, but was potentiated by spantide at a high concentration range of SP (Fig. 2). Similar results were obtained with GR71251. The mechanism of the potentiating effect of the tachykinin antagonists on the depolarizing action of SP is unknown and will not be further discussed here. However, if we assume that the difference between the concentration-response curves of SP in the absence and presence of spantide partly represents the action of SP on NKA-preferring NK_1 receptors in the neonatal rat spinal cord, then the results shown in Fig. 2 suggest that SP exerts only a weak depolarizing action via NKA-preferring NK_1 receptors on motoneurons.

The results presented above suggest that a subpopulation of NK_1 receptors, i.e., NKA-preferring NK_1 receptors, on which NKA has a potent action, whereas SP has only a weak action, is involved in the tachykinin-induced depolarization of motoneurons. Similar results were obtained with respect to the tachykinin-induced ACh release from the isolated spinal cord of the newborn rat (Otsuka et al., 1991).

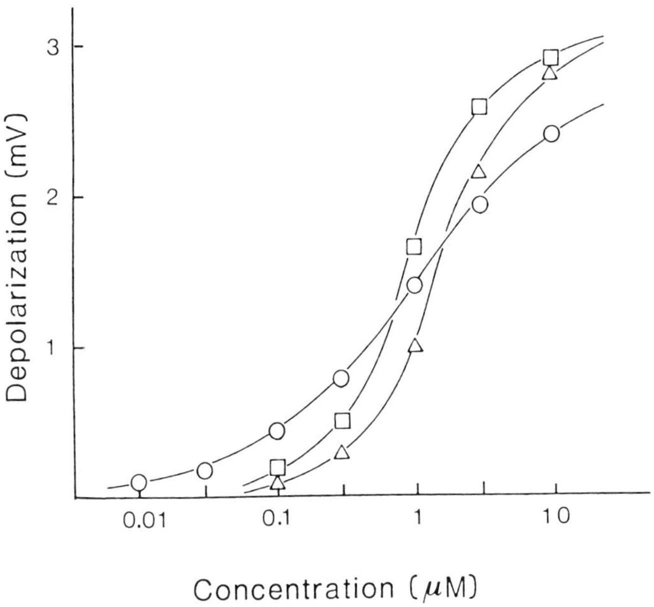

Fig. 2. Effects of spantide on the concentration-response curve to SP. SP was applied by perfusion for 30 s at 10-min intervals. Potentials were recorded extracellularly from the L4 ventral root of a hemisected spinal cord of a 1-d-old rat. Other details are as in Fig. 1. (○) In artificial CSF containing TTX (0.3 μ*M*); (□) and (△) after addition of spantide at 4 and 8 μ*M*, respectively, in the presence of TTX (0.3 μ*M*) (Modified from Fig. 2 in Yanagisawa and Otsuka, 1990).

Namely, as shown in Fig. 3, both SP and NKA evoked an increase in the release of ACh from neonatal rat spinal cords. The action of NKA was antagonized by spantide, whereas that of SP was not. Acetyl-Arg^6-septide at 0.3 μ*M* also evoked an increase in the ACh release, and its action was antagonized by spantide at 10 μ*M*. Likewise, GR71251 antagonized the actions of NKA and acetyl-Arg^6-septide, but not that of SP, to evoke an increase of ACh release (Suzuki et al., 1993). These results suggest that NKA-preferring NK_1 receptors are also involved in the tachykinin-evoked ACh release from the neonatal rat spinal cord.

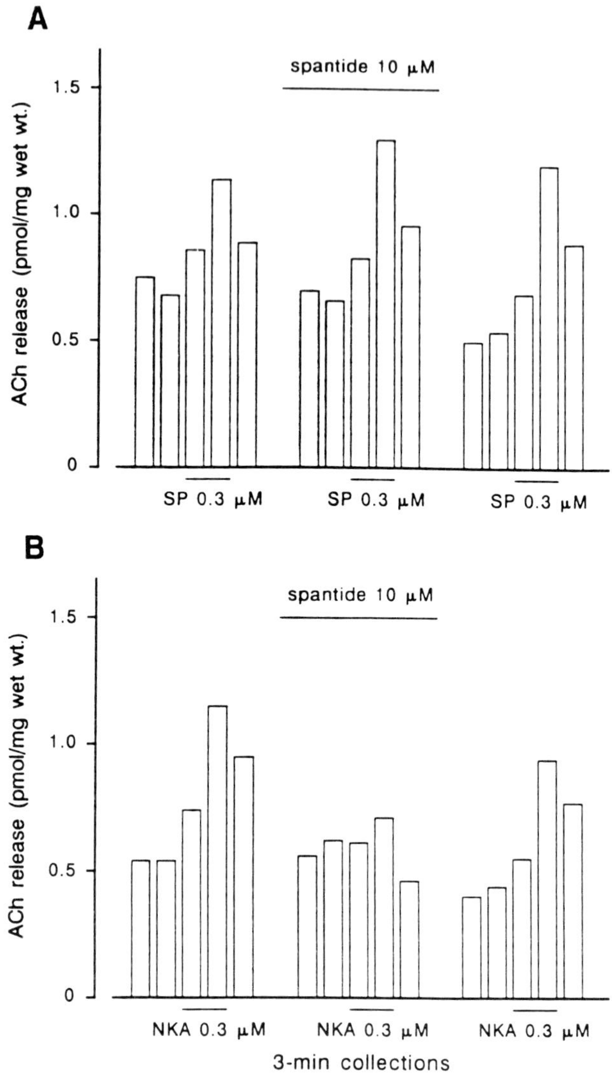

Fig. 3. Effects of spantide on the ACh release evoked by SP (**A**) and NKA (**B**). In each experiment, two hemisected spinal cords were perfused with normal artificial CSF containing 5 μ*M* eserine for more than 1 h, and then three series of collections were carried out with intervals of 40–60 min. In each series, five successive 3-min samples were collected. The first two 3-min fractions represented precontrols. Then, the perfusion medium was switched to a solution containing 0.3 μ*M* SP (A) or NKA (B) for 6 min during the period indicated by the horizontal bar under the columns. The ACh release was first

4. Discussion

Reports from other laboratories also suggest the existence of nonclassical NK_1 receptors. Chassaing et al. (1992) and Petitet et al. (1992) suggested the presence of two subtypes of NK_1 receptors in the guinea pig ileum, one of which is labeled by ^{125}I-Bolton-Hunter (BH)-SP, but has low affinities to septide and GR71251, and another which has high affinities for septide and GR71251. Itoi et al. (1992) showed that cardiovascular and behavioral actions of NKA, but not of SP, were antagonized by an NK_1 antagonist, L-668,169. Kunitomo et al. (1992) also showed that spantide attenuated the contractile response to NKA more effectively than that to SP in the rabbit iris sphincter muscle and suggested the presence of NK_1 receptors that are activated by NKA.

The existence in the CNS of NKA-preferring NK_1 receptors, which may not be labeled by ^{125}I-BH-SP (Chassaing et al., 1992), might provide a partial explanation for the apparent mismatch between tachykinin contents and tachykinin binding sites in some regions. For example, in the substantia nigra and ventral tegmental area, where the contents of SP and NKA are high (Lindefors et al., 1986; Takano et al., 1986; Nagashima et al., 1987), autoradiographic studies revealed low densities of NKA binding sites and almost no SP binding sites (Quirion et al., 1983; Mantyh et al., 1984a, 1989; Rothman et al., 1984a,b). Physiological studies, however, showed that both SP and NKA have potent excitatory actions and that NKA is more potent than SP on some neurons (Kalivas et al., 1985; Tan and Tsou, 1988).

---

observed in normal artificial CSF and then, during the second series of collections, the spinal cords were perfused with the solution containing spantide (10 μM) as shown by the horizontal bars above the columns. After finishing the second series of collections, the perfusion medium was returned to normal artificial CSF. After 40–60 min of perfusion, the third series of collections was started. Each column represents the amount of ACh in pmol/mg wet weight in a 3-min fraction. Reproduced with permission from Fig. 3 in Otsuka et al., 1991.

If NKA acts on NKA-preferring NK_1 receptors on which SP has only a weak action, then the next question is on which subtype(s) of receptors does SP act in the neonatal rat spinal cord. Distributions of tachykinin binding sites in the rat spinal cord were studied by autoradiography (Ninkovic et al., 1984; Charlton and Helke, 1985a, 1985b, 1986; Buck et al., 1986; Danks et al., 1986; Helke et al., 1986; Hunter et al., 1987; Mantyh et al., 1989). NK_1 binding sites, labeled by ^{125}I-BH-SP, are distributed throughout the spinal cord of the neonatal rat, but become organized into specific nuclei in adult rat spinal cord (Charlton and Helke, 1986). NK_3 binding sites, labeled with ^{125}I-BH-eledoisin, are likewise distributed throughout the dorsal and ventral horn of neonatal rat spinal cord, but are confined to superficial layers of the dorsal horn of adult rat spinal cord (Beresford et al., 1992). Yashpal et al. (1990) recently described the presence of (2-^{125}I-iodohistidyl^1)-NKA binding sites in adult rat spinal cord, suggesting that NK_2 receptors may also be present in this tissue.

Recently, Suzuki et al. (1993) examined the nucleotide sequences of tachykinin receptor cDNAs obtained from the neonatal rat spinal cord by using reverse transcription and polymerase chain reaction amplification methods. They found that nucleotide sequences of most of amplified DNA fragments were identical with the sequence of the NK_1 or NK_3 receptor already cloned. Only a very small number, if any, of amplified DNA fragments displayed the nucleotide sequence of the NK_2 receptor. This is consistent with the observations in adult rat spinal cord (Tsuchida et al., 1990). SP, therefore, may act on classical NK_1 receptors. However, the receptors on which SP acts to induce depolarization of neonatal rat spinal motoneurons are distinct from the classical NK_1 receptors of guinea pig ileum, because the action of SP was depressed by the NK_1 antagonists, spantide and GR71251, with lower pA_2 values (Table 2). Another possibility is that SP acts on NK_3 receptors. In this respect, it is interesting to note that the depolarizing action of SP on motoneurons observed in the absence of TTX, which represents mostly transynaptic action via interneurons (Yanagisawa and Otsuka, 1990), is antagonized by spantide and GR71251 with relatively low pA_2 values, and that in rat portal vein, an NK_3 model tissue, the action of

SP is also antagonized by spantide and GR71251 with low pA_2 values (Table 2). A selective NK_3 agonist, senktide, however, evoked a depolarization of motoneurons of neonatal rat spinal cord in the absence of TTX, but not in its presence, suggesting that NK_3 receptors are present on interneurons, but not on motoneurons (Yanagisawa and Otsuka, 1990; Ireland et al., 1992).

In the mammalian CNS, the amount of NK_3 tachykinin receptor mRNA is comparable to that of NK_1 mRNA (Tsuchida et al., 1990), whereas the amount of neurokinin B (NKB) is much smaller than that of SP or NKA (Kanazawa et al., 1984; Arai and Emson, 1986; Nagashima et al., 1989; Tateishi et al., 1989), raising the possibility that NKB may not be the main endogenous ligand for NK_3 receptors in the CNS. Characterization of tachykinin receptors in the CNS using physiological means will be important for the future development of tachykinin-related drugs and for understanding the mechanism of action of drugs, e.g., nonpeptide tachykinin antagonists.

References

Akagi, H., Konishi, S., Otsuka, M., and Yanagisawa, M. (1985) The role of substance P as a neurotransmitter in the reflexes of slow time courses in the neonatal rat spinal cord. *Br. J. Pharmacol.* **84,** 663–673.

Arai, H. and Emson, P.C. (1986) Regional distribution of neuropeptide K and other tachykinins (neurokinin A, neurokinin B and substance P) in rat central nervous system. *Brain Res.* **399,** 240–249.

Barthó, L., Holzer, P., Leander, S., and Lembeck, F. (1989) Evidence for an involvement of substance P, but not cholecystokinin-like peptides, in hexamethonium-resistant intestinal peristalsis. *Neuroscience* **28,** 211–217.

Beresford, I. J. M., Ireland, S. J., Stables, J., and Hagan, R. M. (1992) Ontogeny and characterization of [^{125}I]Bolton Hunter-eledoisin binding sites in rat spinal cord by quantitative autoradiography. *Neuroscience* **46,** 225–232.

Buck, S. H. and Shatzer, S. A. (1988) Agonist and antagonist binding to tachykinin peptide NK-2 receptors. *Life Sci.* **42,** 2701–2708.

Buck, S. H., Helke, C. J., Burcher, E., Shults, C. W., and O'Donohue, T. L. (1986) Pharmacologic characterization and autoradiographic distribution of binding sites for iodinated tachykinins in the rat central nervous system. *Peptides* **7,** 1109–1120.

Charlton, C. G. and Helke, C. J. (1985a) Characterization and segmental distribution of ^{125}I-Bolton-Hunter-labeled substance P binding sites in rat spinal cord. *J. Neurosci.* **5,** 1293–1299.

Charlton, C. G. and Helke, C. J. (1985b) Autoradiographic localization and characterization of spinal cord substance P binding sites: high densities in sensory, autonomic, phrenic, and Onuf's motor nuclei. *J. Neurosci.* **5,** 1653–1661.

Charlton, C. G. and Helke, C. J. (1986) Ontogeny of substance P receptors in rat spinal cord: quantitative changes in receptor number and differential expression in specific loci. *Dev. Brain Res.* **29,** 81–91.

Chassaing, G., Lavielle, S., Brunissen, A., Carruette, A., Garret, C., Petitet, F., Saffroy, M., Beaujouan, J. C., Torrens, Y., and Glowinski, J. (1992) [Pro^9]SP and [$pGlu^6$,Pro^9]SP(6–11) interact with two different receptors in the guinea-pig ileum as demonstrated with new SP antagonists. *Neuropeptides* **23,** 73–79.

Danks, J. A., Rothman, R. B., Cascieri, M. A., Chicchi, G. G., Liang, T., and Herkenham, M. (1986). A comparative autoradiographic study of the distributions of substance P and eledoisin binding sites in rat brain. *Brain Res.* **385,** 273–281.

Fujimori, K. and Yamamoto, K. (1987) Determination of acetylcholine and choline in perchlorate extracts of brain tissue using liquid chromatography-electrochemistry with an immobilized-enzyme reactor. *J. Chromatogr.* **414,** 167–173.

Guo, J.-Z., Yoshioka, K., Yanagisawa, M., Hosoki, R., Hagan, R. M., and Otsuka, M. (1993) Depression of primary afferent-evoked responses by GR71251 in the isolated spinal cord of the neonatal rat. *Br. J. Pharmacol.* **110,** 1142–1148.

Helke, C. J., Charlton, C. G., and Wiley, R. G. (1986) Studies on the cellular localization of spinal cord substance P receptors. *Neuroscience* **19,** 523–533.

Hunter, J. C., Kilpatrick, G. J., and Brown, J. R. (1987) Pharmacological analysis of ^{125}I-Bolton and Hunter labelled eledoisin binding sites in rat spinal cord by quantitative autoradiography. *Neurosci. Lett.* **78,** 12-16.

Ingi, T., Kitajima, Y., Minamitake, Y., and Nakanishi, S. (1991) Characterization of ligand-binding properties and selectivities of three rat tachykinin receptors by transfection and functional expression of their cloned cDNAs in mammalian cells. *J. Pharmacol. Exp. Ther.* **259,** 968-975.

Innis, R. B., Andrade, R., and Aghajanian, G. K. (1985) Substance K excites dopaminergic and non-dopaminergic neurons in rat substantia nigra. *Brain Res.* **335,** 381–383.

Ireland, S. J., Wright, I. K., and Jordan, C. C. (1992) Characterization of tachykinin-induced ventral root depolarization in the neonatal rat isolated spinal cord. *Neuroscience* **46,** 217–223.

Itoi, K., Tschöpe, C., Jost, N., Culman, J., Lebrun, C., Stauss, B., and Unger, T. (1992) Identification of the central tachykinin receptor subclass involved in substance P-induced cardiovascular and behavioral responses in conscious rats. *Eur. J. Pharmacol.* **219,** 435–444.

Kalivas, P. W., Deutch, A. Y., Maggio, J. E., Mantyh, P. W., and Roth, R. H. (1985) Substance K and substance P in the ventral tegmental area. *Neurosci. Lett.* **57,** 241–246.

Kanazawa, I., Ogawa, T., Kimura, S., and Munekata, E. (1984) Regional distribution of substance P, neurokinin α and neurokinin β in rat central nervous system. *Neurosci. Res.* **2,** 111–120.

Kobayashi, N., Sakuma, M., Yoshioka, K., Onishi, Y., Yanagisawa, M., Kawashima, K., and Otsuka, M. (1991) Substance P-evoked release of acetylcholine from isolated spinal cord of the newborn rat. *Neuroscience* **45,** 331–337.

Kunitomo, M., Imaizumi, N., and Fujiwara, M. (1992) Effects of selective tachykinin antagonists on the tachykininergic contraction induced by electrical transmural stimulation in the rabbit iris sphincter muscle. *Regul. Pept.* **Suppl. 1,** S93.

Lindefors, N., Brodin, E., and Ungerstedt, U. (1986) Neurokinin A and substance P in striato-nigral neurons in rat brain. *Neuropeptides* **8,** 127–132.

Mantyh, P. W., Hunt, S. P., and Maggio, J. E. (1984a) Substance P receptors: localization by light microscopic autoradiography in rat brain using [^{3}H]SP as the radioligand. *Brain Res.* **307,** 147–165.

Mantyh, P. W., Maggio, J. E., and Hunt, S. P. (1984b) The autoradiographic distribution of kassinin and substance K binding sites is different from the distribution of substance P binding sites in rat brain. *Eur. J. Pharmacol.* **102,** 361–364.

Mantyh, P. W., Gates, T., Mantyh, C. R., and Maggio, J. E. (1989) Autoradiographic localization and characterization of tachykinin receptor binding sites in the rat brain and peripheral tissues. *J. Neurosci.* **9,** 258–279.

Matsuto, T., Yanagisawa, M., Otsuka, N., Kanazawa, I., and Munekata, E. (1984) The excitatory action of the newly-discovered mammalian tachykinins, neurokinin α and neurokinin β, on neurons of the isolated spinal cord of the newborn rat. *Neurosci. Res.* **2,** 105–110.

Nagashima, A., Takano, Y., Masui, H., and Kamiya, H. (1987) Evidence that neurokinin A (substance K) neurons project from the striatum to the substantia nigra in rats. *Neurosci. Lett.* **77,** 103–108.

Nagashima, A., Takano, Y., Tateishi, K., Matsuoka, Y., Hamaoka, T., and Kamiya, H. (1989) Central pressor actions of neurokinin B: increases in neurokinin B contents in discrete nuclei in spontaneously hypertensive rats. *Brain Res.* **499,** 198–203.

Ninkovic, M., Beaujouan, J. C., Torrens, Y., Saffroy, M., Hall, M. D., and Glowinski, J. (1984) Differential localization of tachykinin receptors in rat spinal cord. *Eur. J. Pharmacol.* **106,** 463–464.

Otsuka, M. and Yanagisawa, M. (1988) Effect of a tachykinin antagonist on a nociceptive reflex in the isolated spinal cord-tail preparation of the newborn rat. *J. Physiol. Lond.* **395,** 255-270.

Otsuka, M., Sakuma, M., Kobayashi, N., Onishi, Y., Yanagisawa, M., Suzuki, H., and Yoshioka, K. (1991) Tachykinin-evoked release of neurotransmitters from isolated spinal cord of the newborn rat. *Ann. NY Acad. Sci.* **632,** 212–219.

Papir-Kricheli, D., Frey, J., Laufer, R., Gilon, C., Chorev, M., Selinger, Z., and Devor, M. (1987) Behavioural effects of receptor-specific substance P agonists. *Pain* **31,** 263–276.

Petitet, F., Saffroy, M., Torrens, Y., Lavielle, S., Chassaing, G., Loeuillet, D., Glowinski, J., and Beaujouan, J. C. (1992) Possible existence of a new tachykinin receptor subtype in the guinea-pig ileum. *Peptides* **13,** 383–388.

Quirion, R. and Dam, T.-V. (1985) Multiple tachykinin receptors in guinea pig brain. High densities of substance K (neurokinin A) binding sites in the substantia nigra. *Neuropeptides* **6,** 191–204.

Quirion, R., Shults, C. W., Moody, T. W., Pert, C. B., Chase, T. N., and O'Donohue, T. L. (1983) Autoradiographic distribution of substance P receptors in rat central nervous system. *Nature Lond.* **303,** 714–716.

Rothman, R. B., Danks, J. A., Herkenham, M., Cascieri, M. A., Chicchi, G. G., Liang, T., and Pert, C. B. (1984a) Autoradiographic localization of a novel peptide binding site in rat brain using the substance P analog, eledoisin. *Neuropeptides* **4,** 343–349.

Rothman, R. B., Herkenham, M., Pert, C. B., Liang, T., and Cascieri, M. A. (1984b) Visualization of rat brain receptors for the neuropeptide, substance P. *Brain Res.* **309,** 47–54.

Suzuki, H., Yoshioka, K., Maehara, T., Hagan, R. M., Nakanishi, S., and Otsuka, M. (1993) Pharmacological characteristics of tachykinin receptors mediating acetylcholine release from neonatal rat spinal cord. *Eur. J. Pharmacol.* **241,** 105–110.

Takano, Y., Nagashima, A., Masui, H., Kuromizu, K., and Kamiya, H. (1986) Distribution of substance K (neurokinin A) in the brain and peripheral tissues of rats. *Brain Res.* **369,** 400–404.

Tan, D.-P. and Tsou, K. (1988) Differential effects of tachykinins injected intranigrally on striatal dopamine metabolism. *J. Neurochem.* **51,** 1333–1337.

Tateishi, K., Matsuoka, Y., and Hamaoka, T. (1989) Establishment of highly specific radioimmunoassays for neurokinin A and neurokinin B and determination of tissue distribution of these peptides in rat central nervous system. *Regul. Pept.* **24,** 245–257.

Tsuchida, K., Shigemoto, R., Yokota, Y., and Nakanishi, S. (1990) Tissue distribution and quantitation of the mRNAs for three rat tachykinin receptors. *Eur. J. Biochem.* **193,** 751–757.

Ward, P., Ewan, G. B., Jordan, C. C., Ireland, S. J., Hagan, R. M., and Brown, J. R. (1990) Potent and highly selective neurokinin antagonists. *J. Med. Chem.* **33,** 1848–1851.

Xu, X.-J., Maggi, C. A., and Wiesenfeld-Hallin, Z. (1991) On the role of NK-2 tachykinin receptors in the mediation of spinal reflex excitability in the rat. *Neuroscience* **44,** 483–490.

Yanagisawa, M. and Otsuka, M. (1990) Pharmacological profile of a tachykinin antagonist, spantide, as examined on rat spinal motoneurones. *Br. J. Pharmacol.* **100,** 711–716.

Yashpal, K., Dam, T.-V., and Quirion, R. (1990) Quantitative autoradiographic distribution of multiple neurokinin binding sites in rat spinal cord. *Brain Res.* **506,** 259–266.

CHAPTER 17

Receptor Function in the Periphery

Judith M. Hall

1. Introduction

The localization of tachykinin receptors to discrete areas, coupled with the identification of structures immunoreactive for tachykinins, has led to proposals of roles for these receptors in physiological control processes. The aim of this chapter is to outline our current understanding of the functional significance of tachykinin receptors in the periphery and to highlight the possible involvement of these receptors in peripheral disorders. Discussion is limited to where it is felt that a physiological or pathological function for specific receptor types has been established with some degree of certainty. In peripheral systems where tachykinins actions have been demonstrated, evidence was sought for the concomitant presence of tachykinin receptors and tachykininergic innervation or local storage and, in particular, inhibition of effects of endogenous tachykinins with receptor antagonists in vivo.

The use of peripheral preparations for isolated tissue and radioligand binding studies to determine receptor selectivity of new ligands

The Tachykinin Receptors Ed.: S. H. Buck © 1994 Humana Press Inc., Totowa, NJ

has been extensively documented in other chapters of this volume (*see* Section 2.). Consequently, this aspect of tachykinin receptor pharmacology will not be discussed. Similarly, many excellent reviews are available describing receptor function in specific peripheral systems. Rather than duplicating these accounts, such reviews will be referred to in the appropriate sections of this chapter. Studies prior to 1983 have been reviewed (Pernow, 1983).

1.1. Source of Tachykinins in the Periphery

Tachykinins have been localized to, and shown to be released from, several distinct neuronal and other cell types. Of the mammalian tachykinins identified, SP and NKA mediate most of the peripheral actions of tachykinins and there is little evidence for a peripheral role for NKB. The majority of effects of endogenous tachykinins in the periphery are a consequence of local effector actions of tachykinins following their release from the peripheral termini of capsaicin-sensitive primary afferents. In these neurons, tachykinins are costored and coreleased with calcitonin gene-related peptide (CGRP) (*see* Holzer, 1988; Maggi, 1991a). In the gastrointestinal tract, however, tachykinin-containing primary afferent neurons make up only a small proportion of immunoreactive tachykinin content. Here, intrinsic neurons of the enteric nervous system account for most of the demonstrated tachykinin-like immunoreactivity (TK-LI), and the tachykinins normally coexist with other transmitters, in particular acetylcholine and 5-HT (*see* Barthó and Holzer, 1985). Intrinsic neurons probably also account for some TK-LI in other hollow organs. Further cell sources of tachykinins include vascular endothelium cells, enterochromaffin paracrine cells of the gastrointestinal tract, and amacrine cells of the retina.

1.2. Tachykinin Receptors in the Periphery

The classification of tachykinin receptors into NK_1, NK_2, and NK_3 types, and their possible subclassification have been discussed in other chapters of this volume (*see* Section 2.). All three types of tachykinin receptors have been demonstrated to be present in mammalian peripheral tissues.

2. Visual System

In mammals, symptoms of ocular inflammation include miosis and breakdown of the blood-aqueous barrier. Following stimulation of the trigeminal nerve, SP-like immunoreactivity (SP-LI) is released into the aqueous humor (Bill et al., 1979), and exogenous SP or infrared irradiation of the iris produce symptoms of inflammation that are inhibited by tachykinin receptor antagonists (Holmdahl et al., 1981). Consequently, tachykinins may play a role in the response to ocular injury, and this has led to speculation of a therapeutic use of tachykinin receptor antagonists in eye disease where neurogenic inflammation is a characteristic.

Excellent discussions of earlier literature relating to tachykinins and their receptors in structures of the eye may be found in Pernow (1983) and Stjernschantz (1984).

2.1. Iris Smooth Muscles

Of all structures of the peripheral visual system, iris smooth muscle has been the subject of the most extensive histological, pharmacological, and biochemical investigation in relation to tachykinin function. Miosis, a consistent observation following ocular injury, is thought to be a consequence of constriction of the iris sphincter resulting from released mediators, including endogenous tachykinins, from sensory neurons (Fig. 1).

2.1.1. Innervation

The tissue content of tachykinin-like peptides has been estimated using radioimmunoassay and HPLC in the iris or iris–ciliary complex of many species. The majority of studies in the rabbit identify the presence of SP, NKA, and neuropeptide K-like immunoreactivity (e.g., Beding-Barnekow et al., 1988). Although one study reported the additional presence of NKB-like immunoreactivity (Taniguchi et al., 1986), later studies, using different extraction methods, failed to confirm the presence of NKB in this tissue (Beding-Barnekow and Brodin, 1989).

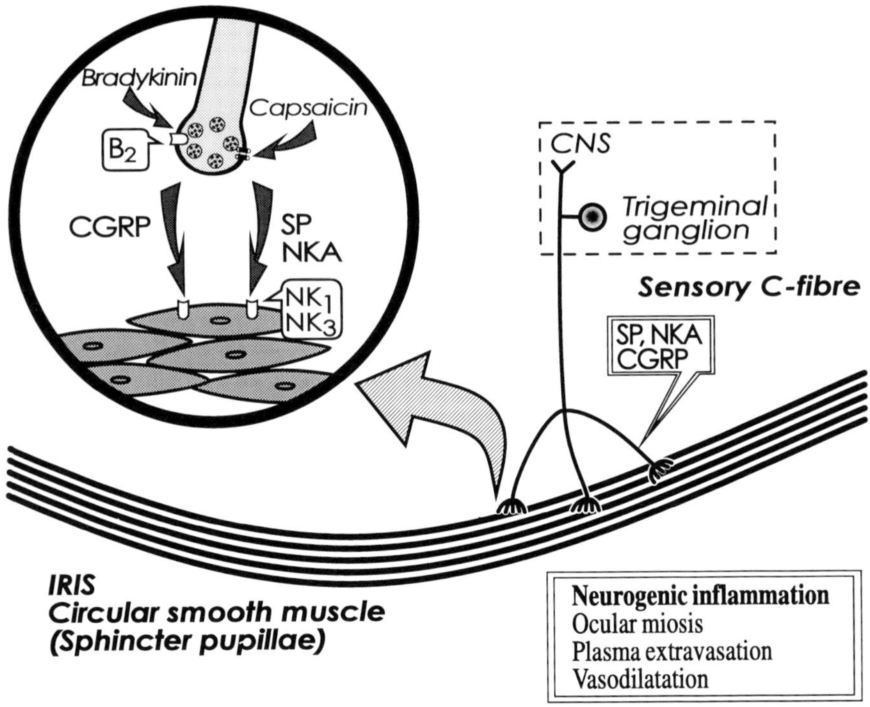

Fig. 1. Schematic diagram of neurons and tachykinin receptors within the iris of the eye. The data relate mainly to the rabbit, and show a cross-section of the iris, with the emphasis on the innervation of the circular smooth muscle (the sphincter pupillae) by sensory afferent C-fibers that contain colocalized substance P (SP), neurokinin A (NKA), and calcitonin gene-related peptide (CGRP). Stimulation of these sensory fibers results in nonadrenergic, noncholinergic (NANC) efferent effects on electrical or chemostimulation (e.g., by capsaicin or bradykinin). The postjunctional contractile effects are mediated mainly through activation of tachykinin receptors in the rabbit. Agonist and antagonist data suggest involvement of both NK_1 and NK_3 receptors, and there is evidence for an NK_1 variant with atypical properties *(see text)*. Hyperreactivity of this system underlies neurogenic inflammation, where important efferent effects of these sensory fibers cause several pathological symptoms (examples boxed in diagram).

In the iris of the rabbit, guinea pig, and rat, tachykinin-containing fibers predominate in the sphincter muscle and stroma, with fibers running close to the dilator muscle (e.g., Terenghi et al., 1982; Ehinger et al., 1983). The localization of the tachykinins seems to be predominantly to sensory neurons originating in the trigeminal nerve and having their cell bodies in the trigeminal ganglia (Unger et al., 1981; Ehinger et al., 1983).

2.1.2. Exogenous Tachykinins and Receptors In Vitro

Irides of the pig and cow (Unger and Tighe, 1984; Tachado et al., 1991), rabbit (Butler and Hammond, 1980), and rat (Banno et al., 1985) respond to tachykinins with contraction of the sphincter smooth muscle. In the human iris, tachykinins were previously considered to have no effect (Unger and Tighe, 1984), but recently have been shown to cause either relaxation (Tachado et al., 1991) or contraction (Anderson et al., 1990). The contractile response of the human iris may be modulated by endogenous enkephalinase (Anderson et al., 1990). In those species where tachykinins cause contraction, receptors are coupled to phospholipase C, whereas in dog, cat, and human irides, where SP caused relaxation or no effect, an increase in cAMP levels was demonstrated (Tachado et al., 1991). Binding sites for tachykinins have been demonstrated in iris smooth muscle membrane preparations of the cow (Tachado et al., 1991), rabbit (Too et al., 1988; Denis et al., 1991; Tachado et al., 1991), dog (Tachado et al., 1991), and rat (Denis et al., 1991). Autoradiographic studies in rabbit and rat irides identified high densities of ^{125}I-Bolton Hunter SP (^{125}I-BHSP) labeled NK_1 binding sites in the muscular area of the sphincter (Denis et al., 1991). The iris dilator smooth muscle seems insensitive to tachykinins and, consistent with this, no binding sites for ^{125}I-BHSP could be demonstrated in this structure in the rat (Denis et al., 1991).

The isolated iris sphincter preparation of the rabbit has been extensively studied with regard to the tachykinin receptors mediating smooth muscle contraction. Contraction in response to exogenous tachykinins results from a direct action on the smooth muscle (Too et al. 1988; Hall et al., 1991). Contractile responses clearly result from both NK_1 and NK_3 receptor activation (Wang and Håkanson, 1992a; Hall et al., 1991, 1993a,b; Fig. 2). However, responses to certain tachykinins, notably SP, exhibit unusual response characteristics that may not simply result from an interaction with either of these two receptor types. Thus, contractile responses exhibit very

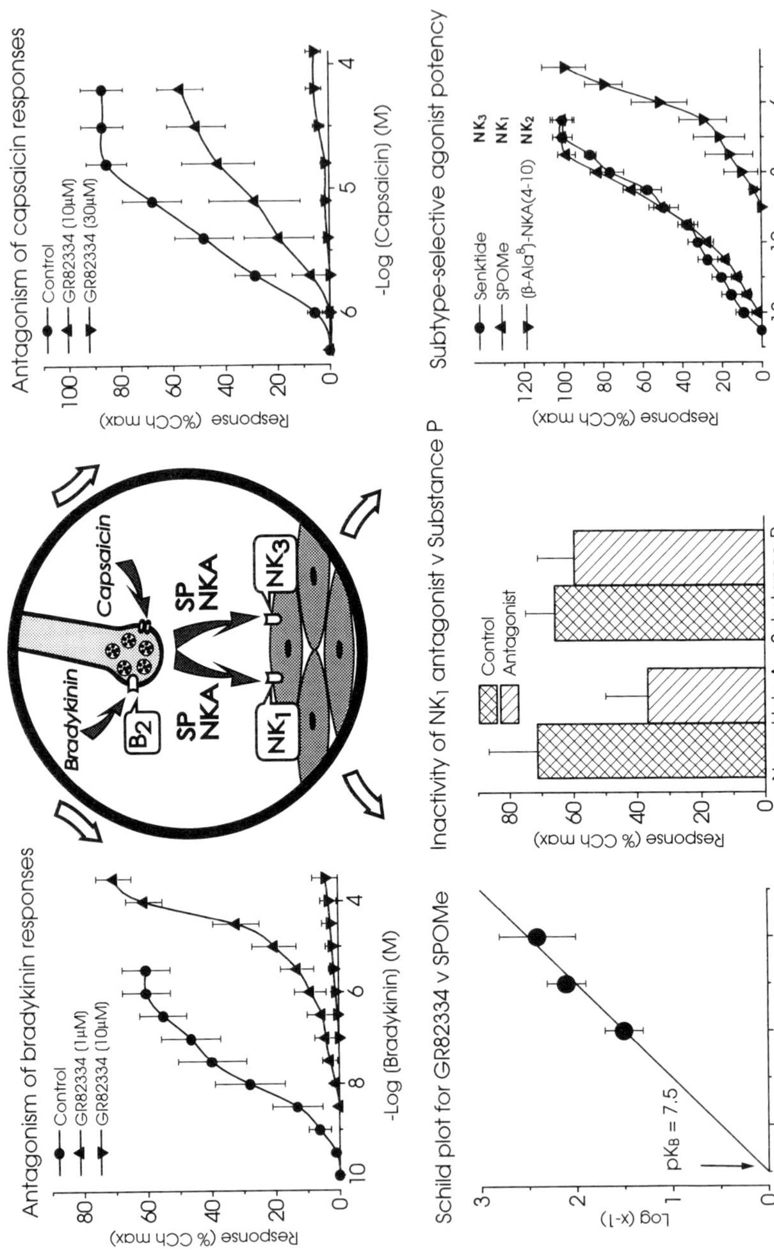
Antagonism of bradykinin responses
Control
GR82334 (1μM)
GR82334 (10μM)
Response (% CCh max)
-Log (Bradykinin) (M)
Bradykinin
Capsaicin
B2
SP NKA
NK1
NK3
Antagonism of capsaicin responses
Control
GR82334 (10μM)
GR82334 (30μM)
Response (%CCh max)
-Log (Capsaicin) (M)
Schild plot for GR82334 v SPOMe
Log (x-1)
pKB = 7.5
-Log GR82334 (M)
Inactivity of NK1 antagonist v Substance P
Control
Antagonist
Response (% CCh max)
Neurokinin A
Substance P
Antagonism by GR82334 (10 μM)
Subtype-selective agonist potency
Senktide
SPOMe
(β-Ala8)-NKA(4-10)
NK3
NK1
NK2
Response (%CCh max)
-Log (Tachykinin) (M)

Fig. 2. NK_1-antagonist blockade of the effects of endogenous and exogenous tachykinins in the rabbit iris. The figure contrasts direct and indirect pathways of excitation involving tachykinins in the rabbit iris sphincter. Prejunctional excitation of the sensory C-fiber is believed to release neurotransmitters, including substance P (SP) and neurokinin A (NKA), which then subserve an efferent function. In the two panels at the top of the figure, chemostimulation by bradykinin, acting via B_2 receptors, or capsaicin to act on an ion channel receptor complex, cause concentration-dependent contractions. These contractions are inhibited by the NK_1 antagonist, GR82334 (1–30 μ*M*), suggesting that tachykinins are involved. Postjunctional effects of tachykinins are shown in the lower three panels, and the relatively high activity of the NK_1-selective agonist, SP methyl ester, and the NK_3-selective agonist, senktide, suggest the presence of these two receptor types on the smooth muscle, whereas the NK_2-selective agonist, [β-Ala^8]-NKA(4–10), had only very low activity. In confirmation, the NK_1-selective antagonist, GR82334, tested against SP methyl ester, showed a competitive interaction as shown in the Schild plot, which yields a pK_B estimate for this antagonist or 7.5. However, this and other (*see* Hall et al., 1991, 1993a,b) NK_1 antagonists do not inhibit SP contractions even at high concentrations *(see figure)*, though they do inhibit neurokinin A contractions. Substance P also has unusually slow offset kinetics, so these data suggest the additional presence of an atypical NK_1 receptor variant. All data are from the author's laboratory.

slow offset kinetics, and NK_1 selective antagonists are considerably less effective at inhibiting responses to SP as compared to those of other tachykinins (Ueda et al., 1986; Muramatsu et al., 1987; Too et al., 1988; Hall et al., 1993b). Recently, it has been shown that the unusual properties of SP are shared by several tachykinins and analogs, including those proposed as being highly selective for the NK_1 receptor, such as [Pro^9]-SP(1-11). The various hypotheses raised to explain these interesting observations are discussed by Hall et al. (1993b). It is not known whether a similar complex system occurs in the iris sphincter of other species.

2.1.3. *Endogenous Tachykinins and Receptors In Vitro*

In the rabbit isolated iris sphincter preparations in vitro, responses to sensory nonadrenergic, noncholinergic nerve stimulation by electrical field-stimulation, or by bradykinin, have long been considered to result from the release of tachykinins from the peripheral termini of sensory nerves. The main body of evidence for such proposals comes from studies involving capsaicin (Ueda et al., 1984; Håkanson et al., 1987) and, more recently, resiniferatoxin (Wang and Håkanson, 1992b), which, following repeated application, completely inhibit responses to nonadrenergic, noncholinergic electrical field-stimulation and to bradykinin (Fig. 2). The various methods of sensory nerve activation are considered to differentially release tachykinins from multiple tachykinin stores (Håkanson et al., 1987). In the rabbit, following unilateral trigeminal nerve denervation, iris sphincter preparations from the denervated eye failed to respond to electrical field-stimulation, bradykinin, and capsaicin, whereas preparations from the control eye responded to these agents (Ueda et al., 1982), thereby providing direct evidence for a role of locally released sensory nerve mediators in these responses. In the human iris, however, Geppetti et al. (1992b) failed to demonstrate an increase in SP-LI release in response to capsaicin. Interestingly, tachykinin-mediated responses to nonadrenergic, noncholinergic electrical field stimulation in the rabbit iris sphincter are inhibited by opioids (Ueda et al., 1983; 1985a) and enhanced by prostaglandins (Ueda et al., 1985b).

In the rabbit iris sphincter, tachykinin receptors mediating smooth muscle contraction following sensory nerve stimulation are predominantly of the NK_1 type, since effects are inhibited by the NK_1 receptor antagonists, GR82334, CP-96,345, and RP67580 (Wang and Håkanson, 1992b; Hall et al., 1993a) and by spantide II (Wang and Håkanson, 1992a). Since responses to SP are relatively resistant to tachykinin receptor antagonists, whereas responses to NKA are readily inhibited, it has been suggested that the contractile response following sensory nerve stimulation in the rabbit iris sphincter is due predominantly to release of NKA (Muramatsu et al., 1987; Wang and Håkanson, 1992b; Fig. 2).

2.1.4. Endogenous and Exogenous Tachykinin Receptors In Vivo

In the rabbit, when injected into the anterior chamber (Bill et al., 1979; Butler and Hammond, 1980; Bynke et al., 1983a; Beding-Barnekow et al., 1988), SP causes a potent miotic effect, inhibited by tachykinin receptor antagonists (Bill et al., 1979), though no effect was demonstrated in the cat (Mandahl et al., 1980). In the rabbit, SP is still miotic following trigeminal nerve denervation (Butler and Hammond, 1980), suggesting a direct action on the pupillary smooth muscle, as indicated by the studies in vitro described above. In the rabbit in vivo, CGRP was shown to potentiate SP-evoked miosis (Wahlstedt et al., 1986).

2.2. Retina

2.2.1. Innervation

Neurons in the retina of many species have been shown to contain TK-LI. These include the chick, bullfrog, guinea pig, dog, cow, monkey, and human (Brecha, 1983; *see* Pernow, 1983; Sternschantz, 1984; Tornqvist and Ehinger, 1988). In contrast to other structures of the peripheral visual system, in most species sensory nerve denervation does not reduce tachykinin levels. Thus, TK-LI is not thought to originate in sensory nerves, but, instead, evidence suggests a localization to amacrine cells (Zhang and Yeh, 1992). Recently, cellular expression of mRNA encoding for SP was demonstrated in rat retina (Brecha et al., 1989).

2.2.2. *Tachykinin Receptors*

The pharmacological analysis of tachykinin receptors in the retina has been restricted in the main to autoradiographic studies. SP-binding sites in the bovine (Osborne, 1984), rat (Denis et al., 1991), and human (Kieselbach et al., 1991a) retina have been described. Recently, the autoradiographic localization and characterization of ^{125}I-BHSP binding sites was elegantly described in detail in the retina of the rat and also the rabbit (Denis et al., 1991). In the rat retina, high-affinity, reversible specific binding (K_D 0.13 n*M*) was reported. Competition studies demonstrated an inhibition profile of endogenous mammalian tachykinins, SP > NKA > NKB, consistent with an interaction with NK_1 receptors. ^{125}I-BHSP binding was also inhibited by the NK_1-selective agonist, SP methyl ester, though surprisingly, not by the NK_1 selective agonist, [Glp^6, L-Pro^9]-SP(6–11) (septide). In a further study, specific binding for ^{125}I-BHSP was demonstrated in the human retina with similar affinity (K_D = 0.27 n*M*; Kieselbach et al., 1990). In rat retinal cultures, SP increased inositol phosphate accumulation (Osborne and Ghasi, 1989), suggesting a link of these receptors with phosphatidylinositol hydrolysis as demonstrated for NK_1 receptors in other tissues (Chapter 10).

In the rat retina, binding of ^{125}I-BHSP was not uniformly distributed, with binding localized principally in the inner plexiform and ganglian cell layers, with lower levels demonstrated in the outer plexiform layer and proximal inner nuclear layer, and with very low binding in other regions, such as the outer retina (Denis et al., 1991). This binding profile is similar to the pattern of tachykinin-immunoreactive cells described previously. Furthermore, degeneration of the inner retinal layers of the neonatal rat by monosodium glutamate treatment results in a marked reduction in ^{125}I-BHSP binding in areas of the retina (Lee and Cheung, 1988), where the majority of specific binding sites have been described (Denis et al., 1991).

2.2.3. *Function*

Despite the identification of tachykinin-containing neurons and the widespread autoradiographic demonstration of tachykinin binding sites, the function of tachykinins in the retina is not established, though a neuromodulatory role in some aspect of visual processing has been suggested. Thus, in fish in vitro, local application of SP increases spontaneous spike activity and enhances light-evoked excitation of retinal ganglian cells (Glickman and Adolph, 1982). A role in modulation of dopamine neurons has also been suggested, since SP increases ^{3}H-dopamine release from rabbit retina (Tsang, 1986).

2.3. *Cornea*

Using monoclonal antibodies, Ehinger et al. (1983) demonstrated SP-immunoreactive fibers subepithelially, intraepithelially, and in the corneal stroma of the rabbit cornea. The SP content of the adult mouse cornea is significantly reduced following trigeminal nerve denervation, and in this species (Keen et al., 1982) and also in the rabbit (Bynke et al., 1984), by neonatal capsaicin treatment, so confirming a sensory origin of these fibers. Using single-unit electrical activity recordings, capsaicin was found to excite polymodal nociceptors in the cat cornea, and this manipulation abolished responses to chemical and thermal sensitivity (Belmonte et al., 1991). However, in functional terms, tachykinins do not appear to contribute to corneal sensitivity (Lembeck and Donnerer, 1981; Bynke et al., 1984).

2.4. *Other Ocular Structures*

SP contracts the human ciliary muscle in vitro (Lograno and Daniele, 1988). In rat and rabbit ciliary smooth muscle, no binding sites for ^{125}I-BHSP were found, though specific binding was demonstrated in rat and rabbit choroid and corneal epithelium (Denis et al., 1991).

2.5. *Pathophysiological Considerations*

Of greatest interest is the possibility that endogenous tachykinins contribute to miosis and ocular hypertension following application of irritant chemicals to the eye. Bradykinin injected into the vitreous humor in the rabbit causes a tetrodotoxin-sensitive miosis that was also inhibited by tachykinin receptor antagonists (Bynke et al., 1983b; Wahlstedt et al., 1985). Capsaicin, injected into the vitreous humor of the eye, causes miosis (Andersson and Almegörd, 1991;

Andersson and Le Greves, 1991), an effect in the rabbit attenuated by omega-conotoxin GVIA and tachykinin receptor antagonists (Andersson and Le Greves, 1991). At certain low doses of capsaicin, the miotic response is inhibited by tetrodotoxin, suggesting the involvement of axon reflexes (Wahlstedt et al., 1985). The miotic effect of intracameral administration of PGE_1, capsaicin, and bradykinin was reduced or abolished 6–8 days after diathermic destruction of the trigeminal nerve (Butler and Hammond, 1980). Miosis in vivo in the rabbit induced by the anticholinesterase inhibitor, echothiophate iodide (Mandahl, 1985), has also been attributed to release of tachykinins from sensory nerves.

Increased intraocular pressure may accompany miosis in the response to several agents, and this may also be caused by the release of tachykinins from sensory nerves. Thus, intracameral injection of SP at high doses in the rabbit eye causes a strong and long-lasting miosis accompanied by an increase in intraocular pressure. The increase in intraocular pressure was prevented by iridectomy, suggesting that it results from miosis-induced papillary block (Stjernschantz et al., 1981,1984). Ruthenium Red at high doses also causes miosis when injected into the anterior chamber of the rabbit eye, with concomitant release of SP-like immunoreactivity and increased intraocular pressure, suggesting an atypical effect on sensory nerves by this agent (Andersson and Le Greves, 1991).

A role for tachykinins has been suggested in the inflammation-induced breakdown of the blood aqueous barrier. Thus, following local injury by infrared irradiation of the iris (Bynke et al., 1984), or capsaicin or PGE_2 administration to the vitreous humor (Bynke et al., 1983b), disruption of the blood-aqueous barrier in the rabbit is prevented by tachykinin receptor antagonists. However, a direct action of tachykinins is unlikely, since SP itself has relatively weak and inconsistent effects on the blood-aqueous barrier in the rabbit (Beding-Barnekow et al., 1988); in the rabbit iris and ciliary epithelium, the major site of the blood-aqueous barrier, no binding sites for ^{125}I-BHSP could be demonstrated (Denis et al., 1991). SP does not affect outflow facility in the cynomolgus monkey (Almegård and Andersson, 1990).

Interestingly, elevated levels of SP have been found in the intraocular fluid of patients having proliferative vitreoretinopathy, the commonest reason for failure in retinal attachment surgery (Kieselbach et al., 1991b).

It seems clear that tachykinins contribute to miosis following ocular injury, and here, NK_1 receptors are the predominant receptor type involved. Although TK-LI and receptors have been demonstrated in various other structures of the peripheral visual system, their role is as yet unclear.

3. Cardiovascular System

Substance P is one of the most potent hypotensive agents known. Indeed, hypotension in the rabbit was one of the earliest bioassays used to identify SP-like material (von Euler and Gaddum, 1931). Practically all vascular beds studied have been shown to be innervated by tachykinin-containing fibers and to respond to exogenous tachykinins, though there is little evidence for a functional role for tachykinins in cardiac muscle (e.g., Lundberg et al., 1984b). Individual studies of the effects of tachykinins on the peripheral cardiovascular system are too numerous to be cited here. Instead, the reader is invited to consult the excellent reviews devoted to this area of research (Pernow, 1983; Mione et al., 1990; Holzer, 1992). An overview only is given in this section concentrating on recent studies relating to tachykinin receptor function, particularly in regard to neurogenic inflammation.

3.1. Innervation

Immunocytochemical studies show extensive innervation by primary afferent C-fibers of the vascular system, particularly in the adventitia and at the adventitial-medial border (Furness et al., 1982; Holzer et al., 1982; Barja et al., 1983). These C-fibers have their cell bodies in the dorsal root ganglia and in the jugular, nodose, and trigeminal ganglia, where tachykinins coexist with calcitonin gene-related peptide (CGRP) (Gibbins et al., 1985; Lundberg et al., 1985). Coexistence with other agents, such as ATP, has also been described (Burnstock, 1977).

Tachykinin-immunoreactive perivascular fibers originating from capsaicin-insensitive neurons have been described in some peripheral tissues, particularly in the gastrointestinal tract (*see* Holzer, 1992). However, whether these fibers represent neurons intrinsic to blood vessels, or whether they have their cell bodies in the plexi associated with the organs they innervate, remains to be established (*see* Section 6).

In addition to neuronally-derived tachykinins, electron microscopic studies have demonstrated that endothelial cells may be a source of SP in femoral and mesenteric arteries (Loesch and Burnstock, 1988). There is some evidence for the release of endothelial-derived tachykinins activated by increased flow through the rat hind limb vasculature (Ralevic et al., 1990) and from rat coronary artery during hypoxia (Milner et al., 1989).

3.2. Large Blood Vessels

3.2.1. Tachykinin Effects and Receptors In Vivo

Intravenous administration of tachykinins induces transient hypotension in many animal species (*see* Pernow, 1983), considered to be a consequence of peripheral vasodilatation mediated by NK_1 receptors. Thus, in the dog (Constantine et al., 1991) and rabbit (Griesbacher et al., 1992), tachykinin effects are inhibited by the NK_1-selective antagonist, CP-96,345. The NK_1 receptors are located on endothelial cells (Saito et al., 1990) and may mediate dilation via release of nitric oxide (Whittle et al., 1989). It should be noted that CP-96,345 has non-specific cardiac effects independent of NK_1 receptor blockade (Donnerer et al., 1992). In view of the paucity of evidence that endogenous tachykinins modulate basal blood flow in the microvasculature under normal conditions *(see below)*, peripheral hypotension in response to administration of tachykinins may be considered to result from predominantly NK_1-mediated tachykinin vasorelaxation of large blood vessels.

3.2.2. Tachykinin Effects and Receptors In Vitro

In large blood vessels, the pharmacological effects of tachykinins, involvement of endothelium-derived factors (e.g., nitric oxide), and the receptor types involved in these effects, depend both on species and tissue source.

3.2.2.1. NK_1 RECEPTORS. NK_1 receptors have been shown by autoradiographic studies to occur on the endothelium of major arteries (e.g., Stephenson et al., 1986), where their activation causes vasodilation (D'Orléans-Juste et al., 1986). The NK_1 antagonist, CP-96,345, inhibits tachykinin-mediated relaxation of the precontracted rabbit aorta (Beresford et al., 1991; Rubino et al., 1992), whereas NK_2 and NK_3 selective agonists are low in activity (Rubino et al., 1992). The dog carotid artery also relaxes in response to NK_1 receptor activation, and CP-96,345 has high affinity in this preparation (Snider et al., 1991). The relaxation of the rabbit pulmonary artery is also mediated by NK_1 receptors (D'Orléans-Juste et al., 1986).

NK_1 selective agonists administered intraluminally to rat perfused venous vasculature are inactive (D'Orléans-Juste et al., 1991). However, stimulation of perivascular nerves causes venoconstriction, and this is inhibited by the NK_1-selective antagonist, CP-96,345. This suggests that endogenously-released tachykinins act on NK_1 receptors located intraluminally in the mesenteric venous vasculature to cause venoconstriction (Claing et al., 1992). Tachykinins contract the rabbit jugular vein (Nantel et al., 1990) and vena cava via NK_1 receptor activation (Rouissi et al., 1991).

3.2.2.2. NK_2 RECEPTORS. Tachykinin-evoked contraction of the endothelium-denuded rabbit pulmonary artery via NK_2 receptor activation (D'Orléans-Juste et al., 1986; Constantine et al., 1990); however, autoradiographic studies have failed to demonstrate the presence of NK_2 receptors in this tissue despite demonstrating functional responses (Xiau et al., 1992). The NK_2 receptors in this tissue are of the proposed NK_{2A} subtype (*see* Chapter 14) for which the antagonist, MEN10207, has high affinity (Maggi et al., 1990b). The NK_2 antagonist, SR48968, also has high affinity for this site (Advenier et al., 1992).

3.2.2.3. NK_3 RECEPTORS. The rat portal vein contracts in response to tachykinins via postjunctional NK_3 receptor activation (Mastrangelo et al., 1983). This preparation has been used extensively as a screen for NK_3 selective ligands. In the rat mesenteric vasculature, intraluminal administration of the NK_3-selective agonist, [Me-Phe^7]-NKB, induces a pronounced venoconstriction (D'Orléans-Juste et al., 1991).

3.3. Microcirculation

Electrical stimulation of dorsal roots results in cutaneous vasodilatation in the skin and plasma extravasation in somatic and visceral tissues (neurogenic inflammation; *see* Lynn, 1988; Holzer, 1992; Fig. 3). These effects are considered to be a consequence of antidromic conduction of impulses in afferent neurons and the local effector action of released mediators on the peripheral vasculature, though evidence indicates that different nerve fiber populations control the two vascular responses (Baylis, 1901; *see* Holzer, 1988, 1992). Release of SP-LI from nerves has been demonstrated in rat skin following noxious thermal stimuli (Helme et al., 1986), and release of SP, as well as CGRP, into the subcutaneous space has been demonstrated in response to various methods of afferent nerve stimulation (Yonehara et al., 1992).

3.3.1. Vasodilatation

Blood vessels of the skin of many species are innervated by tachykinin containing nerve fibers, and capsaicin application results in peripheral vasodilatation (*see* Holzer, 1992). However, although tachykinins have been shown to cause arteriolar dilatation *(see below)* and autoradiographic studies have demonstrated the presence of NK_1 receptors on arterioles (O'Flynn et al., 1989), there is little direct evidence that tachykinins released from primary afferents have an important role in control of normal microvasculature blood flow. Instead, at least in the rat and rabbit, evidence favors a role for CGRP (Delay-Goyet et al., 1992; Hughes and Brain, 1992; Escott and Brain, 1993), coreleased with tachykinins.

There is, however, some limited evidence for a role for tachykinins in control of basal or nerve-evoked arteriolar blood flow. For example, antidromic stimulation of the sectioned sciatic nerve in the rat was shown to induce a biphasic response on cutaneous blood flow, an effect inhibited by the tachykinin receptor antagonist, spantide, which also reduced basal blood flow (Yonehara et al., 1992). These results suggest a tonic regulation of vascular tone by tachykinins in cutaneous microvessels, though the NK_1 selective agonist, GR73632, had no effect on basal blood flow in rat skin (Birch et al., 1992; Escott and Brain, 1993), and RP67580 does not affect the

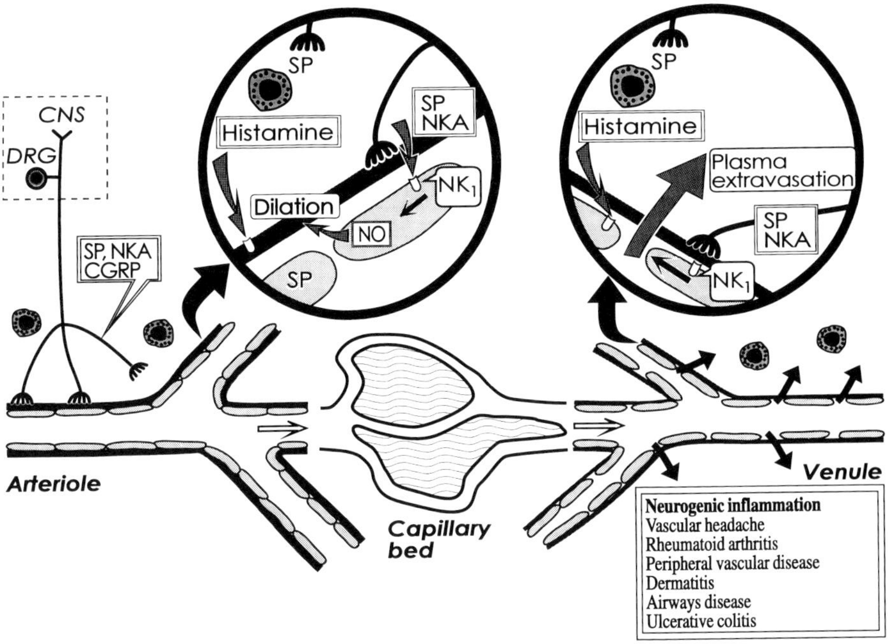

Fig. 3. Schematic diagram of neurons and tachykinin receptors within the microvasculature showing flow from arterioles through the capillary bed to the venules. The three layers of the blood vessel wall represented are the epithelial layer, the smooth muscle of the tunica media, and the adventitia. The innervation shown is that of peripheral sensory afferent C-fibers that contain colocated substance P (SP), neurokinin A (NKA), and calcitonin gene-related peptide (CGRP). There is also evidence for synthesis and release of SP from the endothelial cells themselves. The left of the figure shows the arterioles. Tachykinin receptors (NK_1) are found on the endothelium and mediate relaxation of the media via release of EDRF (endothelial-derived relaxing factor, nitric oxide, NO). Tachykinin receptors (NK_1, NK_2, or NK_3) are on the medial layer of larger vessels, where they act directly to cause contraction. Substance P also has indirect effects through release of histamine from mast cells through a nonreceptor mediated action by virtue of basic amino acid residues. The CGRP also contained in sensory neurons has profound vasodilatory activity. Little is known about tachykinin receptors in the capillary bed, but in the venules, shown on the right of the figure, they have various actions. Here, NK_1 tachykinin receptor stimulation increases endothelial permeability and plasma extravasation. As with the arterioles, there are also indirect actions through release of histamine from mast cells, but here to increase plasma extravasation. Released histamine may itself activate the sensory nerve endings to release further neurotransmitter, both locally and at a distance through activation of the axon reflex. Also, tachykinins have potent effects on several inflammatory cells (e.g., neutrophils, monocytes); however, the receptors involved have not been identified. Inappropriate activation of these mechanisms (neurogenic inflammation) may predispose to inflammatory disease.

increased blood flow induced by saphenous nerve stimulation at stimulation levels that cause edema formation (Delay-Goyet et al., 1992; Shepheard et al., 1993). Using intravital microscopy of the rat cremaster skeletal muscle microvasculature, vasodilation evoked by stimulation of the genito-femoral nerve was inhibited by the tachykinin receptor antagonist, [D-Arg^1,D-Pro^2,D-Trp^{7,9},Leu^{11}]-SP (Brock and Joshua, 1993). In rabbit tenuissimus muscle, nonadrenergic, noncholinergic nerve stimulation results in vasodilation, which is blocked by inhibitors of nitric oxide synthase and also by tachykinin receptor antagonists (Persson et al., 1991). Stimulation of sensory nerves with low doses of capsaicin in rat extrapulmonary airways evokes increased microvascular blood flow, an effect mimicked by SP, but not CGRP, suggesting a role for tachykinins in this effect (Piedimonte et al., 1992). Recently, Lam and Ferrell (1993), using laser-Doppler flowmetry to measure microvascular blood flow in rat hind knee joint, demonstrated that high concentrations of the NK_1 selective antagonist, CP-96,345, partially inhibited blood flow evoked by topically applied SP, though responses to NKA and NKB were unaffected. However, despite these various examples, there is considerably more evidence for a role of CGRP, rather than tachykinins, in control of microvascular blood flow under normal conditions, though tachykinins may contribute to blood flow in certain situations.

3.2.2. *Plasma Extravasation*

Increased permeability of postcapillary venules in the skin, joints, and viscera results in leakage of plasma protein into the interstitial space. Substantial evidence implicates that tachykinins mediate plasma extravasation following antidromic sensory nerve stimulation (for references, *see* Holzer, 1988, 1992; Fig. 3).

In the rat, tachykinin-induced plasma extravasation is mediated via activation of NK_1 receptors. Thus, the NK_1 selective agonists, GR73632 (Birch et al., 1992), septide, [β-Ala^4, Sar^9,Met$(O_2)^{11}$]-SP(4–11)sulfone (Abelli et al., 1991b,c), and [Sar^9 Met$(O_2)^{11}$]-SP (Lam and Ferrell, 1991) were found to potently evoke plasma extravasation in rat skin, upper airways, urinary bladder, and knee joint, respectively. In contrast, NK_2 and NK_3 selective agonists were weakly active or inactive. In the rat, the NK_1 receptor antagonist, CP-96,345, inhibited plasma extravasation in various organs evoked by SP, mustard oil, antidromic stimulation of sensory nerves (Lembeck et al., 1992), capsaicin (Eglezos et al., 1991b), and carrageenin (Birch et al., 1992), as well as that evoked by an NK_1 selective agonist (Eglezos et al., 1991b). It should be noted, however, that the NK_1 receptor inactive enantiomer of CP-96,345 (CP-96,444) also inhibited carrageenin-evoked edema in the rat (Nagahisa et al., 1992a). Of greatest interest is the finding that CP-96,345 (Delay-Goyet and Lundberg, 1991), FR113680, and FK224 (Morimoto et al., 1992b) inhibit edema produced in response to cigarette smoke in the rat. The nonpeptide NK_1 antagonist, RP67580, has been shown to block neurogenic plasma extravasation evoked by electrical stimulation of the trigeminal ganglion in the dura mater of rats (Shepheard et al., 1993) and plasma extravasation evoked by SP in the rat urinary bladder and hind paw following antidromic electrical stimulation of the saphenous nerve (Garret et al., 1991).

In the guinea pig, in the majority of cases, NK_1 receptors also mediate tachykinin-evoked increased plasma extravasation, and specific binding sites for a NK_1 radiolabel (^{125}I-BHSP) have been demonstrated by autoradiographic techniques on postcapillary venules in the guinea pig ureter (Sann et al., 1992). NK_1-selective agonists have been shown to evoke plasma extravasation in guinea pig upper main bronchi (Abelli et al., 1991b), and the NK_1 antagonist, CP-96,345, inhibited plasma extravasation evoked by SP in the airways (Lei et al., 1992); by SP, NKA, and NKB in dorsal skin (Nagahisa et al., 1992a); and by capsaicin in the ureter (Nagahisa et al., 1992b). Edema formation in the guinea pig airways was also inhibited by the NK_1 receptor antagonists, FR113680 (Murai et al., 1992a) and FK888 (Fujii et al., 1992), and the NK_1 receptor agonist, [Sar^9,Met(O$_2$)^{11}]-SP, increased cutaneous plasma extravasation in the guinea pig (Doutremepuich et al., 1992). A recent report, however, demonstrated NK_2 receptor-evoked plasma extravasation in guinea pig lower airways (secondary bronchi and intraparenchymal airways). Thus, the NK_2 selective agonist, [β-Ala[8]]-NKA(4–10), evoked plasma extravasation, and this was inhibited by the NK_2 selective antagonist, SR48968 (Tousignant et al., 1993). It has recently been shown that, although the NK_1 receptor antagonist, CP-96,345, had no effect on the initial neurogenic plasma extravasation evoked by allergen challenge in guinea pig trachea, this antagonist inhibited the subsequent phase of extravasation (Bertrand et al., 1993).

3.3.3. Tachykinin Effects on Inflammatory Cells

In some tissues, mast cell degranulation and prostaglandin release may contribute to tachykinin-evoked plasma extravasation and vasodilation (Foreman and Jordan, 1983; Jacques et al., 1989). Although it is generally considered that histamine release results from a nonreceptor mediated effect of tachykinins having basic N-terminal residues (Bueb et al., 1990), tachykinin-induced histamine release from a cloned murine mast cell line was recently proposed to be mediated predominantly by NK_2 receptors (Krumins and Broomfield, 1992). The relative involvement of mast cells to neurogenic inflammation in different tissues has been discussed (*see* Baraniuk et al., 1990; Kowalski et al., 1990).

Substance P activates guinea pig peritoneal macrophages to release arachidonic acid derivatives and lysosomal enzymes, and binding sites labeled with [^{3}H]SP have been demonstrated on these cells (Hartung et al., 1986). Recently, it has been demonstrated that the NK_1 selective agonist, Ac[Arg^6,Sar^9,Met(O$_2$)^{11}]-SP(6–11), induced shape change of rabbit platelets, an effect that was inhibited by the NK_1 selective antagonist, CP-96,345 (Savi et al., 1992).

Tachykinins have been shown to influence the activity of leukocytes and to release cytokines, prostaglandins, and leukotrienes from monocytes. These various effects all contribute to and perpetuate neurogenic inflammation (*see* Holzer, 1992). However, the receptor types involved in these actions have not been determined and will not be discussed further. Microglia have been proposed to release IL-1 via stimulation of nonclassical NK_1 receptors (Martin et al., 1993), and NK_2 receptors have been proposed to modulate the primary antibody response in rats (Eglezos et al., 1991a).

3.4. Pathophysiological Considerations

Neurogenic inflammation may contribute to the pathophysiology of several chronic inflammatory peripheral disorders. Consequently, tachykinin receptor antagonists have potential as antiinflammatory agents. The involvement of tachykinin receptors in neurogenic inflammation is shown schematically in Fig. 3. Inflammatory mediators released following tissue damage may also sensitize afferent nerve endings, which facilitates further release of tachykinins and CGRP, and thus exacerbate inflammation (Lynn, 1988). The resultant hyperemia, increased vascular permeability, and stimulation of immunocompetent cells will enhance the host defense reaction by facilitating the accumulation of immunocompetent cells to the area of tissue damage and thereby perpetuate inflammation. Hyperreactivity of primary afferents may therefore contribute to the symptoms associated with chronic inflammatory disease. Impairment of primary afferent function, in contrast, may prevent normal functioning of the microcirculation and result in associated symptoms.

3.4.1. Inflammatory Disease of the Joints

TK-LI is increased in the joint fluid of patients with rheumatoid inflammatory diseases (e.g., Devillier et al., 1986; Marabini et al., 1991; *see* Garrett et al., 1992). Peripherally, a significant increase in nerves containing immunoreactive tachykinins was demonstrated in joints that developed severe arthritis in adjuvant-induced arthritis in rats (Levine et al., 1984). Tachykinin receptor antagonists inhibit plasma extravasation in response to antidromic nerve stimulation in the knee of the cat (Ferrell and Russell, 1986) and rat (Lam and Ferrell, 1991). NK_1 sites labeled with ^{125}I-BHSP were found to be localized to vascular endothelial cells in human synovial tissue (Walsh et al., 1992). It may therefore be possible that tachykinins derived from hyperactive afferent nerve endings in the joints contribute to the development of chronic inflammatory processes, and that tachykinin receptor antagonists have therapeutic potential in the treatment of inflammatory diseases of the joints (*see* Kidd et al., 1990).

3.4.2. Skin

Normal functioning of the microcirculation may be impaired as a result of afferent nerve malfunction; for example, neonatal capsaicin pretreatment results in the appearance of cutaneous lesions and aggravation of acid-induced skin lesions in rats (Maggi et al., 1987a), and afferent nerve-mediated axon reflex and vasodilation can be impaired in patients with sensory neuropathies associated with atopic dermatitis (Heyer et al., 1991) and diabetes (Aronin et al., 1987). Furthermore, since healing of injured tissue requires an intact microcirculation, impairment may predispose to trophic disorders; for example, sensory neuropathies may be associated with persistent skin ulcers and connective tissue diseases (*see* Holzer, 1992). Because sensory neuropeptides, including SP and NKA, stimulate cell proliferation, a role in the angiogenesis of healing tissue has been suggested (e.g., Ziche et al., 1990). In contrast, tachykinin release from hyperactive primary afferents in the skin may contribute to the vascular changes in acquired cold and heat urticaria (e.g., Jancsó et al., 1985).

3.4.3. Inflammatory Diseases of the Gastrointestinal Tract

The colonic mucosa of patients with ulcerative colitis contains high levels of tachykinins (e.g., Goldin et al., 1989), and quantitative autoradiographic studies indicated increased NK_1 receptor expression on arterioles, venules, and also lymph nodes in the colon of patients with ulcerative colitis and Crohn's disease (Mantyh et al., 1988). Tachykinin receptor antagonists may therefore represent a therapeutic approach to chronic inflammatory disease of the gastrointestinal tract.

3.4.4. Vascular Headache

Cerebral blood vessels have a particularly rich supply of tachykinin-containing sensory neurons, and neurogenic inflammation within cranial tissues may be involved in the pathogenesis of vascular headaches, including migraine (Moskowitz, 1992). Certain NK_1-selective agonists (SP methyl ester and GR73632) have been shown to cause vasodilation of guinea pig pial artery in vivo, and these effects were inhibited by the NK_1 antagonist, GR82334 (Beattie et al., 1993). However, in canine middle cerebral artery, although NK_1 receptors mediate relaxation, relaxation in response to nonadrenergic, noncholinergic nerve stimulation is unaffected by the NK_1 antagonist, GR82334 (Stubbs et al., 1992). Neurogenic plasma extravasation evoked by electrical stimulation of the trigeminal ganglion in the dura mater of rats has been shown to be inhibited by the nonpeptide NK_1 antagonist, RP67580 (Shepheard et al., 1993), and the NK_1 selective antagonist, GR82334, inhibited plasma protein extravasation evoked by the NK_1-selective agonist, GR73632, in guinea pig dura (O'Shaughnessy and Connor, 1993). These observations suggest a role for tachykinin receptor antagonists in vascular headache.

4. Urinary Tract

The urinary tract receives a dense innervation of capsaicin-sensitive afferent neurons that run in the pelvic and hypogastric nerves. Tachykinin regulation of the urinary tract has been the subject of extensive investigation and exhaustive discussion (*see* Maggi and Meli, 1986; Maggi, 1991b).

4.1. Urinary Bladder

Extensive information is available on the functional role of tachykinins and their receptors in the urinary bladder, particularly in the transmission of sensory information centrally relating to the micturition reflex (Maggi and Meli, 1986; Maggi, 1991b). The isolated urinary bladder of several species contracts in response to tachykinins (Falconieri-Erspamer et al., 1973; Sjögren et al., 1982), and evidence is available suggesting a role for peripheral tachykinin receptors in the control of the micturition reflex, though the local motor response produced by tachykinins released from sensory nerves may be more important during inflammation than during normal micturition (Maggi and Meli, 1986; Maggi et al., 1987b; Maggi, 1991b).

4.1.1. Innervation

TK-LI has been demonstrated in the urinary bladder of many species, almost all of which is present within capsaicin-sensitive primary afferent fibers (*see* Maggi and Meli, 1986; Maggi, 1991b). However, intramural neurons expressing SP-LI have been demonstrated in the guinea pig (Crowe et al., 1986).

4.1.2. Tachykinin Receptors in Rat Urinary Bladder In Vitro

Both NK_1 and NK_2 postjunctional tachykinin receptors mediate the phasic contractile response in the rat isolated bladder (Watson et al., 1983; Maggi et al., 1988; Hall et al., 1992), and NK_1 and NK_2 receptors have also been demonstrated by radioligand binding studies (Burcher and Buck, 1986; Nimmo et al., 1992). Autoradiographic studies confirm the presence of NK_1 receptors located predominantly on the smooth muscle (Burcher and Buck, 1986; Burcher et al., 1986). Capsaicin elicits a biphasic response in the rat isolated urinary bladder. The initial phasic response is sensitive to inhibition by NK_1 receptor antagonists, whereas the later tonic component of the response is inhibited predominantly by NK_2 antagonists, suggesting that endogenous tachykinins with preferential activity at NK_1 receptors (SP) and NK_2 receptors (NKA) are involved in the two phases, respectively (Maggi et al., 1991b).

4.1.3. Peripheral Tachykinin Receptors and the Micturition Reflex

A role has been proposed for tachykinin receptors in the activation of the micturition reflex. In a fixed-volume model in the anesthetized rat, application of tachykinins in vivo to the bladder containing a subthreshold volume of fluid activated micturition (Maggi et al., 1987b). The activation of NK_3 receptors on sensory nerves in the bladder wall was considered to sensitize the sensory fibers to mechanical stimuli and/or initiate afferent impulses and therefore activate micturition. The ability of tachykinins to activate the micturition reflex was impaired in animals treated neonatally with capsaicin. The direct NK_1 and NK_2 mediated contractile effect of higher concentrations of tachykinins described above was thought not to contribute to activation of reflex micturition. These conclusions were substantiated with studies on the voiding-cycle of the rat bladder (Maggi et al., 1987b). Recently, it was demonstrated that intravesical installation of the NK_2 selective agonist, [β-Ala^8]-NKA(4–10), stimulated voiding function of rat urinary bladder by decreasing bladder capacity and residual volume (Maggi et al., 1991b).

4.1.4. Tachykinin Receptors in the Urinary Bladder of Other Species

Early functional and radioligand binding studies indicated the presence of NK_2 receptors in the hamster urinary bladder (Watson et al., 1983; Burcher and Buck, 1986; Lee et al., 1986; Buck and Krstenansky, 1987). This conclusion has recently been confirmed (Van Giersbergen et al., 1992; Guard et al., 1993). In this preparation, receptor activation was found to be associated with inositol phospholipid breakdown (Bristow et al., 1987; Van Giersbergen et al., 1992). In human urinary bladder, ^{125}I-BHSP was shown to bind to a single class of binding sites, and the potency order of displacement of ^{125}I-BHSP was characteristic of an NK_1 receptor, although recently introduced heterosteroid NK_1 antagonists were weaker at displacing than in tissues from other species (Appell et al., 1992). This is in contrast to previous conclusions that the tachykinin receptors in this tissue were of the NK_2 type (Dion et al., 1990). Tachykinins contract the guinea pig bladder via an interaction with NK_1 and NK_2 receptors (Carruette et al., 1992; Longmore and Hill, 1992), and ^{125}I-BHSP binding sites have been demonstrated in the bladder of this species (Burcher and Buck, 1986; Lee et al., 1986). The presence of NK_2 receptors has been demonstrated by radioligand binding studies on rabbit urinary bladder (Guard et al., 1993), and autoradiographic studies have demonstrated specific binding sites for ^{125}I-Bolton Hunter NKA (^{125}I-BHNKA) and ^{125}I-BHSP in the mouse urinary bladder (Burcher et al., 1986).

4.2. Ureter

4.2.1. Innervation

The mammalian ureter exhibits high levels of TK-LI that is of almost entirely sensory origin (Amann et al., 1988b; Dray et al., 1989; *see* Amann, 1993). In the guinea pig, tachykinin-like immunoreactive nerves are localized principally to the adventitia, smooth muscle layer, and submucosa and often on blood vessels where tachykinins are colocalized with CGRP (Sann et al., 1992).

4.2.2. Tachykinin Receptors and Ureteric Motility

Tachykinins have a role in control of ureteric motility either owing to local effects at the site of release (Hua and Lundberg, 1986) or by facilitation of sympathetic transmission via release of tachykinins from collaterals at prevertebral ganglia (*see* Amann et al., 1993). Thus, capsaicin perfused intraluminally in the guinea pig ureter stimulates peristaltic activity, and capsaicin or ureteric distension elicits capsaicin-sensitive slow depolarization in inferior mesenteric ganglian cells (Amann et al., 1988a). In the middle sections of the ureter of several species, tachykinins induce regular contractions, considered to be a result of activation of latent pacemakers (*see* Amann, 1993), and in the renal pelvis neuropeptide release reinforces spontaneous activity sustained by natural pacemakers (Maggi et al., 1992b). It might be of relevance therefore that the renal pelvis is more sensitive to tachykinins than other parts of the ureter (Sann et al., 1992).

In the guinea pig ureter, tachykinin-evoked contractions result from NK_2 receptor activation, and few binding sites for the NK_1 radiolabel, ^{125}I-BH SP, were detected on ureter smooth muscle (Sann et al., 1992), and in guinea pig renal pelvis NK_2 receptors were found to be responsible for contractile responses to endogenous tachykinins (Maggi et al., 1992b). In rat isolated ureter, NK_2 and NK_3 selective agonists potently activated rhythmic contractions, whereas NK_1 selective agonists were less active (Maggi et al., 1988).

4.2.3. Tachykinin Receptors and Epithelial Function

Binding sites for SP are also located in cell layers close to the submucosa in the guinea pig ureter, suggesting a role for tachykinins in transmembrane water and electrolyte movement (Sann et al., 1992).

4.3. Urethra

Tachykinins contract urethra smooth muscle. In rats, NK_1 and NK_2 receptors are present on proximal urethra (Maggi et al., 1988), and in human (Parlani et al., 1991) and guinea pig (Maggi and Patacchini, 1992a), NK_2 and NK_1 receptors, respectively, mediate urethral contraction.

4.4. Pathophysiological Considerations

Since hypermobility of the bladder may be a consequence of the release of tachykinins from peripheral endings of sensory nerves, NK_2 antagonists may have therapeutic potential where bladder hypermobility is a characteristic. NK_2 receptor antagonists have been shown to be active in animal models of cystitis (e.g., Pietra et al., 1992).

Conversely, tachykinins promote bladder emptying and may therefore be used to promote emptying in patients with urinary retention (Maggi et al., 1991). Indeed, the NK_2-selective agonist, [β-Ala^8]-NKA(4–10), reduced bladder capacity and residual volume (Maggi et al., 1991b). Acrylamide intoxication results in severe urinary retention, and this may be a result of impairment of peripheral sensory nerve function. For example, in rats, subcutaneous acrylamide monomer impaired the micturition reflex resulting in urinary retention, and acrylamide depletes capsaicin-sensitive nerves of TK-LI and CGRP-like immunoreactivity (Abelli et al., 1991a). A role for endogenous tachykinins in the genesis of abnormal urethral contractions (urethral instability) has been suggested (Maggi and Patacchini, 1992).

5. Genital Tract

5.1. Vas Deferens

5.1.1. Innervation

A functional role for tachykinin receptors in the male genital tract has not been established (*see* Sastry et al., 1991); however, the vas deferens in particular has been used extensively to investigate the role of tachykinins in the modulation of sympathetic neurotransmission. Tachykinin-immunoreactive fibers of sensory origin have been identified in the vas deferens of several species (Stjernquist et al., 1983). In the guinea pig, these fibers run sparsely in the lamina propria, adventitia, outer longitudinal smooth muscle, and also around small arteries (Alm et al., 1978; Gibbins et al., 1987).

5.1.2. Receptor Studies In Vitro

Contractile responses to sympathetic field-stimulation of the vas deferens in the guinea pig (von Euler and Hedqvist, 1974), rat (Okasada et al., 1986), rabbit (Dion et al., 1987), and mouse (Blackwell et al., 1978; Kastin et al., 1978) are potentiated by tachykinins, though the receptors involved show species dependency. In the guinea pig vas deferens, tachykinins interact with NK_1 receptors (Watson et al., 1983; Mussap et al., 1989; Hall and Morton, 1991a). In contrast, the predominant receptors in rat (Lee et al., 1982; Dion et al., 1987) and rabbit (Dion et al., 1987) vas deferens are of the NK_2 type. Thus, NK_2 selective agonists have high activity (Dion et al., 1987; Tousignant et al., 1987), and in the rat the NK_2 antagonist, L-659, 877 (McKnight et al., 1991), and the NK_1/NK_2 antagonist, FK224 (Morimoto et al., 1992a), have high affinity. Radioligand binding studies confirm a predominant NK_2 receptor population in the rat preparation (Lee et al., 1986; Bergström et al., 1987).

It is unclear, however, whether tachykinin receptors in the vas deferens have a pre-and/or postjunctional localization. In the mouse, postjunctional actions of SP were considered to account for tachykinin actions (Blackwell et al., 1978), and both pre-and postjunctional sites in rat vas deferens (Tousignant et al., 1987; Nagata et al., 1991). However, direct studies of neurotransmitter release were not carried out in these investigations. In the rat, NKA suppressed a voltage-gated outward K^+-current in isolated smooth muscle cells (Nakazawa et al., 1990), suggesting the presence of receptors on the smooth muscle. In the rat vas deferens, tachykinin-evoked contraction of the epididymal section involves only extracellular Ca^{2+}, whereas contraction of the prostatic section relies on both extra-and intracellular Ca^{2+} (Nagata et al., 1991). In the guinea pig, tachykinins do not increase release of ATP (Ellis and Burnstock, 1989) or [^{3}H]-noradrenaline (Ellis and Burnstock, 1989; Hall and Morton, 1991a). However, a postjunctional action, as indicated by potentiation of responses to exogenous α,β-methylene ATP, was demonstrated by one group (Ellis and Burnstock, 1989), but this was not confirmed (Hall and Morton, 1991a).

Autoradiographic studies in the guinea pig vas deferens have demonstrated NK_1 receptors both on sensory nerve endings within the outer longitudinal muscle layer and also on the inner longitudinal smooth muscle (Mussap et al., 1989). It is therefore unresolved whether pre-(on sensory or sympathetic) or postjunctional receptors are functionally involved in potentiation of sympathetic responses in the vas deferens. In relation to the presence of the dense sensory innervation in the vas deferens and tachykinin receptors on sensory nerve endings, capsaicin potentiates sympathetic twitch responses in guinea pig isolated vas deferens through release of tachykinins and inhibition via release of CGRP (Ellis and Burnstock, 1989). In the guinea pig vas deferens, capsaicin pretreatment resulted in a marked increase in ^{125}I-BHSP binding sites over the inner longitudinal muscle and inner part of the circular muscle layer, whereas binding was abolished over the outer longitudinal muscle. The authors suggested that the capsaicin-sensitive binding sites were located presynaptically on primary afferent neurons. The upregulation of binding sites on the inner longitudinal muscle was due to postsynaptic receptor upregulation resulting from loss of sensory innervation (Mussap et al., 1989).

5.2 *Uterus*

Immunohistochemical studies have demonstrated that the human myometrium is innervated by tachykinin-containing fibers (Heinrich et al., 1986). NK_3 receptors have been identified by functional and radioligand binding studies to be present in the rat uterus, and radioligand binding studies identified the additional presence of NK_1 and NK_2 sites (Barr et al., 1991). The NK_3 receptors were coupled to phosphatidylinositol hydrolysis. No contractions were evoked in human myometrium obtained from caesarian section at term, which does not support a physiological role of tachykinins in parturition (Barr et al., 1991).

5.3. *Conclusions*

Few conclusions can be drawn from these studies in the genital tract in relation to the function of tachykinins and their receptors. However, in view of the sensitivity to tachykinins and diversity of receptor subtypes in the vas deferens of various species, the future may well reveal a physiological role for tachykinin receptors in this structure.

6. Gastrointestinal Tract

Gastrointestinal preparations played an essential part in the original assay of SP extracted from gut and brain (Gaddum and Schild, 1934). Substance P is now thought to be a major excitatory transmitter located in enteric neurons that have their cell bodies in the intestinal wall and may be especially important in the coordination of the peristaltic reflex. This area of tachykinin pharmacology has been extensively reviewed (*see* Barthó and Holzer, 1985), so discussion is restricted to recent studies using selective receptor ligands.

6.1. *Innervation*

Direct measurement of tachykinin release and its modulation have been shown by both pharmacological and immunochemical techniques from regions of the gastrointestinal tract of many species (*see* Barthó and Holzer, 1985). It is now clear that tachykinins found in the gastrointestinal tract derive from three sources: intrinsic neurons of the enteric nervous system, extrinsic afferent neurons with sensory and local effector function, and paracrine cells of the diffuse neuroendocrine system found in the mucosa (Fig. 4).

6.1.1. *Intrinsic Neurons*

Immunohistochemical and radioimmunological studies show most neuronal SP to be contained within intrinsic neurons of the intestine (e.g., Franco et al., 1979). There is also NKA-like immunoreactivity, but little if any, NKB-like immunoreactivity (Deacon et al., 1987). In the main, tachykinins are colocalized with other transmitters including ACh (Furness et al., 1984), 5-HT (Legay et al., 1984), and Met-enkephalin (Domoto et al., 1984). Intrinsic neurons have their cell bodies largely confined to the myenteric and submucous plexi, where neurons are highly ordered and form distinct pathways (*see* Barthó and Holzer, 1985). Substance P-positive fibers in myenteric ganglia project to the longitudinal and circular smooth muscle, whereas fibers from the submucous plexus project to the mucosa (*see* Fig. 4). The organization of these neurons suggests a role in control of motor activity, for instance, in the peristaltic reflex (*see* Barthó and Holzer, 1985).

6.1.2. Extrinsic Afferent Neurons

Not all the fibers within the wall of the gut that show TK-LI are intrinsic. There is a small population of unmyelinated extrinsic fibers that appear to be sensory afferents having their cell bodies largely within dorsal root ganglia and show colocalization with CGRP (Gibbins et al., 1985). Following extrinsic denervation or chronic capsaicin pretreatment, there is a loss of TK-LI (Furness et al., 1982). These afferents have both local efferent functions and allow integration of the CNS in gut function.

6.1.3. Paracrine Cells

These cells are regarded as part of the diffuse neuroendocrine system, are widespread in the gut mucosa, and can be considered to be undifferentiated neurons. As well as containing tachykinins, these cells also contain 5-HT (Pearse and Polak, 1975). Their function remains obscure, though when disrupted by radiation or chemotherapy they release 5-HT, which leads to nausea and vomiting. Whether SP also has such a role remains to be determined.

6.2. Motility and Peristalsis

There is now a large body of evidence concerning the innervation, release, and effects of tachykinins in the small intestine. Early studies showed that in the presence of an SP-containing extract subthreshold intraluminal pressure elicited peristaltic waves in rabbit intestine *in situ* (Gernandt, 1942), and, later, release of tachykinins during experimental intestinal peristalsis was demonstrated (e.g., Donnerer et al., 1984). Tachykinins are considered to be essential in the high-threshold noncholinergic pathway involved in the ascend-

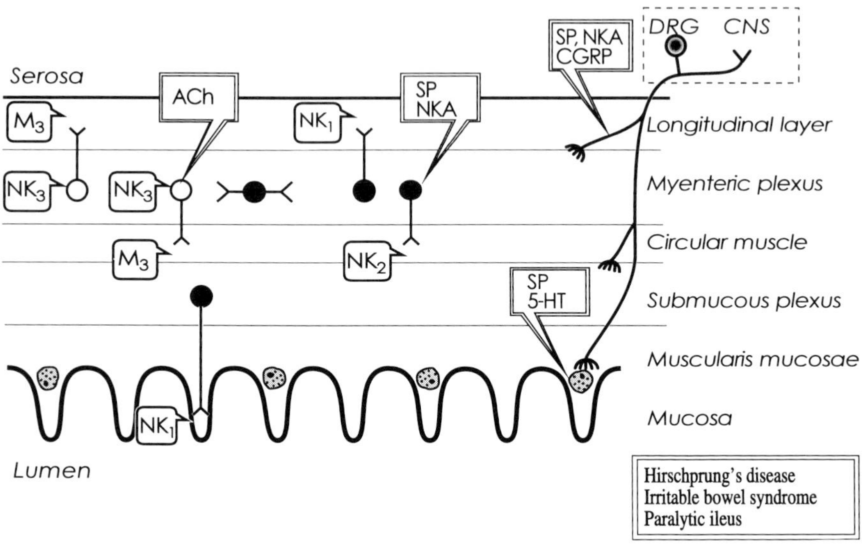

Fig. 4. Schematic diagram of neurons and tachykinin receptors within the gastrointestinal tract, constructed from histochemical and functional information. The data relate mainly to the guinea pig, with the diagram representing a cross-section of the wall of the ileum. Tachykinins are established as having a role in the peristaltic reflex pathways, NANC contractile responses on field stimulation, and indirect contractile responses to other agents (e.g., capsaicin). At the left of the figure, cholinergic neurons (ACh) are drawn with unfilled cell bodies. Excitation of these with tachykinins, acting mainly through NK_3 receptors, causes a well-documented indirect contraction of the longitudinal muscle layer through release of acetylcholine to excite muscarinic M_3 receptors. There is also some evidence for an excitatory or inhibitory role for NK_1 receptors on these neurons *(see text)*. A number of types of tachykinin-immunoreactive enteric (intrinsic) neurons have been demonstrated (shown with filled cell bodies) that contain SP and neurokinin A. Some run transversely, whereas others innervate smooth muscle, plexi, or the mucosa. It should be noted that few enteric interconnections are well understood; further, a number of neurotransmitters are colocalized (e.g., acetylcholine and SP). On NANC stimulation, there is muscle contraction mediated via NK_1 receptors (longitudinal smooth muscle) or NK_2 receptors (circular muscle layer; but with some evidence for NK_1 receptors), and these may be involved in the peristaltic reflex. Also, some of the neurons innervating the mucosal layer and epithelium appear to have a function in modulating fluid transport. Malfunction of certain of these pathways may have a role in the etiology of

ing reflex contraction. Although it is not possible to dissect individual roles for tachykinin receptors in peristalsis, which results from coordinated neuronal pathways, multiple transmitters, and activity of the circular and longitudinal smooth muscles, the receptors present in the various structures of the small intestine will be considered. Certainly, when the tachykinin-containing enteric ganglion cells are not present (e.g., in Hirschprung's disease or aganglionic megacolon), there is no propulsive activity in that segment of the gut.

6.2.1. Guinea Pig Small Intestine

The majority of studies have been carried out on the guinea pig small intestine, particularly the ileum, where tachykinins contract the isolated smooth muscle myenteric plexus preparation (Bury and Mashford, 1977). Early studies indicated that the initial phase of the biphasic contraction to SP is a result of a direct action on the longitudinal smooth muscle; the later tonic phase, in contrast, was proposed to result from tachykinin-mediated ACh release (Holzer and Lembeck, 1980). The dual action of tachykinins on the guinea pig ileum have now been repeatedly confirmed, and the receptors involved in these actions established as being of the NK_1 and NK_3 types for the direct and neuronal effects, respectively (Figs. 4 and 5).

---

several pathological states (examples boxed on diagram). There are extrinsic sensory afferent neurons innervating several layers; these neurons have their cell bodies in the dorsal root ganglia (DRG) and are characterized by colocalized SP, neurokinin A (NKA), and calcitonin gene-related peptide (CGRP). They can have an efferent role, for instance when excited by capsaicin to cause tachykinin-mediated contractions of both muscle layers. They have an afferent reflex physiological role in signaling the state of the gastrointestinal tract, and this may be pathological (e.g., on disruption of the paracrine cells). The paracrine or neuroendocrine cells of the mucosal layers (shown with granulated bodies) contain large amounts of SP coexisting with 5-HT. Though the physiological role of these cells is not clear, they may be disrupted in pathology (e.g., radiation and chemotherapy), when the large amounts of mediators liberated have adverse effects (e.g., nausea through activation of $5\text{-}HT_3$ receptors).

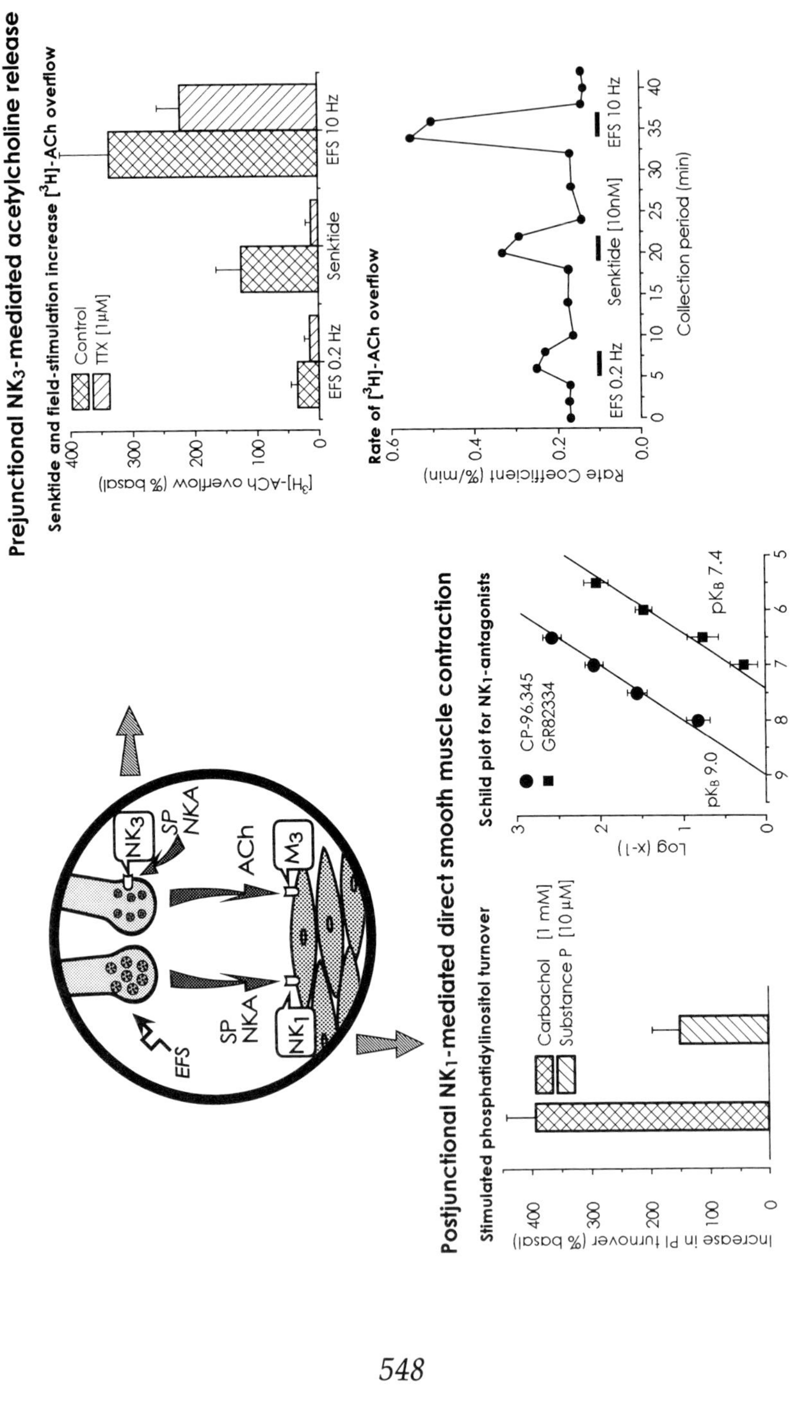

Prejunctional NK3-mediated acetylcholine release
Senktide and field-stimulation increase [3H]-ACh overflow
Control
TTX [1μM]
[3H]-ACh overflow (% basal)
400
300
200
100
0
EFS 0.2 Hz
Senktide
EFS 10 Hz
Rate of [3H]-ACh overflow
Rate Coefficient (%/min)
0.6
0.4
0.2
0.0
EFS 0.2 Hz
Senktide [10nM]
EFS 10 Hz
0
5
10
15
20
25
30
35
40
Collection period (min)
EFS
SP
NKA
NK1
NK3
SP
NKA
ACh
M3
Postjunctional NK1-mediated direct smooth muscle contraction
Stimulated phosphatidylinositol turnover
Carbachol [1 mM]
Substance P [10 μM]
Increase in PI turnover (% basal)
400
300
200
100
0
Agonist
Schild plot for NK1-antagonists
CP-96,345
GR82334
Log (x-1)
3
2
1
0
pKB 9.0
pKB 7.4
9
8
7
6
5
-Log [Antagonist] (M)

Fig. 5. The figure contrasts direct and indirect pathways by which tachykinins act to contract the longitudinal smooth muscle of the guinea pig ileum. Prejunctional NK_3 receptors are located on cell bodies, and probably terminals, of cholinergic neurons within the myenteric plexus that, when activated by tachykinins, release acetylcholine (ACh) to act postjunctionally on muscarinic M_3 receptors to cause contraction. Following activation of NK_3 receptors by the nonselective endogenous neurokinins, substance P (SP) and neurokinin (NKA), or the selective NK_3 ligand, senktide, there is an increase in the rate of release of Ach. The figure shows the effect of 10 n*M* senktide on Ach release, measured here directly as overflow of [^{3}H]-ACh (after prelabeling with [^{3}H]-choline) measured with a superfusion system with collection over 2-min periods (for methods *see* Fox and Morton, 1990). This is compared with the effect of electrical field-stimulation (EFS), in terms of increase in rate coefficient for release and sensitivity to block by 1 μ*M* tetrodotoxin (TTX). Postjunctional contractile effects of tachykinins on the longitudinal muscle are mediated via NK_1 receptors, and these may be activated by exogenous tachykinins or endogenous tachykinin release, as with NANC contractions after stimulation of the myenteric plexus-longitudinal muscle preparation with electrical field-stimulation (EFS). The figure shows affinity estimates for two competitive NK_1 antagonists tested against SP methyl ester contractions, yielding pK_B estimates of 9.0 and 7.4 for CP-96,345 and GR82334, respectively (for methods *see* Hall et al., 1992). The coupling of NK_1 receptors to calcium mobilization and contraction is via increased phosphatidylinositol hydrolysis, and the data shown in the figure compare the magnitude of the effect of carbachol (1 m*M*) and SP (10 μ*M*) on turnover, measured in terms of accumulation of total inositol phosphates in the presence of lithium, after prelabeling with [^{3}H]-inositol (for methods, *see* Hall and Morton, 1991b). All data are from the author's laboratory.

Thus, the atropine and tetrodotoxin-insensitive, tachykinin-mediated contraction is competitively antagonized by numerous NK_1 selective antagonists, including GR82334 and CP-96,345 (Hall et al., 1992). Specific NK_1 binding sites have been demonstrated in guinea pig ileum longitudinal muscle myenteric plexus preparations (Watson and Iversen, 1984; Burcher et al., 1986; Tousignant et al., 1991) and are the only tachykinin receptors found on isolated smooth muscle cells (Souquet et al., 1987). In contrast, contractions evoked by NK_3-selective agonists, such as senktide, are unaffected by NK_1 antagonists, but are inhibited by atropine and tetrodotoxin, and by NK_3-receptor desensitization (Guard and Watson, 1987; Laufer et al., 1988), suggesting contraction results from neuronally-derived ACh. Direct evidence for tachykinin-evoked ACh release was originally provided by neurochemical studies by Yau and Youther (1982), and it was later demonstrated that release was mediated by a novel (NK_3) tachykinin receptor (Fosbraey et al., 1984; Featherstone et al., 1986; Laufer et al. 1985; Wormser et al., 1986; Fox and Morton, 1990). Specific binding sites for the NK_3 selective agonist, senktide, have been demonstrated in the guinea pig ileum myenteric plexus preparation (Guard et al., 1990). NK_3 receptors on enteric neurons have been proposed to also evoke the release of tachykinins (Guard and Watson, 1987) and prejunctional NK_1 receptors to inhibit Ach release (Kilbinger et al., 1986). Both NK_1 and NK_3 receptors in the guinea pig ileum myenteric plexus preparation couple to phosphatidylinositol hydrolysis (Watson and Iversen, 1984; Guard et al., 1988; Hall and Morton, 1990; Fig. 5). In contrast, in the anatomically adjacent guinea pig taenia caeci, both NK_1 and NK_3 receptors are present on the smooth muscle, and their activation results in smooth muscle contraction, also associated with increased hydrolysis of phosphatidylinositol (Hall and Morton, 1991b).

Tachykinins mediate slow synaptic excitation within the myenteric plexus primarily by inactivation of the resting K^+ conductance (Katayama and North, 1978), and although excitation, surprisingly, was inhibited by the NK_1 antagonist, CP-96,345, inhibition was considered to be a result of a local anesthetic effect (Tamura et al., 1993). In myenteric neurons of the guinea pig duodenum, the NK_3 agonist, senktide, evoked depolarization through the closure of K^+ channels (Hanani et al., 1988), again consistent with results from functional studies that the neuronal tachykinin receptor is of the NK_3 type.

The circular muscle is important in the genesis of propulsive activity and peristalsis. In the isolated preparation of the guinea pig ileum, NK_1 and NK_2 receptors contract the smooth muscle via a direct action, and NK_3 receptors stimulate the release of ACh (Barthó et al., 1992). Recently, evidence has been provided for an involvement of NK_2 receptors in excitatory transmission to the circular muscle of the guinea pig ileum. Thus, NK_2 selective antagonists block the atropine-resistant ascending enteric reflex, activated through balloon distension produced by stimulation of intramural neuronal pathways that subserve the atropine-resistant ascending enteric reflex in this tissue (Barthó et al., 1992). NK_1 receptor antagonists, in contrast, are ineffective (Barthó et al., 1992; Holzer et al., 1993). Tachykinin receptors identified as influencing gastrointestinal function are shown in Fig. 4.

6.2.2. Other Gastrointestinal Tissues

Contractile responses to tachykinins have been demonstrated in intestinal preparations from numerous species. Contraction of the human ileum results from activation of NK_1 and NK_2 receptors, with no involvement of cholinergic neurons. The contractile response to electrical field-stimulation is preferentially blocked by NK_2 antagonists (Maggi et al., 1990c, 1992a). Tachykinin binding sites have been demonstrated with radioligand binding studies in human gastrointestinal tract (Gates et al., 1989). Contractile responses of the rat small intestine activation result from NK_2 receptor activation (Hall et al., 1992; Emonds-Alt et al., 1992), and NK_2 binding sites have been demonstrated in rat small intestine, stomach, and colon (Burcher et al., 1986; Bergström et al., 1987).

6.3. Mucosal Ion Transport

It has been suggested that tachykinins have a physiological role in Cl^- secretion in the large and small intestine of several species (e.g., Kachur et al., 1982). In the guinea pig small intestine, tachykinin receptors mediate some of these effects via a direct action on epithelial cells, but predominantly via an indirect action via the stimulation of cholinergic and noncholinergic submucosal neurons (Keast et al., 1985). SP and NK_1 selective agonists evoke a transient increase in short-circuit current (owing to Cl^- secretion) in guinea pig intestine, an effect inhibited by the NK_1 antagonists, GR82334 and CP-96,345 (Reddix and Cooke, 1992).

6.4. Pathophysiological Considerations

The main point of note regarding tachykinins in the gastrointestinal tract is that, unlike the majority of other peripheral organs, the major tachykinin-containing neurons are intrinsic enteric motor fibers rather than C-fiber afferents. The distribution of fibers in the gut plexi suggest roles both in modulation of motility and secretion. With respect to tachykinin receptors, it is of the greatest interest that the receptor types involved in indirect actions via ACh release generally differ from those responsible for direct actions. This fact offers great hope for useful therapeutic ligand selectivity. However, it should be noted that the lack until very recently of antagonist ligands of reasonably high affinity and stability has greatly hindered investigation, and the total lack of NK_3 antagonists has been crucial in this respect. It may confidently be expected that new ligands will make good these deficiencies, and that real progress will be made in further investigating the roles of tachykinin receptors in peristalsis and secretion.

In view of the potent actions of tachykinins in gastrointestinal motility, tachykinin receptor antagonists may have therapeutic applications in diseases such as irritable bowel syndrome, where excessive gut motility is a characteristic; in stasis (e.g., paralytic ileus), agonists may prove to be useful as stimulants.

7. Respiratory Tract

Tachykinins cause airway smooth muscle contraction, increase microvascular permeability, are potent stimulants of the release of macromolecules from airway secretory cells, and may contribute to bronchial hyperreactivity. Overexcitation of capsaicin-sensitive afferent sensory fibers may therefore play a role in the pathophysiology of inflammatory airway disease (*see* Barnes et al., 1988, 1990). Identification of the tachykinin receptors mediating these effects may lead to an understanding of the role of tachykinins in normal airway function and identify targets for the development of therapeutically active agents in airway disease (Maggi, 1990; Frossard and Advenier, 1991; *see* Fig. 6).

7.1. Innervation

Tachykinins have been localized to capsaicin-sensitive, C-fiber afferents having their cell bodies in the nodose and jugular ganglia in the airways in many species (Lundberg and Saria, 1987; Martling, 1987). Tachykinin immunoreactive fibers lie within and around the surface epithelium and are in close proximity to blood vessels, mucus glands, blood vessels, and in a lesser extent to airway smooth muscle (Lundberg et al., 1984a). Release of tachykinins in the airways has been demonstrated in response to capsaicin (Saria et al., 1984), bradykinin, histamine, and nerve stimulation (Saria et al., 1988).

7.2 Airway Smooth Muscle

Tachykinins contract isolated airway smooth muscle and cause bronchoconstriction in vivo in many species, including humans; bronchoconstriction can result from both direct actions on the smooth muscle, via local indirect neuronal effects, or via autonomic reflexes (*see* Maggi, 1990; Frossard and Advenier, 1991; Fig. 6). Tachykinins may therefore have a role in the regulation of bronchomotor tone. General opinion favors a predominant role for NK_2 receptors in control of airway smooth muscle tone, although NK_1 receptors have been demonstrated on some airway smooth muscle and may contribute in some species (Maggi, 1990; Frossard and Advenier, 1991).

7.2.1. Tachykinin Receptors In Vitro

In the guinea pig, NK_2 and NK_1 receptors mediate tracheal smooth muscle contraction (Ireland et al., 1991). Thus, numerous NK_2 antagonists, including R396 (Ireland et al., 1991), GR83074, and GR94800 (McElroy et al., 1992) and NK_1 receptor antagonists, such as GR71251 (Ireland et al., 1991), FR113680 (Hagiwara et al.,

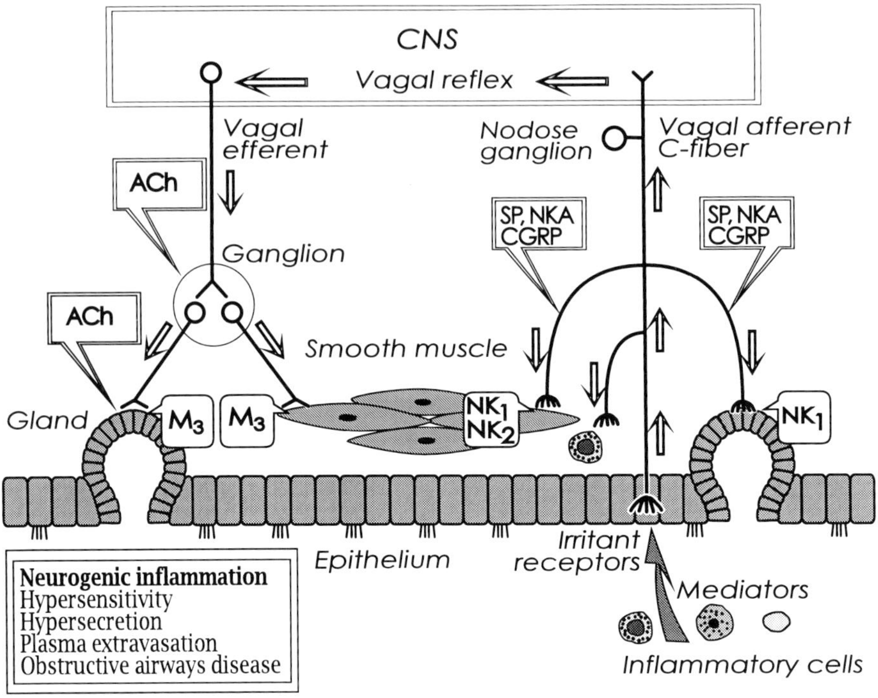

Fig. 6. Schematic diagram of neurons and tachykinin receptors within the airways. A number of structures may influence, or be influenced by, sensory afferent C-fibers that contain colocalized substance P (SP), neurokinin A (NKA), and calcitonin gene-related peptide (CGRP). Stimulation of these sensory fibers by chemostimulation (e.g., by capsaicin, bradykinin, or mediators released from inflammatory cells) can result in local efferent effects through neurotransmitter release or indirect effects through activation of afferent pathways, including a vagal reflex involving the central nervous system. Important effects of these reflex efferent pathways include cholinergically-mediated bronchoconstriction and hypersecretion, but local release of tachykinins may also cause such effects, and here both NK_1 and NK_2 receptors seem involved depending on species. For actions on the microvasculature, *see* Fig. 3. Hyperreactivity of the system shown underlies neurogenic inflammation, where important efferent effects of these sensory fibers underlie several pathological symptoms and conditions (examples boxed on diagram).

1992), FK888 (Murai et al., 1993), and RP67580 (Carruette et al., 1992), inhibit tachykinin-evoked contraction of the isolated preparation. Both NK_1 and NK_2 receptors mediate contraction of isolated guinea pig bronchi (Maggi et al., 1991a), and NK_1 and especially NK_2 receptor antagonists inhibit responses to nonadrenergic, noncholinergic nerve stimulation in the bronchus in this species (Renzetti et al., 1992; Martin et al., 1992). Radioligand binding studies in the guinea pig lung, although demonstrating the presence of NK_1 sites, failed to identify NK_2 binding sites (Geraghty et al., 1992). In radioligand binding studies, NK_1 sites were labeled with ^{125}I-BHSP (Carstairs and Barnes, 1986), and the mixed NK_1/NK_2 receptor antagonist, FK224, displaced [^{3}H]-SP binding and inhibited constriction in the guinea pig lung (Morimoto et al., 1992a; Murai et al., 1992b).

In rabbit airways, both NK_1 and NK_2 receptors have been demonstrated by functional and autoradiographic studies (Black et al., 1992; Ireland et al., 1991). The density of NK_1 receptors was found to be greatest in the peripheral airways, whereas NK_2 receptors were uniformly distributed throughout the airways (Black et al., 1992). The hamster trachea contains only NK_2 receptors that mediate contraction (Maggi et al., 1990b).

In the precontracted isolated mouse main bronchus, tachykinins, capsaicin, and electrical field-stimulation all cause smooth muscle relaxation mediated by NK_1 receptors, thought to result from the release of prostanoids (Manzini, 1992). Tachykinins also relax the precontracted rat trachea, and this effect is dependent on an intact epithelium and is due to NK_1 receptor-mediated release of prostaglandin E_2 (Devillier et al., 1992).

NK_1 sites labeled with ^{125}I-BHSP have been demonstrated in human lung predominantly in the smooth muscle and to a lesser extent on epithelium and submucosal glands (Carstairs and Barnes, 1986). In human isolated bronchi, contraction results predominantly from NK_2 receptor activation (Naline et al., 1989; Dion et al., 1990). In this tissue, removal of the epithelium or preincubation with the neutral endopeptidase inhibitor, phosphoramidon, enhances the contractile response to tachykinins (Naline et al., 1989).

7.2.1. Tachykinin Receptors and Bronchoconstriction In Vivo

Both NK_1 and NK_2 receptors contribute to tachykinin mediated bronchoconstriction in vivo. In the anesthetized guinea pig, the NK_2-selective agonist, [β-Ala^8]-NKA(4–10), is a potent bronchoconstrictor (Maggi et al., 1990a), and NK_2 selective antagonists, including GR83074 (McElroy et al., 1992) and SR48969 (Emonds-Alt et al., 1992, 1993), are potent inhibitors of tachykinin-evoked bronchoconstriction. NK_1 receptor agonists also cause bronchoconstriction in vivo, and NK_1 receptor antagonists, such as FR113680 (Hagiwara et al., 1992), FK888 (Murai et al., 1993), and CP-96,345 (Griesbacher et al., 1992), inhibit bronchoconstrictor responses to intravenous SP. Capsaicin-evoked bronchoconstriction in the guinea pig, which involves vagal and/or cholinergic reflex and noncholinergic pathways (Buchan and Adcock, 1992), is also inhibited by the mixed NK_1/NK_2 receptor antagonist, FK224 (Murai et al., 1992), and by the NK_1 receptor antagonist, FK888 (Murai et al., 1993). Several studies have indicated that SP-evoked bronchoconstriction in the guinea pig in vivo is mediated predominantly by NK_1 receptors, whereas that evoked by NKA is mediated predominantly by NK_2 receptors (e.g., Murai et al., 1993).

7.3. Mucus Secretion and Mucociliary Clearance

Tachykinins induce mucus secretion from the airways of several species (*see* Maggi, 1990; Frossard and Advenier, 1991), including humans (Rogers et al., 1989). In ferret trachea and ferret tracheal explants, this effect has been attributed to NK_1 receptor activation (Gentry, 1991; Geppetti et al., 1992a). Secretion of mucus from goblet cells in guinea pig trachea is capsaicin-sensitive, suggesting a role for C-fibers in this effect (Kuo et al., 1990).

Mucociliary activity is increased after exposure to airway irritants, such as cigarette smoke and capsaicin. In the upper airways, [Sar^9, Met$(O_2)^{11}$]-SP dose-dependently stimulated mucociliary activity in rabbit maxillary sinus, suggesting that tachykinin release from antidromic stimulation of the maxillary nerve may modulate mucociliary activity via activation of NK_1 receptors in upper airways (Lindberg and Dolata, 1993).

7.4. Pathophysiological Considerations

Various lines of evidence from both animal and human studies implicate tachykinins in inflammatory airway disease. For example, tachykinins are released by antidromic stimulation of vagal afferent C-fibers in the airways by pathologically relevant stimuli (Saria et al., 1988), and capsaicin desensitization in the conscious guinea pig was shown to protect against allergic bronchospasm (Manzini et al., 1987). In the guinea pig, tachykinin release was evoked by toluene diisocyanate. This toxic compound is considered to be a causative agent in occupational asthma (*see* Mapp et al., 1991).

Airway epithelial damage is a characteristic of asthma and may expose sensory nerve endings. It may be relevant that epithelial removal from guinea pig trachea enhanced sensitivity to certain tachykinins, considered to result from the removal of tachykinin degrading enzymes, such as neutral endopeptidase, present in the epithelium (e.g., Frossard et al., 1989). The lipoxygenase metabolites, the hepoxilins, sensitize the guinea pig trachea to NKA (Laneuville et al., 1992), and repeated antigen challenge selectively increases responsiveness to NKA and vagal nonadrenergic, noncholinergic nerve mediated bronchospasm in the guinea pig in vivo (Ballati et al., 1992). Tachykinins stimulate alveolar macrophages via NK_2 receptor stimulation (Brunelleschi et al., 1990, 1992a), and macrophages from sensitized guinea pigs show an increased responsiveness to NK_2 receptor stimulation, suggesting a role in lung disease (Brunelleschi et al., 1992b).

In humans, increased SP levels in bronchoalveolar lavage fluid following allergen provocation was demonstrated (Nieber et al., 1991). Inhaled NKA, though not SP, is bronchoconstrictor in asthmatic, but not in normal subjects (Joos et al., 1987; Fuller et al., 1987). In human airways, capsaicin causes intense coughing, but only transient bronchoconstriction. The latter is considered to be cholinergically mediated, because of the vagal reflex, rather than because of the peripheral release of sensory neuropeptides (Fuller et al., 1985). These various findings support the proposal that capsaicin-sensitive sensory nerves are involved in the pathogenesis of inflammatory disease of the airways.

The proinflammatory mediator, bradykinin, has also been implicated in the etiology of airway inflammatory disease. Bradykinin is a prime stimulant of C-fibers, causing the release of tachykinins and CGRP (*see* Hall, 1992; Geppetti, 1993). Recently, bronchoconstriction and coughing evoked by bradykinin inhalation in asthmatic subjects was reported to be inhibited by the NK_1/NK_2 tachykinin receptor antagonist, FK-224 (Ichinose et al., 1992).

Excessive plasma extravasation is also a characteristic of inflammatory airway disease. The receptors involved in neurogenic plasma extravasation in the airways are discussed in Section 3.3.

7.4.2. Upper Airways and Rhinitis

Tachykinin-containing nerves have been demonstrated, tachykinin binding sites have been demonstrated by autoradiography, and tachykinins induce secretion in the nasal mucosa (*see* Baraniuk and Kaliner, 1991). Intranasal administration of SP induces nasal obstruction in allergic rhinitis patients (though not in nonrhinitics) accompanied by recruitment of inflammatory cells and plasma extravasation that may involve NK_1 and NK_2 receptors (Braunstein et al., 1991).

8. Overview

After some sixty years of study, it is hardly surprising that so many receptor-mediated effects of SP and other tachykinins are defined and catalogued. However, it should be evident from the above account that, in terms of defined function, the situation is less satisfactory. As was pointed out, where there is a coincidence in a particular organ or tissue of demonstrable receptor-mediated responses to the exogenous tachykinin coupled with a potential source of endogenous tachykinins, there is every reason to search for a physiological or pathophysiological role. Here, two main experimental approaches are of value: the first is to look for evidence of release on appropriate stimulation, and the second is to attempt to block the studied physiological manifestations with antagonists.

Direct evidence for the release of TK-LI has generally only been obtained under favorable technical conditions, such as in organs with a dense innervation by sensory afferents, and with powerful stimulants of release, such as capsaicin. In contrast, demonstration of release of measurable amounts of SP-LI on field-stimulation of intrinsic neurons within the myenteric plexus is technically very demanding, though theoretically there is every reason to expect such release *(see above)*.

Clearly, the second approach, that of inhibiting effects of released materials, has much to commend it. There seems little documentation of experiments with antibodies to SP used as chemical antagonists, so the reliance has been on receptor antagonists. In practice, the literature has left much to be desired in terms of clarity of interpretation, and there seem to have been two main problems. One has been the low affinity (and stability) of all but very recently developed tachykinin receptor antagonists; the other is the existence of multiple types of receptor and recently described species variants (Hall et al., 1993c), only some of which are blocked by available ligands.

Consequently, the ambiguities inherent in the use of analytical tools of such limited suitability has not markedly enhanced our understanding of the central or peripheral physiological function for what is, in fact, the earliest discovered neuropeptide. Recently introduced more active ligands, although clearly offering great hope, have not been without problems. First, some have had unexpected additional actions, and it is unfortunate that CP-96,345 is a quite potent calcium channel blocker, since this action alone might be responsible for blockade of some pathophysiological phenomena *(see above)*. Second, this and other NK_1 and NK_2 antagonists have very different affinities at different species-related isoforms of these receptors (*see* Hall et al., 1993c), thus greatly complicating choice of suitable concentrations or doses in in vivo experiments. Third, the continuing lack of active or selective NK_3 receptor antagonists is a serious impairment to progress, and in the meantime only highly selective agonists, such as senktide, allow some progress in our understanding of the effects of NK_3 receptor stimulation, though not of the consequences of their blockade.

Notwithstanding these reservations, there do seem to be some areas where many workers hope that antagonists may prove to be of value in treating pathophysiological manifestations. The possibilities for use of NK_1 or NK_2 receptor ligands in diseases of the airways has raised much interest *(see above)*, and the wide variety of experimental models that exist in this area certainly increases the chances of finding a worthwhile application. It is noteworthy that current ideas regarding the role of tachykinins in airway function involve a quite complex organized response to pathophysiological challenge involving local efferent effects of afferent neurons, local reflexes, and reflexes mediated through other neurotransmitters involving the central nervous system. Perhaps other systems should be approached in the expectation of a multifaceted response phenomenon.

Much earlier hope for therapeutic applications of tachykinin antagonists rested on the anticipation they would prove to be analgesics acting at central sites. The emphasis now seems ready to turn to their use as peripherally-acting analgesics, in that the neurogenic inflammatory response is seen as part of a general peripheral hypersensitivity and hyperreactivity, so that antagonism of the proinflammatory actions of tachykinins released from the peripheral termini of sensory nerves may be, in effect, analgesic.

In conclusion, there have been some advances in our understanding of the roles and actions of tachykinins released from capsaicin-sensitive afferent neurons and capsaicin-insensitive enteric neurons within the gut, and here it may be confidently predicted that new antagonists will allow rapid breakthroughs. For other areas, for instance regarding the role of SP contained in neuroendocrine cells and claimed to exist in endothelial cells, the story is only just beginning.

Acknowledgments

The author wishes to thank Ian K. M. Morton for helpful discussion and for preparing the illustrations, Debra Mitchell for expert technical assistance, and the Wellcome Trust for support.

References

Abelli, L., Ferri, G.-L., Astolfi, M., Conte, B., Geppetti, P., Parlani, M., Dahl, D., Polak, J. M., and Maggi, C. (1991a) Acrylamide-induced visceral neuropathy: Evidence for the involvement of capsaicin-sensitive nerves of the rat urinary bladder. *Neuroscience* **41,** 311–321.

Abelli, L, Maggi, C. A., Rovero, P., Del Bianco, E., Regoli, D., Drapeau, G., and Giachetti, A. (1991b) Effect of synthetic tachykinin analogs on airway microvascular leakage in rats and guinea pigs: Evidence for the involvement of NK-1 receptors. *J. Auton. Pharmacol.* **11,** 267–275.

Abelli, , Nappi, F., Maggi, C. A., Rovero, P., Astolfi, M., Regoli, D., Drapeau, G., and Giachetti, A. (1991c) NK-1 receptors and vascular permeability in rat airways. *Ann. NY Acad. Sci.* **632,** 358,359.

Advenier, C., Naline, E., Toty, L., Bakdach, H., Emonds-Alt, X., Vilain, P., Breliere, J.-C., and Le Fur, G. (1992) Effects on the isolated human bronchus of SR 48968, a potent and selective nonpeptide antagonist of the neurokinin A (NK_2) receptors. *Am. Rev. Respir. Dis.* **146,** 1177–1181.

Alm, P., Alumets, J., Brodin, E., Håkanson, R., Nilsson, G., Sjöberg, N. O., and Sundler, F. (1978) Peptidergic (substance P) nerves in the genito-urinary tract. *Neuroscience* **3,** 419–425.

Almegard, B. and Andersson, S. E. (1990) Outflow facility in the monkey eye: Effects of calcitonin gene-related peptide, cholecystokinin, galanin, substance P and capsaicin. *Exp. Eye Res.* **51,** 685–689.

Amann, R. (1993) Neural regulation of ureteric motility, in *Autonomic Control of the Urogenital System* (Maggi, C. A., ed.), Harwood, Reading, UK, pp. 209–225.

Amann, R., Dray, A., and Hankins, M. W. (1988a) Stimulation of afferent fibers of the guinea pig ureter evokes potentials in inferior mesenteric ganglian neurons. *J. Physiol. (Lond.)* **402,** 543–553.

Amann, R., Skofitsch, G., and Lembeck, F. (1988b) Species-related differences in the capsaicin-sensitive innervation of the rat and guinea pig ureter. *Naunyn Schmiedeberg's Arch. Pharmacol.* **338,** 407–410.

Anderson, J. A., Malfroy, B., Richard, N. R., Kullerstrand, L., Lucas, C., and Binder, P. S. (1990) Substance P contracts the human iris sphincter: possible modulation by endogenous enkephalinase. *Regul. Pept.* **29,** 49–58.

Andersson, S. E. and Almegård, B. (1991) The capsaicin-induced inflammatory reaction in the cat eye: Antagonism by ruthenium red. *Exp. Eye Res.* **52,** 519–523.

Andersson, S. E. and Le Greves, P. (1991) Ruthenium red and capsaicin induce a neurogenic inflammatory response in the rabbit eye: Effects of omega-conotoxin GVIA and tetrodotoxin. *Eur. J. Pharmacol.* **209,** 175–183.

Appell, K. C., Fragale, B. J., Loscig, J., Singh, S., and Tomczuk, B. E. (1992) Antagonists that demonstrate species differences in neurokinin-1 receptors. *Mol. Pharmacol.* **41,** 772–778.

Aronin, N., Leeman, S. E., and Clements, R. S. Jr. (1987) Diminished flare response in neuropathic diabetic patients. Comparison of effects of substance P, histamine, and capsaicin. *Diabetes* **36,** 1139–1143.

Ballati, L, Evangelista, S., and Manzini, S. (1992) Repeated antigen challenge induced airway hyperresponsiveness to neurokinin A and vagal nonadrenergic, noncholinergic (NANC) stimulation in guinea pigs. *Life Sci.* **51,** PL119–PL124.
Baraniuk, J. N. and Kaliner, M. (1991) Neuropeptides and nasal secretion. *Am. J. Physiol. Lung Cell. Mol. Physiol.* **261,** L223–L235.
Baraniuk, J. N., Kowalski, M. L., and Kaliner, M. (1990) Relationships between permeable vessels, nerves, and mast cells in rat cutaneous neurogenic inflammation. *J. Appl. Physiol.* **68,** 2305–2311.
Barja, F., Mathison, R., and Huggel, H. (1983) Substance P-containing nerve fibers in large peripheral blood vessels of the rat. *Cell Tissue Res.* **229,** 411–422.
Barnes, P. J., Belvisi, M. G., and Rogers, D. F. (1990) Modulation of neurogenic inflammation: novel approaches to inflammatory disease. *Trends Pharmacol. Sci.* **11,** 185–189.
Barnes, P. J., Chung, K. F., and Page, C. P. (1988) Inflammatory mediators and asthma. *Pharmacol. Rev.* **40,** 49–84.
Barr, A. J., Watson, S. P., Bernal, A. L., and Nimmo, A. J. (1991) The presence of NK_3 tachykinin receptors on rat uterus. *Eur. J. Pharmacol.* **203,** 287–290.
Barthó, L. and Holzer, P. (1985) Search for a physiological role of substance P in gastrointestinal motility. *Neuroscience* **10,** 1–32.
Barthó, L., Santicioli, P., Patacchini, R., and Maggi, C. (1992) Tachykininergic transmission to the circular muscle of the guinea pig ileum: Evidence for the involvement of NK_2 receptors. *Br. J. Pharmacol.* **105,** 805–810.
Baylis, W. M. (1901) On the origin from spinal cord of the vaso-dilator fibers of the hind-limb, and on the nature of these fibers. *J. Physiol. (Lond.)* **26,** 173–209.
Beattie, D. T., Stubbs, C. M., Conner, H. E., and Feniuk, W. (1993) Neurokinin-induced changes in pial artery diameter in the anesthetized guinea pig. *Br. J. Pharmacol.* **108,** 146–149.
Beding-Barnekow, B., Brodin, E., and Håkanson, R. (1988) Substance P, neurokinin A and neurokinin B in the ocular response to injury in the rabbit. *Br. J. Pharmacol.* **95,** 259–267.
Beding-Barnekow, B. and Brodin, E. (1989) Neurokinin A, neurokinin B and neuropeptide K in the rabbit iris: a study comparing different extraction methods. *Regul. Pept.* **25,** 199–206.
Belmonte, C., Gallar, J., Pozo, M. A., and Rebollo, I. (1991) Excitation by irritant chemical substances of sensory afferent units in the cat's cornea. *J. Physiol. (Lond.)* **437,** 709–725.
Beresford, I. J. M., Birch, P. J., Hagan, R. M., and Ireland, S. J. (1991) Investigation into species variants in tachykinin NK_1 receptors by use of the nonpeptide anatgonist, (±)-CP-96,345. *Br. J. Pharmacol.* **104,** 292,293.
Bergström, L, Beaujouan, J. C., Torrens, Y., Saffroy, M., Glowinski, J., Lavielle, S., Chassaing, G., Marquet, A., D'Orléans-Juste, P., Dion, S., and Regoli, D. (1987) ^{3}H-Neurokinin A labels a specific tachykinin-binding site in the rat duodenal smooth muscle. *J. Pharmacol. Exp. Ther.* **32,** 764–771.
Bertrand, C., Geppetti, P., Baker, J., Yamawaki, I., and Nadel, J. A. (1993) Role of neurogenic inflammation in antigen-induced vascular extravasation in guinea-pig trachea. *J. Immunol.* **150,** 1479–1485.

Bill, A., Stjernschantz, J., Mandahl, A., Brodin, E., and Nilsson, G. (1979) Substance P: Release on trigeminal nerve stimulation, effects in the eye. *Acta Physiol. Scand.* **106,** 371–373.

Birch, P. J., Harrison, S. M., Hayes, A. G., Rogers, H., and Tyers, M. B. (1992) The nonpeptide NK_1 receptor antagonist, (±)-CP-96,345, produces antinociceptive and anti-edema effects in the rat. *Br. J. Pharmacol.* **105,** 508–510.

Black J. L., Diment, M., Alouan, L. A., Johnson, P. R. A., Armour, C., Badgery-Parker, T., and Burcher, E. (1992) Tachykinin receptors in rabbit airways-characterization by functional, autoradiographic and binding studies. *Br. J. Pharmacol.* **107,** 429–436.

Blackwell, M., James, T. A., and Starr, M. S. (1978) Potentiation by substance P of contractions of the isolated vas deferens of the mouse elicited by electrical field stimulation and by drugs. *Br. J. Pharmacol.* **63,** 323–329.

Braunstein, G., Fajac, I., Lacronique, J., and Frossard, N. (1991) Clinical and inflammatory responses to exogenous tachykinins in allergic rhinitis. *Am. Rev. Respir. Dis.* **144,** 630–635.

Brecha, N. (1983) Retinal neurotransmitters, in *Chemical Neuroanaomy* (Emson, P. C., ed.), Raven, New York, pp. 85–129.

Brecha, N. C., Sternini, C., Anderson, K., and Krause, J. E. (1989) Expression and cellular localization of substance P/neurokinin A and neurokinin B mRNAs in the rat retina. *Visual Neurosci.* **3,** 527.

Bristow, D. R., Curtis, N. R., Suman-Chauhan, N., Watling, K. J., and Williams, B. J. (1987) Effects of tachykinins on inositol phospholipid hydrolysis in slices of hamster urinary bladder. *Br. J. Pharmacol.* **90,** 211–217.

Brock, J. W. and Joshua, I. G. (1993) Neural release of substance P causes dilation of arterioles in rat striated muscle. *Regul. Pept.* **43,** 65–71.

Brunelleschi, S., Ceni, E., Fantozzi, R., and Maggi, C. A. (1992a) Evidence for tachykinin NK-2B-like receptors in guinea pig alveolar macophages. *Life Sci.* **51,** PL177–PL181.

Brunelleschi, S., Parenti, A., Ceni, E., Giotti, A., and Fantozzi, R. (1992b) Enhanced responsiveness of ovalbumin-sensitized guinea pig alveolar macophages to tachykinins. *Br. J. Pharmacol.* **107,** 964–969.

Brunelleschi, S., Vanni, L, Ledda, F., Giotti, A., Maggi, C. A., and Fantozzi, R. (1990) Tachykinins activate guinea pig alveolar macrophages: Involvement of NK_2 and NK_1 receptors. *Br. J. Pharmacol.* **100,** 417–420.

Buchan, P. and Adcock, J. J. (1992) Capsaicin-induced bronchoconstriction in the guinea pig: contribution of vagal cholinergic reflexes, local axon reflexes and their modulation by BW443C81. *Br. J. Pharmacol.* **105,** 448–452.

Buck, S. H. and Krstenansky, J. L. (1987) The dogfish peptides scyliorhinin I and scyliorhinin II bind with differential selectivity to mammalian tachykinin receptors. *Eur. J. Pharmacol.* **144,** 109–111.

Bueb, J. L., Mousli, M., Bronner, C., Rouot, B., and Landry, Y. (1990) Activation of Gi-like proteins, a receptor-independent effect of kinins in mast cells. *Mol. Pharmacol.* **38,** 816–822.

Burcher, E. and Buck, S. H. (1986) Multiple tachykinin binding sites in hamster, rat and guinea pig urinary bladder. *Eur. J. Pharmacol.* **128,** 165–177.

Burcher, E., Buck, S. H., Lovenberg, W., and O'Donohue, T. L (1986) Characterization and autoradiographic localization of multiple tachykinin binding sites in gastrointestinal tract and bladder. *J. Pharmacol. Exp. Ther.* **236,** 819–831.

Burnstock, G. (1977) Autonomic neuroeffector junctions-reflex vasodilatation of the skin. *J. Invest. Dermatol.* **69,** 47–57.

Bury, R. W. and Mashford, M. (1977) A pharmacological investigation of synthetic substance P on the isolated guinea pig ileum. *Clin. Exp. Pharmacol. Physiol.* **4,** 453–461.

Butler, J. M. and Hammond, B. R. (1980) The effects of sensory denervation on the responses of the rabbit eye to prostaglandin E_1, bradykinin and substance P. *Br. J. Pharmacol.* **69,** 495–502.

Bynke, G., Håkanson, R., and Horig, J. (1983a) Ocular responses evoked by capsaicin and prostaglandin E_2 are inhibited by a substance P antagonist. *Experientia* **39,** 996–998.

Bynke, G., Håkanson, R., Horig, J., and Leander, S. (1983b) Bradykinin contracts the pupillary sphincter and evokes ocular inflammation through release of neuronal substance P. *Eur. J. Pharmacol.* **91,** 469–475.

Bynke, G., Håkanson, R., and Sundler, F. (1984) Is substance P necessary for corneal nociception? *Eur. J. Pharmacol.* **101,** 253–258.

Carruette, A., Moussaoui, S. M., Champion, A., Cottez, D., Goniot, P., and Garret, C. (1992) Comparison in different tissue preparations of the in vitro pharmacological profile of RP 67580, a new nonpeptide substance P antagonist. *Neuropeptides* **23,** 245–250.

Carstrairs, J. R. and Barnes, P. J. (1986) Autoradiographic mapping of substance P receptors in lung. *Eur. J. Pharmacol.* **127,** 295,296.

Claing, A., Telemaque, S., Cadieux, A., Fournier, A., Regoli, D., and D'Orléans-Juste, P. (1992) Nonadrenergic and noncholinergic arterial dilation and venoconstriction are mediated by calcitonin gene-related peptidel and neurokinin-1 receptors, respectively, in the mesenteric vasculature of the rat after perivascular nerve stimulation. *J. Pharmacol. Exp. Ther.* **263,** 1226–1232.

Constantine, J. W., Lebel, W. S., and Woody, H. A. (1990) Smooth muscle of rabbit isolated aorta contains the NK-2 tachykinin receptor. *Naunyn Schmiedeberg's Arch. Pharmacol.* **342,** 722–724.

Constantine, J. W., Lebel, W. S. and Woody, H. A. (1991) Inhibition of tachykinin-induced hypotension in dogs by CP-96,345, a selective blocker of NK-1 receptors. *Naunyn Schmiedeberg's Arch. Pharmacol.* **344,** 471–477.

Crowe, R., Haven, A. J., and Burnstock, G. (1986) Intramural neurons of the guinea pig urinary bladder: histochemical localization of putative neurotransmitters in cultures and newborn animals. *J. Auton. Nerv. System* **15,** 319–339.

D'Orléans-Juste, P., Claing, A., Telemaque, S., Warner, T. D., and Regoli, D. (1991) Neurokinins produce selective venoconstriction via NK-3 receptors in the rat mesenteric vascular bed. *Eur. J. Pharmacol.* **204,** 329–334.

D'Orléans-Juste, P., Dion, S., Mizrahi, J., and Regoli, D. (1986) Effect of peptides and nonpeptides on isolated arterial smooth muscles: role of endothelium. *Eur. J. Pharmacol.* **114,** 9–21.

Deacon, C. F., Agoston, D. V., Nau, R., and Conlon, J. M. (1987) Conversion of neuropeptide K to neuropeptide A and vesicular colocalization of neurokinin A and substance P in neurons of the guinea pig small intestine. *J. Neurochem.* **48,** 141–146.

Delay-Goyet, P. and Lundberg, J. M. (1991) Cigarette smoke-induced airway edema is blocked by the NK_1 antagonist, CP-96,345. *Eur. J. Pharmacol.* **203,** 157,158.

Delay-Goyet, P., Satoh, H., and Lundberg, J. M. (1992) Relative involvement of substance P and CGRP mechanisms in antidromic vasodilation in the rat skin. *Acta Physiol. Scand.* **146,** 537,538.

Denis, P., Fardin, V., Nordmann, J.-P., Elena, P.-P., Laroche, L., Saraux, H., and Rostene, W. (1991) Localization and characterization of substance P binding sites in rat and rabbit eyes. *Invest. Ophthalmol. Vis. Sci.* **32,** 1894–1902.

Devillier, P., Acker, G. M., Advenier, C., Marsac, J., Regoli, D., and Frossard, N. (1992) Activation of an epithelial neurokinin NK-1 receptor induces relaxation of rat trachea through release of prostaglandin E_2. *J. Pharmacol. Exp. Ther.* **263,** 767–772.

Devillier, P., Weill, B., Renoux, M., Menkes, C., and Pradelles, P. (1986) Elevated levels of TK-LI in joint fluids from patients with rheumatic inflammatory diseases. *N. Engl. J. Med.* **314,** 1323.

Dion, S., D'Orléans-Juste, P., Drapeau, G., Rhaleb, N.-E., Rouissi, N., Tousignant, C., and Regoli, D. (1987) Characterization of neurokinin receptors in various isolated isolated organs by use of selective agonists. *Life Sci.* **41,** 2269–2278.

Dion, S., Rouissi, N., Nantel, F., Drapeau, G., Regoli, D., Naline, E., and Advenier, C. (1990) Receptors for neurokinins in human bronchus and urinary bladder are of the NK-2 type. *Eur. J. Pharmacol.* **178,** 215–219.

Domoto, T., Gonda, T., Oki, M., and Yanaihara, N. (1984) Coexistence of substance P-and methionine 5-enkephalin-like immunoreactivity in nerve cells of the myenteric ganglia in the cat ileum. *Neurosci. Lett.* **47,** 9–13.

Donnerer, J., Barthó, L., Holzer, P., and Lembeck, F. (1984) Intestinal peristalsis associated with release of immunoreactive substance P. *Neuroscience* **11,** 913–918.

Donnerer, J., Stark, U., Tritthart, H. A., and Lembeck, F. (1992) CP-96,345, a nonpeptide antagonist of substance P. III. Cardiovascular effects in mammals unrelated to actions on substance P receptors. *Naunyn Schmiedeberg's Arch. Pharmacol.* **346,** 328–332.

Doutremepuich, J. D., Barbier, A., Vilain, P., and Lacheretz, F. (1992) Effect of substance P and $Sar^9Met(O_2)^{11}$-substance P on cutaneous capillary permeability in guinea pig skin. *Skin Pharmacol.* **5,** 171–176.

Dray, A., Hankins, M. W., and Yeats, J. C. (1989) Desensitization and capsaicin-induced release of substance P-like immunoreactivity from guinea pig ureter in vitro. *Neuroscience* **31,** 479–483.

Eglezos, A., Andrews, P. V., Boyd, R., and Helme, R. D. (1991a) Tachykinin-mediated modulation of the primary antibody response in rats: Evidence for mediation by an NK-2 receptor. *J. Neuroimmunol.* **32,** 11–18.

Eglezos, A., Giuliani, S., Viti, G., and Maggi, C. A. (1991b) Direct evidence that capsaicin-induced plasma protein extravasation is mediated through tachykinin NK_1 receptors. *Eur. J. Pharmacol.* **209,** 277–279.

Ehinger, B., Sundler, F., Tervo, K., Tervo, T., and Tornqvist, K. (1983) Substance P fibers in the anterior segmant of the rabbit eye. *Acta Physiol. Scand.* **118,** 215–218.

Ellis, J. L. and Burnstock, G. (1989) Modulation of neurotransmission in the guinea pig vas deferens by capsaicin: involvement of calcitonin gene-related peptide and substance P. *Br. J. Pharmacol.* **98,** 707–713.

Emonds-Alt, X., Golliot, F., Pointeau, P., Le Fur, G., and Breliere, J. C. (1993) Characterization of the binding sites of [^{3}H]SR48968, a potent nonpeptide radioligand antagonist of the neurokinin-2 receptor. *Biochem. Biophys. Res. Commun.* **191,** 1170–1177.

Emonds-Alt, X., Vilain, P., Goulaouic, P., Proietto, V., Van Broeck, D., Advenier, C., Naline, E., Neliat, G., Le Fur, G., and Breliere, J. C. (1992) A potent and selective nonpeptide antagonist of the neurokinin A (NK_2) receptor. *Life Sci.* **50,** PL101–PL106.

Escott, K. J. and Brain, S. D. (1993) The inhibitory effect of a calcitonin generelated peptide receptor antagonist ($CGRP_{8-37}$) on the local increased blood flow and edema formation induced in rat skin after electrical stimulation of the saphenous nerve. *Br. J. Pharmacol.* In press.

Evans, T. W., Dixon, C. M., Clarke, B., Conradson, T. B., and Barnes, P. J. (1988) Comparison of neurokinin A and substance P on cardiovascular and airway function in man. *Br. J. Clin. Pharmacol.* **25,** 273–275.

Falconieri-Erspamer, G., Negri, L., and Piccinelli, D. (1973) The use of preparations of urinary bladder smooth muscle for bioassays of and discrimination between polypeptides. *Naunyn Schmiedeberg's Arch. Pharmacol.* **279,** 61–74.

Featherstone, R. L, Fosbraey, P., and Morton, I. K. M. (1986) A comparison of the effects of three substance P antagonists on tachykinin-stimulated [^{3}H]acetylcholine release in the guinea pig ileum. *Br. J. Pharmacol.* **87,** 73–78.

Ferrell, W. R. and Russell, N. J. W. (1986) Extravasation in the knee induced by antidromic stimulation of articular C fibre afferents of the anesthetized cat. *J. Physiol. (Lond.)* **379,** 407–416.

Foreman, J. and Jordan, C. (1983) Histamine release and vascular changes induced by neuropeptides. *Agents Actions* **13,** 105–116.

Fosbraey, P., Featherstone, R. L., and Morton, I. K. M. (1984) Comparison of potency of substance P and related peptides on [^{3}H]-acetylcholine release, and contractile actions, in guinea-pig ileum. *Naunyn Schmiedeberg's Arch. Pharmacol.* **326,** 111–115.

Fox, A. J. and Morton, I. K. M. (1990) An examination of the 5-HT_3 receptor mediating contraction and evoked [^{3}H]-acetylcholine release in the guinea pig ileum. *Br. J. Pharmacol.* **101,** 553–558.

Franco, R., Costa, M., and Furness, J. B. (1979) Evidence that axons containing substance P in the guinea pig ileum are of intrinsic origin. *Naunyn Schmiedeberg's Arch. Pharmacol.* **307,** 57–63.

Frossard, N. and Advenier, C. (1991) Tachykinin receptors and the airways. *Life Sci.* **49,** 1941–1953.

Frossard, N., Rhoden, K. J., and Barnes, P. J. (1989) Influence of epithelium on guinea pig airway responses to tachykinins: role of endopeptidase and cyclooxygenase. *J. Pharmacol. Exp. Ther.* **248,** 292–298.

Fujii, T., Murai, M., Morimoto, H., Maeda, Y., Yamaoka, M., Hagiwara, D., Miyake, H., Ikari, N. and Matsuo, M. (1992) Pharmacological profile of a high affinity dipeptide NK_1 receptor antagonist, FK888. *Br. J. Pharmacol.* **107,** 785–789.

Fuller, R. W., Dixon, C. M. S., and Barnes, P. J. (1985) Bronchoconstrictor response to inhaled capsaicin in humans. *J. Appl. Physiol.* **58,** 1080–1084.

Fuller, R. W., Maxwell, D. L., Dixon, C. M. S., McGregor, G. P., Barnes, V. F., Bloom, S. R., and Barnes, P. J. (1987) Effect of substance P on cardiovascular and respiratory function in subjects. *J. Appl. Physiol.* **62,** 1473–1479.

Furness, J. B., Costa, M., and Keast, J. R. (1984) Choline acetyltransferase and peptide immunoreactivity of submucous neurons in the small intestine of the guinea pig. *Cell Tissue Res.* **237,** 329–336.

Furness, J. B., Papka, R. E., Della, N. G., Costa, M., and Eskay, R. L. (1982) Substance P-like immunoreactivity in nerves associated with the vascular system of guinea pigs. *Neuroscience* **7,** 447–459.

Gaddum, J. H. and Schild, H. (1934) Depressor substances in extracts of intestine. *J. Physiol. (Lond.)* **83,** 1–14.

Garret, C., Carruette, A., Fardin, V., Moussaoui, S., Peyronel, J.-F., Blanchard, J.-C., and Laduron, P. M. (1991) Pharmacological properties of a potent and selective nonpeptide substance P antagonist. *Proc. Natl. Acad. Sci. USA* **88,** 10,208–10,212.

Garrett, N. E., Mapp, P. I., Cruwys, S. C., Kidd, B. L., and Blake, D. R. (1992) Role of substance P in inflammatory arthritis. *Ann. Rheum. Dis.* **51,** 1014–1018.

Gates, T. S., Zimmermann, R. P., Mantyh, C. R., Vigna, S. R., Maggio, J. E., Welton, M. L, Passaro, E. P. Jr., and Mantyh, P. W. (1989) Substance P and substance K receptor binding sites in the human gastrointestinal tract: localization by autoradiography. *Peptides* **9,** 1207–1219.

Gentry, S. E. (1991) Tachykinin receptors mediating airway macromolecular secretion. *Life Sci.* **48,** 1609–1618.

Geppetti, P. (1993) Sensory neuropeptide release by bradykinin: mechanisms and pathophysiological implications. *Regul. Pept.* **47,** 1–23.

Geppetti, P., Bertrand, C., Bacci, E., Snider, R. M., Maggi, C. A. and Nadel, J. (1992a) The NK-l receptor antagonist, CP96,345 blocks substance P-evoked mucus secretion from the ferret trachea. *Neuropeptides* **22,** 25.

Geppetti, P., Del Bianco, E., Cecconi, R., Tramontana, M., Romani, A., and Theodorsson, E. (1992b) Capsaicin releases calcitonin gene-related peptide from the human iris and ciliary body in vitro. *Regul. Pept.* **41,** 83–92.

Geraghty, D. P., Mussap, C. J., and Burcher, E. (1992) Radioiodinated substance P, neurokinin A, and eledoisin bind predominantly to NK_1 receptors in guinea pig lung. *Mol. Pharmacol.* **41,** 147–153.

Gernandt, B. (1942) Untersuchung uber die biologische Wirkung der Substanz P. *Acta Physiol. Scand.* **3,** 270–274.

Gibbins, I. L, Furness, J. B., Costa, M., MacIntyre, I., Hillyard, C., and Girgis, S. (1985) Co-localization of calcitonin gene-related peptide with substance P in cutaneous, vascular and visceral sensory neurons of guinea-pigs. *Neurosci. Lett.* **57,** 125–130.

Gibbins, I. L, Furness, J. B., and Costa, M. (1987) Pathway-specific patterns of the co-existence of substance P, calcitonin gene-related peptide, cholecystokinin and dynorphin in neurons of the dorsal root ganglian of the guinea pig. *Cell Tissue Res.* **248,** 417–437.

Glickman, R. D. and Adolph, A. R. (1982) Acetylcholine and substance P: action via distinct receptors on carp retinal ganglian cells. *Invest. Ophthalmol. Vis. Sci.* **22,** 804–808.

Goldin, E., Karmeli, F., Selinger, Z., and Rachmilewitz, D. (1989) Colonic substance P levels are increased in ulcerative colitis and decreased in chronic severe constipation. *Dig. Dis. Sci.* **34,** 754–757.

Griesbacher, T., Donnerer, J., Legat, F. J., and Lembeck, F. (1992) CP-96,345, a nonpeptide antagonist of SP: II. Actions on substance P-induced hypotension and bronchoconstriction, and on depressor reflexes in mammals. *Naunyn Schmiedeberg's Arch. Pharmacol.* **346,** 323–327.

Guard, S., Pain, D., Franks, R., and Watling, K. J. (1993) [^{125}I]Neurokinin A labels pharmacologically distinct populations of NK_2 binding sites in hamster and rabbit urinary bladder. *Eur. J. Pharmacol.* **232,** 287–290.

Guard, S., Watling, K. J., and Watson, S. P. (1988) Neurokinin-3-receptors are linked to inositol phospholipid hydrolysis in the guinea pig ileum longitudinal muscle-myenteric plexus preparation. *Br. J. Pharmacol.* **94,** 148–154.

Guard, S. and Watson, S. P. (1987) Evidence for neurokinin-3 receptor-mediated tachykinin release in the guinea pig ileum. *Eur. J. Pharmacol.* **144,** 409–412.

Guard, S., Watson, S. P., Maggio, J. E., Too, H. P., and Watling, K. J. (1990) Pharmacological analysis of [^{3}H]-senktide binding to NK_3 tachykinin receptors in guinea pig ileum longitudinal muscle-myenteric plexus and cerebral cortex membranes. *Br. J. Pharmacol.* **99,** 767–773.

Hagiwara, D., Miyake, H., Morimoto, H., Murai, M., Fujii, T., and Matsuo, M. (1992) Studies on neurokinin antagonists. 2. Design and structure-activity relationships of novel tripeptide substance P antagonists, *N*a-[*N*a-(*N*a-acetyl-L-threonyl)-*N*1-formyl-D-tryptophyl]-*N*-methyl-*N*-(phenylmethyl)-L-phenylalaninamide and its related compounds. *J. Med. Chem.* **35,** 3184–3191.

Håkanson, R., Beding, B., Ekman, R., Heilig, M., Wahlstedt, C., and Sundler, F. (1987) Multiple tachykinin pools in sensory nerve fibers in the rabbit iris. *Neuroscience* **21,** 943–950.

Hall, J. M. (1992) Bradykinin receptors: pharmacological properties and biological roles. *Pharmacol. Ther.* **56,** 131–190.

Hall, J. M., Flowers, J. M., and Morton, I. K. M. (1992) A pharmacological study of NK_1 and NK_2 tachykinin receptor characteristics in the rat isolated urinary bladder. *Br. J. Pharmacol.* **107,** 777–784.

Hall, J. M., Mitchell, D., and Morton, I. K. M. (1991) Neurokinin receptors in the rabbit iris sphincter characterised by novel agonist ligands. *Eur. J. Pharmacol.* **199,** 9–14.

Hall, J. M., Mitchell, D., and Morton, I. K. M. (1993a) Tachykinin receptors mediating responses to sensory nerve stimulation and exogenous tachykinins and analogs in the rabbit isolated iris sphincter. *Br. J. Pharmacol.* **109,** 1008–1013.

Hall, J. M., Mitchell, D., and Morton, I. K. M. (1993b) Tachykinin NK_1 receptor ligands show two distinct pharmacological profiles in rabbit iris sphincter. *Br. J. Pharmacol.* In press.

Hall, J. M., Caulfield, M. P., Watson, S. P. and Guard, S. (1993c) Receptor subtypes or species homologues: Revelance to drug discovery. *Trends Pharmacol. Sci.* **14,** 376–383.

Hall, J. M. and Morton, I. K. M. (1990) Mechanism of action of substance P in guinea pig ileum: a re-evaluation. *J. Physiol. (Lond.)* **431,** 529–541.

Hall, J. M. and Morton, I. K. M. (1991a) Novel selective agonists and antagonists confirm neurokinin NK_1 receptors in guinea pig vas deferens. *Br. J. Pharmacol.* **102,** 511-517.

Hall, J. M. and Morton, I. K. M. (1991b) Subtypes and excitation-contraction coupling mechanisms for neurokinin receptors in smooth muscle of the guinea pig *Taenia caeci. Naunyn Schmiedeberg's Arch. Pharmacol.* **344,** 225–234.

Hanani, M., Chorev, M., Gilon, C., and Selinger, Z. (1988) The actions of receptor-selective substance P analogs on myenteric neurons: an electrophysiological investigation. *Eur. J. Pharmacol.* **153,** 247–253.

Hartung, H.-P., Wolters, K., and Toyka, K. V. (1986) Substance P binding properties and studies on cellular responses in guinea pig macrophages. *J. Immunol.* **136,** 3856–3863.

Heinrich, D., Reinecke, M. and Forssmann, W. G. (1986) Peptidergic innervation of the guinea pig uterus. *Arch Gynecol.* **237,** 213.

Helme, R. D., Koschorke, G. M., and Zimmermann, M. (1986) Immunoreactive substance P release from skin nerves in the rat by noxious thermal stimulation. *Neurosci. Lett.* **63,** 295–299.

Holmdahl, G., Håkanson, R., Leander, S., Rosell, S., Folkers, K., and Sundler, F. (1981) A substance P antagonist (D-Pro^2,D-$Trp^{7,9}$)SP, inhibits inflammatory responses in the rabbit eye. *Science* **214,** 1029–1031.

Holzer, P. (1988) Local effector functions of capsaicin-sensitive sensory nerve endings: Involvement of tachykinins, calcitonin gene-related peptide and other neuropeptides. *Neuroscience* **24,** 739–768.

Holzer, P. (1992) Peptidergic sensory neurons in the control of vascular functions: mechanisms and significance in the cutaneous and splanchnic vascular beds. *Rev. Physiol. Biochem. Pharmacol.* **121,** 49–146.

Holzer, P., Bucsics, A., and Lembeck, F. (1982) Distribution of capsaicin sensitive nerve fibers containing immunoreactive substance P in cutaneous and visceral tissues of the rat. *Neurosci. Lett.* **31,** 253–257.

Holzer, P. and Lembeck F. (1980) Neurally mediated contraction of ileal longitudinal muscle by substance P. *Neurosci. Lett.* **17,** 101–105.

Holzer, P., Schluet, W., and Maggi, C. A. (1993) Ascending enteric reflex contraction: roles of acetylcholine and tachykinins in relation to distension and propagation of excitation. *J. Pharmacol. Exp. Ther.* **264,** 391–396.

Hua, X.-Y. and Lundberg, J. M. (1986) Dual capsaicin effects on ureteric motility: low dose inhibition mediated by calcitonin gene-related peptide and high dose stimulation by tachykinins? *Acta Physiol. Scand.* **128,** 453–465.

Hughes, S. R. and Brain, S. D. (1991) A calcitonin gene-related peptide (CGRP) antagonist ($CGRP_{8-37}$) inhibits microvascular responses induced by CGRP and capsaicin in skin. *Br. J. Pharmacol.* **104,** 738–742.

Ichinose, M., Nakajima, N., Takahashi, T., Yamauchi, H., Inoue, H., and Takishima, T. (1992) Protection against bradykinin-induced bronchoconstriction in asthmatic patients by neurokinin receptor antagonist. *Lancet* **340,** 1248–1251.

Ireland, S. J., Bailey, F., Cook, A., Hagan, R. M., Jordan, C. C., and Stephens-Smith, M. L. (1991) Receptors mediating tachykinin-induced contractile responses in guinea pig trachea. *Br. J. Pharmacol.* **103,** 1463–1469.

Jacques, L, Couture, R., Drapeau, G., and Regoli, D. (1989) Capillary permeability induced by intravenous neurokinins. Receptor characterization and mechanism of action. *Naunyn Schmiedeberg's Arch Pharmacol.* **30,** 170–179.

Jancsó, G., Obal, F. Jr., Toth-Kasa, I., Katona, M., and Husz, S. (1985) The modulation of cutaneous inflammatory reactions by peptide-containing sensory nerves. *Int. J. Tissue React.* **7,** 449–457.

Joos, G., Pauwels, R., and van der Straeton, M. (1987) Effect of inhaled substance P and neurokinin A on the airways of normal and asthmatic subjects. *Thorax* **42,** 779–783.

Kachur, J. F., Miller, H. J., Field, M., and Rivier, J. (1982) Neurohormonal control of ileal electrolyte transport II, neurotensin and substance P. *J. Pharmacol. Exp. Ther.* **220,** 456–463.

Kastin, A. J., Coy, D. H., Schally, A. V., and Meyers, C. A. (1978) Activity of VIP, somatostatin and other peptides in the mouse vas deferens assay. *Pharmacol. Biochem. Behav.* **9,** 673–676.

Katayama, Y. and North, R. A. (1978) Does substance P mediate slow synaptic excitation within the myenteric plexus? *Nature* **274,** 387,388.

Keast, J. R., Furness, J. B., and Costa, M. (1985) Different substance P receptors are found on mucosal epithelial cells and submucous neurons of the guinea pig small intestine. *Naunyn Schmiedeberg's Arch. Pharmacol.* **329,** 382–387.

Keen, P., Tullo, A. B., Blyth, W. A. and Hill, T. J. (1982) Substance P in the mouse cornea: effects of chemical and surgical denervation. *Neurosci. Lett.* **29,** 231–235.

Kidd, B. L, Mapp, P. I., Blake, D. R., Gibson, S. J., and Polak, J. M. (1990) Neurogenic influences in arthritis. *Ann. Rheumatic Dis.* **49,** 649–652.

Kieselbach, G. F., Ragaut, R., Knaus, H. G., Konig, P., and Wiedermann, C. J. (1990) Autoradiographic analysis of binding sites for Bolton-Hunter-substance P in the human eye. *Peptides* **11,** 655–659.

Kieselbach, G. F., Ragaut, R., Knaus, H. G., Konig, P., Saria, A., and Wiedermann, C. J. (1991a) Neurokinin-1 receptors in the human eye: characterization and autoradiographic distribution. *Ann. NY Acad. Sci.* **632,** 466,467.

Kieselbach, G. F., Wiedermann, C., Kunze, C., Gottinger, W., Humpel, C., and Saria, A. (1991b) Elevated levels of substance P in intraocular fluid in proliferative vitreoretinopathy. *Peptides* **12,** 275–278.

Kilbinger, H., Stauss, P., Erlhof, I., and Holzer, P. (1986) Antagonist discrimination between subtypes of tachykinin receptors in the guinea pig ileum. *Naunyn Schmiedeberg's Arch. Pharmacol.* **334,** 181–187.

Kowalski, M. L, Sliwinska-Kowalska, M., and Kaliner, M. A. (1990) Neurogenic inflammation, vascular permeability, and mast cells. 2. Additional evidence that mast cells are not involved in neurogenic inflammation. *J. Immunol.* **145,** 1214–1221.

Krumins, S. A. and Broomfield, C. (1992) Evidence of NK_1 and NK_2 tachykinin receptors and their involvement in histamine release in a murine mast cell line. *Neuropeptides* **21,** 65–72.

Kuo, H.-P., Rohde, J. A. L, Tokuyama, K., Barnes, P. J., and Rogers, D. F. (1990) Capsaicin and sensory neuropeptide stimulation of goblet cell secretion in guinea pig trachea. *J. Physiol. (Lond.)* **431,** 629–641.

Lam, F. Y. and Ferrell, W. R. (1991) Specific neurokinin receptors mediate plasma extravasation in the rat knee joint. *Br. J. Pharmacol.* **103,** 1263–1267.

Lam, F. Y. and Ferrell, W. R. (1993) Effects of interactions of naturally-occurring neuropeptides on blood flow in the rat knee joint. *Br. J. Pharmacol.* **108,** 694–699.

Laneuville, O., Couture, R., and Pace-Asciak, C. R. (1992) Neurokinin A-induced contraction of guinea pig isolated trachea: potentiation by hepoxilins. *Br. J. Pharmacol.* **107,** 808–812.

Laufer, R., Gilon, C., Chorev, M., and Selinger, Z. (1988) Desensitization with a selective agonist discriminates between multiple tachykinin receptors. *J. Pharmacol. Exp. Ther.* **245,** 639–643.

Laufer, R., Wormser, U., Friedman, Z. Y., Gilon, C., Chorev, M., and Selinger, Z. (1985) Neurokinin B is a preferred agonist for a neuronal substance P receptor and its action is antagonized by enkephalin. *Proc. Natl. Acad. Sci. USA* **82,** 7444–7448.

Lee, C.-M., Campbell, N. J., Williams, B. J., and Iversen, L. L. (1986) Multiple tachykinin binding sites in peripheral tissues and in brain. *Eur. J. Pharmacol.* **130,** 209–217.

Lee, C.-M., Iversen, L. L., Hanley, M. R., and Sandberg, B. E. B. (1982) The possible existence of multiple receptors for substance P. *Naunyn Schmiedeberg's Arch. Pharmacol.* **318,** 281–287.

Lee, C. M. and Cheung, W. T. (1988) Effects of neonatal monosodium glutamate treatment on substance P binding sites in the rat retina. *Neurosci. Lett.* **92,** 310.

Legay, C., Saffrey, M. J., and Burnstock, G. (1984) Coexistence of immunoreactive substance P and serotonin in neurons of the gut. *Brain Res.* **302,** 379–382.

Lei, Y.-H., Barnes, P. J., and Rogers, D. F. (1992) Inhibition of neurogenic plasma exudation in guinea pig airways by CP-96,345, a new nonpeptide NK_1 receptor antagonist. *Br. J. Pharmacol.* **105,** 261,262.

Lembeck, F. and Donnerer, J. (1981) Timecourse of capsaicin-induced functional impairments in comparison with changes in neuronal substance P content. *Naunyn Schmiedeberg's Arch. Pharmacol.* **316,** 240–243.

Lembeck, F., Donnerer, J., Tsuchiya, M., and Nagahisa, (1992) The non-peptide tachykinin antagonist, CP-96,345, is a potent inhibitor of neurogenic inflammation. *Br. J. Pharmacol.* **105,** 527–530.

Levine, J. D., Clark, R., Devor, M., Helms, C., Moskowitz, M., and Basbaum, A. I. (1984) Intraneuronal substance P contributes to the severity of experimental arthritis. *Science* **226,** 547–549.

Lindberg, S. and Dolata, J. (1993) NK_1 receptors mediate the increase in mucociliary activity produced by tachykinins. *Eur. J. Pharmacol.* **231,** 375–380.
Loesch, A. and Burnstock, G. (1988) Ultrastructural localisation of serotonin and substance P in vascular endothelial cells of rat femoral and mesenteric arteries. *Anat. Embryol. (Berl)* **178,** 137–142.
Lograno, M. and Daniele, E. (1988) Substance P as a transmitter in the human ciliary muscle. *Pharmacol. Res. Commun.* **20,** 901–905.
Longmore, J. and Hill, R. G. (1992) Characterisation of NK receptors in guinea pig urinary bladder smooth muscle: Use of selective antagonists. *Eur. J. Pharmacol.* **222,** 167–170.
Lundberg, J. M., Franco-Cereceda, A., Hua, X.-Y., Hokfelt, T., and Fischer, J. (1985) Co-existence of substance P and calcitonin gene-related peptide immunoreactivities in sensory nerves in relation to cardiovascular and bronchoconstrictor effects of capsaicin. *Eur. J. Pharmacol.* **108,** 315–319.
Lundberg, J. M., Hokfelt, T., Martling, C. R., Saria, and Cuello, C. (1984a) Substance P-immunoreactive sensory nerves in the lower respiratory tract of various mammals including man. *Cell Tissue Res.* **235,** 251–261.
Lundberg, J. M., Hua, X.-Y., and Fredholm, B. B. (1984b) Capsaicin-induced stimulation of the guinea pig atrium. Involvement of a novel sensory transmitter or a direct action on myocytes? *Naunyn Schmiedeberg's Arch. Pharmacol.* **325,** 176–182.
Lundberg, J. M. and Saria, A. (1987) Polypeptide-containing neurons in airway smooth muscle. *Annu. Rev. Physiol.* **49,** 557–572.
Lynn, B. (1988) Neurogenic inflammation. *Skin Pharmacol.* **1,** 217–224.
Maggi, C. A. (1990) Tachykinin receptors in the airways and lung: what should we block? *Pharmacol. Res.* **22,** 527–540.
Maggi, C. A. (1991a) The pharmacology of the efferent function of sensory nerves. *J. Auton. Pharmacol.* **11,** 173–208.
Maggi, C. A. (1991b) The role of peptides in the regulation of the micturition reflex: An update. *Gen. Pharmacol.* **22,** 1–24.
Maggi, C. A., Borsini, F., Santicioli, P., Geppetti, P., Abelli, L, Evangelista, S., Manzini, S., Theodorsson-Norheim, E., Somma, V., Amenta, F., Bacciarelli, C., and Meli, A. (1987a) Cutaneous lesions in capsaicin-pretreated rats. A trophic role of capasicin-sensitive afferents? *Naunyn Schmiedeberg's Arch. Pharmacol.* **336,** 538–545.
Maggi, C. A., Giuliani, S., Santicioli, P., Abelli, L, Regoli, D., and Meli, A. (1987b) Further studies on the mechanisms of the tachykinin-induced activation of micturition reflex in rats: evidence for the involvement of the capsaicin-sensitive bladder mechanoreceptors. *Eur. J. Pharmacol.* **136,** 189–205.
Maggi, C. A., Giuliani, S., Ballati, L, Rovero, P., Abelli, L., Manzini, S., Giachetti, A., and Meli, A. (1990a) *In vivo* pharmacology of [β-Ala^8]neurokinin A-(4–10), a selective NK-2 tachykinin receptor agonist. *Eur. J. Pharmacol.* **177,** 81–86.
Maggi, C. A., Giuliani, S., Santicioli, P., Abelli, L., and Giachetti, A. (1991) Facilitation of reflex micturition by intravesical administration of [β-Ala^8]-neurokinin A (4–10), a selective NK-2 tachykinin receptor agonist. *J. Urol.* **145,** 184–187.

Maggi, C, Giuliani, S., Patacchini, R., Santicioli, P., Theodorsson, E., Barbanti, G., Turini, D., and Giachetti, A. (1992a) Tachykinin antagonists inhibit nerve-mediated contractions in the circular muscle of the human ileum. Involvement of neurokinin-2 receptors. *Gastroenterology* **102,** 88–96.

Maggi, C. A. and Meli, A. (1986) The role of neuropeptides in the regulation of the micturition reflex. *J. Auton. Pharmacol.* **6,** 133–162.

Maggi, C. A., Parlani, M., Astolfi, M., Santicioli, P., Rovero, P., Abelli, L., Somma, V., Giuliani, S., Regoli, D., Patacchini, R., and Meli, A. (1988) Neurokinin receptors in the rat lower urinary tract. *J. Pharmacol. Exp. Ther.* **246,** 308–315.

Maggi, C. A., Patacchini, R., Giuliani, S., Rovero, P., Dion, S., Regoli, D., Giachetti, A., and Meli, A. (1990b) Competitive antagonists discriminate between NK_2 tachykinin receptor subtypes. *Br. J. Pharmacol.* **100,** 588–592.

Maggi, C. A., Patacchini, R., Santicioli, P., Giuliani, S., Turini, D., Barbanti, G., Giachetti, A., and Meli, A. (1990c) Human isolated ileum: Motor responses of the circular muscle to electrical field stimulation and exogenous neuropeptides. *Naunyn Schmiedeberg's Arch. Pharmacol.* **341,** 256–261.

Maggi, C. A., Patacchini, R., Quartara, L, Rovero, P., and Santicioli, P. (1991a) Tachykinin receptors in the guinea pig isolated bronchi. *Eur. J. Pharmacol.* **197,** 167–174.

Maggi, C. A., Patacchini, R., Santicioli, P., and Giuliani, S. (1991b) Tachykinin antagonists and capsaicin-induced contraction of the rat isolated urinary bladder: Evidence for tachykinin-mediated cotransmission. *Br. J. Pharmacol.* **103,** 1535–1541.

Maggi, C. A. and Patacchini, R. (1992a) Tachykinin NK_1 receptor in the guinea pig isolated proximal urethra: Characterization by receptor selective agonists and antagonists. *Br. J. Pharmacol.* **106,** 888–892.

Maggi, C. A., Patacchini, R., Eglezos, A., Quartara, L, Giuliani, S., and Giachetti, A. (1992b) Tachykinin receptors in the guinea pig renal pelvis: Activation by exogenous and endogenous tachykinins. *Br. J. Pharmacol.* **107,** 27–33.

Maggi, C. A., Theodorsson, E., Santicioli, P., and Giuliani, S. (1992c) Tachykinins and calcitonin gene-related peptide as co-transmitters in local motor responses produced by sensory nerve activation in the guinea pig isolated renal pelvis. *Neuroscience* **46,** 549–559.

Mandahl, A. (1985) Effects of the substance P antagonist, [D-Arg^1,D-Pro_2,D-$Trp^{7,9}$,Leu^{11}] SP on miosis caused by echothiophate iodide or pilocarpine hydrochloride. *Eur. J. Pharmacol.* **114,** 121–127.

Mandahl, A., Brodin, E., Nilsson, G., and Bill, A. (1980) Substance P, release and effects in the eye. *Acta Physiol. Scand.* **108,** 18A.

Mantyh, C. R., Gates, T. S., Zimmerman, R. P., Welton, M. L, Passaro, E. P. Jr., Vigna, S. R., Maggio, J. E., Kruger, L., and Mantyh, P. W. (1988) Receptor binding sites for substance P, but not substance K or neuromedin K, are expressed in high concentrations by arterioles, venules, and lymph nodules in surgical specimens obtained from patients with ulcerative colitis and Crohn disease. *Proc. Natl. Acad. Sci. USA* **85,** 3235–3239.

Manzini, S. (1992) Bronchodilatation by tachykinins and capsaicin in the mouse main bronchus. *Br. J. Pharmacol.* **105,** 968–972.

Manzini, S., Maggi, C. A., Geppetti, P., and Bacciarelli, C. (1987) Capsaicin desensitization protects from antigen-induced bronchospasm in conscious guinea pigs. *Eur. J. Pharmacol.* **138,** 307,308.

Mapp, C. E., Graf, P. D., Boniotti, A., and Nadel, J. A. (1991) Toluene diisocyanate contracts guinea pig bronchial smooth muscle by activating capsaicin-sensitive sensory nerves. *J. Pharmacol. Exp. Ther.* **256,** 1082–1085.

Marabini, S., Matucci-Cerinic, M., Geppetti, P., Del Bianco, E., Marchesoni, A., Tosi, S., Cagnoni, M., and Partsch, G. (1991) Substance P and somatostatin levels in rheumatoid arthritis, osteoarthritis, and psoriatic arthritis synovial fluid. *Ann. NY Acad. Sci.* **632,** 435,436.

Martin, C. A. E., Naline, E., Emonds-Alt, X., and Advenier, C. (1992) Influence of (±)-CP-96,345 and SR 48968 on electrical field stimulation of the isolated guinea pig main bronchus. *Eur. J. Pharmacol.* **224,** 137–143.

Martin, F. C., Anton, P. A., Gornbein, J. A., Shanahan, F., and Merrill, J. E. (1993) Production of interleukin-1 by microglia in response to substance P: role for a non-classsical NK-1 receptor. *J. Neuroimmunol.* **42,** 53–60.

Martling, C. R. (1987) Sensory nerves containing tachykinins and CGRP in the lower airways: functional implications for bronchoconstriction, vasodilatation, and protein extravasation. *Acta Physiol. Scand.* **563,** 1–57.

Mastrangelo, D., Mathison, R., and Huggel, H. (1983) Postjunctional localization of substance P receptors on the rat portal vein. *Pharmacology* **27,** 305–318.

McElroy, A. B., Clegg, S. P., Deal, M. J., Ewan, G. B., Hagan, R. M., Ireland, S. J., Jordan, C. C., Porter, B., Ross, B. C., Ward, P., and Whittington, A. R. (1992) Highly potent and selective heptapeptide antagonists of the neurokinin NK-2 receptor. *J. Med. Chem.* **35,** 2582–2591.

McKnight, A. T., Maguire, J. J., Elliott, N. J., Fletcher, A. E., Foster, A. C., Tridgett, R., Williams, B. J., Longmore, J., and Iversen, L. L. (1991) Pharmacological specificity of novel, synthetic, cyclic peptides as antagonists at tachykinin receptors. *Br. J. Pharmacol.* **104,** 355–360.

Milner, P., Ravelic, V., Hopwood, M., Feher, E., Lincoln, J., Kirkpatrick, K. A., and Burnstock, G. (1989) Ultrastructural localisation of substance P and choline acetyltransferase in endothelial cells of rat coronary artery and release of substance P and acetylcholine during hypoxia. *Experientia* **45,** 121–125.

Mione, M. C., Ralevic, V., and Burnstock, G. (1990) Peptides and vasomotor mechanisms. *Pharmacol. Ther.* **46,** 429–468.

Morimoto, H., Murai, M., Maeda, Y., Yamaoka, M., Nishikawa, M., Kiyotoh, S., and Fujii, T. (1992a) FK 224, a novel cyclopeptide substance P antagonist with NK_1 and NK_2 receptor selectivity. *J. Pharmacol. Exp. Ther.* **262,** 398–402.

Morimoto, H., Yamashita, M., Matsuda, A., Miyake, H., and Fujii, T. (1992b) Effects of FR 113680 and FK 224, novel tachykinin receptor antagonists, on cigarette smoke-induced rat tracheal plasma extravasation. *Eur. J. Pharmacol.* **224,** 1–5.

Moskovitz, M. A. (1992) Neurogenic versus vascular mechanisms of sumatriptan and ergot alkaloids in migraine. *Trends Pharmacol. Sci.* **13,** 307–311.

Murai, M., Maeda, Y., Hagiwara, D., Miyake, H., Ikari, N., Matsuo, M., and Fujii, T. (1993) Effects of an NK_1 receptor antagonist, FK888, on constriction and plasma extravasation induced in guinea pig airway by neurokinins and capsaicin. *Eur. J. Pharmacol.* **236,** 7–13.

Murai, M., Morimoto, H., Maeda, Y., and Fujii, T. (1992a) Effects of the tripeptide substance P antagonist, FR113680, on airway constriction and airway edema induced by neurokinins in guinea pigs. *Eur. J. Pharmacol.* **217,** 23–29.

Murai, M., Morimoto, H., Maeda, Y., Kiyotoh, S., Nishikawa, M., and Fujii, T. (1992b) Effects of FK224, a novel compound NK_1 and NK_2 receptor antagonist, on airway constriction and airway edema induced by neurokinins and sensory nerve stimulation in guinea pigs. *J. Pharmacol. Exp. Ther.* **262,** 403–408.

Muramatsu, I., Nakanishi, S. and Fujiwara, M. (1987) Comparison of the responses to the sensory neuropeptides, substance P, neurokinin A, neurokinin B and calcitonin gene-related peptide and to trigeminal nerve stimulation in the iris sphincter muscle of the rabbit. *Jpn. J. Pharmacol.* **44,** 85–92.

Mussap, C. J., Lew, R. and Burcher, E. (1989) The autoradiographic distribution of SP binding sites in guinea pig vas deferens is altered by capsaicin pretreatment. *Eur. J. Pharmacol.* **168,** 337–345.

Nagahisa, A., Asai, R., Kanai, Y., Murase, A., Tsuchiya-Nakagaki, M., Nakagaki, T., Shieh, T.-C., and Taniguchi, K (1992a) Nonspecific activity of (±)-CP96,345 in models of pain and inflammation. *Br. J. Pharmacol.* **107,** 273–275.

Nagahisa, A., Kanai, Y., Suga, O., Taniguchi, K, Tsuchiya, M., Lowe, J. A., III, and Hess, H.-J. (1992b) Antiinflammatory and analgesic activity of a nonpeptide substance P receptor antagonist. *Eur. J. Pharmacol.* **217,** 191–195.

Nagata, K., Saito, H., and Matsuki, N. (1991) Efficient Ca^{2+} mobilization induced by neurokinin A in rat vas deferens. *Eur. J. Pharmacol.* **204,** 295–300.

Nakazawa, K., Inoue, K., Fujimori, K., and Takanaka, A. (1990) Neurokinin A suppresses a voltage-gated K^+ current in smooth muscle cells from rat vas deferens. *Eur. J. Pharmacol.* **182,** 189–192.

Naline, E., Devillier, P., Drapeau, G., Toty, L., Bakdach, H., Regoli, D., and Advenier, C. (1989) Characterization of neurokinin effects and receptor selectivity in human isolated bronchi. *Am. Rev. Respir. Dis.* **140,** 679–686.

Nantel, F., Rouissi, N., Rhaleb, N.-E., Dion, S., Drapeau, G., and Regoli, D. (1990) The rabbit jugular vein is a contractile NK-1 receptor system. *Eur. J. Pharmacol.* **179,** 457–462.

Nieber, K, Baumgarten, C., Witzel, A., Rathsack, R., Oehme, P., Brunnee, T., Kleine-Tebbe, J., and Kunkel, G. (1991) The possible role of substance P in the allergic reaction, based on two different provocation models. *Int. Arch. Allergy Appl. Immunol.* **94,** 334-338.

Nimmo, A. J., Carstairs, J. R., Maggi, C. A., and Morrison, J. F. B. (1992) Evidence for the co-existence of multiple NK_2 tachykinin receptor subtypes in rat bladder. *Neuropeptides* **22,** 48.

O'Flynn, N. M., Helme, R. D., Watkins, D. J., and Burcher, E. (1989) Autoradiographic localization of substance P binding sites in rat footpad skin. *Neurosci. Lett.* **106,** 43–48.

O'Shaughnessy, C. T. and Connor, H. E. (1993) Neurokinin NK_1 receptors mediate protein extravasation in guinea pig dura. *Eur. J. Pharmacol.* **236,** 319–321.

Okasada, F., Kubo, K., Goto, K., Kanazawa, I., and Munekata, E. (1986) The contractile activities of neurokinin A, B and related peptides on smooth muscle. *Eur. J. Pharmacol.* **120,** 201–208.

Osborne, N. N. (1984) Substance P in the bovine retina: localization, identification, release, uptake and receptor analysis. *J. Physiol. (Lond.)* **349,** 83–93.

Osborne, N. N. and Ghasi, H. (1989) The effect of substance P and other tachykinins on inositol phospholipid hydrolysis in rabbit retina, superior colliculus and retinal cultures. *Vision Res.* **29,** 754–764.

Parlani, M., Conte, B., Majmore, S., Maggi, C. A., Rovero, P., and Giachetti, A. (1991) The contractile effect of tachykinins on human prostatic urethra: involvement of NK-2 receptors. *J. Urol.* **144,** 1543–1545.

Pearse, A. G. and Polak, J. M. (1975) Immunocytochemical localization of substance P in mammalian intestine. *Histochemistry* **41,** 373–375.

Pernow, B. (1983) Substance P. *Pharmacol. Rev.* **35,** 85–141.

Persson, M. G., Hedqvist, P., and Gustafsson, L. E. (1991) Nerve-induced tachykinin-mediated vasodilatation in skeletal muscle is dependent on nitric oxide formation. *Eur. J. Pharmacol.* **205,** 295–301.

Piedimonte, G., Hoffman, J. I. E., Husseini, W. K, Hiser, W. L., and Nadel, J. (1992) Effect of neuropeptides released from sensory nerves on blood flow in the rat airway microcirculation. *J. Appl. Physiol.* **72,** 1563–1570.

Pietra, C., Bettelini, L, Hagan, R. M., Ward, P., McElroy, A., and Trist, D. G. (1992) Effect of selective antagonists at tachykinin NK-1 and NK-2 receptors on xylene-induced cystitis in rats. *Neuropeptides* **22,** 32.

Ralevic, V., Milner, P., Hudlicka, O., Kristek, F., and Burnstock, G. (1990) Substance P is released from the endothelium of normal and capsaicin-treated rat hind-limb vasculature, in vivo, by increased flow. *Circ. Res.* **66,** 1178–1183.

Reddix, R. A. and Cooke, H. J. (1992) Neurokinin 1 receptors mediate substance P-induced changes in ion transport in guinea pig ileum. *Regul. Pept.* **39,** 215–225.

Renzetti L. M., Shenvi, A., and Buckner, C. K. (1992) Nonadrenergic, noncholinergic contractile responses of the guinea pig hilar bronchus involve the preferential activation of tachykinin $neurokinin_2$ receptors. *J. Pharmacol. Exp. Ther.* **262,** 957–963.

Rogers, D. F., Aursudkij, B., and Barnes, P. J. (1989) Effects of tachykinins on mucus secretion in human bronchi in vitro. *Eur. J. Pharmacol.* **174,** 283–286.

Rouissi, N., Gitter, B. D., Waters, D. C., Howbert, J. J., Nixon, J. A., and Regoli, D. (1991) Selectivity and specificity of new, non-peptide, quinuclidine antagonists of substance P. *Biochem. Biophys. Res. Commun.* **176,** 894–901.

Rubino, A., Thomann, H., Henlin, J.-M., Schilling, W., and Criscione, L. (1992) Endothelium-dependent relaxant effect of neurokinins on rabbit aorta is mediated by the NK_1 receptor. *Eur. J. Pharmacol.* **212,** 237–240.

Saito, R., Konishi, H., Takano, Y., Nonaka, S., Sakaguchi, K., Shimohigashi, Y., and Kamiya, H. (1990) Characterization of tachykinin receptors in endothelial cells of porcine artery. *Neurosci Lett.* **110,** 337–342.

Sann, H., Rossler, W., Hammer, K., and Pierau, F.-K. (1992) Substance P and calcitonin gene-related peptide in the ureter of chicken and guinea pig: Distribution, binding sites and possible functions. *Neuroscience* **49,** 699–713.

Saria, A., Martling, C. R., Yan, Z., Theodorsson-Norheim, E., Gamse, R., and Lundberg, J. M. (1988) Release of multiple tachykinins from capsaicin-sensitive sensory nerves in the lung by bradykinin, histamine, dimethylphenyl piperazinium, and vagal nerve stimulation. *Am. Rev. Respir. Dis.* **137,** 1330–1335.

Saria, A., Theodorsson-Norheim, E., Gamse, R., and Lundberg, J. M. (1984) Release of substance P-and substance K-like immunoreactivities from the isolated perfused guinea pig lung. *Eur. J. Pharmacol.* **106,** 207,208.

Sastry, B. V. R., Janson, V. E., and Owens, L. K. (1991) Significance of substance P- and enkephalin-peptide systems in the male genital tract. *Ann. NY Acad. Sci.* **632,** 339–353.

Savi, P., Laplace, M. C. L., Le Fur, G., Emonds-Alt, X., and Herbert, J. M. (1992) The NK_1 receptor is involved in the neurokinin-induced shape change of rabbit platelets. *FEBS Lett. 312,* 200–202.

Shepheard, R. J., Williamson, D. J., Hill, R. G., and Hargreaves, J. M. (1993) The nonpeptide $neurokinin_1$, receptor antagonist, RP 67580, blocks neurogenic plasma extravasation in the dura mater of rats. *Br. J. Pharmacol.* **108,** 11,12.

Sjögren, C., Andersson, K.-E., and Husted, S. (1982) Contractile effects of some polypeptides on the isolated urinary bladder of guinea pig, rabbit, and rat. *Acta Pharmacol. Toxicol.* **50,** 175–184.

Snider, R. M., Constantine, J. W., Lowe, J. A., III, Longo, K. P., Lebel, W. S., Woody, H. A., Drozda, S. E., Desai, M. C., Vinick, F. J., Spencer, R. W., and Hess, H.-J. (1991) A potent nonpeptide antagonist of the substance P (NK_1) receptor. *Science* **251,** 435–437.

Souquet, J.-C., Bitar, K. N., Grider, J. R., and Makhlouf, G. M. (1987) Receptors for substance P on isolated intestinal smooth muscle cells of the guinea pig. *Am. J. Physiol.* **253,** G666–G672.

Stephenson, J. A., Burcher, E., and Summers, R. J. (1986) Autoradiographic demonstration of endothelium-dependent Bolton-Hunter substance P binding to dog carotid artery. *Eur. J. Pharmacol.* **124,** 377,378.

Stjernquist, M., Håkanson, R., Leander, S., Owman, C., Sundler, F., and Uddman, R. (1983) Immunohistochemical localization of substance P, vasoactive intestinal polypeptide, and gastrin-releasing peptide in vas deferens and seminal vesicle, and the effect of these and eight other neuropeptides on resting tension and neurally evoked contractile activity. *Regul. Pept.* **7,** 67–86.

Stjernschantz, J. (1984) Autacoids and neuropeptides, in *Pharmacology of the Eye* (Sears, M. L, ed.), Springer-Verlag, New York, pp. 311–365.

Stjernschantz, J., Sears, M., and Stjernschantz, L (1981) Intraocular effects of substance P in the rabbit. *Invest. Ophthalmol. Vis. Sci.* **20,** 53–60.

Stubbs, C. M., Waldron, G. J., Connor, H. E., and Feniuk, W. (1992) Characterization of the receptor mediating relaxation to substance P in canine middle cerebral artery: No evidence for involvement of substance P in neurogenically mediated relaxation. *Br. J. Pharmacol.* **105,** 875–880.

Tachado, S. D., Akhtar, R. A., Yousufzai, S. Y. K., and Abdel-Latif, A. A. (1991) Species differences in the effects of substance P on inositol trisphosphate accumulation and cyclic AMP formation, and on contraction in isolated iris sphincter of the mammalian eye: differences in receptor density. *Exp. Eye Res.* **53,** 729–739.

Tamura, K, Mutabagani, K., and Wood, J. D. (1993) Analysis of a nonpeptide antagonist for substance P on myenteric neurons of guinea pig small intestine. *Eur. J. Pharmacol.* **232,** 235–239.

Taniguchi, T., Fujiwara, M., Masuo, Y., and Kanazawa, I. (1986) Levels of neurokinin A, neurokinin B and substance P in rabbit iris sphincter muscle. *Jpn. J. Pharmacol.* **42,** 590–593.

Terenghi, G., Polak, J. M., Probert, L, McGregor, G. P., Ferri, G., Blank, M. A., Unger, W. G., Zhang, S., Cole, D. F., and Bloom, S. R. (1982) Mapping, quantitative distribution and origin of substance P and VIP containing nerves in the uvea of guinea pig eye. *Histochemistry* **75,** 399–417.

Too, H. P., Unger, W. C., and Hanley, M. R. (1988) Evidence for multiple tachykinin receptor substyes on the rabbit iris sphincter muscle. *J. Pharmacol. Exp. Ther.* **33,** 64–71.

Tornqvist, K. and Ehinger, B. (1988) Peptide immunoreactive neurons in the human retina. *Invest. Ophthalmol. Vis. Sci.* **29,** 680.

Tousignant, C., Chan, C.-C., Guevremont, D., Brideau, C., Hale, J. J., MacCoss, M., and Rodger, I. W. (1993) NK_2 receptors mediate plasma extravasation in guinea pig lower airways. *Br. J. Pharmacol.* **108,** 383–386.

Tousignant, C., Dion, S., Drapeau, G., and Regoli, D. (1987) Characterization of pre- and postjunctional receptors for neurokinins and kinins in the rat vas deferens. *Neuropeptides* **9,** 333–343.

Tousignant, C., Guillemette, G., and Regoli, D. (1991) Binding sites for $[^3H][Sar^9,Met(O_2)^{11}]SP$ in rat brain and guinea pig ileum. *Brain Res.* **560,** 1–11.

Tsang, D. (1986) Effect of substance P on dopamine release in rat retina, in *Contemporary Themes in Biochemistry* (Kon, O. L., Chung, M. C. M., Hwang, P. H. et al., eds.), Cambridge University Press, London, UK, pp. 588,589.

Ueda, N., Muramatsu, I., Hayashi, H., and Fujiwara, M. (1982) Trigeminal nerve: the possible origin of substance P-nergic response in isolated rabbit iris sphincter muscle. *Life Sci.* **31,** 369–375.

Ueda, N., Muramatsu, I., Hayashi, H., and Fujiwara, M. (1983) Effects of Met-enkephalin on the substance P-ergic and cholinergic responses in the rabbit iris sphincter muscle. *J. Pharmacol. Exp. Ther.* **226,** 507–511.

Ueda, N., Muramatsu, I., and Fujiwara, M. (1984) Capsaicin and bradykinin-induced substance P-ergic responses in the iris sphincter muscle of the rabbit. *J. Pharmacol. Exp. Ther.* **230,** 469–473.

Ueda, N., Muramatsu, I., and Fujiwara, M. (1985a) Dual effects of dynorphin (1–13) on cholinergic and substance P-ergic transmissions in the rabbit iris sphincter muscle. *J. Pharmacol. Exp. Ther.* **232,** 545–550.

Ueda, N., Muramatsu, I., and Fujiwara, M. (1985b) Prostaglandins enhance trigeminal substance P-ergic responses in the rabbit iris sphincter muscles. *Brain Res.* **337,** 349–351.

Ueda, N., Muramatsu, I., Taniguchi, T., Nakanishi, S., and Fujiwara, M. (1986) Effects of neurokinin A, substance P and electrical stimulation on the rabbit iris sphincter muscle. *J. Pharmacol. Exp. Ther.* **239,** 494–499.

Unger, W. G., Butler, J. M., Cole, D. F., Bloom, S. R., and McGregor, G. P. (1981) Substance P, vasoactive intestinal polypeptide (VIP) and somatostatin levels in ocular tissue of normal and sensorily denervated rabbit eyes. *Exp. Eye Res.* **32,** 797–801.

Unger, W. G. and Tighe, J. (1984) The response of the isolated iris sphincter muscle of various marnmalian species to substance P. *Exp. Eye Res.* **39,** 677–684.

Van Giersbergen, P. L. M., Shatzer, S. A., Burcher, E., and Buck, S. H. (1992) Comparison of the effects of neuropeptide K and neuropeptide gamma with neurokinin A at NK_2 receptors in the hamster urinary bladder. *Naunyn Schmiedeberg's Arch. Pharmacol.* **345,** 51–56.

Von Euler, U. S. and Gaddum, J. H. (1931) An unidentified depressor substance in certain tissue extracts. *J. Physio (Lond.)* **72,** 74–87.

Von Euler, U. S. and Hedqvist, P. (1974) Effect of substance P on the response of guinea pig vas deferens to transmural stimulation. *Acta Physiol. Scand.* **90,** 651–653.

Wahlstedt, C., Beding, B., Ekman, R., Oksala, O., Sternschantz, J., and Håkanson, R. (1986) Calcitonin gene-related peptide in the eye: release by sensory nerve stimulation and effects associated with neurogenic inflammation. *Regul. Pept.* **16,** 107.

Wahlstedt, C., Bynke, G., and Håkanson, R. (1985) Pupillary constriction by bradykinin and capsaicin: mode of action. *Eur. J. Pharmol.* **106,** 577–583.

Walsh, D., Mapp, P. I., Wharton, J., Rutherford, R. D., Kidd, B. L., Revell, P. A., Blake, D. R., and Polak, J. M. (1992) Localisation and characterisation of substance P binding to human synovial tissue in rheumatoid arthritis. *Ann. Rheum. Dis.* **51,** 313–317.

Wang, Z. and Håkanson, R. (1992a) The electrically evoked, tachykinin-mediated contractile response of the isolated rabbit iris sphincter muscle involves NK_1 receptors only. *Eur. J. Pharmacol.* **216,** 327–329.

Wang, Z.-Y. and Håkanson, R. (1992b) Effect of resiniferatoxin on the isolated rabbit iris sphincter muscle: Comparison with capsaicin and bradykinin. *Eur. J. Pharmacol.* **213,** 235–241.

Watson, S. P. and Iversen, L. L. (1984) ^{3}H-Substance P binding to guinea pig ileum longitudinal smooth muscle membranes. *Regul. Pept.* **8,** 273–281.

Watson, S. P., Sandberg, B. E. B., Hanley, M. R., and Iversen, L. L. (1983) Tissue selectivity of substance P alkyl esters: suggesting multiple receptors. *Eur. J. Pharmaol.* **87,** 77–84.

Whittle, B. J. R., Lopez-Belmonte, J., and Rees, D. D. (1989) Modulation of the vasodepressor actions of acetylcholine, bradykinin, substance P and endothelin in the rat by a specific inhibitor of nitric oxide formation. *Br. J. Pharmacol.* **98,** 646–652.

Wormser, U., Laufer, R., Hart, Y., Chorev, M., Gilon, C., and Selinger, Z. (1986) Highly selective agonists for substance P receptor subtypes. *EMBO J.* **5,** 2805–2808.

Xiao, X.-H., Mussap, C. J. and Burcher, E. (1992) Characterization of the tachykinin NK_2 receptor subtype in the rabbit pulmonary artery. *Peptides* **13,** 281–285.

Yau, W. M. and Youther, M. (1982) Direct evidence for a release of acetylcholine from the myenteric plexus of the guinea pig small intestine by substance P. *Eur. J. Pharmacol.* **81,** 665–668.

Yonehara, N., Chen, J.-Q., Imai, Y., and Inoki, R. (1992) Involvement of substance P present in primary afferent neurons in modulation of cutaneous blood flow in the instep of rat hind paw. *Br. J. Pharmacol.* **106,** 256–262.

Zhang, D. and Yeh, H. H. (1992) Substance P-like immunoreactive amacrine cells in the adult and the developing rat retina. *Brain Res. Dev. Brain Res.* **68,** 55–65.

Ziche, M., Morbidelli, L., Pacini, M., Geppetti, P., Alessandri, G., and Maggi, C. A. (1990) Substance P stimulates neovascularization in vivo and proliferation of cultured endothelial cells. *Microvasc. Res.* **40,** 264–278.

CHAPTER 18

Receptor Involvement in Pathology and Disease

Patrick W. Mantyh, Steven R. Vigna, and John E. Maggio

1. Introduction

Tachykinins have been hypothesized to play a role in regulating the inflammatory and immune response in such diverse tissues as the lung, joint, colon, and central nervous system (CNS). The present chapter will focus on the involvement of tachykinins in diseases and pathology of the colon/appendix and the CNS. Several other tissues could also have been chosen (e.g., lung or joint), but more is known about the alterations in tachykinins and tachykinin receptors in the colon and the CNS than in these other tissues. More importantly, in both the colon and in the CNS, the individual cell types that express the tachykinins and the tachykinin receptors have been identified, and the functional relevance of the expression and activation of these tachykinin receptors has begun to be addressed. Several reports on the alterations in tachykinin receptors after tissue injury to the lung or joints have indicated substantial similarities to insult in the colon or CNS. Thus, it is assumed that the colon and CNS are models of tachykinin involvement in other tissues.

One ancillary issue we will address is the possible interactions between substance P (SP) and the β-amyloid peptide. Previous reports that SP protects against β-amyloid peptide neurotoxicity have raised considerable interest in the tachykinin in Alzheimer's disease research communities with regard to potential treatments. Using tachykinin binding assays and novel β-amyloid peptide aggregation and deposition assays, we also have examined potential interaction between tachykinins, tachykinin receptors, and β-amyloid peptide.

2. Inflammatory Bowel Disease

Neurons with cell bodies located in dorsal root ganglia are known to convey somatic sensory information from peripheral tissues to the CNS. Recently, several neuropeptides have been identified within subpopulations of these sensory neurons. The most extensively characterized of these sensory neuropeptides is SP. SP and neurokinin A (NKA) are expressed in high concentrations by sensory neurons (Too and Maggio, 1991; Hua et al., 1985; Saria et al., 1989), but the third tachykinin, neurokinin B (NKB), is virtually undetectable (Too and Maggio, 1991) in these same neuronal populations.

Several sensory neuropeptides, most notably SP, have been associated with neurons involved in the conduction of nociceptive information. Thus, intrathecal injection of SP produces biting and scratching behavior, consistent with a role for SP as a peptide neurotransmitter associated with primary afferent nociceptors (Wilcox, 1988). Also, SP release in the spinal cord is inhibited by opiate analgesics (Jessell and Iversen, 1977), depletion of SP by capsaicin (a neurotoxin relatively selective for unmyelinated sensory neurons [Nagy et al., 1981; Nagy, 1982], including those containing SP) is associated with a loss of specific nociceptive responses, and release of SP in the spinal dorsal horn in response to normally innocuous stimuli is enhanced in polyarthritic rats (Oku et al., 1987).

It has become increasingly evident in the last decade that a specific class of dorsal root ganglia neurons conveying *afferent* somatosensory information from peripheral tissues to the spinal cord is also involved in the *efferent* regulation of the peripheral tissues they innervate. Thus, SP-containing dorsal root ganglia neurons have been implicated in both the *afferent* central transmission of nociceptive information and in the *efferent* regulation of inflammation and sensitization of sensory endings in chronic inflammatory pain states such as arthritis, asthma, and inflammatory bowel disease (Lembeck and Holzer, 1979; Lembeck et al., 1981; Levine et al., 1986; Basbaum and Levine, 1991; Lembeck et al., 1992). Support for this concept includes observations that the bulk of the SP synthesized by dorsal root ganglia neurons is transported to the peripheral terminals rather than to the spinal cord (Brimijoin et al., 1980; Hokfelt et al., 1980; Hokfelt et al., 1982), SP is a potent vasodilator in several peripheral tissues (Lundberg et al., 1991), terminals of SP-containing sensory neurons are observed in association with blood vessels (Gibbins et al., 1985), electrical stimulation of these peripheral nerves at intensities that release SP in peripheral tissues reproduces many of the physiological changes seen in acute inflammation, including plasma extravasation (Lembeck and Holzer, 1979; Lembeck et al., 1992), and in experimental animals SP binding sites are expressed by several tissues involved in the inflammatory and immune responses (Mantyh, C. R., et al., 1988; Mantyh, P. W., et al., 1988; Popper et al., 1988; Mantyh et al., 1989a; Helke et al., 1990).

What emerges from these observations is the hypothesis that sensory neurons containing SP and/or other neuropeptides are involved both in conveying nociceptive information to the spinal cord and in regulating the inflammatory and immune responses in the peripheral tissues they innervate (Mantyh et al., 1989a). Thus, SP released by a sensory neuron may signal tissue damage in the spinal cord and may participate in regulating the inflammation and immune responses and, ultimately, wound healing, in the affected peripheral tissue (Fig. 1).

To test whether this hypothesis is applicable to human inflammatory diseases, we examined the changes in tachykinin neuropeptide receptor binding sites in colonic tissue obtained from inflammatory bowel disease patients. Inflammatory bowel disease is a generic

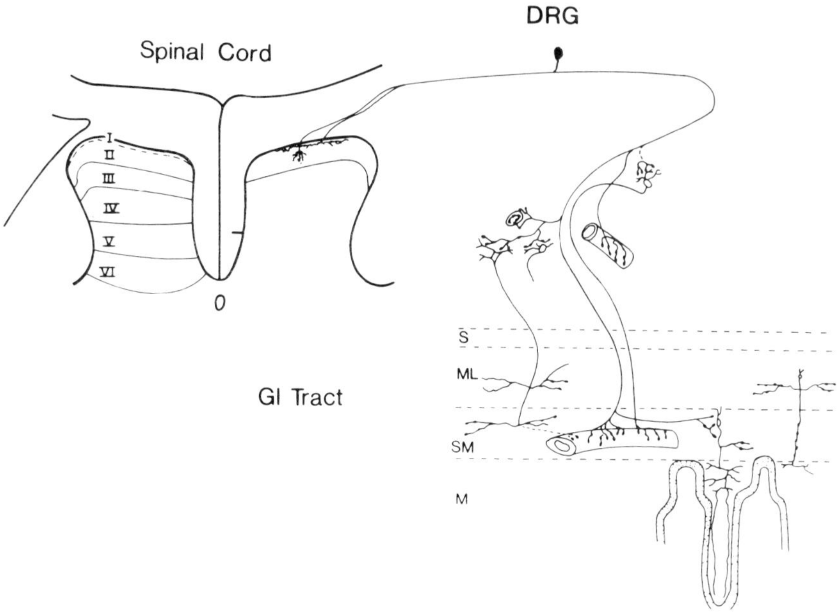

Fig. 1. Schematic diagram of a dorsal root ganglia (DRG) neuron innervating the colon. Note that the DRG neuron is pseudounipolar, with one process projecting to the spinal cord and the other to the colon. Abbreviations: M, mucosa; SM, submucosa; ML, muscle layers; S, serosa.

term that refers to chronic inflammatory diseases of the intestine of unknown etiology, principally ulcerative colitis and Crohn's disease. Ulcerative colitis is an inflammatory, ulcerating process of the colon; Crohn's disease is an inflammation of the intestine characterized by nodular granulomatous inflammatory lesions throughout the entire wall that may involve any part of the intestine, but primarily attacks the distal small intestine and colon. Since surgical removal of the inflamed colon is used to ameliorate the disease in severe cases, we used these surgical specimens to ask whether significant alteration in either the location or levels of NK receptors occur in these human inflammatory diseases.

We found that NK_1 receptors were unique among the several binding sites examined in that they were dramatically upregulated in the inflamed colon tissue (Figs. 2 and 3), and that this dramatic upregulation appeared to be restricted to cells involved in mediating the inflammatory and immune response (Figs. 3 and 4), i.e., small arterioles and lymph nodules. The density of NK_2 receptors did not differ significantly between control in any inflamed tissue, whereas NK_3 receptors were not detected in any peripheral tissue examined. We have also demonstrated that the NK_1 binding sites associated with the small arterioles and lymph nodules of inflammatory bowel disease tissue have the pharmacological characteristics of NK_1 receptors, and that CP-96,345 potently inhibits binding of ^{125}I-SP to these sites (IC_{50} of approximately 2 n*M*).

To test whether this increase in NK_1 receptor density is restricted to inflammatory bowel disease or is characteristic of other inflammatory conditions, we examined the appendix of patients with appendicitis, negative appendicitis (where the patient presents with the symptoms of appendicitis, but where surgical removal and standard pathological stains suggest the tissue is normal), and normals. In normal appendix, similar to normal colon, low levels of NK_1 receptor binding site are largely confined to the smooth muscle of the circular muscle layer (Fig. 5). In the inflamed appendicitis tissue, additional NK_1 binding sites were associated with the small arterioles and lymph nodules (Fig. 6). Interestingly, the negative appendicitis tissue also showed the additional NK_1 binding sites associated with both the blood vessels and the lymph nodules, suggesting that the upregulation of NK_1 receptors occurs prior to the infiltration of mononuclear inflammatory cells characteristic of the highly inflamed appendix. Whether SP and NK_1 receptors initiate the inflammatory changes in inflammatory bowel disease and appendicitis or are a result of the inflammation that occurs in these conditions is unknown, but the data from the negative appendicitis cases suggest that SP receptor expression may actively promote the plasma extravasation and infiltration of mononuclear inflammatory cells observed in these conditions.

An important issue that the above studies was not able to address is identification of the specific cell types exhibiting this upregulation of NK_1 receptors in these human inflammatory states. Previous studies on experimental animals suggested that the endothelial cells of

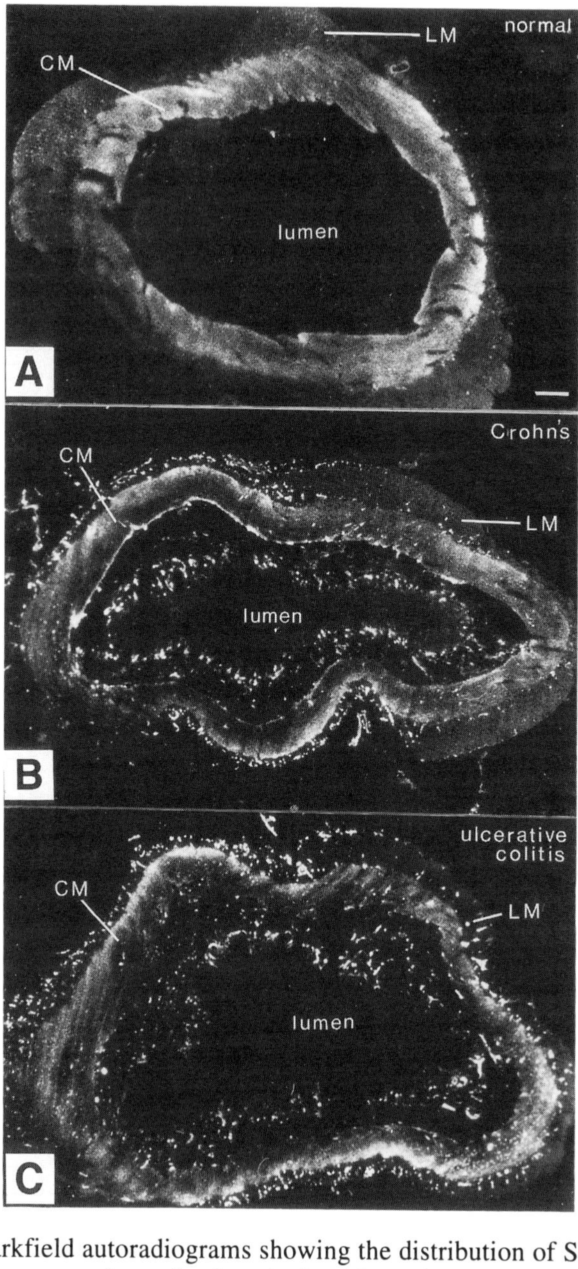

Fig. 2. Darkfield autoradiograms showing the distribution of SP binding sites in transverse sections of colon obtained from the margins of extensive resection for carcinoma **(A)** and from affected regions in patients with Crohn's disease **(B)** and ulcerative colitis **(C)**. In these darkfield autoradiograms, the white silver grains represent areas of high concentration of SP binding sites.

blood vessels expressed high concentrations of NK_1 receptors and suggested that these receptors in large part mediated SP-induced plasma extravasation (Khalil and Helme, 1989; Lembeck et al., 1992). To examine this question, we chose to use endothelial cells from human umbilical cord (Greeno et al., 1993). Initial autoradiographic examination of the human umbilical cord revealed that endothelial cells associated with the umbilical artery, but not the umbilical veins, express high concentrations of NK_1 receptors. Paradoxically, several previous investigators who examined the function of NK_1 receptors expressed by human endothelial cells (Nakagawa et al., 1992) used almost exclusively cultured venous endothelial cells that, at least in vivo, did not appear to express NK_1 receptors. We therefore cultured human umbilical vein endothelial cells to explore what factors might affect NK_1 receptor expression. We found that stress in the form of serum starvation produced a marked upregulation of the NK_1 receptors on these cells. Only 10% of the unstarved endothelial cells responded to SP, but after serum starvation with subsequent refeeding, over 50% of the endothelial cells showed a robust Ca^{2+} mobilization in response to SP. These studies suggest that in inflammatory bowel disease and appendicitis early hypoxic or metabolic stress of endothelial cells may produce an upregulation of NK_1 receptors. On activation by SP, endothelial cells produce a vasodilation of the arteriole smooth muscle by a mechanism consistent with the production of nitric oxide (Khalil and Helme, 1989; Barnes, 1992). This vasodilation in turn results in an increase in the pressure at the level of the capillary bed, resulting in plasma extravasation (Barnes et al., 1990; Lei et al., 1992; Lembeck et al., 1992). Additional experimental data have implicated SP in a number of inflammatory processes including vascular leakage (Khalil and Helme,

---

Notice that, whereas in normal colon (A) a moderate concentration of SP receptor binding sites is expressed by the external circular muscle (CM) and the tunica media of a large artery in the serosa, in both Crohn's disease (B) and ulcerative colitis (C) arterioles and lymph nodules express very high levels of SP receptor binding sites. Abbreviations: CM, external circular muscle; LM, external longitudinal muscle. Line bar = 1.4 mm.

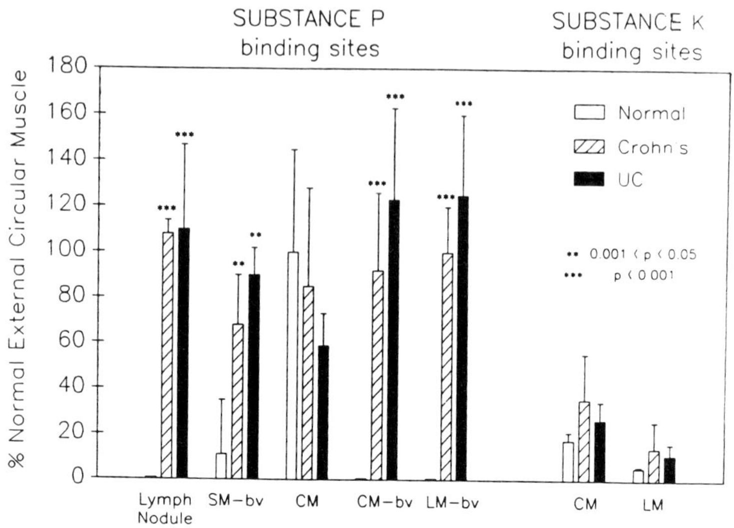

Fig. 3. Histogram showing the changes in the location and concentration of SP and NKA (also known as substance K, SK) binding sites in the surgical specimens of normal colon obtained from carcinoma resection (open bars), Crohn's disease (hatched bars), and ulcerative colitis (dark bars). In this histogram, 100% specific binding is that concentration of specific SP binding sites expressed by the smooth muscle of the normal external circular muscle (CM). Note that, whereas SP receptor binding sites are undetectable in blood vessels (bv) vascularizing the circular (CM) and longitudinal muscle (LM) and in the lymph nodules of normal patients, in the Crohn's disease and ulcerative colitis patients they reach levels that are the highest observed in the human GI tract. Specific binding sites for NKB were not detected in any area of the human GI tract. Abbreviations: bv, blood vessel (arteriole or venule); CM, external circular muscle; LM external longitudinal muscle; SM, submucosa.

1989), induction of ICAM-1 expression (Axelsson and Ahlstedt, 1990; Nakagawa et al., 1992), migration and activation of white blood cells (McGillis et al., 1987; Iwamoto et al., 1990; Parnet et al., 1990), and neovascularization (Ziche et al., 1990a,c), all processes in which endothelial cell interactions are critical.

Do SP-containing sensory neurons have a function other than signaling peripheral tissue injury to the CNS and regulating the

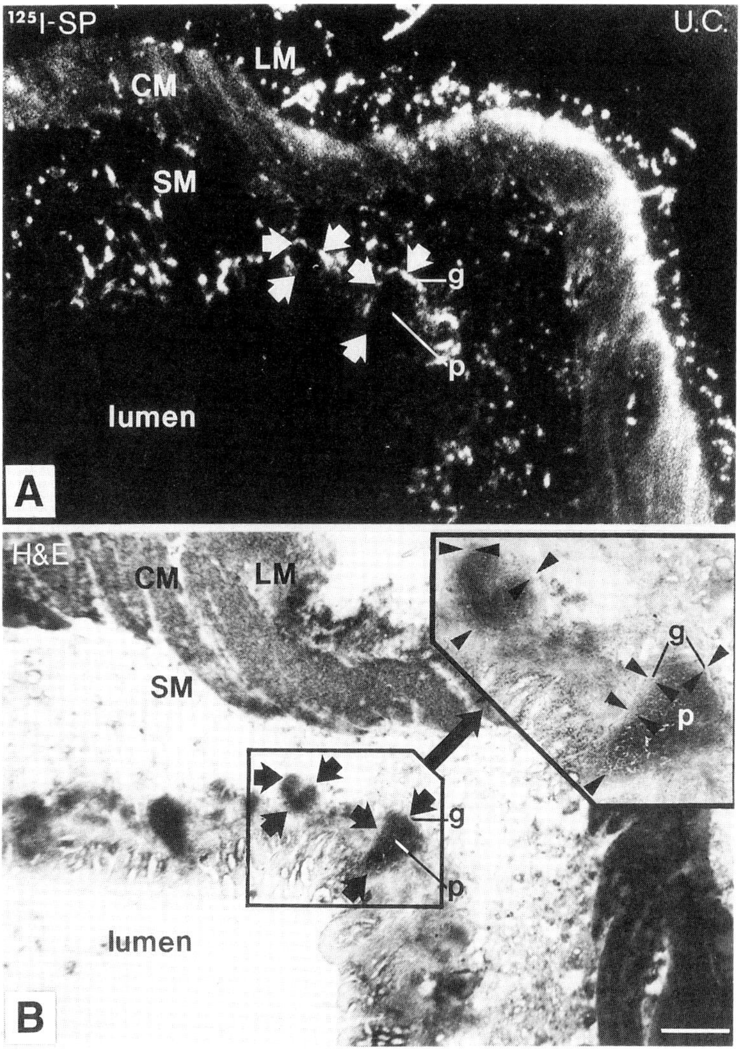

Fig. 4. A series of photomicrographs showing the detailed localization of SP binding sites over the germinal center (g), but not the proliferative zone (p), of lymph nodules (outlined by arrows) in an inflamed colon of an ulcerative colitis patient. **(A)** Darkfield autoradiogram showing the localization of SP binding sites over the lymph nodules. **(B)** Lightfield photomicrograph of the H&E stained section of (A). Inset: high power enlargement of the area outlined in (B). Note that the only part of the lymph nodule that expresses SP binding sites in (A) is the lightly stained germinal center outlined by arrowheads in the inset in (B). Line bar = 0.9 mm (B).

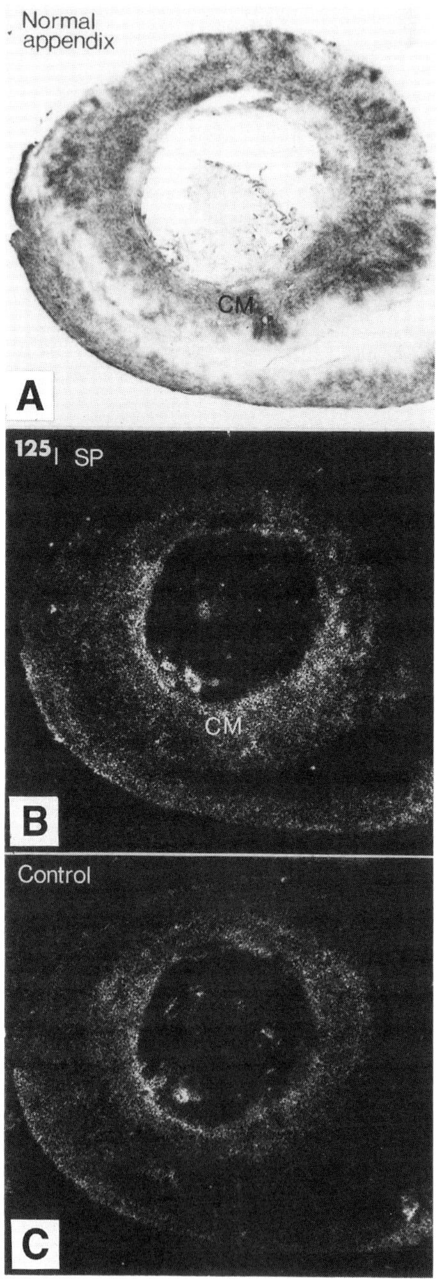
Normal appendix
CM
A
125I SP
CM
B
Control
C

inflammatory and immune responses in peripheral tissues? One observation suggests a normal function for these neuropeptide-containing sensory neurons in that one of the most pronounced deficits in rats treated neonatally with capsaicin (the neurotoxin destroys primarily the neuropeptide containing C-fibers) is that the skin and fur lose their luster and that these animals lack a normal wound healing response; i.e., these animals have numerous small wounds that do not appear to be healing (Nagy, 1982; Buck and Burks, 1986). Such a lack of the normal trophic and/or wound healing response after sensory denervation can also be inferred from the ulceration of the cornea that follows trigeminal deafferentation (Beuerman and Schimmelpfennig, 1980). Together, these observations suggest that in a pathological state SP might be involved in regulating a hyper-inflammatory and immune response, whereas in the normal condition these sensory neuropeptide-containing neurons may have a trophic action on tissues they innervate and also regulate wound healing after injury. Additional support for this hypothesis comes from studies showing that both SP and NKA are potent mitogens for myocytes and fibroblasts in culture (Nilsson et al., 1985; Hultgardh-Nilsson et al., 1988). This hypothesis suggests that SP will be both stimulatory and inhibitory toward the same cells depending on the content of other chemical signals present. This multifunctional role for SP is not unprecedented for other peptide growth factors; it has been shown that another peptide, transforming growth factor-β can stimulate growth of certain fibroblasts in vitro in the presence of platelet-derived growth factor, but inhibits the growth of the same

---

Fig. 5 *(opposite page)*. Photomicrographs showing the distribution of substance P (SP) binding sites in transverse sections of a normal human appendix. The appendix was obtained from the margins of an extensive resection for carcinoma. **(A)** Lightfield photomicrograph of a serially adjacent section to **(B)** stained with H&E, (B) a darkfield photomicrograph of an autoradiogram showing the total ^{125}I-SP binding, and **(C)** a serially adjacent section to (B) showing the nonspecific binding (^{125}I-SP + $10^{-6}M$ cold SP). Note that there is a moderate concentration of SP receptor binding sites expressed by the external circular muscle (CM) and by a large artery in the longitudinal muscle. Line bar = 0.4 mm.

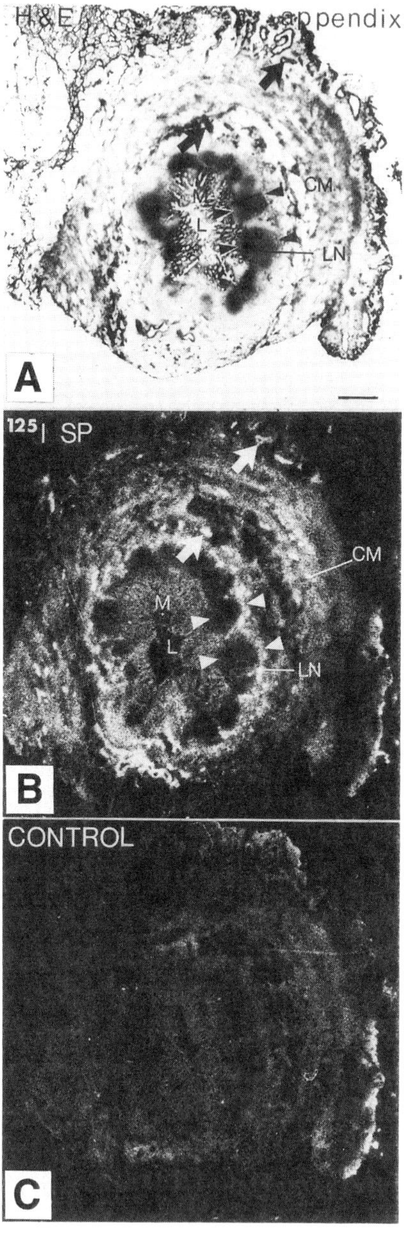
H&E
appendix
CM
M
L
LN
A
125I SP
CM
M
L
LN
B
CONTROL
C

cells if epidermal growth factor is present (Roberts and Sporn, 1988; Logan et al., 1992). In human inflammatory diseases, it may be that ectopic expression of SP or its binding site is not the primary pathology, but rather that the pathology may involve whatever other factors are needed to switch SP action from the "catabolic" mode, where inflammatory and immune responses are promoted, to the anabolic process of tissue growth. It remains to be determined what factors regulate SP binding site expression, the time course of the up- and downregulation of NK_1 receptors, and what the physiological consequences of SP release and SP binding site activation are in both normal and inflamed tissues.

3. Glial Expression of SP Receptors After Neuronal Injury

A major question in neurobiology is why damaged mammalian CNS neurons do not regenerate in vivo. In recent years, the focus of attention has shifted from CNS neurons themselves, which appear to have the capacity to regenerate, to CNS glia, which can either inhibit or promote the regrowth of axons in the CNS. Thus, it has been demonstrated that after injury regenerating axons grow a short distance until they reach the glial scar, at which time they appear to stop growing and degenerate (David and Aguayo, 1985; Bray et al., 1987; Aguayo et al., 1990). Other studies, however, have demonstrated that under a different set of conditions areas of the brain rich in astrocytes support neuronal adhesion and fiber outgrowth (Savio and Schwab, 1989; Walsh et al., 1992), suggesting that astrocytes may support

Fig. 6 *(opposite page)*. Photomicrographs showing the distribution of SP binding sites in transverse sections of an inflamed human appendix. **(A)** Lightfield photomicrograph of a serially adjacent section to **(B)** stained with H&E, (B) a darkfield photomicrograph of an autoradiogram showing the total ^{125}I-SP binding, and **(C)** a serially adjacent section to (B) showing the nonspecific binding (^{125}I-SP + $10^{-6}M$ cold SP). Note the high concentrations of ^{125}I-SP binding sites expressed by the germinal zone of the lymph nodule and the blood vessels throughout the wall of the appendix. Abbreviations: L, lumen; LN, lymph nodule; CM, circular muscle. Line bar = 0.4 mm.

neuronal regeneration. To understand further the factors that promote or inhibit neuronal regeneration, it is important to define the factors that promote the proliferation of astrocytes after injury, the factors that regulate what glia express and release after injury, whether these factors promote or inhibit neuronal regeneration, and just what results obtained from cultured glia tell us about the role glia play in vivo after neuronal injury.

The major cellular constituent of a CNS glial scar is the astrocyte (Miller et al., 1986; Miller et al., 1989; Smith and Miller, 1991). Unlike fibroblasts, which form scars in nonneural tissue by secreting large amounts of collagenous extracellular matrix, astrocytes form scars by extending numerous processes that become packed with intracellular glial filaments (Maxwell and Kruger, 1965). Astrocytes proliferate in response to injury (Miller et al., 1986), and it appears that these "reactive astrocytes" are biochemically different from the major class of astrocytes present in the normal, nonlesioned brain (Miller et al., 1986). Recently, several neuropeptides, including bombesin, NKA, and SP, have been shown to be mitogenic (Rozengurt, 1986; Zachary et al., 1987; Fabregat and Rozengurt, 1990; Ziche et al., 1990b) for several cell types that may be involved in the inflammatory and wound healing responses in peripheral tissues. In vitro studies suggest that glia are potential targets for a variety of neurotransmitters (Cholewinski and Wilkin, 1988; Wilkin et al., 1990; Wilkin et al., 1991) including SP (Hartung et al., 1988; Torrens et al., 1989; Mantyh et al., 1989c; Beaujouan et al., 1990; Beaujouan et al., 1991; Delumeau et al., 1991), somatostatin, vasoactive intestinal peptide (Wilkin et al., 1990; Wilkin et al., 1991), and a host of other growth factors and neurotransmitters. By analogy with the inflammatory and immune responses occurring in response to injury in peripheral tissues, neuropeptides may also regulate glial response in the CNS. However, since glial cells exhibit different functional properties depending on their biochemical environment, it is necessary to show that the receptor binding sites expressed by glia in vitro are also expressed by glia in vivo.

To determine if glia express similar neurotransmitter receptors in vivo as they do in vitro, we examined the glial scar in the transected optic nerve of the albino rabbit using quantitative receptor autoradiography. Receptor binding sites for radiolabeled calcitonin gene-related peptide, cholecystokinin, galanin, somatostatin, SP, and vasoactive intestinal peptide were examined (Mantyh et al., 1989c). Specific receptor binding sites for each of these neurotransmitters were identified in the rabbit forebrain, but were not detected in the normal optic nerve or tract. In the transected optic nerve and tract, only receptor binding sites for SP were expressed at detectable levels (Mantyh et al., 1989c). The density of SP receptor binding sites observed in this glial scar is among the highest observed in the rabbit forebrain (Fig. 7). Ligand inhibition and saturation experiments (Fig. 8) indicate that the SP receptor binding site expressed by the glial scar has pharmacological characteristics similar to those of NK_1 receptors in the rabbit striatum (Mantyh et al., 1989c), rat brain (Mantyh et al., 1989b), and canine chief cells of the GI tract (Vigna et al., 1989). We have also recently demonstrated that CP-96,345 potently displaces ^{125}I-SP from this binding site, further suggesting that these astrocytes are expressing NK_1 receptors. These results demonstrate that glial cells in vivo express high concentrations of SP receptor binding sites after transection of retinal ganglion cell axons.

Several studies have examined whether the NK_1 receptors expressed by cultured astrocytes are functionally coupled. SP receptors have been reported to be present in 2- to 3-wk-old primary cultures of cortical astrocytes from newborn mice, and addition of SP to these astrocyte cultures stimulates phosphatidylinositol turnover (Torrens et al., 1989; Beaujouan et al., 1990; Beaujouan et al., 1991). Substance P has also been shown to stimulate the cyclooxygenase pathway of arachidonic acid metabolism in 2- to 3-wk-old cultures of rat astrocytes and to evoke the formation of prostaglandin E and thromboxane B in a dose-dependent manner (Hartung et al., 1988; Hartung and Toyka, 1989). Last, SP has been shown to cause a dose dependent release of IL-6 from the astrocytoma cell line, U373 (Gitter et al., 1992), suggesting that SP may modulate the synthesis and release of cytokines and growth factors from astrocytes. These studies, together with the present results, are consistent with the hypothesis that astrocytes that proliferate after neuronal injury express

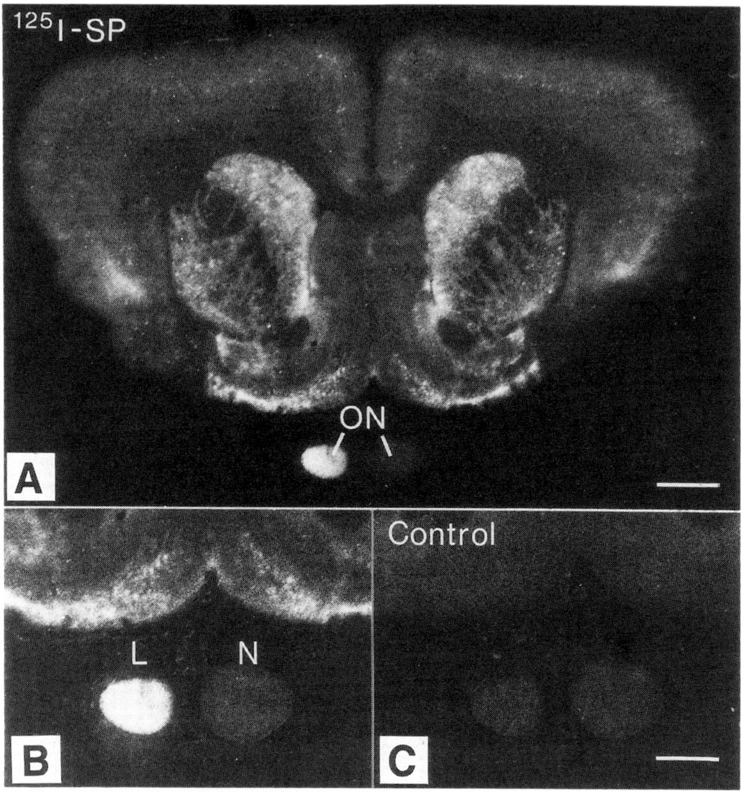

Fig. 7. A series of darkfield photomicrographs showing the autoradiographic localization of SP receptor binding sites in a coronal section of the rabbit brain (animal #05) 99 d after unilateral transection of the optic nerve (ON). Autoradiograms (**A** and **B**) show the total binding, whereas (**C**) is the nonspecific binding. The control section (C) illustrating the nonspecific binding was treated identically to the adjacent section, which shows the total binding (B), except that nonradioactive SP ($10^{-6}M$) was added to the incubation medium. In all the darkfield autoradiograms, the highest density of white silver grains represents the highest concentration of binding sites. The specific binding is obtained by subtracting the binding in (C) from (B). Note that, whereas the lesioned (L) optic nerve is reduced in size, it expresses a very high density of specific receptor binding sites relative to the larger normal (N) optic nerve in which specific binding sites are not detectable. Line bar (A), 1.9 mm; (B and C), 1.2 mm.

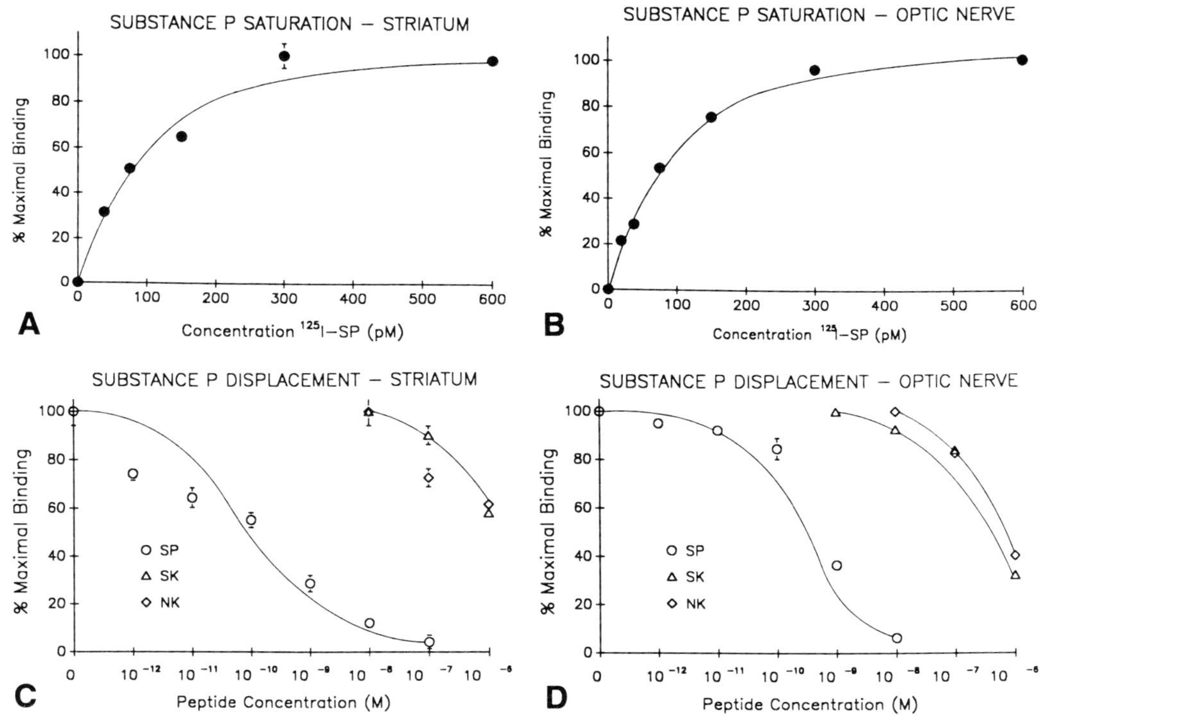

Fig. 8. Saturation **(A,B)** and competition **(C,D)** curves for SP receptor binding sites expressed by the striatum (A,C) and the lesioned optic nerve (B,D) 99 d after transection (rabbit #04). Note that both the saturation and competition curves for SP receptor binding sites in the normal striatum (A,C) and the transected optic tract (B,D) are similar, suggesting that the SP receptor binding site expressed by the lesioned optic nerve is similar to the SP receptor binding site expressed by the normal striatum. The data in Fig. 8 are from a representative experiment. Each point represents the mean of triplicate determinations.

functional SP receptors. These data suggest that when glia are separated from neurons either in vitro or in vivo after the transection and degeneration of retinal ganglion cell axons, the glia that survive and proliferate express SP receptors.

Unlike peripheral tissues, such as the GI tract or skin, where there is a dense innervation by SP-containing dorsal root ganglia neurons, the brain lacks such a sensory innervation. This important difference raises the question of the origin of the SP that could occupy the SP receptors expressed by the CNS glia after neuronal injury. To address this question, we and others examined the expression of mRNA coding for SP in mononuclear inflammatory cells, as previous studies have reported that these cells can synthesize several other neuropeptides including ACTH and opiates. In these studies, we were able to show that in the encephalitic rabbit brain, mRNA coding for the preprotachykinin precursor of SP could be detected in infiltrates of mononuclear inflammatory cells that are characteristic of these brain infections (Boehmer et al., 1989). These data, along with the work showing similar SP mRNA expression in eosinophils (Weinstock et al., 1988) and release of immunoreactive SP from eosinophils (Weinstock and Blum, 1990) in a model of schistosomiasis, suggest that one possible source of SP after brain injury is the mononuclear inflammatory cells that invade the CNS after local tissue injury. Key questions that remain to be answered are the functional relevance of this NK_1 receptor expression, the factors that control this NK_1 receptor expression in astrocytes, and whether activation of these astrocytic NK_1 receptors promotes or inhibits neuronal regeneration.

4. Interactions Between β-Amyloid, Tachykinins, and Tachykinin Receptors

Alzheimer's disease, a widespread progressive dementia affecting a significant fraction of the elderly, is characterized by the presence in brain of numerous extracellular proteinaceous deposits called amyloid on the basis of their staining properties (Selkoe, 1991). The extracellular amyloid is deposited both at neuropil and vascular sites, and the density of these deposits in the cerebral cortex and blood vessels correlates positively with the degree of dementia. The principal component of both neuritic and vascular plaques is βA4, a hydrophobic peptide of 39–43 amino acids that is encoded by a gene for a much larger protein termed the amyloid precursor protein. Neither the processing of amyloid precursor protein to βA4 nor the genesis of the amyloid deposits is well understood. The characteristics of βA4 deposition and the factors that affect it remain key questions in the pathology of Alzheimer's disease.

Recent work has suggested that the amyloid peptide may have biological activities of its own, and the 42-mer peptide or its fragments have been reported to be trophic and/or toxic at various concentrations (Whitson et al., 1989; Koh et al., 1990; Whitson et al., 1990; Flood et al., 1991). The amyloid peptide has a considerable tendency to form insoluble aggregates under various conditions (Pike and Snyderman, 1988; Pike et al., 1991a; Burdick et al., 1992; Knauer et al., 1992), and evidence has been presented that the neurotoxicity of βA4 is related to the aggregation process (Pike et al., 1991a; Pike et al., 1991b). A sequence similarity (Fig. 9) between residues 25–35 of βA4 and the tachykinin neuropeptides led to an investigation of the latter and the finding that tachykinins can block and tachykinin antagonists can mimic some of the biological activities of βA4 on rat embryonic hippocampal neurons in primary culture (Yankner et al., 1990). Furthermore, direct injection of $\beta A4^{1-40}$ into the cerebral cortex or hippocampus of adult rats was reported (Kowall et al., 1991; Yankner and Mesulam, 1991) to have neurotoxic effects that are blocked by coadministration of the mammalian tachykinin, SP, in a dose-dependent manner. The most parsimonious explanation of these data was that βA4 was exerting its neurotoxic actions via an interaction with tachykinin receptors.

In several experiments, we failed to detect any conventional interaction between βA4 and tachykinin receptors. Saturable binding sites for tachykinin radioligands selective for NK_1 and NK_3 tachykinin receptor types were present at similar densities in both

Human Amyloid Peptide

```
            10        20        30        40
H-DAEFRHDSGYEVHHQKLVFFAEDVGSNKGAIIGLMVGGVVIAT-OH
```

Rat Amyloid Peptide

```
            10        20        30        40
H-DAEFGHDSGFEVRHQKLVFFAEDVGSNKGAIIGLMVGGVVIAT-OH
```

Dutch Amyloid Peptide

```
            10        20        30        40
H-DAEFRHDSGYEVHHQKLVFFAQDVGSNKGAIIGLMVGGVVIAT-OH
```

βAP (25-35)-OH H-GSNKGAIIGLM-OH

βAP (25-35)-NH_2 H-GSNKGAIIGLM-NH_2

Substance P H-RPKPQQFFGLM-NH_2

Neurokinin A H-HKTDSFVGLM-NH_2

Neurokinin B H-DMHDFFVGLM-NH_2

Fig. 9. Sequences of amyloid and tachykinin peptides. Abbreviations for the amino acids are as follows: A, Ala; D, Asp; E, Glu; F, Phe; G, Gly; H, His; I, Ile; K, Lys; L, Leu; M, Met; N, Asn; P, Pro; Q, Gln; R, Arg; S, Ser; T, Thr; V, Val; and Y, Tyr. Abbreviations for peptide termini are as follows: H–, free amino group; –OH, free carboxyl group; –NH_2, carboxyamide. Sequences are numbered from the amino termini with positions indicated by superscripts.

Alzheimer's disease and normal brain tissue (Maggio et al., 1992), and unlabeled tachykinins inhibited binding of their radiolabeled analogs at nanomolar concentrations in both tissues. In contrast, there was no displacement of any of the tachykinin radioligands by $\beta A4^{1-40}$ at concentrations up to 30 μ*M*, suggesting that the amyloid peptide does not interact with tachykinin receptors under these conditions (Mitsuhashi et al., 1991; Maggio et al., 1992). $\beta A4^{1-40}$ also failed to inhibit radiolabeled tachykinin binding to receptors in rat brain homogenates and sections. These results are consistent with the reported structure–activity studies among the tachykinin peptide family, i.e., that a carboxyl-terminal amide is required for activity, and argue against the hypothesis (Yankner et al., 1990; Kowall et al., 1991) that the reported biological activity of βA4 results from its interaction with NK_1 or NK_3 tachykinin receptors.

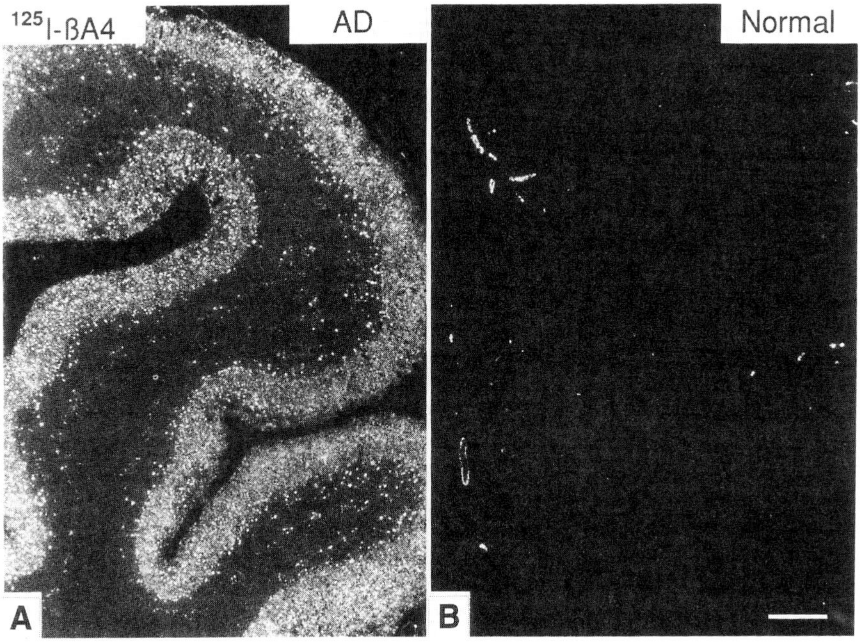

Fig. 10. Autoradiographic localization of ^{125}I-βA4$^{1-40}$ binding sites in sections of human Alzheimer's disease (AD) brain and in a brain from a normal age matched control. Tissue from areas without plaques showed negligible displaceable binding of ^{125}I-βA4$^{1-40}$. **(A)** Darkfield photomicrograph autoradiogram showing distribution of ^{125}I-βA4$^{1-40}$ binding sites in Alzheimer's disease temporal cerebral cortex reveals numerous plaques through the grey matter, **(B)** Darkfield photomicrograph autoradiogram showing the distribution of ^{125}I-βA4$^{1-40}$ binding sites in temporal cortex of an age-matched control. Note that there is very little deposition of radioligand under conditions identical to those used in (A). Line bar, 2 mm.

Another possible mechanism by which SP might interact with β-amyloid would be to interfere with the aggregation of β-amyloid, since it has been reported that β-amyloid must first aggregate to be neurotoxic. To explore the aggregation properties of human βA4, the sites at which the peptide is bound, and what factors may affect this binding, we have synthesized a radiolabeled amyloid peptide, ^{125}I-βA4$^{1-40}$, and developed a method (Fig. 10) to characterize and localize the sites with which it interacts in homogenates or thin sections of normal or Alzheimer's disease tissue (Maggio et al., 1992). Radioiodinated human βA4 is rapidly deposited in vitro from dilute ($<10^{-11}M$) solution onto neuritic and diffuse plaques and cerebrovascular amyloid in Alzheimer's disease brain tissue, whereas no deposition is detectable in tissue without preformed plaques. This growth of plaques by deposition of ^{125}I-βA4 to plaques is reversible, with a dissociation half-time of approximately 1 h. Using the ^{125}I-βA4, we also demonstrated that ^{125}I-βA4 has a considerable tendency to self-aggregate and that aluminum, iron, and zinc under the same conditions strongly promoted ^{125}I-βA4 aggregation, whereas several other metal cations had no significant effect.

Using the above deposition and aggregation assays, we exam ined whether any of the mammalian tachykinins had an effect on ^{125}I-βA4 deposition or aggregation. The mammalian tachykinins, SP, NKA, and NKB, and the nonpeptide NK_1 tachykinin receptor antagonist, CP-96,345, did not inhibit the deposition of radiolabeled βA4 onto Alzheimer's disease plaques over a broad range of concentrations (Fig. 11). We also examined the possibility that tachykinins may inhibit β-amyloid aggregation. Similar to the results we observed with β-amyloid deposition, SP, NKA, NKB, and the nonpeptide NK_1 tachykinin receptor antagonist, CP-96,345, inhibited neither metal-induced aggregation nor the self-aggregation of β-amyloid (Fig. 12). These results suggest that if the tachykinins do block β-amyloid neurotoxicity it is probably not a tachykinin receptor-mediated mechanism nor is it by NKA, NKB, or SP blocking the aggregation or deposition of β-amyloid peptide. One possible, although unproven, mechanism by which SP may play a role in regulating βA4 neurotoxicity is by regulating the amount of βA4 released by astrocytes. Evidence supporting this hypothesis is that in response to neuronal injury astrocytes have been shown to express NK_1 receptors (Mantyh et al., 1989c) and βA4 (Siman et al., 1989), and that neurotransmitters have been shown to modulate the release of βA4 from cells (Nitsch et al., 1992).

5. Summary

Tachykinins appear to be involved in the response to tissue injury in both peripheral tissues and the CNS. In the inflamed colon

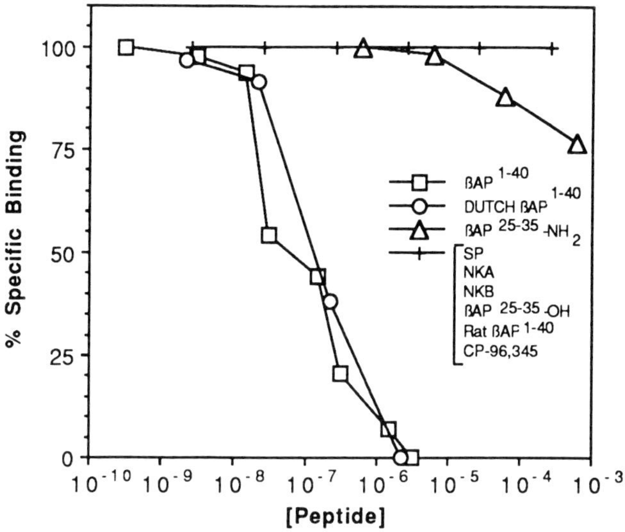

Fig. 11. Inhibition of specific binding of ^{125}I-βA4$^{1-40}$ to homogenates of AD temporal cortex by amyloid and tachykinin peptides. Densitometric analysis of autoradiograms of similar experiments performed on sections of intact tissue ($N = 9$) gave similar results. Each point is an average of at least three experiments, each done in triplicate; standard errors were less than 10% of the mean for all points. Because concentrations of βA4$^{1-40}$ higher than $10^{-5}M$ can form filterable aggregates during the course of the incubation, the asymptote of the binding curve is estimated by extrapolation. Peptide sequences are defined in Fig. 9.

and appendix, a major cell type that displays an upregulation of NK_1 receptors appears to be the endothelial cells. On activation by SP, endothelial cells appear to release nitric oxide that in turn produces a vasodilation of the smooth muscle of the arteries. This vasodilation in turn results in an increase in the blood pressure at the level of the capillary bed resulting in plasma extravasation. Other experiments have suggested that SP may regulate the expression of adhesion molecules by the endothelial cell, thereby regulating mononuclear cell migration into the damaged tissue.

In the CNS, tissue injury appears to promote NK_1 receptor upregulation in astrocytes. Thus, after optic nerve lesion in the rat or

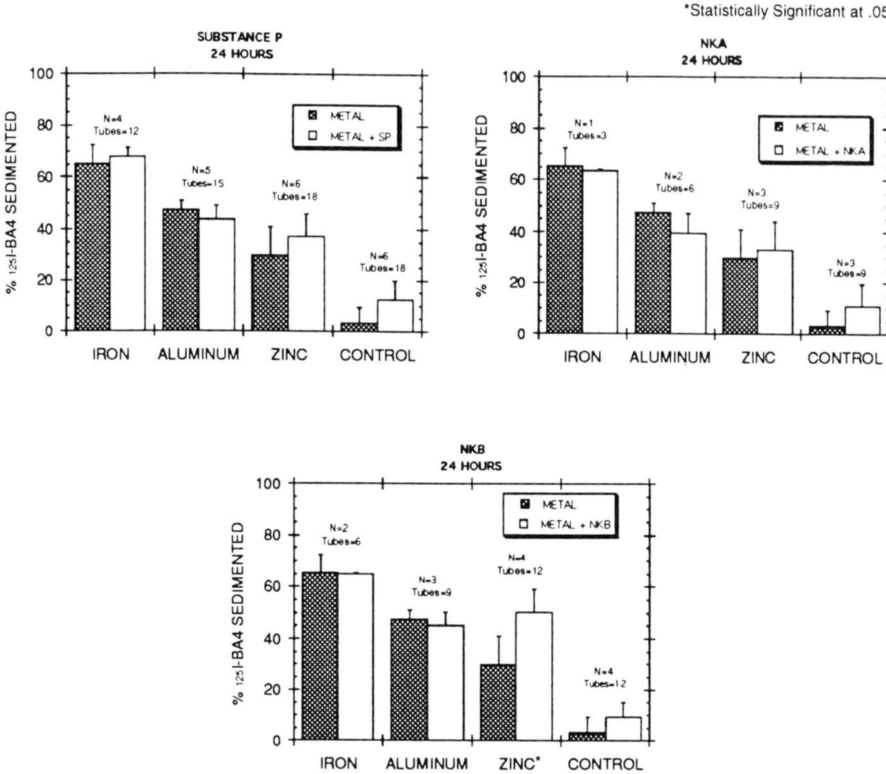

Fig. 12. Failure of mammalian tachykinins (SP, NKA, or NKB, all at $10^{-5}M$) to inhibit either self-aggregation or metal (Al^{2+}, Fe^{2+}, or Zn^{2+})-induced aggregation of ^{125}I-βA4. All aggregation was carried out as described in Fig. 11 and text, with the exception that the experiments were carried out over a 24-h period at room temperature.

rabbit, NK_1 receptors are expressed by the astrocytes that form the glial scar. In vitro a similar upregulation of NK_1 receptors is also seen after glial cells are plated and begin to grow in culture. Thus, astrocytes in vitro express NK_1 receptors, and addition of SP to these cultures has been shown to stimulate phosphatidylinositol turnover, stimulate the cyclooxygenase pathway of arachidonic acid metabolism, evoke the formation of prostaglandin E and thromboxane B_2 in dose-dependent manners, and stimulate IL-6 release. The source of the ligand that occupies these astrocytic NK_1 receptors and the functional consequences of NK_1 receptor activation in these cells, especially with regard to promoting or inhibiting neuronal regeneration, requires further clarification.

The mechanism by which tachykinins influence endothelial cells and astrocytes appears to be via the NK_1 receptor. In contrast, the mechanism by which tachykinins might block the β-amyloid peptide-induced neurotoxicity is unclear. To date, we and others have not observed any interaction between tachykinins, tachykinin receptors, and the β-amyloid peptide at physiologically relevant concentrations.

References

Aguayo, A. J., Bray, G. M., Rasminsky, M., Zwimpfer, T., Carter, D., and Vidal-Sanz, M. (1990) Synaptic connections made by axons regenerating in the central nervous system of adult mammals. *J. Exp. Biol.* **153,** 199–224.

Axelsson, L.-G. and Ahlstedt, S. (1990) A fluorescence method for the visualization of inflammatory cells in intestinal mucosa. *J. Immunol. Meth.* **127,** 271–277.

Barnes, P. J., Belvisi, M. G., and Rogers, D. F. (1990) Modulation of neurogenic inflammation: Novel approaches to inflammatory disease. *TIPS* **11,** 185–189.

Barnes, P. J. (1992) Neurogenic inflammation and asthma. *J. Asthma* **29,** 165–180.

Basbaum, A. I. and Levine, J. D. (1991) The contribution of the nervous system to inflammation and inflammatory disease. *Can. J. Physiol. Pharmacol.* **69,** 647–651.

Beaujouan, J.-C., Teutsch, B., Saffroy, M., Petitet, F., Torrens, Y., and Glowinski, J. (1991) NK-1 receptors are the only class of tachykinin receptors found on mouse cortical astrocytes. *Peptides* **12,** 813–820.

Beaujouan, J. C., Daguet de Montety, M. C., Torrens, Y., Saffroy, M., Dietl, M., and Glowinski, J. (1990) Marked regional heterogeneity of ^{125}I-Bolton Hunter substance P binding and substance P-induced activation of phospholipase C in astrocyte cultures from the embryonic or newborn rat. *J. Neurochem.* **54,** 669–675.

Beuerman, R. W. and Schimmelpfennig, B. (1980) Sensory denervation of the rabbit cornea affects epithelial properties. *Exp. Neurol.* **69,** 196–201.

Boehmer, C. G., Norman, J., Catton, M., Fine, L.G., and Mantyh, P. W. (1989) High levels of mRNA coding for substance P, somatostatin and b-tubulin are expressed by rat and rabbit dorsal root ganglia neurons. *Peptides* **10,** 1179–1194.

Bray, G. M., Villegas-Perez, M. P., Vidal-Sanz, M., and A. Aguayo, J. (1987) The use of peripheral nerve grafts to enhance neuronal survival, promote growth and permit terminal reconnections in the central nervous system of adult rats. *J. Exp. Biol.* **132,** 5–19.

Brimijoin, S., Lundberg, J. M., Brodin, E., Hokfelt, T., and Nilsson, G. (1980) Axonal transport of substance P in the vagus and sciatic nerves of the guinea pig. *Brain Res.* **191,** 443–457.

Buck, S. H. and Burks, T. F. (1986) The neuropharmacology of capsaicin: review of some recent observations. *Pharmacol. Rev.* **38,** 179–250.

Burdick, D., Soreghan, B., Kwon, M., Kosmoski, J., Knauer, M., Henschen, A., Yates, J., Cotman, C., and Glabe, C. (1992) Assembly and aggregation properties of synthetic Alzheimer's A4/b amyloid peptide analogs. *J. Biol. Chem.* **267,** 546–554.

Cholewinski, A. J. and Wilkin, G. P. (1988) Peptide receptors on astrocytes: Evidence for regional heterogeneity. *Biochem. Soc. Trans.* **16,** 429–432.

David, S. and Aguayo, A. J. (1985) Axonal regeneration after crush injury of rat central nervous system fibres innervating peripheral nerve grafts. *J.Neurocytol.* **14,** 1–12.

Delumeau, J. C., Petitet, F., Cordier, J., Glowinski, J., and Premont, J. (1991) Synergistic regulation of cytosolic Ca^{2+} concentration in mouse astrocytes by NK1 tachykinin and adenosine agonists. *J. Neurochem.* **57,** 2026–2035.

Fabregat, I. and Rozengurt, E. (1990) [D-Arg^{1},D-Phe^{5},D-Trp^{7,9},Leu^{11}] substance P, a neuropeptide antagonist, blocks binding, C^{a2}-mobilizing, and mitogenic effects of endothelin and vasoactive intestinal contractor in mouse 3T3 cells. *J. Cell. Physiol.* **15,** 88–94.

Flood, J. F., Morley, J. E., and Roberts, E. (1991) Amnestic effects in mice of four synthetic peptides homologous to amyloid b protein from patients with Alzheimer disease. *Proc. Natl. Acad. Sci. USA* **88,** 3363–3366.

Gibbins, I. L., Furness, J. B., Costa, M., MacIntyre, I., Hillyard, C. J., and Girgis, S. (1985) Co-localization of calcitonin gene-related peptide-like immunoreactivity with substance P in cutaneous, vascular and visceral sensory neurons of guinea pigs. *Neurosci. Lett.* **57,** 125–130.

Gitter, B. D., Waters, D. C., Regoli, D., Glasebrook, A. L., and Howbert, J. J. (1992) Interleukin-6 secretion from the human astrocytoma cell line U373 induced by substance P. *Regul. Pept.* **Suppl. 1,** s62 (abstract).

Greeno, E. W., Mantyh, P. W., Vercellotti, G., and Moldow, C. F. (1993) Functional NK-1 receptors for substance P are expressed by human vascular endothelium. *J. Exp. Med.* **177,** 1269–1276.

Hartung, H. P., Heininger, K., Schafer, B., and Toyka, K. V. (1988) Substance P and astrocytes: stimulation of the cyclooxygenase pathway of arachidonic acid metabolism. *FASEB J.* **2,** 48–51.

Hartung, H. P. and Toyka, K. V. (1989) Substance P, the immune system and inflammation. *Int. Rev. Immunol.* **4,** 229–249.

Helke, C. J., Krause, J. E., Mantyh, P. W., Couture, R., and Bannon, M. J. (1990) Diversity in mammalian tachykinin peptidergic neurons: multiple peptides, receptors, and regulatory mechanisms. *FASEB J.* **4,** 1606–1615.

Hokfelt, T., Johansson, O., Ljungdahl, A., Lundberg, J. M., and Schultzberg, M. (1980) Peptidergic neurones. *Nature* **28,** 515–521.

Hokfelt, T., Vincent, S., Dalsgaard, C. J., Skirboll, L., Johansson, O., Schultzberg, M., Lundberg, J. M., Rosell, S., Pernow, B., and Jancso, G. (1982) Distribution of substance P in brain and periphery and its possible role as a co-transmitter. *Ciba. Found. Symp.* **24,** 84–106.

Hua, X. Y., Theodorsson-Norheim, E., Brodin, E., Lundberg, J. M., and Hokfelt, T. (1985) Multiple tachykinins (neurokinin A, neuropeptide K and substance P) in capsaicin-sensitive sensory neurons in the guinea pig. *Regul. Pept.* **13,** 1–19.

Hultgardh-Nilsson, A., Nilsson, J., Jonzon, B., and Dalsgaard, C.-J. (1988) Coupling between inositol phosphate formation and DNA synthesis in smooth muscle cells stimulated with neurokinin A. *J. Cell Physiol.* **137,** 141–145.

Iwamoto, I., Yamazaki, H., Nakagawa, N., Kimura, A., Tomioka, H., and Yoshida, S. (1990) Differential effects of two C-terminal peptides of substance P on human neutrophils. *Neuropeptides* **16,** 103–107.

Jessell, T. M. and Iversen, L. L. (1977) Opiate analgesics inhibit substance P release from rat trigeminal nucleus. *Nature* **268,** 549–551.

Khalil, Z. and Helme, R. D. (1989) Sequence of events in substance P-mediated plasma extravasation in rat skin. *Brain Res.* **500,** 256–264.

Knauer, M. F., Soreghan, B., Burdick, D., Kosmoski, I., and Glabe, C. G. (1992) Intracellular accumulation and resistance to degradation of the Alzheimer amyloid A4/b protein. *Proc. Natl. Acad. Sci. USA* **89,** 7437–7441.

Koh, J. Y., Yang, L. L., and Cotman, C. W. (1990) Beta-amyloid protein increases the vulnerability of cultured cortical neurons to excitotoxic damage. *Brain Res.* **533,** 315–320.

Kowall, N. W., Beal, M. F., Busciglio, J., Duffy, L. K., and Yankner, B. A. (1991) An *in vivo* model for the neurodegenerative effects of b amyloid and protection by substance P. *Proc. Natl. Acad. Sci. USA* **88,** 7247–7251.

Lei, Y.-H., Barnes, P. J., and Rogers, D. F. (1992) Inhibition of neurogenic plasma exudation in guinea-pig airways by CP-96,345, a new non-peptide NK1 receptor antagonist. *Br. J. Pharmacol.* **105,** 261,262.

Lembeck, F., Donnerer, J., and Colpaert, F. C. (1981) Increase of substance P in primary afferent nerves during chronic pain. *Neuropeptides* **1,** 175–180.

Lembeck, F., Donnerer, J., Tsuchiya, M., and Nagahisa, A. (1992) The non-peptide tachykinin antagonist, CP-96,345, is a potent inhibitor of neurogenic inflammation. *Br. J. Pharmacol.* **105,** 527–530.

Lembeck, F. and Holzer, P. (1979) Substance P as a neurogenic mediator of antidromic vasodilatation and neurogenic plasma extravasation. *Naunyn Schmiedeberg's Arch. Pharmcol.* **310,** 175–183.

Levine, J. D., Dardick, S. J., Roizen, M. F., Helms, C., and Basbaum, A. I. (1986) Contribution of sensory afferents and sympathetic efferents to joint injury in experimental arthritis. *J. Neurosci.* **6,** 3423–3429.

Logan, A., Frautschy, S. A., Gonulez, A.-M., Sporn, M. B., and Baird, A. (1992) Enhanced expression of transforming growth factor 1 in the rat brain after a localized cerebral injury. *Brain Res.* **587,** 216–225.

Lundberg, J. M., Franco-Cereceda, A., Lacroix, J. S., and Pernow, J. (1991) Release of vasoactive peptides from autonomic and sensory nerves. *Blood Vessels* **28,** 27–34.

Maggio, J. E., Stimson, E. R., Ghilardi, J. R., Allen, C. J., E. Dahl, C., Whitcomb, D. C., Vigna, S. R., Vinters, H. V., E. Labenski, M., and Mantyh, P. W. (1992)

Reversible *in vitro* growth of Alzheimer disease β-amyloid plaques by deposition of labeled amyloid peptide. *Proc. Natl. Acad. Sci. USA* **89,** 5462–5466.

Mantyh, C. R., Gates, T. S., Zimmerman, R. P., Welton, M. L., Passaro, Jr., E. P., Vigna, S. R., Maggio, J. E., Kruger, L., and Mantyh, P. W. (1988) Receptor binding sites for substance P, but not substance K or neuromedin K, are expressed in high concentrations by arterioles, venules, and lymph nodules in surgical specimens obtained from patients with ulcerative colitis and Crohn disease. *Proc. Natl. Acad. Sci. USA* **85,** 3235–3239.

Mantyh, P. W., Mantyh, C. R., Gates, T., Vigna, S. R., and Maggio, J. E. (1988) Receptor binding sites for substance P and substance K in the canine gastrointestinal tract and their possible role in inflammatory bowel disease. *Neuroscience* **25,** 817–837.

Mantyh, P. W., Catton, M. D., Boehmer, C. G., Welton, M. L., Passaro, Jr., E. P., Maggio, J. E., and Vigna, S. R. (1989a) Receptors for sensory neuropeptides in human inflammatory diseases: implications for the effector role of sensory neurons. *Peptides* **10,** 627–645.

Mantyh, P. W., Gates, T., Mantyh, C. R., and Maggio, J. E. (1989b) Autoradiographic localization and characterization of tachykinin receptor binding sites in the rat brain and peripheral tissues. *J. Neurosci.* **9,** 258–279.

Mantyh, P. W., Johnson, D. J., Boehmer, C. G., Catton, M. D., Vinters, H. V., Maggio, J. E., Too, H. P., and Vigna, S. R. (1989c) Substance P receptor binding sites are expressed by glia in vivo after neuronal injury. *Proc. Natl. Acad. Sci. USA* **86,** 5193–5197.

Macwell, D. S. and Kruger, L. (1965) The fine structure of astrocytes in the cerebral cortex and their response to focal injury produced by heavy ionising particles. *J. Cell Biol.* **25,** 141–157.

McGillis, J. P., Organist, M. L., and Payan, D. G. (1987) Substance P and immunoregulation. *Fed. Proc.* **46,** 196–199.

Miller, R. H., Abney, E. R., David, S., French-Constant, C., Lindsay, R., Patel, R., Stone, J., and Raff, M. C. (1986) Is reactive gliosis a property of a distinct subpopulation of astrocytes? *J. Neurosci.* **6,** 22–29.

Miller, R. H., French-Constant, C., and Raff, M. C. (1989) The macroglial cells of the rat optic nerve. *Annu. Rev. Neurosci.* **12,** 517–534.

Mitsuhashi, M., Akitaya, T., Turk, C. W., and Payan, D. G. (1991) Amyloid beta protein substituent peptides do not interact with the substance P receptor expressed in cultured cells. *Mol. Brain Res.* **11,** 177–180 (abstract).

Nagy, J. I., Hunt, S. P., Iversen, L. L., and Emson, P. C. (1981) Biochemical and anatomical observations on the degeneration of peptide-containing primary afferent neurons after neonatal capsaicin. *Neuroscience* **6,** 1923–1934.

Nagy, J. I. (1982) Capsaicin A chemical probe for sensory neuron mechanisms, in *Handbook of Psychopharmacology*, vol. 6 (Iversen, L. L., Iversen S. D., and Snyder, S. H., eds.), Elsevier, Amsterdam, pp. 185–235.

Nakagawa, N., Iwamoto, I., and Yoshida, S. (1992) Effect of substance P on the epression of an adhesion molecule ICAM-1 in human vascular endothelial cells. *Regul. Pept.* **Suppl. 1,** s124.

Nilsson, J., von Euler, A. M., and Dalsgaard, C. J. (1985) Stimulation of connective tissue cell growth by substance P and substance K. *Nature* **315,** 61–63.

Nitsch, R. M., Slack, B. E., Wurtman, R. J., and Growdon, J. H. (1992) Release of Alzheimer amyloid precursor derivatives stimulated by activation of muscarinic acetylcholine receptors. *Science* **258,** 304–307 (abstract).

Oku, R., Saton, M., and Takagi, H. (1987) Release of substance P from the spinal dorsal horn is enhanced in polyarthritic rats. *Neurosci. Lett.* **74,** 315–319.

Parnet, P., Payan, D. G., Kerdelhué, B., and Mitsuhashi, M. (1990) Neuro-endocrine interaction on lymphocytes. Testosterone-induced modulation of the lymphocyte substance P receptor. *J. Neuroimmunol.* **28,** 185–188.

Pike, C. J., Walencewicz, A. J., Glabe, C. G., and Cotman, C. W. (1991a) Aggregation-related toxicity of synthetic β-amyloid protein in hippocampal cultures. *Eur. J. Pharmacol. Mol. Pharmacol.* **207,** 367,368.

Pike, C. J., Walencewicz, A. J., Glabe, C. G., and Cotman, C. W. (1991b) In vitro aging of β-amyloid protein causes peptide aggregation and neurotoxicity. *Brain Res.* **563,** 311–314.

Pike, M. C. and Snyderman, R. (1988) Leukocyte chemoattractant receptors. *Methods Enzymol.* **162,** 236–245.

Popper, P., Mantyh, C. R., Vigna, S. R., Maggio, J. E., and Mantyh, P. W. (1988) The localization of sensory nerve fibers and receptor binding sites for sensory neuropeptides in canine mesenteric lymph nodes. *Peptides* **9,** 257–267.

Roberts, A. B. and Sporn, M. B. (1988) Transforming growth factor β. *Adv. Cancer Res.* **51,** 107–146.

Rozengurt, E. (1986) Early signals in the mitogenic response. *Science* **234,** 161–166.

Saria, A., Ma, R. C., Dun, N. J., Theodorson-Norheim, E., and Lundberg, J. M. (1989) Neurokinin A in capsaicin-sensitive neurons of the guinea-pig inferior mesenteric ganglia: An additional putative mediator for the non-cholinergic exitatory postsynaptic potential. *Neuroscience* **211,** 951–958.

Savio, T. and Schwab, M. E. (1989) Rat central nervous system white matter, but not gray matter, is nonpermissive for neuronal cell adhesion and fiber outgrowth. *J. Neurosci.* **9,** 1126–1133.

Selkoe, D. J. (1991) The molecular pathology of Alzheimer's disease. *Neuron* 487–498.

Siman, R., Card, J. P., Nelson, R. B., and Davis, L. G. (1989) Expression of beta-amyloid precursor protein in reactive astrocytes following neuronal damage. *Neuron* **3,** 275–285 (abstract).

Smith, G. M. and Miller, R. H. (1991) Immature type-1 astrocytes suppress glial scar formation, are motile and interact with blood vessels. *Brain Res.* **543,** 111–122.

Too, H. P. and Maggio, J. E. (1991) Immunocytochemical localization of neuromedin K (neurokinin B) in rat spinal ganglia and cord. *Peptides* **12,** 431–443.

Torrens, Y., Daguet de Montety, M. C., el Etr, M., Beaujouan, J. C., and Glowinski, J. (1989) Tachykinin receptors of the NK1 type (substance P) coupled positively to phospholipase C on cortical astrocytes from the newborn mouse in primary culture. *J. Neurochem.* **52,** 1913–1918.

Vigna, S. R., Mantyh, C. R., Soll, A. H., Maggio, J. E., and Mantyh, P. W. (1989) Substance P receptors on canine chief cells: localization, characterization, and function. *J. Neurosci.* **9,** 2878–2886.

Walsh, E., Ueda, Y., Nakanishi, H., and Yoshida, K. (1992) Neuronal survival and neurite extension supported by astrocytes co-cultured in transwells. *Neurosci. Lett.* **138,** 103–106.

Weinstock, J. V., Blum, A., Walder, J., and Walder, R. (1988) Eosinophils from granulomas in murine schistosomiasis mansoni produce substance P. *J. Immunol.* **141,** 961–966.

Weinstock, J. V. and Blum, A. M. (1990) Release of substance P by granuloma eosinophils in response to secretagogues in murine schistosomiasis mansoni. *Cell. Immunol.* **125,** 380–385.

Whitson, J. S., Selkoe, D. J., and Cotman, C. W. (1989) Amyloid beta protein enhances the survival of hippocampal neurons in vitro. *Science* **23,** 1488–1490.

Whitson, J. S., Glabe, C. G., Shintani, E., Abcar, A., and Cotman, C. W. (1990) Beta-amyloid protein promotes neuritic branching in hippocampal cultures. *Neurosci. Lett.* **110,** 319–324.

Wilcox, G. L. (1988) Pharmacological studies of grooming and scratching behavior elicited by spinal substance P and excitatory amino acids. *Ann. NY Acad. Sci.* **525,** 228–236.

Wilkin, G. P., Marriott, D. R., and Cholewinski, A. J. (1990) Astrocyte heterogeneity. *Trends Neurosci.* **13,** 43–46.

Wilkin, G. P., Marriott, D. R., Cholewinski, A. J., Wood, J. N., Taylor, G. W., Stephens, G. J., and Djamgoz, M. B. (1991) Receptor activation and its biochemical consequences in astrocytes. *Ann. NY Acad. Sci.* **633,** 475–488.

Yankner, B. A., Duffy, L. K., and Kirschner, D. A. (1990) Neurotrophic and neurotoic effects of amyloid β protein: Reversal by tachykinin neuropeptides. *Science* **250,** 279–282.

Yankner, B. A. and Mesulam, M.-M. (1991) Seminars in medicine of the Beth Israel Hospital, Boston: β-Amyloid and the pathogenesis of Alzheimer's disease. *N. Engl. J. Med.* **325,** 1849–1857.

Zachary, I., Woll, P. J., and Rozengurt, E. (1987) A role for neuropeptides in the control of cell proliferation. *Dev. Biol.* **124,** 295–308.

Ziche, M., Morbidelli, L., Geppetti, P., Maggi, C. A., and Dolara, P. (1990a) Substance P induces migration of capillary endothelial cells: A novel NK_1 selective receptor mediated activity. *Life Sci.* **8,** PL7–PL11.

Ziche, M., Morbidelli, L., Pacini, M., Dolara, P., and Maggi, C.A. (1990b) NK1-receptors mediate the proliferative response of human fibroblasts to tachykinins. *Br. J. Pharmacol.* **100,** 11–14.

Ziche, M., Morbidelli, L., Pacini, M., Geppetti, P., Alessandri, G., and Maggi, C. A. (1990c) Substance P stimulates neovascularization in vivo and proliferation of cultured endothelial cells. *Microvasc. Res.* **40,** 264–278.

PART V

SUMMARY AND FUTURE PERSPECTIVES

Chapter 19

Summary, Thoughts, and Future Perspectives

Fred Lembeck and Stephen H. Buck

As all of the chapters in this book demonstrate, there has been much recent progress in the tachykinin receptor field and many exciting scientific discoveries have appeared. The report by Masu et al. in 1987 of the structure of the bovine stomach NK_1 receptor marked the first neuropeptide receptor of any kind to have its primary structure determined. Since then, many other neuropeptide receptor structures have been reported, including what seem to be mammalian NK_2 and NK_3 receptors (Nakanishi, 1991). It is not surprising that these tachykinin receptor types are all G-protein-linked since the natural agonists of these receptors have long been known to be efficient stimulators of phosphatidylinositol hydrolysis in the appropriate tissues. The regulation of the expression of these tachykinin receptors can now be studied, and hopefully this will provide some explanation of the peptide-receptor mismatch that has been discussed in this volume. And, it should now be possible to address the issue of potential receptor subtypes and whether there are just NK_1, NK_2, and

The Tachykinin Receptors Ed.: S. H. Buck © 1994 Humana Press Inc., Totowa, NJ

NK_3 receptors or if the tachykinin receptor story is even more complicated. The growing application of computerized tachykinin receptor structure analysis, coupled with the production of mutant receptors for biological analysis, will no doubt hasten our understanding of all of the tachykinin receptors and allow for the design and development of specific ligands for each receptor type. Of course, this progress will be aided by the similarity in receptor structure and function of the tachykinin receptors to other G-protein-linked receptors, both peptidergic and nonpeptidergic.

One of the most exciting aspects of receptor pharmacology is to see laboratory science lead to the development of receptor-active molecules that can be effective in human therapeutics. There are now several peptide antagonists of some tachykinin receptors, and these will contribute greatly to the study and understanding of the nature of the receptors. However, peptides as potential therapeutic agents have many limitations working against them, and it is unlikely that a useful systemic tachykinin antagonist that itself is a peptide will ever have major therapeutic value. For nonsystemic applications (e.g., topical), it is possible that a peptide that acts on tachykinin receptors might have some utility. Therefore, some of the major targets of tachykinin-receptor-active molecules, such as the CNS and the GI tract, would most likely not be addressable with such peptide molecules. This raises the potential therapeutic importance of nonpeptide tachykinin-receptor-active agents. For the tachykinin receptor field, the era of such molecules has begun. As discussed in several chapters of this volume, interesting nonpeptide tachykinin antagonists have now been discovered and seem to be very specific for tachykinin receptors. Work in this area of the tachykinin story is growing and is likely to produce even more remarkable compounds. Nonpeptide tachykinin agonists have yet to be conclusively demonstrated, but it is likely that these will appear in the near future. However, the existence of the nonpeptide tachykinin antagonists promises the greatest potential therapeutic benefit, just as has happened for many nonpeptide biological mediators.

In general, antagonists are of greater significance than agonists in pharmacology, as well as in applied therapeutics. *Antagonists* are defined as compounds that bind to the recognition site of the receptor, but lack intrinsic activity. *Agonists* can usually bind to and activate several types of receptors in the same organism, of which only one or another can be blocked by a specific antagonist. Antagonists are therefore useful to distinguish between different receptors. Acetylcholine can be used as an example: A small dose causes vasodilatation that is of no physiological significance, since blood vessels have no cholinergic innervation. A larger dose of acetylcholine causes bradycardia, mimicking the natural vagal effect on the heart and, therefore, this effect is of physiological significance. After blockade of these receptors by atropine, a very large dose of acetylcholine stimulates the ganglionic cholinergic receptors that can be blocked by a large dose of nicotine or another similar antagonist. Likewise, antagonists of the biogenic amines, which were first discovered among plant alkaloids (e.g., atropine, ergotoxin), showed receptor selectivity. Ergotoxin, for example, only blocked the effect of adrenaline on blood pressure and not on heart rate, thus giving the first hint of two types of adrenergic receptors.

It is interesting to note that the β-adrenergic agonist, isoproterenol, which is not found in the body, gave the first hint of the possibility of selectively stimulating the β-type of adrenergic receptors. Likewise, some of the nonmammalian and synthetic tachykinin agonists have been of great value in tachykinin receptor differentiation. Some of them have also had the advantage of increased resistance to enzymatic degradation.

Some amine receptor antagonists were discovered by modification of the structure of the agonist, which led to antagonists that were specific for receptor subtypes, such as the H_1- and H_2-histamine antagonists. By modifying the amino acid sequence of tachykinins, the peptidic tachykinin antagonists were discovered. All of these findings were based on recording biological intrinsic activity, i.e., the action evoked by an agonist and its inhibition by an antagonist. On the other hand, the search for novel, nonpeptide tachykinin antagonists was based on the screening of inhibition of specific tachykinin binding to receptors using automated procedures that allowed the screening of vast chemical "libraries."

In the area of biogenic amines, pharmacological influences can also be achieved by stimulation or inhibition of agonist biosynthesis, release, storage, reuptake, or degradation by specific enzymes. Some of these routes, but not all of them, may be manipulatable for the tachykinins as well. Tachykinin biosynthesis is influenced by up- or downregulation of the expression of preprotachykinin mRNA. An upregulation can be mediated by NGF (Donnerer et al., S3*; Harmar et al., S9), whereas a downregulation can be observed after axotomy (Hokfelt et al., S11; Noguchi et al., S127) or after capsaicin desensitization (for review, *see* Holzer, 1991). Colchicine and lidocaine blocked the substance P (SP) supply to nerve terminals (Moussaoui et al., S113), morphine inhibited SP release (Moussaoui et al., S113; Lembeck and Donnerer, 1985; Donnerer, 1989), and dexamethasone reduced SP receptor synthesis (Ihara and Nakanishi, 1990). The receptor interaction of agonists and of antagonists is clearly similar for amines and tachykinins in that in both cases this interaction is competitive.

There is a wide spectrum of intrinsic activities of tachykinins that can be used to evaluate tachykinin antagonists. The influence of tachykinins on phosphatidylinositol hydrolysis (Cascieri et al., S1; Krause et al., S16) and activation of phospholipase C and adenylate cyclase (Nakanishi et al., S22) can best be studied on certain cell lines. Channel-mediated actions (Nakajima et al., S21, and this volume) are easily seen on smooth muscle contraction or relaxation or by electrophysiological recording of nerve and muscle cells. The sialogogic effect can be studied in vitro and in vivo. Other effects, such as those on blood pressure or bronchoconstriction, can require anesthetized animals (Griesbacher et al., 1992; Lembeck et al., 1992). Whereas isolated cells or organs have the advantage of analytical precision and ease, effects in whole animals can reveal unspecific

*For brevity, citations of reports from the 1992 International Symposium on Substance P and Related Peptides (abstracts in *Regulatory Peptides* (1992) Suppl. 1, 1–194) are indicated in the text by the author's name and the abstract page number (e.g., Garret et al., S5) and are not included in the list of references of this chapter.

effects of antagonists (Nagahisha et al., S121), such as the effect of CP-96,345 and its inactive enantiomer on calcium channels (Donnerer et al., 1992; Nagahisha et al., 1992). Whole animal studies can also provide information about the duration of antagonist action and pharmacokinetic properties.

Receptor binding studies have shown considerable species differences for nonpeptide tachykinin antagonists; i.e., CP-96,345 was more potent in guinea pig and man compared to rat and mouse, whereas RP-67,580 was more potent in rat and mouse (Garret et al., S5; Watling and Guard, 1992). Is this because of variations in the structure of the tachykinin receptor? There is also a marked difference in the affinity of some tachykinin agonists between tissues of the same species (Fardin et al., S53). Might this point to different receptor types or subtypes in the same species or simply to different modes of receptor binding in different tissues?

In a strict sense, tachykinin receptors can be regarded as relevant in terms of physiology only if (1) tachykinins are released from nerve terminals in close proximity to the receptor (e.g., in the spinal dorsal horn or near cutaneous blood vessels), and (2) the released tachykinin exerts a relevant physiological effect (e.g., neurotransmission, cutaneous vasodilatation, or bronchoconstriction). The known release of tachykinins from cells is usually defined as neurotransmission. However, there is one exception and that is when neurokinin A (NKA) is released into the circulation from carcinoid tumors, where the tachykinin coexists with serotonin. NKA then distributes in the blood stream like a hormone and causes skin flushes, i.e., cutaneous vasodilatation (Norheim et al., 1987). Many effects of tachykinins on visceral or vascular smooth muscle are not physiologically meaningful; they are pharmacological. The same is true for effects of biogenic amines; the vasodilatation produced by acetylcholine has been mentioned earlier in this regard. In addition, the isolated rat uterus is very sensitive to serotonin, bradykinin, and oxytocin, although for the first two of these there is no known physiological significance.

For SP, the marked stimulation of salivation (Takeda et al., S156), which was the critical activity for Susan Leeman's discovery of the structure of SP, has no bearing on the physiological roles of SP. The amount of the tachykinin required to induce salivation is far in excess of that released from nerves in the salivary glands and very much larger, for example, than the dose that causes vasodilatation. In addition, no sialogogic response to SP is seen in the rabbit or the cat in contrast to all other laboratory animal species. Might there be lack of salivary NK_1 receptors in these two species? Or, are the receptors in the other species only a genetic relic without physiological relevance?

SP can also release histamine from mast cells. In contrast, this effect is not seen with NKA, eledoisin, or kassinin (Holzer-Petsche et al., 1985). SP, containing two N-terminal basic amino acids, is a much more basic peptide than other tachykinins. In addition, other peptides with at least two basic charges, some of them devoid of biological activity, release histamine from mast cells. Therefore, it is unlikely that there are specific SP receptors on mast cells (Schmidt et al., S144). Another instance where receptors for SP may be of questionable physiological significance is the guinea pig ileum. The possible presence of two different receptors for the peptide in this tissue has been reported (Chassaing et al., S44) and is discussed elsewhere in this volume. The receptor on the smooth muscle underwent rapid desensitization following a large dose of SP, whereas the other receptor (on cholinergic structures) remained responsive (Holzer and Lembeck, 1980). However, no effect of the nonpeptide antagonists has been observed on the peristaltic reflex (Legat et al., 1992), although SP was thought to be involved in this reflex (Bartho and Holzer, 1985). The ascending enteric reflex contraction, i.e., that part of the peristaltic reflex that mediates the propulsion of gut contents, may be mediated in part by NK_2 receptors at the neuromuscular junction, whereas neuroneuronal transmission may involve primarily muscarinic and nicotinic acetylcholine receptors (Holzer et al., 1993).

As for evaluating tachykinins and antagonists in tissues, there exists a large number of preparations that have been and are being used (Legat et al., 1992). Smooth muscle preparations of the guinea pig and the rat have been used most frequently instead of the more sensitive rabbit jejunum on which Von Euler and Gaddum (1931) first characterized SP actions. The relative potencies of NKA, NKB, eledoisin, and kassinin were generally smaller than that of SP on the guinea pig ileum, rabbit jejunum, rat salivary glands, and rabbit vasculature. The rat vas deferens was an exception in that all of these former peptides were one to two orders of magnitude more potent than SP (Holzer-Petsche et al., 1985). However, the rat vas deferens is known to respond to many types of compounds in unspecific ways, and it is doubtful that such a preparation should be used any further. The new nonpeptide tachykinin antagonists have been tested in many different bioassay preparations, which makes comparison of potencies difficult. Perhaps again certain preparations, such as the rat vas deferens and the barely accessible dog carotid artery, should be excluded and a consensus about the most useful and relevant preparations should be reached. In this respect, it seems that the guinea pig ileum and the rabbit jejunum are the most suitable visceral smooth muscles. For vasodilatation, the perfused rat hind leg is a preparation that is sensitive to SP, NKA, and NKB in the picomole range (Holzer-Petsche et al., 1985). Isolated vessels should probably only be used from rats, guinea pigs, and rabbits. For in vivo tests, the hypotensive effects of tachykinins in rat and rabbit are invaluable as is bronchoconstriction in the guinea pig (Griesbacher et al., 1992; Lembeck et al., 1992). Protein extravasation can easily be tested by the Evans blue leakage method in these same species. This effect, however, is not specific as it can also be evoked by histamine, serotonin, bradykinin, and antigen challenge. SP injected, or released by antidromic nerve stimulation or by capsaicin from peripheral nerve terminals, can give valuable information about the localization of neurogenic inflammation (Saria et al., 1983). The salivation evoked by SP and NKA, but not by NKB, is not very sensitive, but is highly specific for tachykinins (Takeda et al., S156; Holzer-Petsche et al., 1985). There are also other preparations that are especially selective for one or another of the tachykinins (for a review, *see* Watling and Guard, 1992). Of course, what will ultimately be the most valuable information will be that generated from investigations in which human tissues can be obtained and used.

More difficult is the evaluation of actions of neuropeptides and their antagonists in the CNS. Spinal cord slices and isolated dorsal root ganglia can be used to measure the stimulated release of neuropeptides, but effects of antagonists cannot be tested in these preparations. However, the SP-induced stimulation of locus coeruleus cells in guinea pig brain slices was found to be a useful test of tachykinin antagonists (McLean et al., S120). M. Otsuka's spinal cord preparations are also of outstanding analytical value for this testing (Guo et al., S65; Otsuka and Yanagisawa, 1990; and this volume). In vivo central actions of a tachykinin antagonist will depend largely on penetration into the CNS. Additionally, reflexes evoked by intra-arterial injection of capsaicin or by afferent nerve stimulation do not allow one to conclude that the reflex is mediated exclusively by release of tachykinins from central terminals; other peptides could just as easily be involved. SP is certainly important in such reflexes and also in neuroendocrine regulation, which is blocked after sensory denervation by capsaicin (for a review, *see* Lembeck, 1988). It should also be kept in mind that the usual tests for transmission of noxious stimuli, such as paw pressure, tail flick, writhing response, and formalin irritation, are behavioral measures that are of limited value if the drug tested has a sedative action or an influence on blood pressure, thermoregulation, or the motor system.

In dreaming about a universally potent, *non plus ultra* tachykinin receptor antagonist, what could one expect to see it do? The experimental animal would probably show at least a partial loss of peristalsis and a diminished response to peripheral and visceral noxious stimuli and to airway irritation. The micturition reflex and other important autonomic reflexes (e.g., heat dissipation, hypoxic drive of respiration) might be greatly disturbed. Impairment of motor function might also occur. All of these hypothetically possible actions of a powerful tachykinin antagonist concern physiological functions, but any or all of them might also be extremely beneficial in correcting pathophysiological conditions.

Concluding remarks can be like the last course of a dinner, either colorful sweetness or harsh cheese with pepper. The impact of the new nonpeptide tachykinin antagonists on CNS functions is likely to be more complex than in the periphery, in congruence with the greater neuronal complexity of the CNS (Bower, 1992). Topics limited to the classical "SP and pain" are surely an oversimplification of facts. Unexpected CNS actions of the new antagonists are likely and thus should not be ignored. Nevertheless, in the CNS as in the periphery, the new tachykinin antagonists promise to be valuable scientific and therapeutic tools for use in investigating the known and the unknown. Visions like these should provide some outlook for the future, a future that will not be based only on the "sweetness" of easily achieved progress, but also on the more complicated unexpected aspects, tasting more like "peppered cheese" at the end of a dinner.

References

Bartho, L. and Holzer, P. (1985) Search for a physiological role of substance P in gastrointestinal motility. *Neuroscience* **16,** 1–32.

Bower, J. M. (1992) Modeling the nervous system. *Trends Neurosci.* (Special Issue) **15,** 411,412.

Donnerer, J. (1989) Primary sensory neurones and naloxone-precipitated morphine withdrawal. *Br. J. Pharmacol.* **96,** 767–772.

Donnerer, J., Stark, U., Tritthart, H. A., and Lembeck, F. (1992) CP-96,345, a non-peptide antagonist of substance P. III. Cardiovascular effects in mammals unrelated to actions on substance P receptors. *Naunyn Schmiedeberg's Arch. Pharmacol.* **346,** 328–332.

Griesbacher, T., Donnerer, J., Legat, F.J., and Lembeck, F. (1992) CP-96,345, a non-peptide antagonist of substance P. II. Actions on substance P-induced hypotension and bronchoconstriction and on depressor reflexes in mammals. *Naunyn Schmiedeberg's Arch. Pharmacol.* **346,** 323–327.

Holzer, P. (1991) Capsaicin: cellular targets, mechanisms of action, and selectivity for thin sensory neurons. *Pharmacol. Rev.* **43,** 143–201.

Holzer, P. and Lembeck, F. (1980) Neurally mediated contraction of ileal longitudinal muscle by substance P. *Neurosci. Lett.* **17,** 101–105.

Holzer, P., Schluet, W., and Maggi, C. A. (1993) Ascending enteric reflex contraction: roles of acetylcholine and tachykinins in relation to distension and propagation of excitation. *J. Pharmacol.* **264,** 391–396.

Holzer-Petsche, U., Schimek, E., Amann, R., and Lembeck, F. (1985) In vivo and in vitro actions of mammalian tachykinins. *Naunyn Schmiedeberg's Arch. Pharmacol.* **330,** 130–135.

Ihara, H. and Nakanishi, S. (1990) Selective inhibition of expression of the substance P receptor mRNA in pancreatic acinar AR42J cells by glucocorticoids. *J. Biol. Chem.* **265,** 10,441–10,445.

Legat, F. J., Griesbacher, T., and Lembeck, F. (1992) CP-96,345, a non-peptide antagonist of substance P. I. Effects on the actions mediated by substance P and related tachykinins on the guinea pig ileum and rabbit jejunum. *Naunyn Schmiedeberg's Arch. Pharmacol.* **346,** 315–322.

Lembeck, F. (1988) The Ulf von Euler Lecture. Substance P: from extract to excitement. *Acta. Physiol. Scand.* **133,** 435–454.

Lembeck, F. and Donnerer, J. (1985) Opioid control of the function of primary afferent substance P fibres. *Eur. J. Pharmacol.* **114,** 241–246.

Lembeck, F., Donnerer, J., Tsuchiya, M., and Nagahisha, A. (1992) The non-peptide tachykinin antagonist, CP-96,345, is a potent inhibitor of neurogenic inflammation. *Br. J. Pharmacol.* **105,** 527–530.

Masu, Y., Nakayama, K., Tamaki, H., Harada, Y., Kuno, M., and Nakanishi, S. (1987) cDNA cloning of bovine substance K-receptor through oocyte expression system. *Nature* **329,** 836–838.

Nagahisha, A., Asai, R., Kanai, Y., Murase, A., Tsuchiya-Nakagaki, M., Nakagaki, T., Shieh, T.-C., and Tanaguchi, K. (1992) Non-specific activity of (+/-)-CP-96,345 in models of pain and inflammation. *Br. J. Pharmacol.* **107,** 273–275.

Nakanishi, S. (1991) Mammalian tachykinin receptors. *Annu. Rev. Neurosci.* **14,** 123–136.

Norheim, I., Oberg, K., Theodorsson-Norheim, E., Lindgren, P. C., Lundqvist, G., Magnusson, A., Wide, L., and Wilander, L. (1987) Malignant carcinoid tumors: an analysis of 103 patients with regard to tumor localisation, hormone production and survival. *Ann. Surg.* **206,** 115–125.

Otsuka, M. and Yanagisawa, M. (1990) Pain and neurotransmitters. *Cell. Mol. Neurobiol.* **10,** 293–302.

Saria, A., Lundberg, J. M., Skofitsch, G., and Lembeck, F. (1983) Vascular protein leakage in various tissues induced by substance P, capsaicin, bradykinin, serotonin and histamine and by antigen challenge. *Naunyn Schmiedeberg's Arch. Pharmacol.* **324,** 212–218.

Von Euler, U. S. and Gaddum, J. H. (1931) An unidentified depressor substance in certain tissue extracts. *J. Physiol. (Lond.)* **72,** 74–87.

Watling, K. J. and Guard, S. (1992) Tachykinin receptor agonists and antagonists. *Neurotransmission* **8,** 1–8.

Index

N

P

T

U

V